ANALYSIS OF STAINLESS CHIP

FORMED BY CIRCULAR SEGMENTAL SAWS

ANATOLY ROZENBLAT

authorHOUSE

1663 LIBERTY DRIVE, SUITE 200
BLOOMINGTON, INDIANA 47403
(800) 839-8640
www.authorhouse.com

First published by AuthorHouse 09/22/04

ISBN: 1-4184-7881-4 (e)
ISBN: 1-4184-7880-6 (sc)

Library of Congress Control Number: 2004095763

Printed in the United States of America
Bloomington, Indiana

This book is printed on acid-free paper.

*Dedicated with love to my
dear parents-father, Rozenblat
Isaac Samoilovich and mother,
Fedorisheva Natalia Ivanovna*

ABOUT THE AUTHOR

Anatoly Rozenblat is an Independent author, Scientist and Inventor, specializing in Manufacturing (Mechanical) Engineering. He is based in Chicago, Illinois.

Mr. Rozenblat has been active in the Engineering for over 25 years. He has published more than 50 technical articles and 30 innovations in various aspects of technology.

He is the author of such books as *" Advanced machining problem solving "* and other six books. As the Independent Scientist he was invited on 26th, 27th,28th and 29th Israel Conference on Mechanical Engineering for presentation of 8 technical papers. Anatoly Rozenblat has received the numerous awards from International Biographical Center , American Biographical Center and Marquis " Who's who".

PREFACE

This book is about of metal cut-off processes mainly for the big rolled and other products from stainless steels with using of cold circular segmental saws.

The main objective of this book is to define and analyze the geometrical parameters of stainless chip (external and internal its diameters, number of chip wraps and clearance between these wraps, and also its width and thickness) which formed in cutting processes. The topics presented in this book have been organized into 8 chapters. The chapter one presents chip-formation in cut-off processes. The chapter two shows the main characteristics of polygon distribution for stainless chip. In chapter three is shown the functional oriented graph in evaluation of geometrical parameters for stainless chip. In chapter four is given the applied statistics in evaluation of correlation between two parameters of stainless chip. In chapter five is shown the correlation between three parameters of stainless chip for internal and external diameters and other parameters of stainless chip. In chapter six is shown the multiple regression analysis for external diameter of stainless chip in dependence from some general its parameters. In chapter seven is shown the multiple regression analysis for internal diameter of stainless chip in dependence from some its parameters. In chapter eight is given the multiple regression analysis of the main parameters of stainless chip and shown the empirical formula for calculation of external and internal diameters of stainless chip in dependence from other its parameters.

And besides were evaluated many functions characterizing these dependencies on the main parameters of stainless chip in question of finally definition regression model(linear or non-linear) on the basis using such statistical characteristics as coefficient determination, coefficient of correlation, standard deviation, minimization of the mean square error (min MSE) and minimization of the mean absolute deviation (min MAD) for each given function. This book is intended for use in industry, government, professional in business and also by the students of undergraduate or graduate level.

August 25,2003   Anatoly Rozenblat, Independent Scientist
Chicago, Illinois USA

TABLE OF CONTENTS

CHAPTER FOUR APPLIED STATISTICS IN EVALUATION OF CORRELATION BETWEEN PARAMETERS OF STAINLESS CHIP

CHAPTER FIVE CORRELATION BETWEEN THREE PARAMETERS OF STAINLESS CHIP IN FUNCTION OF $Y_i = \alpha_1(X_{i,1}; X_{i,2})$

CHAPTER SIX   MULTIPLE REGRESSION ANALYSIS FOR EXTERNAL DIAMETER OF STAINLESS CHIP IN DEPENDENCE FROM SOME GENERAL ITS PARAMETERS IN FUNCTION OF $Y_i = \gamma (X_{i,1} ; X_{i,2} ; X_{i,3})$

CHAPTER SEVEN MULTIPLE REGRESSION ANALYSIS FOR INTERNAL DIAMETER AND OTHER PARAMETERS OF STAINLESS CHIP IN DEPENDENCE FROM SOME GENERAL ITS PARAMETERS IN FUNCTION OF $Y_i = \gamma(X_{i,1}; X_{i,2}; X_{i,3})$

CHAPTER EIGHT MULTIPLE REGRESSION ANALYSIS OF THE MAIN PARAMETERS OF STAINLESS CHIP

INTRODUCTION

This book embraces one of the most important problem of metalworking process, as the chip-formation of stainless chip which is formed by cold circular segmental saws in process of cutting mainly the big rolled products from stainless and high-resistance steels.

And besides in this book a big attention is given to the analysis of geometrical parameters of stainless chip and also its form, character of modification in result of the different cutting conditions.

In addition were discovered such important aspects of cutting process such as shape of circular segmental saw and its tool life advantageously in cut-off processes for stainless steels. At present these questions were not discussed widely in many research papers and for this reason the author try to analyze this problem in detail with using of statistical experimental data, on the basis of multiple regression analysis.

The present book has goal to assist many manufacturing companies to improve the cutting process of stainless steels and increase the tool life of cold circular segmental saws, and also to break and remove further the stainless chip from cutting zone by the pneumatic transport.

CHAPTER ONE CHIP-FORMATION IN CUT-OFF PROCESSES

1.1 Industrial experimental part

At present in industry widely uses the stainless and heat-resistant steels [1]. However, the cutting process of blanks, advantageously from stainless and heat-resistance steels for rolled products and forging of a big diameter by the circular segmental saws arise the definite technological difficulties which are joined with low productivity of machining from the insignificant resistance of tool [2] .The utmost importance role in question of increasing of tool life relates for the cold circular segmental saw to the questions of studying chip parameters and also methods of its breaking and removal from the cut-off machines. But these questions advantageously to the circular segmental saws have learned insufficiency particularly in evaluation of types and forms generated chip and also of its constructive sizes [3].

Evaluation of characteristics for stainless chip makes possibility to use the different arrangement for its breaking and removal from the cutting area particularly by systems of pneumatic transportation [4] .For analysis of chip-formation in industrial experiment on the cut-off machines, with using of cold circular segmental saws, were put such important tasks as:

1.Evaluation of tool life for the cold circular segmental saw;

2.Analysis of distribution for the constructive sizes of stainless chip;

3.Designing some new arrangements for breaking and removal of stainless chip from the circular segmental saw by pneumatic transportation.

For analysis of chip-formation in industrial experiment was selected the austenitic chrome-nickel-titanium stainless steel Cr18N9T* in view of round bars with diameter D=180 mm (7.09in).Cutting of this bar has been fulfilled in accordance with scheme which is shown in Figure 1 on the cut-off machine with using of cold circular segmental saw [5].

---

* analogous with stainless steel 302 B AISI [1]

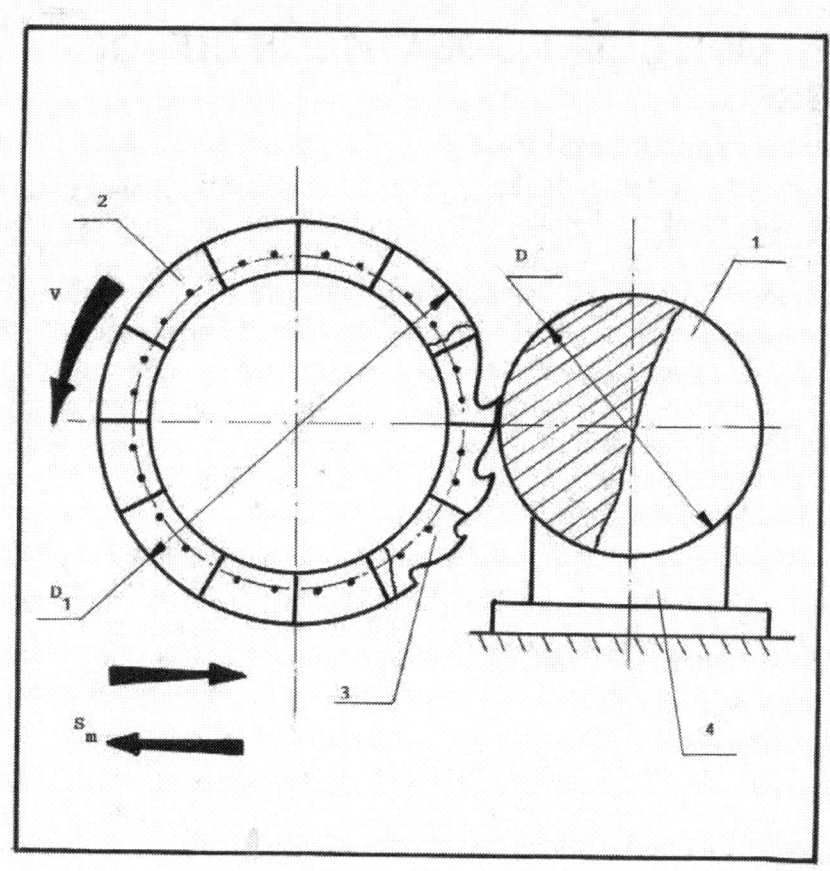

Figure 1 Scheme of cut-off process with using of cold circular segmental saw
(1-bar ;2-cold segmental saw;3-teeth of saw; 4-prism for fixing of bar;
V= cutting speed; S_m= feed of saw; D= diameter of bar; D_1= diameter of saw)

1.2 Shape of circular segmental saw, sharpening of its teeth and some cutting conditions

A. The geometrical parameters of circular segmental saw is used at cutting of stainless steels in process of experiment are the following:
- External diameter of saw D_e = 710 mm (27.95 in)
- Width of saw B=6.5 mm (0.26 in)
- Quantity of saw teeth Z=96

B. Cut-off process of stainless steels is made in industrial cutting conditions:
- Speed V= 13.26 m/min (43.5 fpm)
- Horizontal feed on the teeth of saw F_t =0.03 mm/teeth (0.001 in/teeth)
- Minute of feed S_m =150 mm/min (5.91 ipm)

C. Period of tool life for the cold circular segmental saw has been estimated by the author's method **[6]**.

The period of tool life for circular segmental saw evaluates by method *"Quantity of cut-off billets"* and can be calculated by formula:

2

$$N_g{}^{0.23} = C_v \cdot S_m{}^{0.2} \cdot D_e{}^{0.25} \cdot V^{-1.0} \cdot D^{-0.5} \cdot B^{-0.2} \cdot F_t{}^{-0.2} \cdot Z^{-0.1}$$

where,   (1)

N_g = quantity of cutting billets by saw
C_v = experimental coefficient for cutting material
S_m = minute feed, mm/min
D_e = external diameter of saw, mm
V = cutting speed, m/min
D = diameter of bar (work-piece)
B = width of saw, mm
F_t = horizontal feed on the teeth of saw, mm/teeth
Z = quantity of teeth saw

Sample 1:
To calculate the period of tool life for circular segmental saw with using of formula(1). Assume that we have the following data (in metric unit):
width of saw B= 6.5mm; cutting speed V=13.26 m/min; minute of feed S_m=15mm/min; external diameter of saw D_e=710mm; material of bar- heat –resistance steel ;diameter of bar D=180mm;experimental coefficient for cutting material C_v=45;horizontal feed on the teeth of saw F_t=0.03mm/teeth.

Solution:
The quantity of cutting billets can be evaluated by formula (1)

$$N_g^{0.23} = C_v S_m^{0.2} D_e^{0.25} V^{-1.0} D^{-0.5} B^{-0.2} F_t^{-0.2} Z^{-0.1} = 45(15)^{0.2}(710)^{0.25}(13.26)^{-1.0}(180)^{-0.5}(6.5)^{-0.2}(0.03)^{-0.2}(96)^{-0.1}$$

So, we have the value $N_g^{0.23}$=1.98 or N_g =$(1.98)^{4.35} \approx 20$
Comments:
The quantity of cutting billets in industrial experiment was equal N'_g=18 for these cutting conditions before of re-sharpening circular segmental saw. So, in accordance with experimental data, the average tool life for the cold circular segmental saws for this quantity of cutting billets N'_g=18 was approximately equal T= 140 min before of re-sharpening this circular segmental saw.
So, we see that relative error of calculation the *" Quantity of cutting billets"* by formula (1) is equal

$$\Delta = [(N_g - N'_g)/N_g]100 = [(20-18)/20]100 = 10\% \quad (2)$$

D. Shape of teeth circular segmental saw and sharpening of its has been made in accordance with the recommendations of author **[7]** and **[8]** on the special semi-automatic grinder machine with the following control of sharpening quality, as shown in Figure 2.

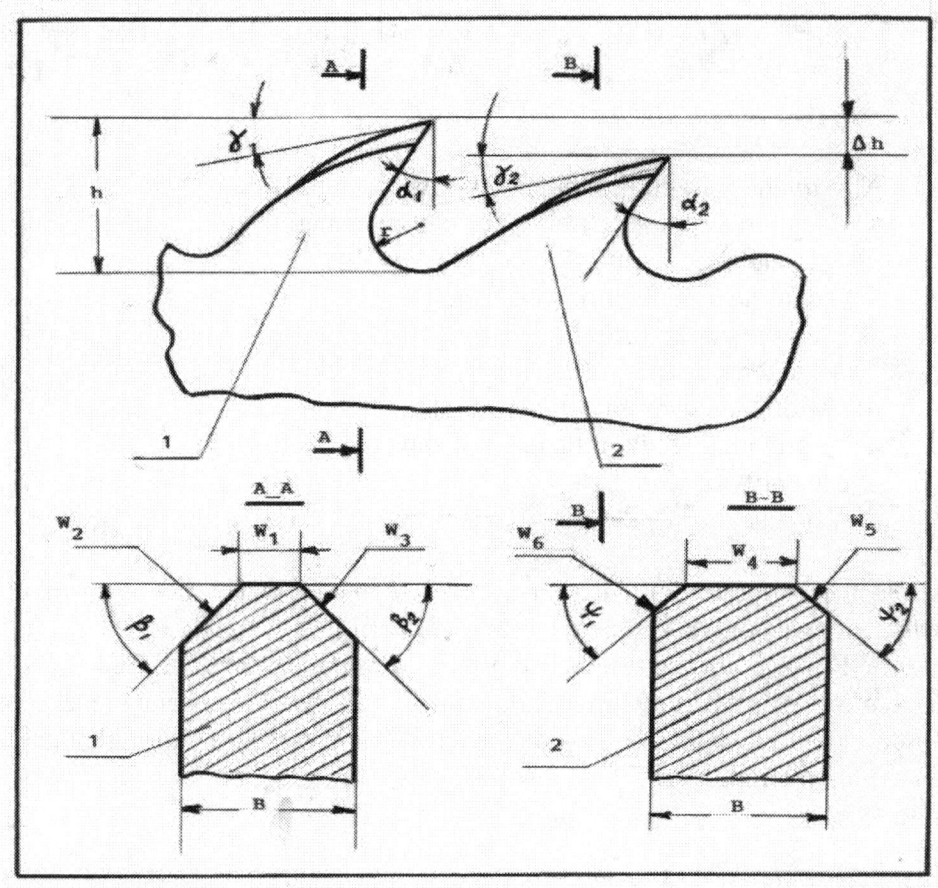

Figure 2 Shape of cold circular segmental saw and sharpening of its teeth:
   1-rough-slotting tooth; 2- finish-scraping tooth

The constructive parameters of circular segmental saw in industrial experiment is the following:

- Clearance angle $\gamma_1=\gamma_2=8°$
- Rake angle $\alpha_1=\alpha_2=12°$
- Difference between of rough-slotting and finish-scraping teeth is equal $\Delta h=3.0$ mm (0.012in)
- Radius between of tooth space r=3.5mm (0.138in)
- Angle of chamber on the rough-slotting tooth $\beta_1=\beta_2=45°$
- Angle of chamber on the finish-scraping tooth $\phi_1=\phi_2=45°$
- Average   width   of   chamber   for   the   rough-slotting   tooth $w_2=w_3=2.5$mm(0.098 in)
- Average width of chamber for the finish-scraping tooth $w_5=w_6=1.0$mm (0.039in)
- Average   width of cutting surface for the rough-slotting tooth $w_1=3.0$mm(0.118in)

- Average width of cutting surface for the finish-scraping tooth $w_4=5.0mm(0.197in)$
- Width of saw $B=6.5mm$ (0.26in)

1.3 The geometrical parameters of stainless chip

In process of industrial experiment is analyzed about of n=156 observation data of the different constructive sizes of stainless chip which is given in **Appendix 1.** The constructive sizes of stainless chip formed by the cold circular segmental saws after of cut-off processes evaluated by measurements with using of universal measure tools.

In Figure 3 is shown the typical form of stainless chip with the main its parameters.

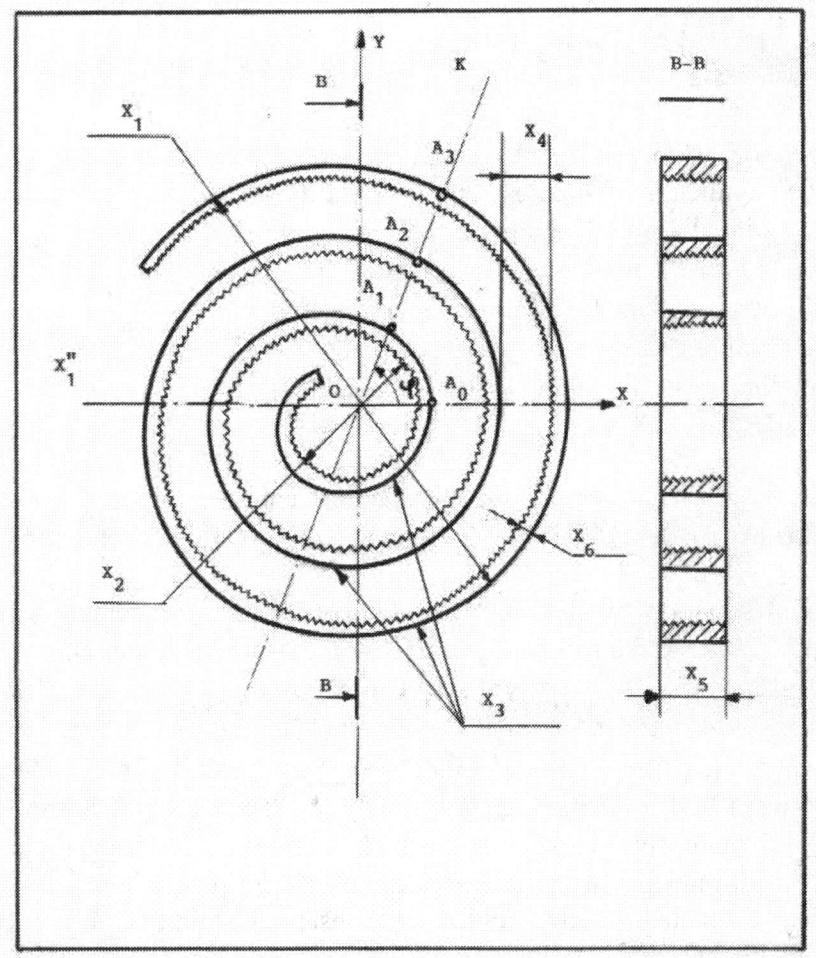

Figure 3 Schematic drawing of constructive parameters of stainless chip formed by the cold circular segmental saws in period of cut-off process
(X_1=average external diameter; X_2=average internal diameter; X_3=number of chip wraps; X_4=clearance between chip wraps; X_5= width of chip; X_6=thickness of chip)

Analyzing the Figure 3 ,we see that stainless chip has view of " Arhimed's spiral which could be described by the mathematical equation view of $\rho= a\varphi$ **[9]** where ρ= polarity radius (length OA_3)

φ= polarity angle ,i.e the angle on which the straight line OK turns from its primary position $XX"_1$

a= the length of way OA_o passed by point in period of turning of straight line OK on the angle 2π.

So, we see that when the value of angle is equal φ=0 , the value of ρ=0.And further with increasing of value φ,the value of polarity radius(ρ) increases proportionally. From Figure 3 we see that variable point $A_1(\rho;\varphi_1)$ of this polarity system moves in positive direction and simultaneous removal from the primary point of this system.

So, the point A_1describes by the Arhimed's spiral and its mathematical equation is ρ=aφ . And besides the straight line OK crosses this spiral curve in the different points A_1,A_2,A_3….. A_n and all these points stand one from another on the equal values, i.e the length $OA_1=A_1A_2=A_2A_3=2\pi a$ (3) .

Sample 2:

To define the parameters of stainless chip in polarity system in presumption that its form submits to " Arhimed's spiral" at data:

- X_1=external diameter of chip , $X_1=\varnothing$ 7.2 mm;
- X_2=internal diameter of chip , $X_2=\varnothing$ 2.2 mm;
- X_4=clearance between of wraps chip,X_4=0.7 mm;
- X_6=thickness of chip ,X_6=0.1mm.

Solution:

1.The clearance between of lengths for this stainless chip (see Figure 3) could be evaluated by the formula (3): $OA_1=A_1A_2=A_2A_3=2\pi a$= ($0.5X_2+X_6$) =1.2mm,where the value a= $0.6\pi^{-1}$.

2.Mathematical equation of this stainless chip describes by formula in polarity system coordinate ρ=aφ and has the following values for the different angles φ:

φ_1=0° ρ_1=0;$\varphi_2=\pi/4$ (45°) $\rho_2=0.6\pi^{-1} (\pi/4)$=0.15mm;$\varphi_3=\pi/2$ (90°) $\rho_3=0.6\pi^{-1}(\pi/2)$=0.30mm; $\varphi_4=3/4\pi$ (135°) $\rho_4=0.6\pi^{-1}(3\pi/4)$ =0.45 mm;$\varphi_5=\pi$(180°) $\rho_5=0.6\pi^{-1} (\pi)$=0.60mm;$\varphi_6=5\pi/4$ (225°) $\rho_6=0.6\pi^{-1} (5\pi/4)$=0.75mm;$\varphi_7=3\pi/2$ (270°) $\rho_7=0.6\pi^{-1} (3\pi/2)$=0.90mm; $\varphi_8= 7\pi/4$ (315 °) $\rho_8=0.6\pi^{-1} (7\pi/4)$=1.05mm; $\varphi_9=2\pi$ (360°) $\rho_9=0.6\pi^{-1} (2\pi)$=1.20mm.

In Figure 4 schematic are shown the different contours of stainless chip evaluated in accordance to formula ρ=aφ in presumption of its form in view of Arhimed's curve (position "a" –evaluated stainless chip) and observed stainless chip (position "b") in industrial experiment.

Comments.

Analysis of Figure 4 shows that there is similarity in both contours for stainless chip (position "a" and "b"). And for this reason we can conclude that the stainless chip formed by the circular segmental saws in cut-off process has the form of "Arhimed's spiral and could be described by the mathematical equation view of ρ=aφ.

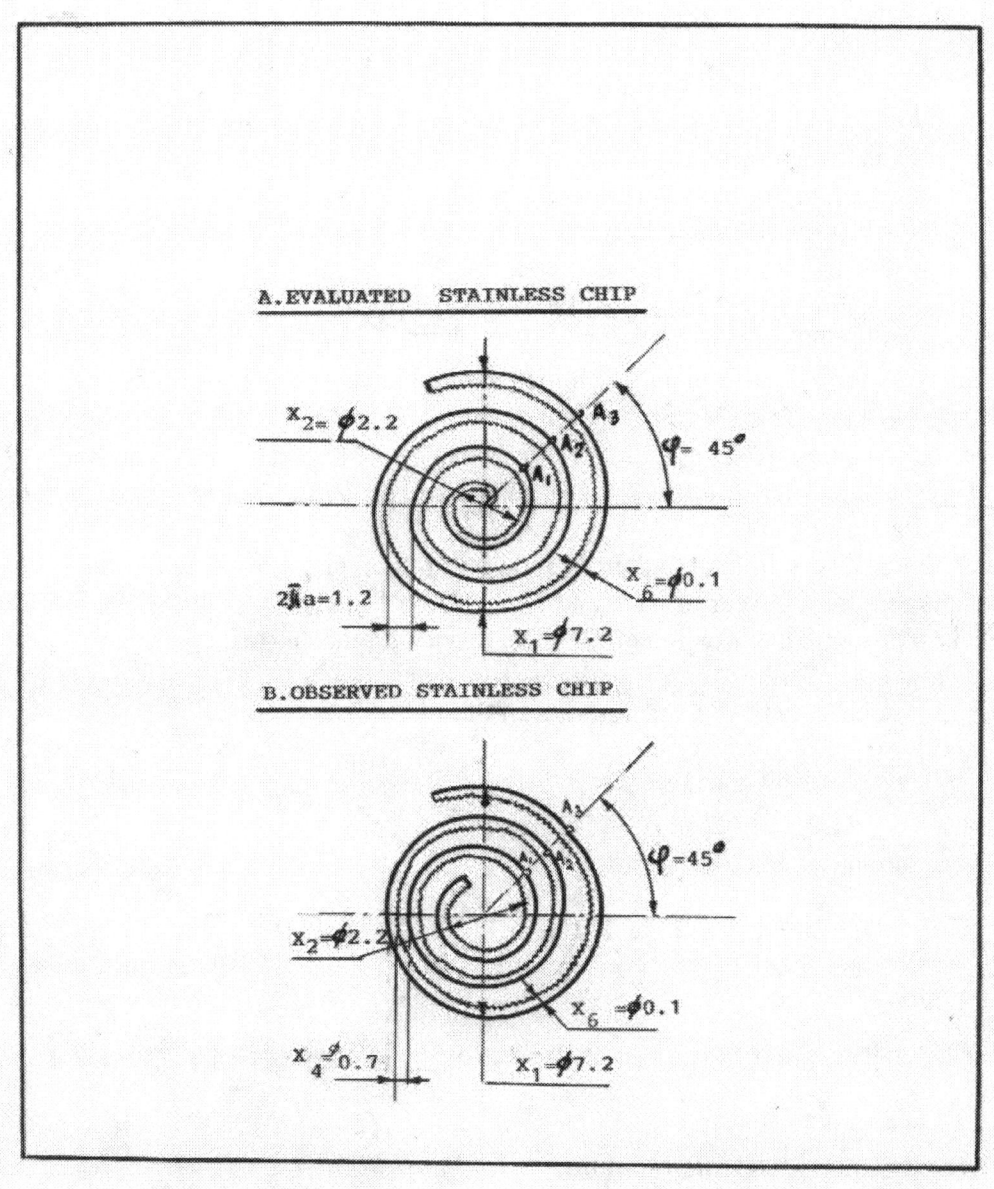

Figure 4 The different contours of stainless chip
(position " a ": evaluated by calculation ; position " b " : observed by experiment)

1.4 Analysis of tool life for the cold segmental circular saw

 Application of cold segmental saws on the cut-off machines in manufacturing production has some advantages (absence of burn material in cut zone, etc.) and the some time has disadvantages. The main parameter of their disadvantages is the low tool life particularly at cut-off process for the billets of big diameter for the different materials, advantageously of stainless and heat-resistance steels.
It is known that period of tool life for the circular saw evaluates by the equation of view

$$T = N_g \, (D/S_m) \quad [10] \qquad\qquad (4)$$

where,

T= period of tool life, in min;

N_g= quantity of cut-off billets by saw for the period of its tool life (this parameter evaluates by formula (1));

D= diameter of bar (work-piece) ,in mm;

S_m= minute feed ,in mm/min.

Analyzing the formula (4) we see that ratio $D/S_m = t_0$ and then we have **$T = N_g t_0$** (**4a**) ,where t_0=machining time for one billet in cut-off process ,in min.

So, we can conclude that tool life (T) of cold circular segmental saw is function of quantity (N_g) of cut-off billets for this cutting process and its machining time (t_0) for one billet ,i.e we have **$T = \varphi \, (N_g; t_0)$** (**4b**)

These conclusions well coordinate with the recommendation of author **[11]** which underlines that: " *Period of tool life of saw also can be evaluated by the quantity (N_g) of cut-off billets…* ".

This is the best way in manufacturing production to evaluate in-time the period of tool life (T) than measure the wear of back and front faces of the teeth, as shown in Figure 2, of the cold circular segmental saw in period of its service on the cut-off machines.

For evaluation of tool life (T) of the cold segmental saw at cut-off process for stainless steels uses such statistical characteristics of distribution as **[12]**:

- Sample mean $\bar{T} = (\sum_{i=1}^{n} T_i \,)/\, n = (1/n) \sum_{i=1}^{n} T_i \quad (5)$, where $\hat{\mu} = \bar{T}_i$

- Sample standard deviation

$$\sigma = S = [\sum_{i=1}^{n} (T_i - T)^2 / n\text{-}1]^{0.5} = [(1/n\text{-}1) \sum_{i=1}^{n} (T_i - \bar{T})^2]^{0.5} \quad (6)$$

- Standard error of the estimate $\bar{T}$ of $\hat{\mu}$ (mean): $\quad \sigma_T = \sigma/ (n)^{0.5} \quad (7)$

where,

T_i= average value i-interval of tool life

n= observation data (n=70)

$\bar{T}$= average value (population mean μ)

S=average deviation of random value T_i from $\bar{T}$ (standard deviation).

In Figure 5 is shown symmetric (normal) distribution of tool life for the cold circular segmental saw at cut-off processes.

8

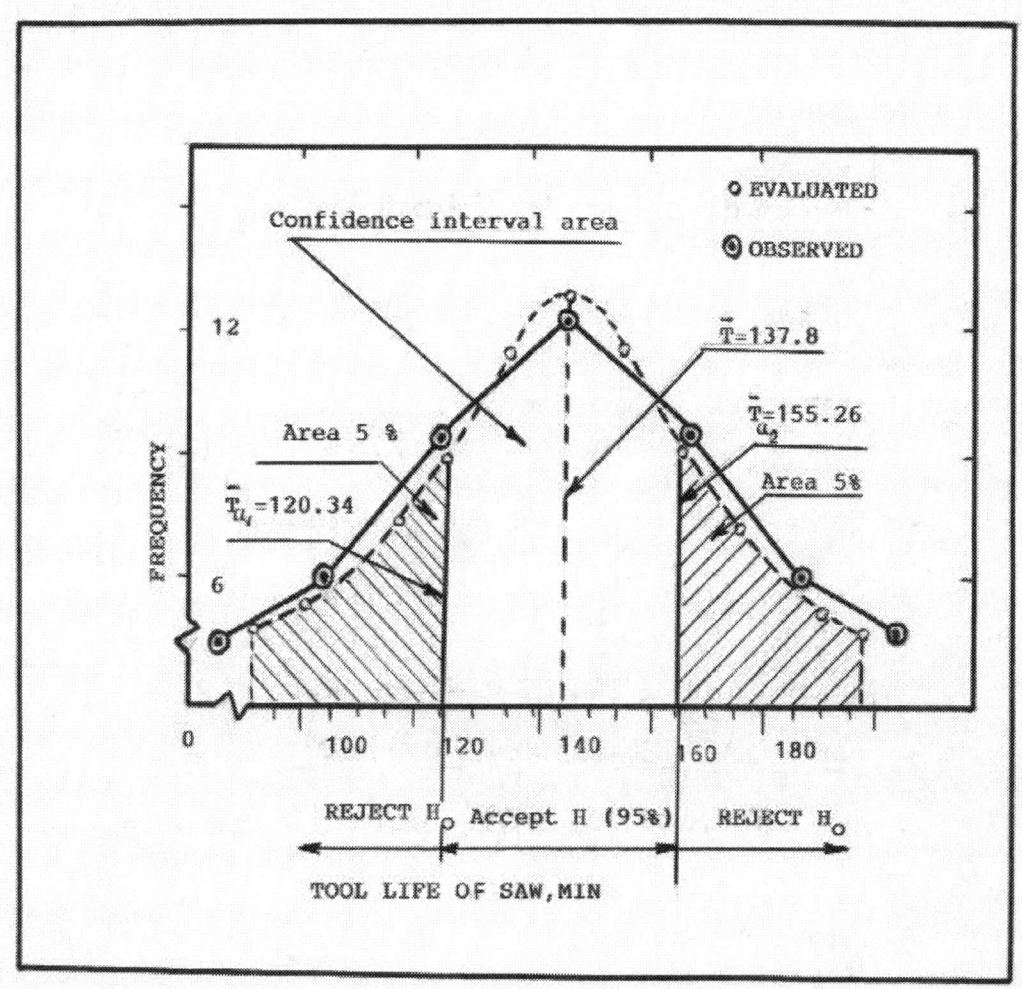

Figure 5 **Histogram of experimental data and evaluated normal (binomial) distribution of tool life of the cold circular segmental saw at cut-off processes (at cutting conditions : diameter of bar D= 70÷80 mm; cutting speed V=25.6 m/min; minute feed S_m=15 m/min)**

In result of calculation we have the following data: $\overline{T}$ =137.8 min; $\overline{S}$ =74.6 min; σ_T =8.91 min.
 As the value $\overline{T}$ normally distributed then 95 percent of the values of $\overline{T}$ will be contained two standard deviations –more correctly 1.96 standard deviations ,as shown in Figure 5. And this conclusion could be expressed by equation view of

$$P_{rob}\left(\overline{T} - 1.96\sigma_T \leq \overset{\wedge}{\mu} \leq \overline{T} + 1.96\sigma_T\right) = 0.95 \qquad (8)$$

9

where

($\overline{T} \pm 1.96\sigma_T$) = confidence interval is equal $T_a = \overline{T} \pm \Delta$ (where $\Delta = 1.96\sigma_T = 17.46$ min). For estimation of confidential value of real interval of tool life and average of its deviation , we use the formula of view

$$\left| T_a - \overline{T} \right| \angle t\, (\, P,k\,)\, [S \cdot n^{0.5}] \quad (9) \quad \text{and} \quad S(1-q) \angle \sigma_s \angle S(1+q) \quad (10)$$

where
T_a = real value of tool life, min
$t(P,k)$ = Student's distribution
σ_s = confidential value of average square error which is equal:
$T_a - 137.8 | \leq 1.995[74.6 \cdot 70^{0.5}]$ and then we have $T_a = (138.8 \pm 17.46)$min

and $120.34 \leq \overline{T} \leq 155.26$ where $T_{a1} = 120.34$ min and $T_{a2} = 155.26$ min .

For examination of conformity of empirical distribution (Figure 5) to the normal distribution for the cold segmental saw, we use the criteria Pearson X^2 as chi-square test which is equal:

$$X^2 = \sum_{i=1}^{n} (\, h_m - n \cdot p_m\,)^2 / (\, n \cdot p_m\,) \quad (11)$$

with degree of freedom equal by $k = m - p_1 - 1$.

Where
m = number of comparative frequencies
np_m = theoretical frequency of I-interval of values T
p_1 = number of theoretical distribution is equal $p_1 = 2$
h_m = empirical frequency of I-interval of the values T_i .

In result of calculation we have $X^2 = 7.24$ and k=5. And with account of inequality

$$P\,[\, X^2 \leq X^{*2}_{0.05;5}\,] = \alpha \quad (12) \quad \text{we have that} \quad X^2 \leq X^{*2}\ (\ 7.24 \leq 11.10\ ) \text{ and}$$

for this reason we accept the null hypothesis (H_0) that empirical and theoretical distribution have the normal law of distribution for the cold segmental saws.
Therefore with probability P=0.95 we can conclude that tool life for the cold circular segmental saw submits to the normal law of distribution in accordance with equation of

$$\varphi\,(\, T\,) = [\ (1/S) \cdot (\, 2\pi\,)^{0.5}\,] \cdot e^{-\,[T_i - \overline{T}]^2 / (2S^2)} \quad (13)$$

The estimation of tool life for the cold segmental saw by formulas (1) has some deviations which do not exceed one percent with comparative observation of data some authors [13] and [7] .

So, the above-named formulas (1) could be used in calculation of tool life for the cold circular segmental saws in manufacturing on the cut-off machines in cutting processes advantageously of the stainless and heat-resistance steels.

Analysis and calculation of tool life for saw permitted in manufacturing objectively to evaluate the total term of service and consumption of cutting tool, using the formula of

$$T_w = (K_1 + 1) \overline{T} \qquad (14)$$

where

T_w = total term of tool life for saw until of its wear in service ,min (for the cold circular segmental saw of diameter D_e =710 mm the value of T_w = 336 hours)

K_1= number of potentialities inherent for cutting tool in question of its re-sharpening

$\overline{T}$ = average tool life for the cold circular segmental saw before of its re-sharpening, min.

For calculation of rate consumption (A) for the cold segmental saw recommends to use the formula of view

$$A = [(t_0 \cdot N_g)/(60 \cdot T_w)] \cdot K_y \qquad (15)$$

where

K_y =coefficient showing the accidental losses of cutting tool ,K_y=1.02.

In Figure 6 is shown the graphic for calculation of machining time for one work-pierce in cut-off process at the different cutting conditions [normative recommendations (1) and experimental statistical (2) observation data].

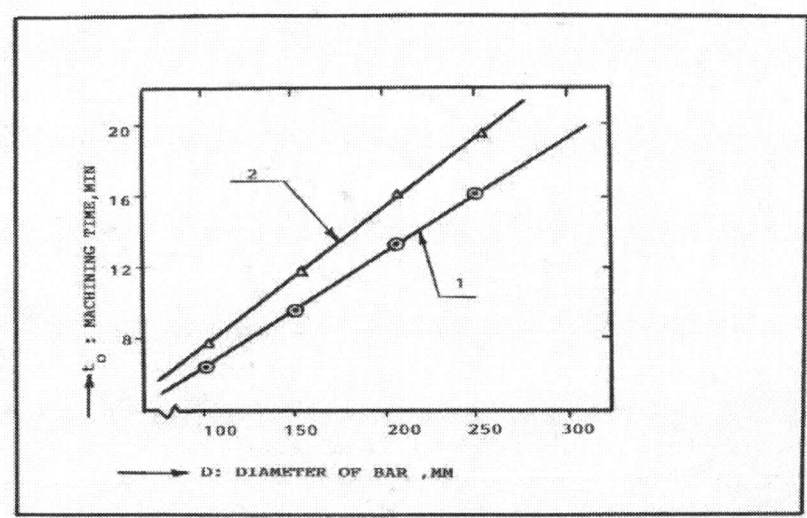

Figure 6 Graph for calculation of machining time for one stainless work-piece in cut-off process by the cold circular segmental saw [1-normative cutting conditions with linear regression equation t_0= -0.05+ 0.067D (16) and 2- experimental statistical cutting conditions with linear regression equation t_0* = - 0.013+0.077D (17)]

From Figure 6 we see that at experimental-statistical cutting conditions (2) , the machining (technological) time submits to linear regression equation t_0= - 0.013 + +0.077D and has the larger value than for the normative cutting conditions (1) and this could be explained by some reasons:

- Presence of imperfect cut-off machine, i.e the absence of good technological accuracy of cut-off machine that forces to decrease the saw feed (S_m) in cutting processes
- There are some negative moments in process of sharpening and re-sharpening of teeth saw
- There are difficulties of removal stainless chip and cleaning of saw teeth in cut-off processes.

In Figure 7 is shown the comparative analysis of required quantity of cold circular segmental saw in the different conditions (normative and experimental-statistical cutting processes).

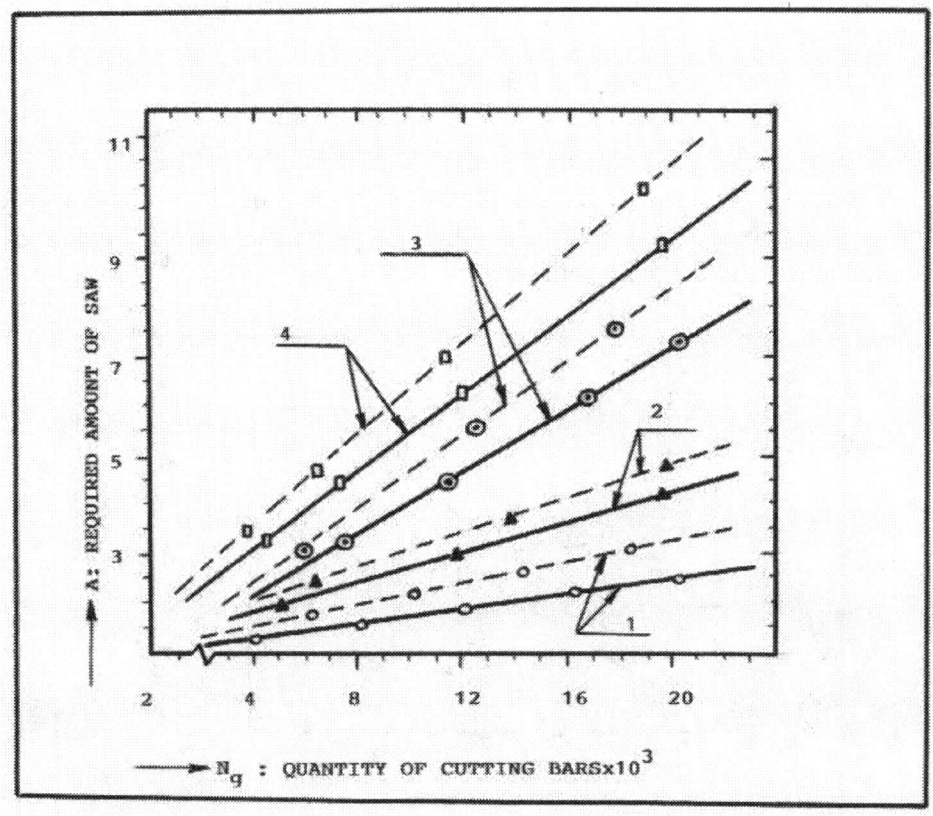

Figure 7 Graph for evaluation of required quantity for the cold circular segmental saws in cut-off process for the different diameters of stainless round bars in dependence of manufacturing program [1-Ø50mm;2-Ø100mm;3-Ø150mm;4-Ø200mm; ——— normative cutting conditions; ---------experimental cutting conditions]

Analysis of Figure 7 shows that required amount of saws increases with increasing of diameter bar and manufacturing program in production and has the linear regression model.

12

And besides we also see that required amount of saws are larger for the experimental-statistical cutting conditions than for the normative conditions.

One obvious advantage in using of multiple linear regression model for estimation of required amount for the cold circular segmental saws at different cutting conditions (1) and (2)was shown in Figure 7.

So, using this multiple regression analysis we have the following model with two independent variables view of $Y_i = b_0 + b_1 x_{i,1} + b_2 x_{i,2}$ (i=1,2..n)   (18)
where
 two independent variables:
$X_{i,1}$ =diameter of cutting bar (D)
$X_{i,2}$ = quantity of cutting bars (N_g)
and dependent variable: Y_i = required amount of saws (A).
Three unknown coefficients b_0 , b_1 and b_2 can be calculated from the normal system equations:

$$\Sigma\,Y = n_1 b_0 + b_1\,\Sigma X_1 + b_2 \Sigma X_2$$
$$\Sigma\,X_1 Y = b_0 \Sigma X_1 + b_1 \Sigma X^2_{\,1} + b_2 \Sigma X_1 X_2 \qquad (19)$$
$$\Sigma X_2 Y = b_0 \Sigma X_2 + b_1 \Sigma X_1 X_2 + b_2 \Sigma X^2_{\,2}$$

A. At normative cutting conditions (1) we have:

$\Sigma Y = 70; n_1 = 20; \Sigma X_1 = 2.5 \cdot 10^3 ; \Sigma X_2 = 2.4 \cdot 10^5 ; \Sigma X_1 Y = 1.1 \cdot 10^4 ; \Sigma X_2 Y = 10^6 ; \Sigma X_1^{\,2} = 3.8 \cdot 10^6$;
$\Sigma X_1 X_2 = 0.3 \cdot 10^8$; $\Sigma X_2^{\,2} = 3.5 \cdot 10^9$.
So, the coefficients are equal: $b_0 = -4.225$; $b_1 = 0.033$ and $b_2 = 0.3 \cdot 10^{-3}$ and fitted regression line for required amount of saw at normative cutting is equal

$$\hat{Y} = -4.225 + 0.033 X_1 + 0.3 \cdot 10^{-3} X_2 \qquad (20)$$

or $A = -4.225 + 0.033 D + 0.3 \cdot 10^{-3} N_g$

The coefficient of determination R^2 for evaluation of formula (20) is equal:

$$R^2 = [\,\Sigma(\,\overset{-}{Y_i - Y}\,)^2 - \Sigma\,(Y_i - \hat{Y}\,)^2\,] / \Sigma\,(Y_i - \overset{-}{Y}\,)^2 \qquad (21)$$

where
$\Sigma(Y_i - \overset{-}{Y}\,)^2 = 129$; $\Sigma\,(\,Y_i - \hat{Y}\,)^2 = 14.169$ and $R^2 = 0.89$.
That is the bivariate regression accounts for 89 percent of the variability in Y.

B. At experimental- statistical cutting conditions(2):

Using the formula (19) we have $\Sigma Y = 85$; $n_1 = 20$; $\Sigma X_1 = 2.5 \cdot 10^4$; $\Sigma X_2 = 2.4 \cdot 10^5$;
$\Sigma X_1 Y = 1.3 \cdot 10^4$; $\Sigma X_2 Y = 1.2 \cdot 10^5$; $\Sigma X_1^{\,2} = 3.8 \cdot 10^5$; $\Sigma X_1 X_2 = 0.3 \cdot 10^8$; $\Sigma X_2^{\,2} = 3.5 \cdot 10^9$.
So, the coefficients are equal : $b_0 = -1.20$; $b_1 = 0.034$ and $b_2 = 0.1 \cdot 10^{-3}$.
Thus, the fitted regression equation for required amount of saws at experimental-statistical cutting conditions is equal
$\overset{\wedge}{}$

13

$$Y = -1.20 + 0.034X_1 + 0.1 \cdot 10^{-3}X_2 \qquad (22)$$

or $\quad A = -1.20 + 0.034D + 0.1 \cdot 10^{-3}N_g$

The coefficient of determination R^2 is computed using the formula (21) is $R^2 = 0.706$

where $\sum(Y_i - \overline{Y})^2 = 141.718$; and $\sum(Y_i - \hat{Y}) = 41.70$.

In Figure 8 is shown separately the scatter plots of diameter bar (D) versus required amount of saws (A) and quantity of cutting bars (N_g) versus required amount of saws (A) for normative cutting conditions.

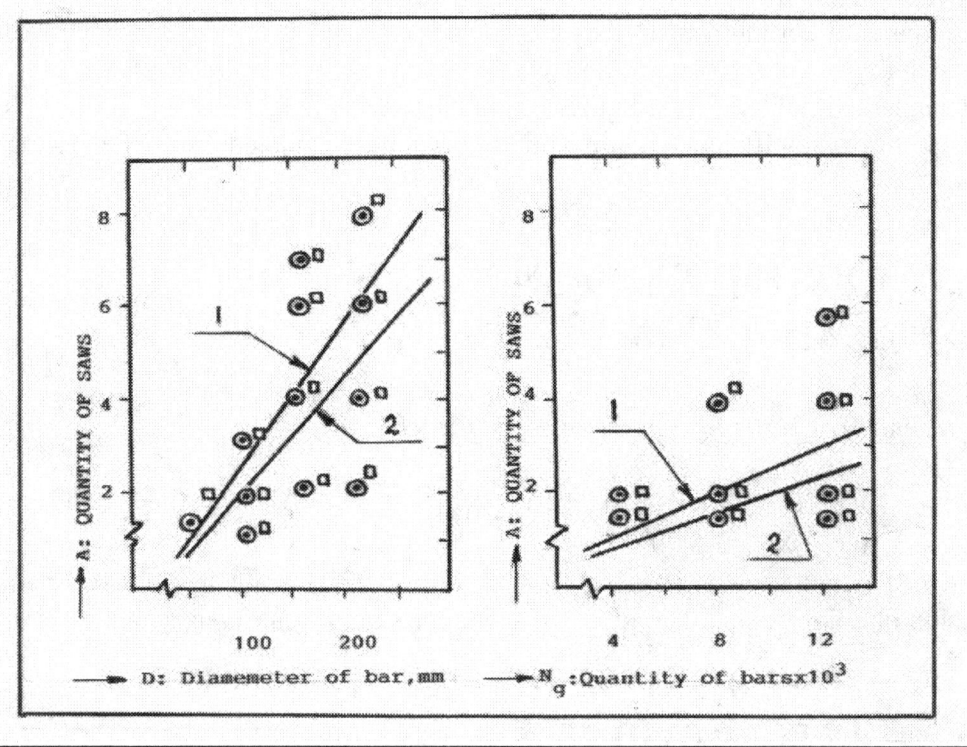

**Figure 8 Scatter plots of diameter bar (D) versus required amount of saws (A)
and quantity of cutting bars (N_g) versus of required amount of saws (A)
[1-at normative cutting conditions (⊕) has the linear regression equations**

$A_1 = -0.2 + 0.03\ D$ **and** $A_1^* = 0.6 + 0.22 \cdot 10^{-3}N_g$ **;2- experimental-statistical cutting
conditions (   ) has the linear regression equations** $A_2 = -1.0 + 0.03\ D$ **and**
$A_2^* = -1.3 + 0.25 \cdot 10^{-3}N_g$ **]**

Figure 8 shows that the first variable X_1 (D) and the second independent variable X_2 (N_g) have a good positive linear relationship with Y.

The residual plots (residual versus X_1 and residual versus X_2) are illustrated in Figure 9.

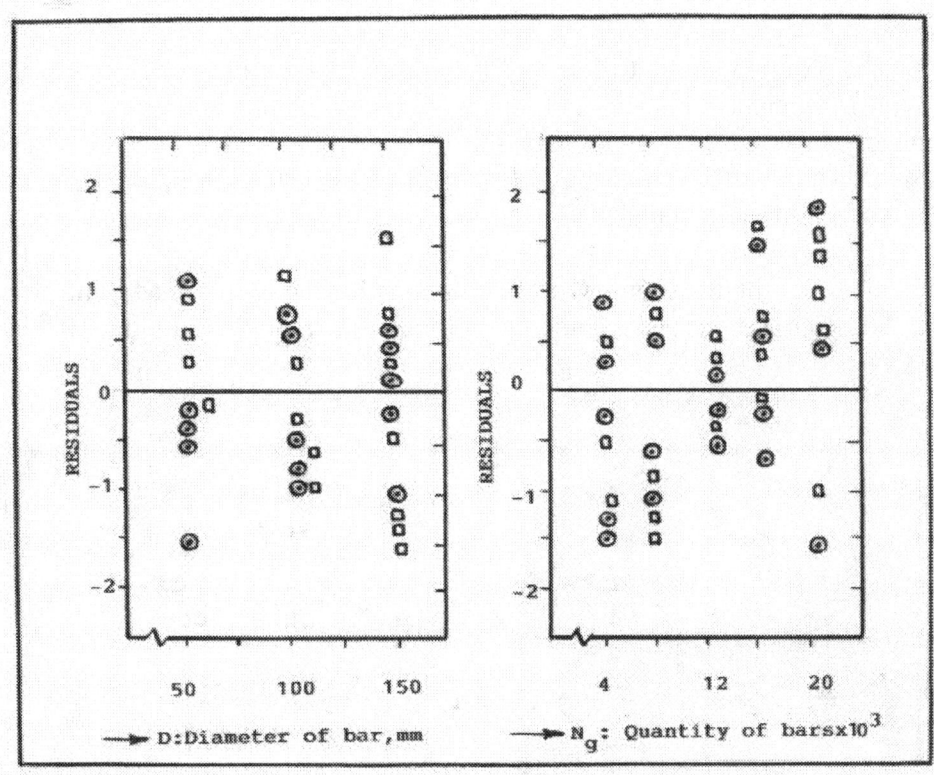

Figure 9 Residual plots for the different cutting conditions in evaluation of amount circular segmental saws [⊕ normative and experimental-statistical)

The residual plots from Figure 9 show that the data have been " rigged " –both plots are too systematic for the Y observation to have occurred at random.

Sample 3
To calculate the required amount of cold circular segmental saws at normative and experimental-statistical cutting conditions:
- Diameter of bar D= 150mm
- Cutting material =stainless steel
- Quantity of cutting bars $N_g = 8 \cdot 10^3$

Solving:
Using the formula (20) for the normative cutting conditions we could calculate the required amount of saws

$A = -4.225 + 0.033 \, D + 0.3 \cdot 10^{-3} N_g = -4.225 + 0.033(150) + 0.3 \cdot 10^{-3} \, (8 \cdot 10^3) \approx 3$

And using the formula (22) we have for the experimental-statistical cutting conditions the required amount of saws:

$A = -1.20 + 0.034D + 0.1 \cdot 10^{-3} N_g = -1.20 + 0.034(150) + 0.10 \cdot 10^{-3} \, (8 \cdot 10^3) \approx 5$

15

Comments:

Analyzing the required amount of saws for the different cutting conditions we see that in industry (experimental-statistical cutting conditions) demands more amount of saws approximately in 1.5 times larger than this by recommended for the normative cutting conditions and this could be explained by result of imperfection of cut-off machines and sharpening of circular segmental saws.

With objective of simplification of the calculations for the required amount of saws at the different cutting conditions is designed the nomogram ,as shown in Figure 10.

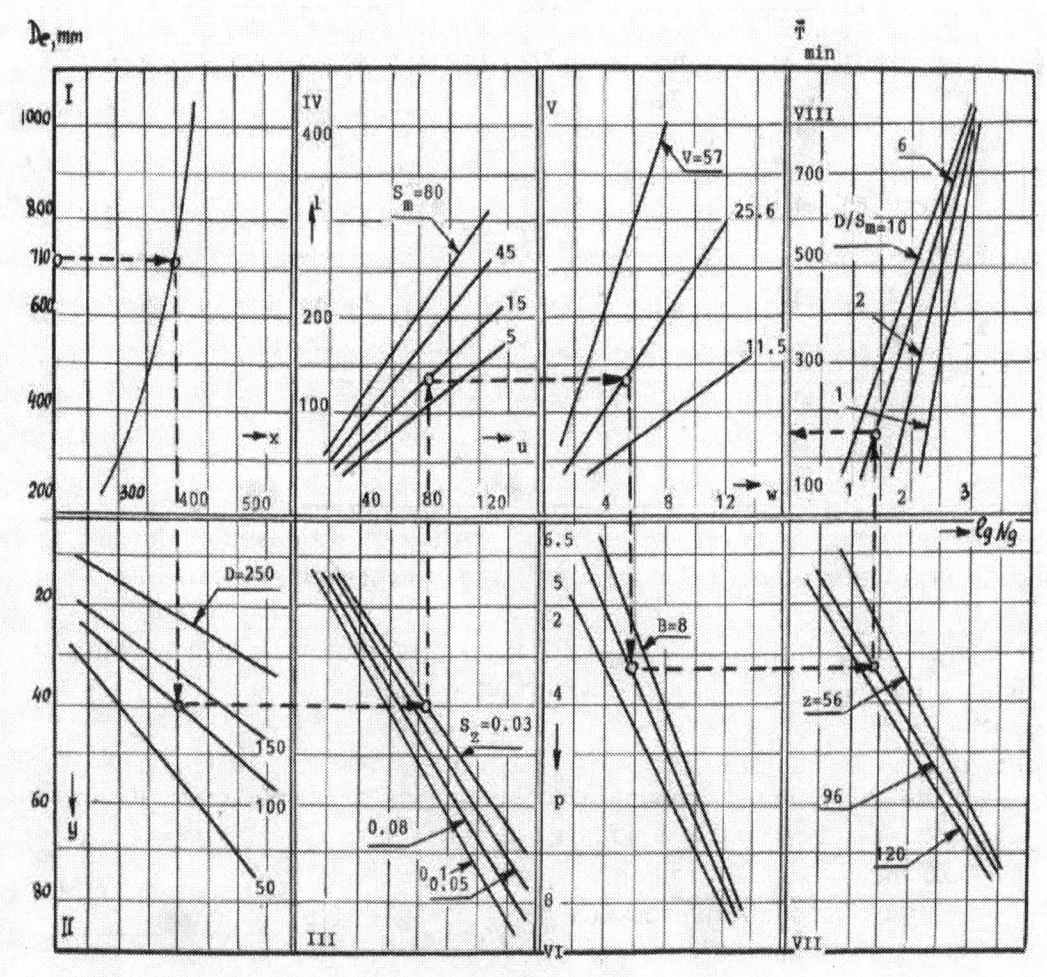

Figure 10 Nomogram for evaluation of tool life for the cold circular segmental saws

In base of this nomogram is put the equations (1) and (4). Designing of nomogram is made into account of introducing some auxiliary variables such as x, y, u, l, w and p in equation (1):

$$x=C_v \cdot D_e^{0.25} ; \quad y=x \cdot D^{-0.5} ; \quad u=y \cdot S_z^{-0.2} ; \quad i=u \cdot S_m^{0.2} ; \quad w=l \cdot v^{-1.0} ; \quad p=w \cdot B^{-0.2} .$$

16

So, for designing of nomogram we have finally expression in view of $N_g = p \cdot z$ for calculation of cut-off bars by the cold segmental saws.

Sample 4

In Figure 10 is shown as the sample (in dash lines) of using nomogram for evaluation of tool life (T_i) in view of circular segmental saw and also calculation of quantity of cut-off stainless and other bars (N_g) at this process at the following data:

1. External diameter of cold circular segmental saw D_e=710mm
2. Diameter of cut-off bar D=100mm
3. Feed on the tooth S_z =0.03mm/tooth
4. Minute feed of saw S_m=15mm/min
5. Cutting speed V= 25.6 m/min
6. Width of saw B=6.5 mm
7. Quantity of saw teeth Z=96

From nomogram ,shown in Figure 10, we find the quantity of cut-off bars (log N_g=1.4 and N_g=25) and also the tool life for circular segmental saw (T =150 min) before of its re-sharpening.

SUMMARY

1. *Cutting of stainless and heat-resistance steels on the cut-off machines by the cold circular segmental saws expediently to make for the bar having the external diameter more than 100 mm*
2. *The tool life of cold circular segmental saw in manufacturing processes should be evaluated better by the quantity of cutting bars on the cut-off machines at experimental-statistical cutting conditions and calculate by the formula (1)*
3. *The tool life of cold circular segmental saw has the normal distribution and could be evaluated by the methods of mathematical statistics with using of regression models*
4. *Evaluation of tool life until of its re-sharpening for the cold circular segmental saw in real manufacturing processes gives the opportunity to evaluate the required amount of saws and tool life of saw at the different cutting conditions*

CHAPTER TWO CHARACTERISTICS OF POLYGON DISTRIBUTION FOR THE STAINLEES CHIP

2.1 Frequency polygon for each parameter of the stainless chip

The experimental data of constructive sizes of chip formed in period of cutting stainless chip by the cold circular segmental saws were obtained from the total number of observations N=42 in process of industrial experiment and have the following parameters:

X_1 =average external diameter

X_2 = average internal diameter

X_3 =number of chip wraps

X_4 =clearance between chip wraps

X_5 =width of chip

X_6 = thickness of chip.

In Figure 1 is shown the polygon of distribution for values displayed in Table 1 for external diameter (X_1) of stainless chip.

Table 1 Grouped frequency distribution of measures of external diameter X_1 for the stainless chip

Values X_1	Taillies	Midpoints	f	rf (%)	cf	Cumulative(%)
11.9÷12.2	/	12.05	1	2.38	42	100
11.6÷11.9	/	11.75	1	2.38	41	97.6
11.3÷11.6	///	11.45	3	7.14	40	95.2
11.0÷11.3	//////	11.15	8	19.06	37	88.0
10.7÷11.0	/	10.85	1	2.38	29	69.1
10.4÷10.7	/	10.55	1	2.38	28	66.6
10.1÷10.4	////	10.25	4	9.52	27	64.3
9.8÷10.1	//////////////	9.95	13	30.96	23	54.7
9.5÷9.8	/	9.65	1	2.38	10	23.8
9.2÷9.5	/	9.35	1	2.38	9	21.4
8.9÷9.2	////	9.05	4	9.52	8	19.0
8.6÷8.9	-	8.75	0	0	4	9.5
8.0÷8.3	//	8.45	2	4.76	4	9.5
8.0÷8.6	-	8.15	0	0	2	4.70
7.7÷8.0	/	7.85	1	2.38	2	4.7
7.4÷7.7	/	7.55	1	2.38	1	2.4

The main formulas for calculation *of cumulative = [(cf) /n]100 %* (1) and *rf = (f /n) 100 %* (2)

18

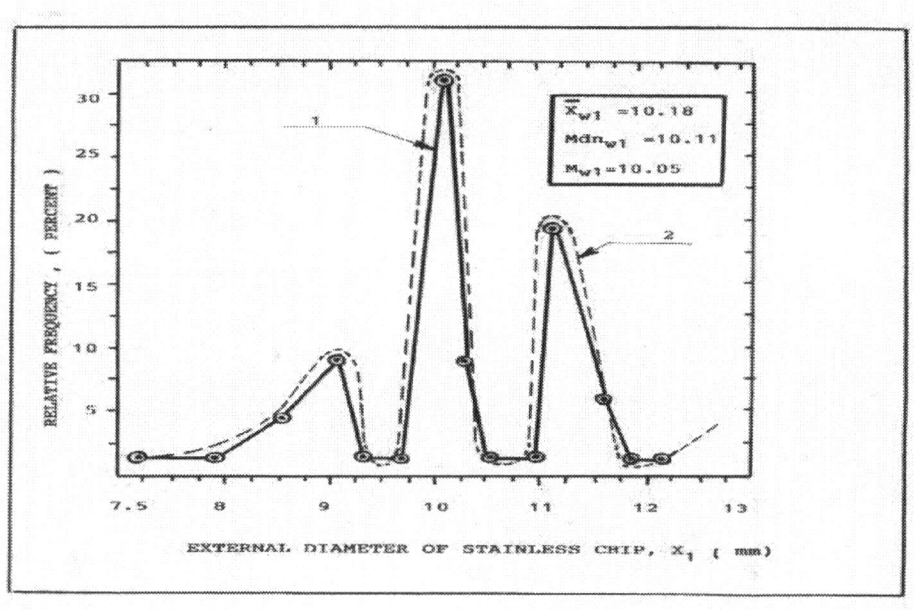

Figure 1 Frequency polygon for external diameter (X_1) of the stainless chip (1-observed;2- evaluated)

In Figure 2 is shown the polygon of distribution of values displayed in Table 2 for internal diameter of the stainless chip.

Table 2 Grouped frequency distribution of measures for internal diameter X_2 of the stainless chip

Values X_2	Tallies	Midpoints	f	rf (%)	cf	Cumulative (%)
8.6÷9.0	/	8.80	1	2.38	42	100
8.2÷8.6	/	8.40	1	2.38	41	97.6
7.8÷8.2	/	8.00	1	2.38	40	95.2
7.4÷7.8	////	7.60	4	9.52	39	92.9
7.0÷7.4	/	7.20	1	2.38	35	83.3
6.6÷7.0	////	6.80	4	9.52	34	80.9
6.2÷6.6	///////	6.40	7	16.70	30	71.4
5.8÷6.2	/	6.00	1	2.38	23	54.7
5.4÷5.8	///	5.60	3	7.14	22	52.4
5.0÷5.4	////	5.20	4	9.52	19	45.2
4.6÷5.0	///	4.80	3	7.14	16	38
4.2÷4.6	/	4.40	1	2.38	13	30.9
3.8÷4.2	//	4.00	2	4.76	11	26.2
3.4÷3.8	///	3.60	3	7.14	9	21.4
3.0÷3.4	///	3.20	3	7.14	6	14.3
2.6÷3.0	/	2.80	1	2.38	3	7.10
2.2÷2.6	/	2.40	1	2.38	2	4.7
1.8÷2.2	/	2.00	1	2.38	1	2.4

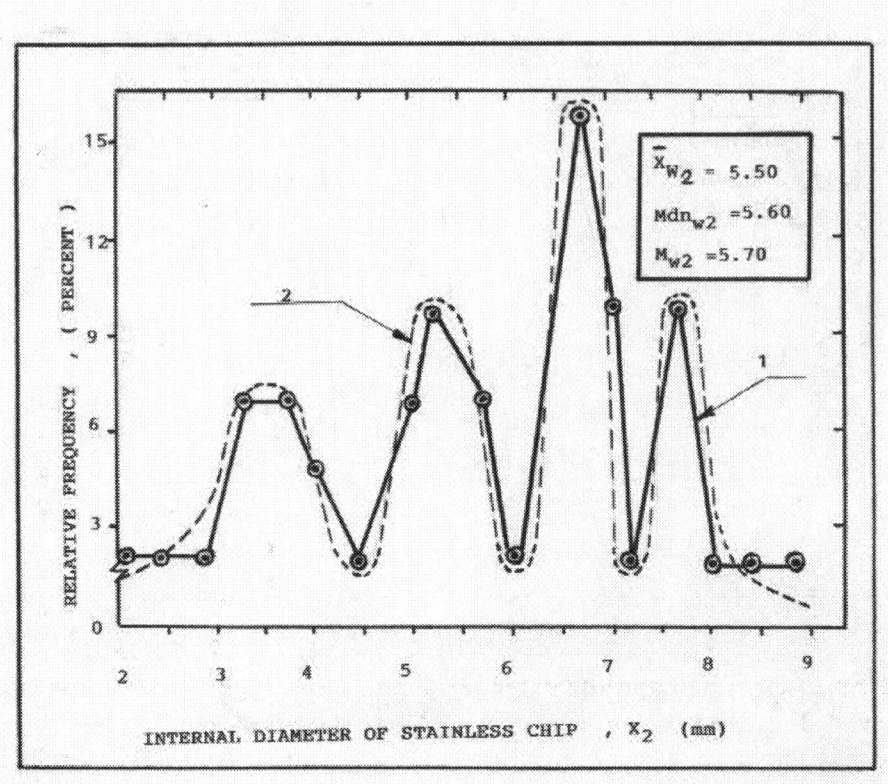

Figure 2 Frequency polygon for internal diameter (X_2) of stainless chip (1- observed ; 2 – evaluated)

In Figure 3 is shown the polygon of distribution of the values displayed in Table 3 for the number of chip wraps (X_3) of stainless chip.

Table 3 Grouped frequency distribution of measures for the number chip wraps X_3 of stainless chip

Values X_3	Tallies	Midpoints	f	rf (%)	cf	Cumulative (%)
5.0÷5.5	/	5.25	1	2.38	42	100
4.5÷5.0	///	4.75	3	7.14	41	97.6
4.0÷4.5	-	4.25	0	0	38	90.5
3.5÷4.0	/////////// ///	3.75	13	30.94	38	90.5
3.0÷3.5	-	3.25	0	0	25	59.5
2.5÷3.0	/////////// ////	2.75	19	45.26	25	59.5
2.0÷2.5	///	2.25	3	7.14	6	14.3
1.5÷2.0	-	1.75	0	0	3	7.2
1.0÷1.5	//	1.25	2	4.76	3	7.2
0.5÷1.0	/	0.75	1	2.38	1	2.4

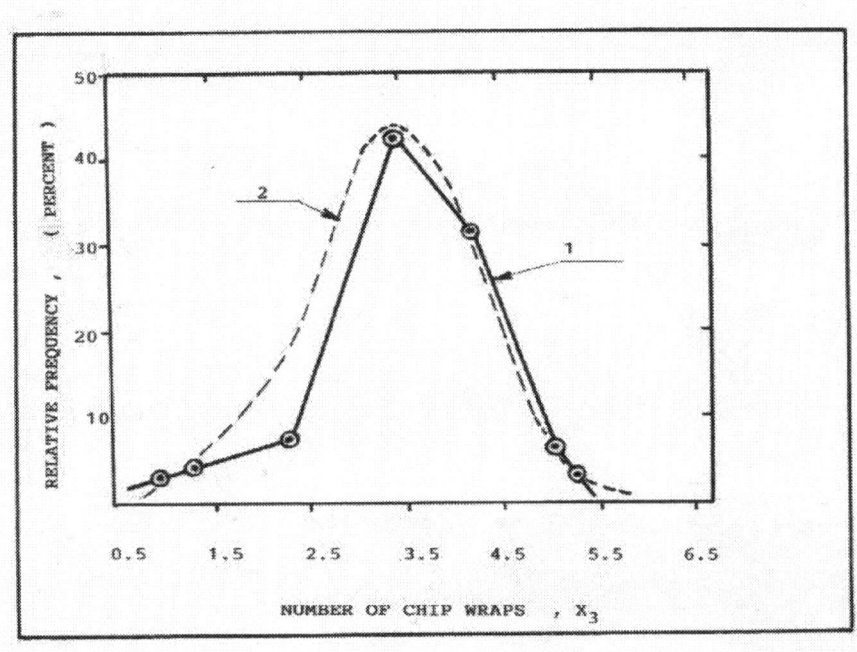

Figure 3   Frequency polygon for the number of chip wraps (X_3) of stainless chip (1 –observed ; 2- evaluated)

In Figure 4 is shown the polygon of distribution of the values displayed in Table 4 for clearance between of chip wraps (X_4).

Table 4 Grouped frequency distribution of measures for clearance between chip wraps X_4 of stainless chip

Values X_4	Tallies	Midpoints	f	rf (%)	cf	Cumulative (%)
0.66÷0.70	/	0.68	1	2.38	42	100
0.62÷0.66	-	0.64	0	0	41	97.6
0.58÷0.62	-	0.60	0	0	41	97.6
0.54÷0.58	-	0.56	0	0	41	97.6
0.50÷0.54	-	0.52	0	0	41	97.6
0.46÷0.50	//////	0.48	7	11.90	41	97.6
0.42÷0.46	-	0.44	0	0	36	85.7
0.38÷0.42	////////	0.40	9	21.44	36	85.7
0.34÷0.38	-	0.36	0	0	27	64.3
0.3÷0.34	-	0.32	0	0	27	64.3
0.26÷0.30	/////////	0.28	10	23.8	27	64.3
0.22÷0.26	-	0.24	0	0	18	42.9
0.18÷0.22	////////	0.20	9	21.44	18	42.9
0.14÷0.18	-	0.16	0	0	7	16.6
0.10÷0.14	/////	0.12	6	14.28	7	16.6

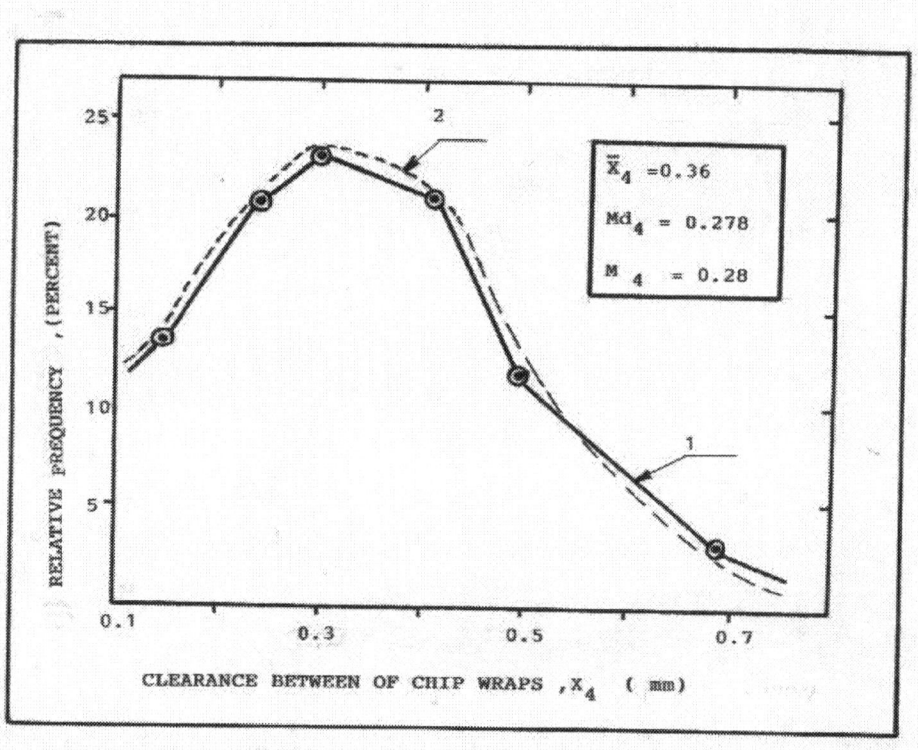

Figure 4 Frequency polygon for clearance between of chip wraps (X_4) of stainless chip (1-observed ; 2 – evaluated)

In Figure 5 presents the polygon of distribution of the values displayed in Table 5 for width of stainless chip.

Table 5 Grouped frequency distribution of measures for width X_5 of stainless chip

Values X_5	Tallies	Midpoints	f	rf (%)	cf	Cumulative (%)
7.0÷7.2	/	7.1	1	2.38	42	100
6.8÷7.0	/	6.9	1	2.38	40	95.2
6.6 ÷6.8	//	6.7	2	4.76	39	92.9
6.4÷6.6	/////	6.5	5	11.9	37	88.1
6.2÷6.4	///	6.3	3	7.14	33	78.5
6.0÷6.2	//////	6.1	6	14.28	30	71.4
5.8÷6.0	//////	5.9	6	14.28	24	57.1
5.6÷5.8	////////	5.7	8	19.08	18	42.9
5.4÷5.6	//////	5.5	6	14.28	10	23.8
5.2÷5.4	//	5.3	2	4.76	4	9.5
5.0÷5.2	/	5.1	1	2.38	2	4.8
4.8÷5.2	/	4.9	1	2.38	1	2.4

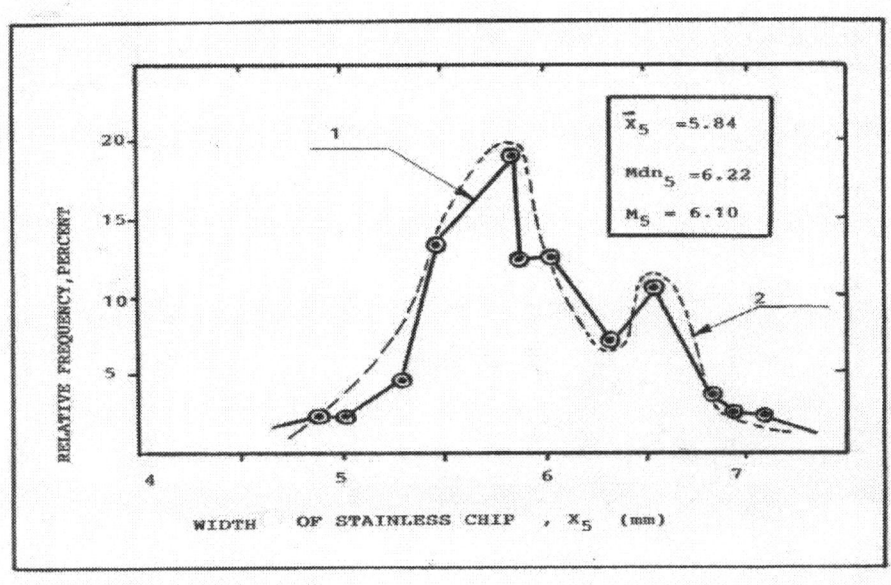

Figure 5 Frequency polygon for width (X_5) of stainless chip (1-observed; 2-evaluated)

Polygon of distribution of the values displayed in Table 6 for thickness of stainless chip is presented in Figure 6.

Table 6 Grouped frequency distribution of measures for thickness X_6 of stainless chip

Values X_6	Tallies	Midpoints	f	rf (%)	cf	Cumulative (%)
0.66÷0.70	/	0.68	1	2.38	42	100
0.62÷0.66	-	0.64	0	0	41	97.6
0.58÷0.62	//	0.60	2	47.6	41	97.6
0.54÷0.58	-	0.56	0	0	39	92.8
0.50÷0.54	/////////// /	0.52	14	33.32	39	92.8
0.46÷0.50	-	0.48	0	0	25	59.5
0.42÷0.46	-	0.44	0	0	25	59.5
0.38÷0.42	////////	0.40	8	19.06	25	59.5
0.34÷0.38	-	0.36	0	0	17	40.5
0.30÷0.34	-	0.30	0	0	17	40.5
0.26÷0.30	/////////	0.28	9	21.44	17	40.5
0.22÷0.26	-	0.24	0	0	8	19.1
0.18÷0.22	//////	0.20	6	14.28	8	19.1
0.14÷0.18	-	0.16	0	0	2	4.7
0.10÷0.14	//	0.12	2	4.76	2	4.7

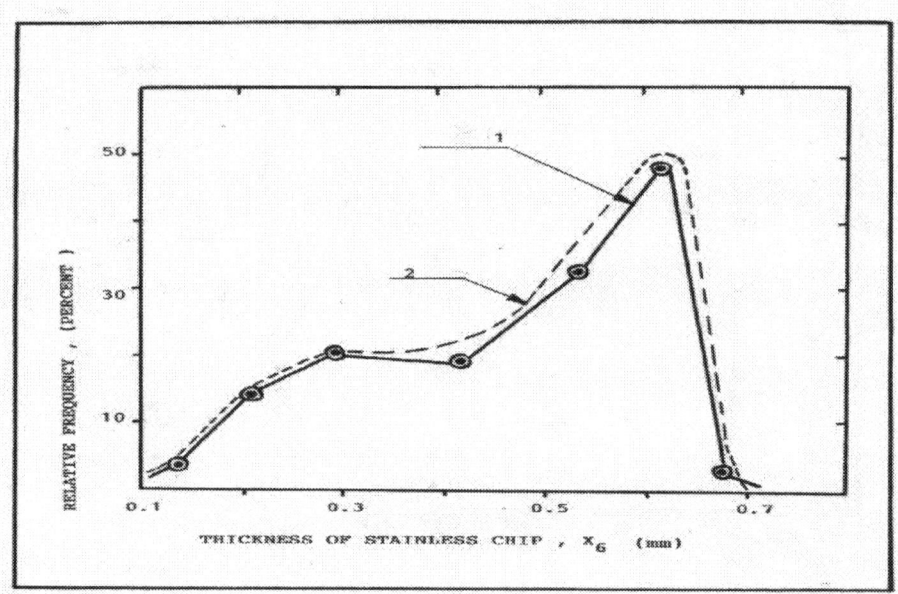

Figure 6 Polygon of distribution for thickness (X_6) of stainless chip (1-observed ;2-evaluated)

And besides in Figure 7 presents the abstract results of statistical analysis in view of frequencies of polygon distribution in evaluation of constructive sizes for stainless chip.

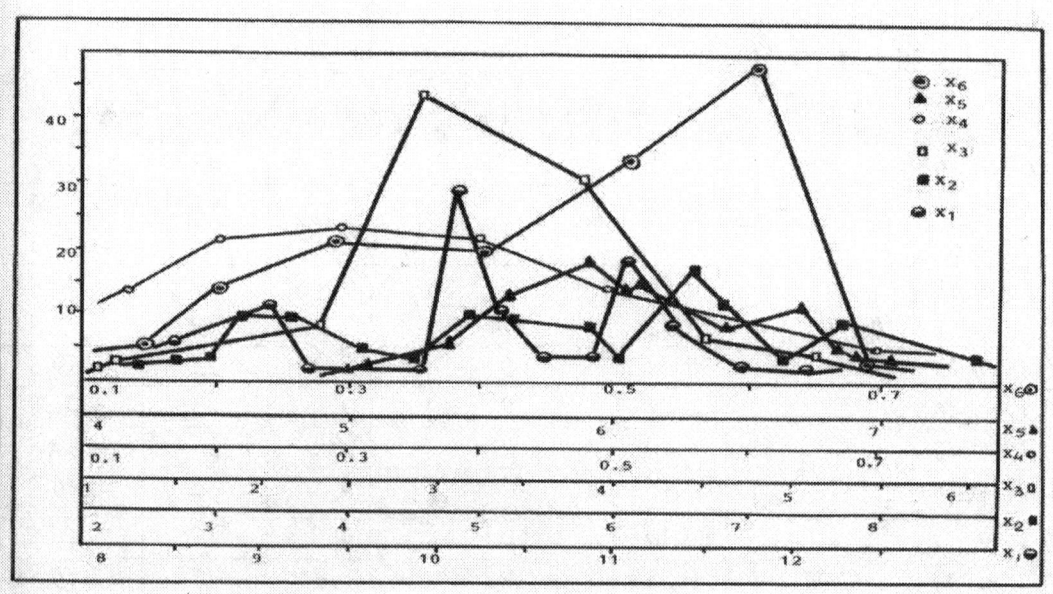

Figure 7 Polygon distribution in evaluation of constructive sizes for stainless chip

In Table 7 is shown the separating and combined description of frequency polygon for external diameter (X_1) of stainless chip as was above shown in Figure 1.

Table 7 Separating and combined description of the values X_1 stainless chip

View to the component curves of distribution	Shape of distribution	Parameter of distribution									
		Mode (M)	Median Mdn	Mean ($\overline{X}$)	Total range	Sum of square (SS)	Sqd deviation (σ^2)	Std deviation (S)	Coefficient of variation (CV)	Skewness (g_1)	Kurtosis (g_2)
	Distribution of values are symmetrical but indicate a slight platykurtic shape distribution	9.05	8.94	8.45	2.80	2.52	0.36	0.65	7	0	-1.3
	Distribution of values are symmetrical but indicate a slight platykurtic shape of distribtion	9.95	10.0	10.1	1.20	0.45	0	0	0	0	-1.4
	Distribution of values are not symmetrical with positive skewness and indicates a slight platykurtic shape of distribution	11.2	11.4	11.5	2.2	1.35	0.27	0.58	4.0	0.71	-2.8
	The shape of frequency of data is distributed as abnormal distribution **PLATY-KURTIC**	10.1	10.1	10.2	-	-	-	0.4	3.7	0.2	-1.8

In Table 8 is shown the separating and combined description of frequency polygon for the internal diameter (X_2) of stainless chip, as this was above-shown in Figure 2.

Table 8 Separating and combined description of the values for stainless chip

View to the component curves of distribution	Shape of distribution	Parameter of distribution									
		Mode (M)	Median Mdn	Mean $\overline{X}$	Total range	Sum of square (SS)	Sqd deviation (σ^2)	Std. deviation (S.D)	Coefcient of variation (CV)	Skewness (g_1)	Kurtosis (g_2)
	Distribution of values is symmetrical but indicate a slight platykurtic shape distribution	3.6	3.33	3.00	3.00	3.80	0.63	0.79	26	0	-2.0
	Distribution of values are symmetrical but indicate a slight platykurtic shape distribution	5.2	4.41	5.20	2.80	1.60	0.32	0.57	10.96	0	-1.3
	Distribution of values are symmetrical but indicate a slight platykurtic shape distribution	6.4	6.35	6.40	1.80	0.32	0.11	0.34	5.30	0	-1.6
	The same	7.6	8.26	8.00	2.60	1.60	0.32	0.57	7.10	0	-1.3
	The shape of frequency polygon of data is distributed as abnormal distribution-PLATYKURTIC	5.7	5.6	5.50	-	-	-	1.86	12.34	0	-1.6

The separating and combined description of frequency polygon for width of stainless chip (X_5) is shown in Table 9, as this was discovered in Figure 5.

26

Table 9 The separating and combined description of frequency polygon for width   (X_5) of stainless chip

View of distribution	Shape of distribution	Parameters of distribution									
		Mode (M)	Median Mdn	Mean $\overline{(X)}$	Total range	Sum of squares (SS)	Sqd deviation (σ^2)	Std deviation (SD)	Coefficient of variation (CV)	Skewness (g_1)	Kurtosis (g_2)
	Distribution of values are symmetrical but indicate a slight platykurtic shape of distribution	5.70	5.82	5.50	2.20	1.12	0.16	0.40	7.0	0	-1.3
	Distribution of values are symmetrical but indicate a slight platy-kurtic shape distribution	6.50	6.62	6.70	1.80	0.40	0.08	0.28	4.2	0	-1.3
	The shape of frequency polygon of data is distributed as abnormal-PLATYKU-RTIC	5.84	6.22	6.10	-	-	-	0.70	5.60	0	-1.3

Analysis of combined description of the values X_1 in Figure 1 and Table 7 show that the shape of frequency polygon of data are distributed as abnormal distribution-*Platykurtic*. And the same picture we see in process of analysis for combined description of the values X_2 in Figure 2 and Table 8 where the shape of frequency polygon of data also is distributed as abnormal distribution – *Platykurtic*. The shape of frequency polygon of data for the values X_5 shown in Table 9 and Figure 5 also is distributed as abnormal – *Platykurtic*. Analysis of data which present the polygon of distribution for the number of chip wraps X_3 of stainless chip in Figure 3 show that this curve has the shape of distribution –*Platykurtic* with the following descriptions:

M=Means=X_3=2.964 ; Mode=2.75 ; Median = Mdn =2.90. The curve of Figure 3 characterizes also by the asymmetrical distribution with positive skew (skewness g_1=0.067) and slight platykurtic (Kurtosis g_2= – 1.39).

Analysis of Figure 6 also characterizes by the asymmetrical positive skew distribution with skewness (g_1=0.42 0 and slight platykurtic distribution (Kurtosis g_2 = – 0.92). And besides analysis of Figure 6 which presents the polygon of distribution the values X_6 show that these data seek to disperse from center and this shape of distribution characterizes as *Platykurtic* with value of Kurtosis g_2= –1.57. This distribution of platykurtic has symmetrical description (skewness g_1=0) with the following parameters of distribution:

Mean = X_6=0.40 ; Mode =M_6= 0.52 ; Median =Mdn=0.36 ; total range 1.56 ; the sum of squares SS = 0.266 ; squared deviation σ^2 =0.038 ; variance S^2 =0.041.

2.2 Summary characteristics for the stainless chip and histogram of its distribution

In Table 10 is given the summary characteristics of distribution for the stainless chip parameters.

Table 10 Summary characteristics of distribution for the stainless chip parameters *

Parameters of distribution	Symbolism	External diameter (X_1)	Internal diameter (X_2)	Number of chip wraps (X_3)	Clearance between of chip wraps (X_4)	Width of chip (X_5)	Thickness of chip (X_6)
Mode	M_w	10.05	5.70	2.75	0.28	6.10	0.52
Median	Mdn_w	10.11	5.60	2.90	0.28	6.22	0.36
Mean	X_w	10.18	5.50	2.96	0.36	5.84	0.40
Standard deviation	$S.D_w$	0.40	1.86	–	–	0.70	0.19
Coefficient of variation	CV_w	3.66	12.34	–	–	5.60	0.21
Skewness	G_1	0.24	0	0.07	0.42	0	0
Kurtosis	G_2	–1.81	–1.57	–1.39	– 0.92	–1.28	–1.57

* sizes in millimeters

Analysis of data shown in Table 10 indicate on fact that there is common tendency of all characteristics of distribution for the stainless chip parameters which characterize as the asymmetrical distribution with positive skewness and slightly platykurtic distribution in view of presented negative kurtosis.

In Figure 8 is shown the histogram for constructive sizes of the stainless chip.

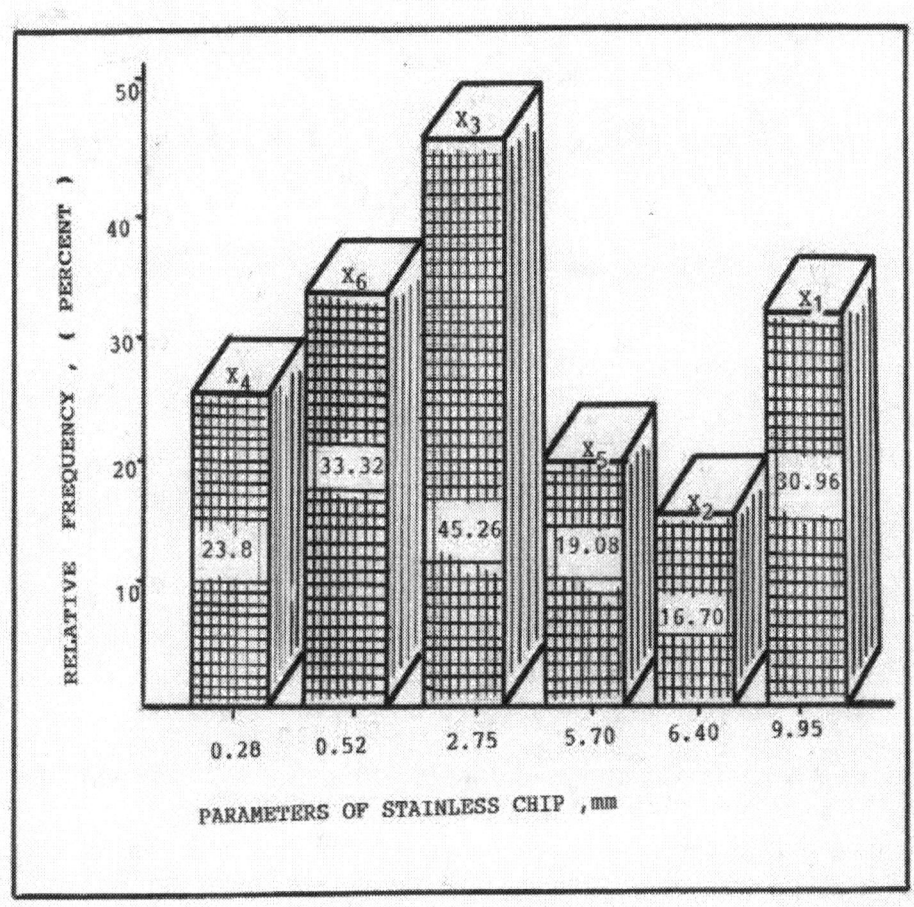

Figure 8 Histogram for constructive sizes of the stainless chip

2.3 Some new methods and innovations for breaking and disposal of chip from the cutting zone and tool

Changing of angles for the cutting tool in machining process considerably improves the break-chips process particularly for the stainless and heat-resistance steels. A variety of this cutting tool is the angular-variable assembling cutting tool of A.I.Rozenblat [**14**] ,as is shown in Figure 9.

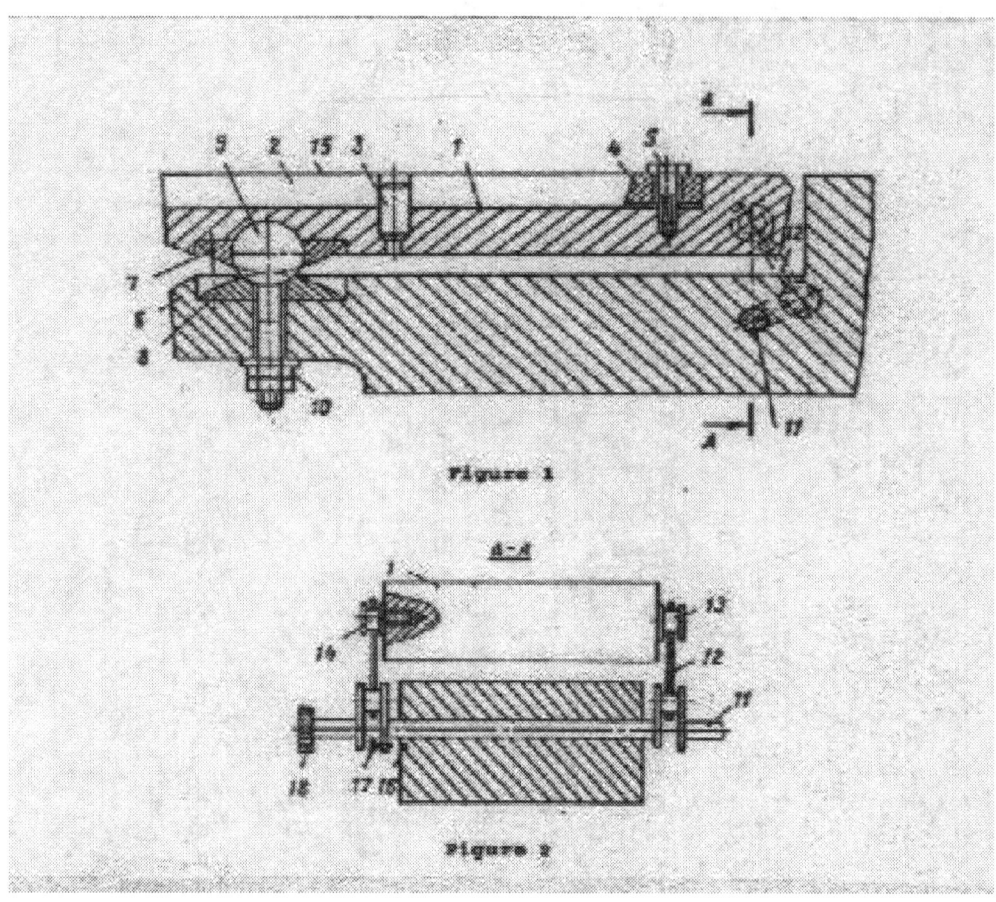

Figure 1

Figure 2

Figure 9 Angular-variable assembling cutting tool of A.I.Rozenblat

This tool is used that to expand it technological possibilities by means of changing of cutting angles in process of machining and also to improve the breaking process particularly for stainless chip with the next it disposal from the cutting zone. Changing of the angles of this tool makes it possible to decrease the wear of front surface for cutting tool and also break the stainless chip that cutting process improves in whole.

The other varieties of cutting tool for improvement of cutting process advantageously of stainless steels are the break-chip inserts as shown in Figure 10 and 11 **[15].**

This multiple break-chip insert (Figure 10) is designed so that its frontal surface made with the variable rake angles. And besides this front surface consists from three separate parts : the first part is made with the value of rake equal to zero ; the second part made with the negative value of rake and the third part made with the positive value of rake. Also the multiple break-chip insert can has the variable front angles with the different other combination of values for this insert.

The other variety of cutting tool in the question of reduction of break chips is the break-chips insert for rotated turning with the variable angles as shown in Figure 11.

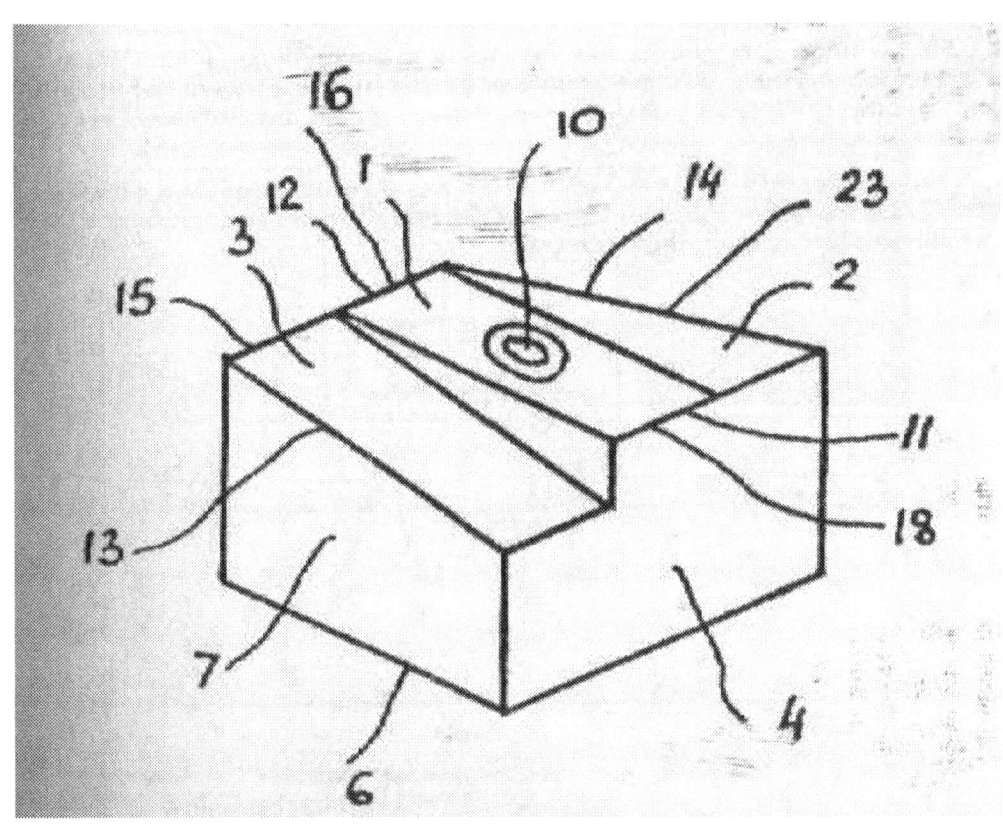

Figure 10 Multiple break-chip insert with the variable front angles

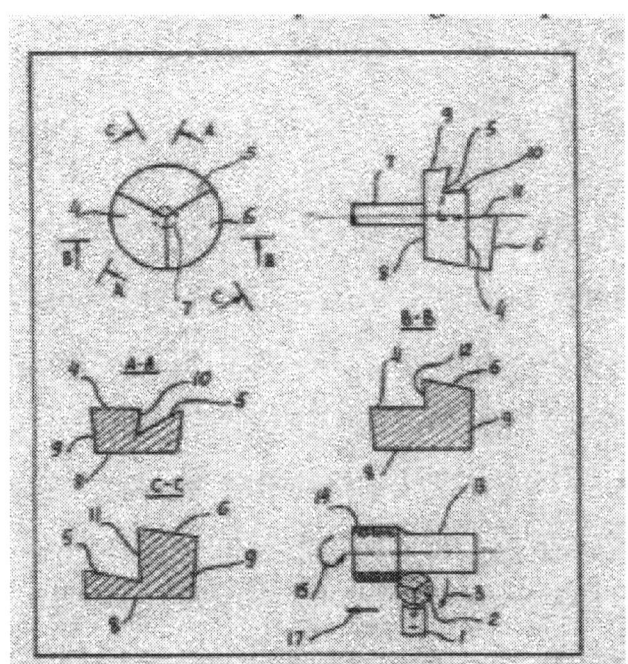

Figure 11 Rotated turning tool with break-chip insert

Analyzing the other innovations in question of breaking stainless chip necessary to consider the tool cleaning device advantageously for the circular segmental saw which is shown in Figure 12.

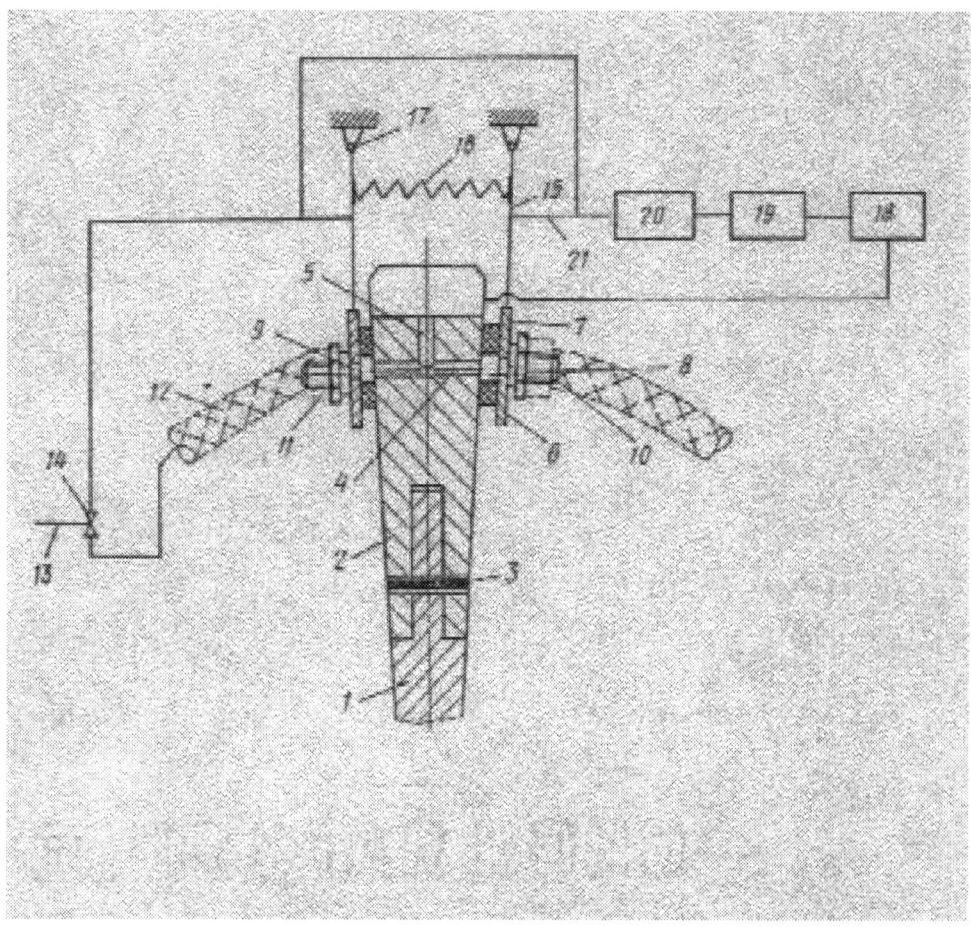

Figure 12 Tool cleaning device for the circular segmental saw

The main objective of tool cleaning device for the circular segmental saw consists in that that automatically disposals chip and other fragments from the zone of saw teeth particularly in operation cut-off process [16]. This device uses two spring-actuated bars in opposing reciprocal motion, placed over the faces of a tool. The tool itself is given specific " cleaning grooves " along its transverse and longitudinal axis.

The cleaning device then sweeps chip and fragment into these " cleaning grooves" and out from the tool. The actuation of pick up motion is set to match the revolution of tool so that it is always synchronized with the appearance of " cleaning grooves". Utilization of the suggested device increases the resistance of cutting tool and efficiency of cut-off process with using of circular segmental saws particularly the stainless steels.

Pneumatic transport widely uses in manufacturing for disposal of ferromagnetic chip and iron dust from the cutting zone of machine-tools as is shown schematically in Figure 13 [17].

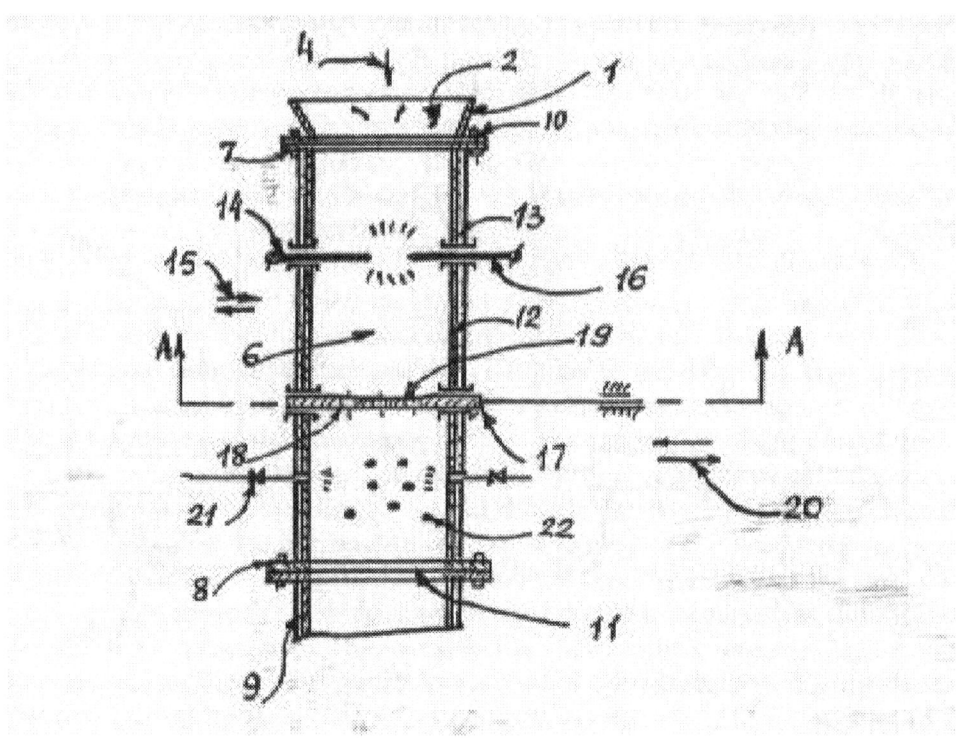

Figure 13 Pneumatic system for disposal of ferromagnetic chip and dust from the cutting zone and tool

SUMMARY

1.Chip formation in cut-off processes advantageously for the stainless and heat-resistance steels by the circular segmental saw has a big value with point view of studying of its shape and geometrical parameters of chip such as external and internal diameters, number and clearance between of chip wraps, width and also the thickness chip;

2. Making a careful study of geometrical parameters of stainless chip in the perspective all this will be useful for designing of special arrangements for breaking and disposal chip from the cutting zone with using of the progressive methods such as the pneumatic transportation;

3. Geometrical parameters of chip is formed in cut-off processes have the accidental characteristics and dependent from the cutting conditions, forms of material and submit to the laws of mathematical statistics;

4.Analysis of frequency polygon descriptions for the geometrical parameters of stainless chip show that this polygon of data is distributed as abnormal distribution – PLATYKURTIC.

CHAPTER THREE   FUNCTIONAL ORIENTED GRAPHS IN EVALUATION OF GEOMETRICAL PARAMETERS FOR THE STAINLESS CHIP

3.1 The general functional dependencies between two variables having the regression model view of $Y_i = \varphi(X_i)$

In Figure 1 is shown the functional oriented graphs between geometrical parameters of the stainless chip.

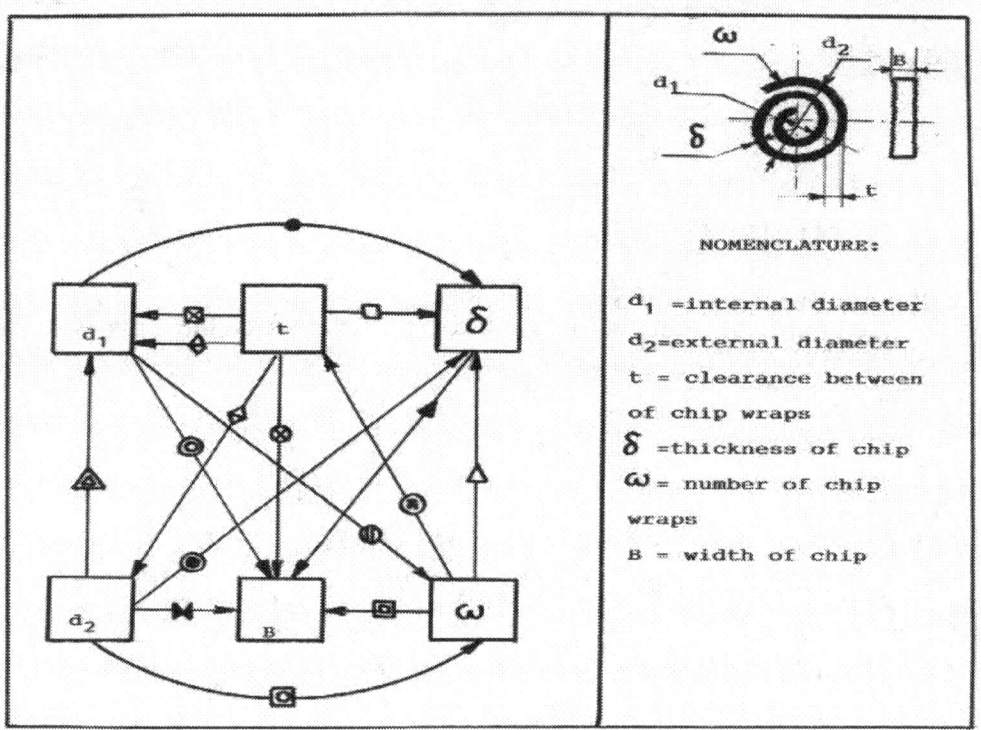

Figure 1   **Functional oriented graphs between geometrical parameters of the stainless chip**

Analysis of Figure 1 shows that we have the following correlations between two variables of the stainless chip having regression model view of $Y_i = \varphi(X_i)$:

1. Dependence of internal diameter (d_1) chip from its thickness (δ),i.e $d_1 = \varphi(\delta)$ ($\bullet$)
2. Dependence of external diameter (d_2) of chip from its thickness (δ),i.e $d_2 = \gamma(\delta)(\beta)$
3. Dependence of clearance (t) between of chip wraps from its thickness (δ) ,i.e $t = \phi(\delta)$ (   )
4. Dependence of number chip wraps (ω) from its thickness (δ) ,i.e $\omega = \alpha(\delta)$ (Δ)
5. Dependence of thickness (δ) chip from its width (B) ,i.e $\delta = f(B)$ (χ)
6. Dependence of internal diameter (d_1) chip from its width (B),i.e $d_1 = \varphi_1(B)$ ($\circledcirc$)

7.Dependence of external diameter (d_2) chip from its width (B) ,i.e $d_2=\varphi_2$ (B) (►◄)

8.Dependence of clearance (t) between of chip wraps from its width (B) ,i.e t=γ_1 (B) ($\otimes$)

9. Dependence of number chip wraps (ω) from its width (B) ,i.e $\omega=\gamma_2$(B) (▣)

10. Dependence of external diameter (d_2) from internal diameter (d_1) of chip ,i.e $d_2=\varphi_3(d_1)$ (A)

11.Dependence of clearance (t) between of chip wraps from its external (d $_2$) and internal (d$_1$) diameters ,i.e t=γ_3 (d$_2$) and t=γ_4(d$_1$) (◁▷)

12. Dependence of external diameter (d_2) from the number of chip wraps (ω) ,i.e $d_2=\phi_1(\omega$) (◈)

13.Dependence of clearance (t) between of chip wraps from its internal diameter (d_1),i.e t=ϕ_2 (d$_1$) (☒)

14.Dependence of internal diameter (d_1) from the number of chip wraps (ω) ,i.e $d_1=\phi_3(\omega)$ (⑪)

15.Dependence of the number chip wraps (ω) from the clearance (t) between of chip wraps, i.e $\omega =\alpha_1$ (t) (*)

3.2 The general functional relations between three variables for the stainless chip having the regression model view of $Y_i= \alpha (X_{i,1} ; X_{i,2})$:
   a) for internal diameter (d_1) of chip
   b) for external diameter (d_2) of chip
   c) for clearance (t) between of chip wraps
   d) for number of chip wraps (ω).

In Figure 2 is shown the general functional relations between of three variables for the stainless chip into account of its internal diameter (d_1).

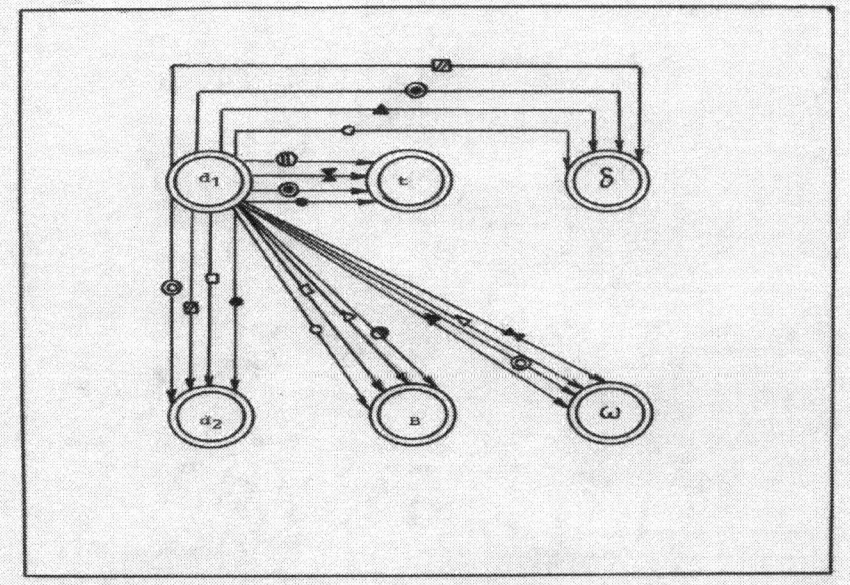

Figure 2 The general functional relations between three variables for the stainless chip into account of its internal diameter (d_1)

Analysis of Figure 2 shows that we have the following functional relations between three variables for the stainless chip having the regression model view of $Y_i=\alpha(X_{i,1};X_{i,2})$ into account:

A .For internal diameter (d_1) of the stainless chip :

1. Dependence of internal diameter (d_1) from external diameter (d_2) and clearance (t) between of chip wraps, i.e $d_1=\varphi(d_2, t)$ (•)

2.Dependence of internal diameter (d_1) from width (B) and thickness (δ) of chip ,i.e $d_1=\varphi_2(B,\delta)$ (○)

3.Dependence of internal diameter (d_1) from the number of chip wraps (ω) and thickness (δ) of stainless chip , $d_1=\varphi_3(\omega,\delta)$ (▲)

4. Dependence of internal diameter (d_1) from external diameter (d_2) and width (B) of this chip ,i.e $d_1=\varphi_4(d_2,B)$ ( )

5.Dependence of internal diameter (d_1) from clearance (t) between of chip wraps and its thickness (δ) ,i.e $d_1=\varphi_5(t,\delta)$ (◉)

6.Dependence of internal diameter (d_1) from width (B) of chip and number of chip wraps (ω) , i.e $d_1=\varphi_6(B,\omega)$ (△)

7. Dependence of internal diameter (d_1) from clearance (t) between of chip wraps and number of chip wraps (ω) ,i.e $d_1=\varphi_7(t,\omega)$ (►◄)

8.Dependence of internal diameter (d_1) from external diameter (d_2) and thickness chip ,i.e $d_1=\varphi_8(d_2,\delta)$ (▧)

9.Dependence of internal diameter (d_1) from clearance (t) between of chip wraps and its width(B) ,i.e $d_1=\varphi_9(t, B)$ (⊘)

10. Dependence of internal diameter (d_1) from external diameter (d_2) and number of chip wraps (ω) ,i.e $d_1=\varphi_{10}(d_2,\omega)$ (◎).

In Figure 3 is shown the general functional relations between three variables for the stainless chip into account of its external diameter.

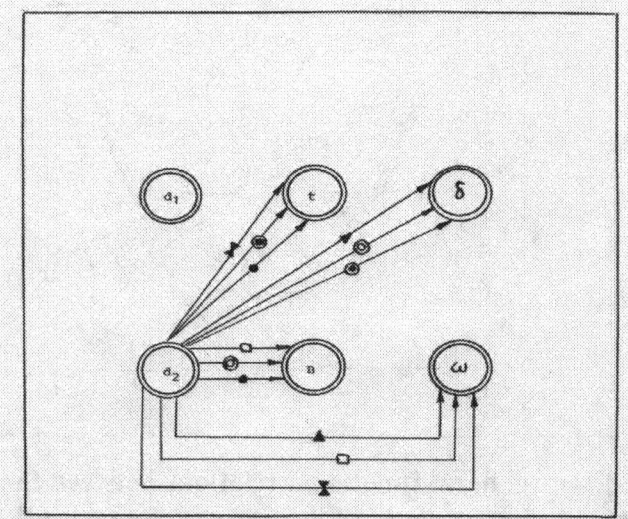

**Figure 3   The general functional relations between three variables
for the stainless chip into account of its external diameter (d_2)**

Analysis of Figure 3 shows that we have the following correlations between three variables for the stainless chip having the regression model view of $Y_i=\alpha\ (X_{i,1};X_{i,2})$ into account:

B. For external diameter (d_2) of the stainless chip:

1.Dependence of external diameter (d_2) from clearance (t) between of chip wraps and its width (B) ,i.e $d_2=\gamma_1(t,B)$ (•)

2.Dependence of external diameter (d_2) from number of chip wraps (ω) and thickness (δ) of this chip ,i.e $d_2=\gamma_2\ (\omega,\delta)$ (▲)

3.Dependence of external diameter (d_2) from width (B) of chip and its thickness (δ) ,i.e $d_2=\gamma_3(B,\delta)$ (◎)

4.Dependence of external diameter (d_2) from clearance (t) between of chip wraps and its thickness (δ) ,i.e $d_2=\gamma_4(t,\delta)$ (◉)

5. Dependence of external diameter (d_2) from width (B) of chip and number of chip wraps (ω) , i.e $d_2=\gamma_5(B,\omega)$ ( )

6. Dependence of external diameter (d_2) from clearance (t) between of chip wraps and number of chip wraps (ω) ,i.e $d_2=\gamma_6\ (t,\omega)$ (►◄).

In Figure 4 is shown the general functional relations between three variables for the stainless chip into account of its clearance between of chip wraps.

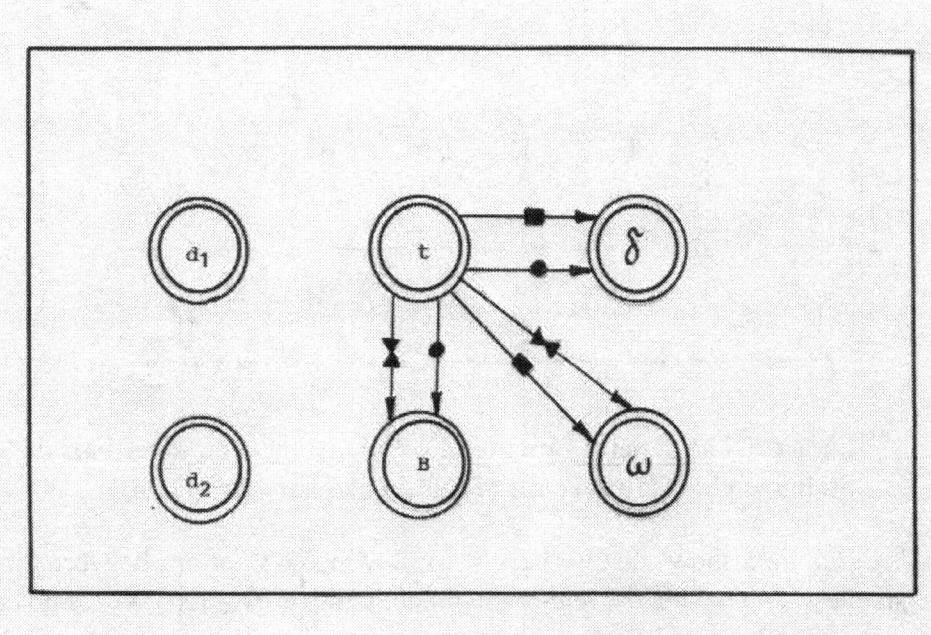

Figure 4 The general functional relations between three variables for the stainless chip into account of its clearance (t) between of chip wraps

Analysis of Figure 4 shows that we have the following correlations between three variables for the stainless chip having the regression model view of $Y_i=\alpha(X_{i,1};X_{i,2})$ into account:

C. Clearance (t) between of chip wraps:

1.Dependence of clearance (t) between of chip wraps from width (B) of chip and its thickness (δ) ,i.e $t=\alpha_1(B,\delta)$ ($\bullet$)
2.Dependence of clearance (t) between of chip wraps from number of chip wraps (ω) and its thickness (δ) ,i.e $t=\alpha_2(\omega,\delta)$ ($\blacksquare$)
3.Dependence of clearance (t) between of chip wraps from width (B) of chip and number of chip wraps (ω), i.e $t=\alpha_3(B,\omega)$ ($\blacktriangleright\blacktriangleleft$).

In Figure 5 is shown the general functional relations between three variables for the stainless chip into account of number of chip wraps (ω).

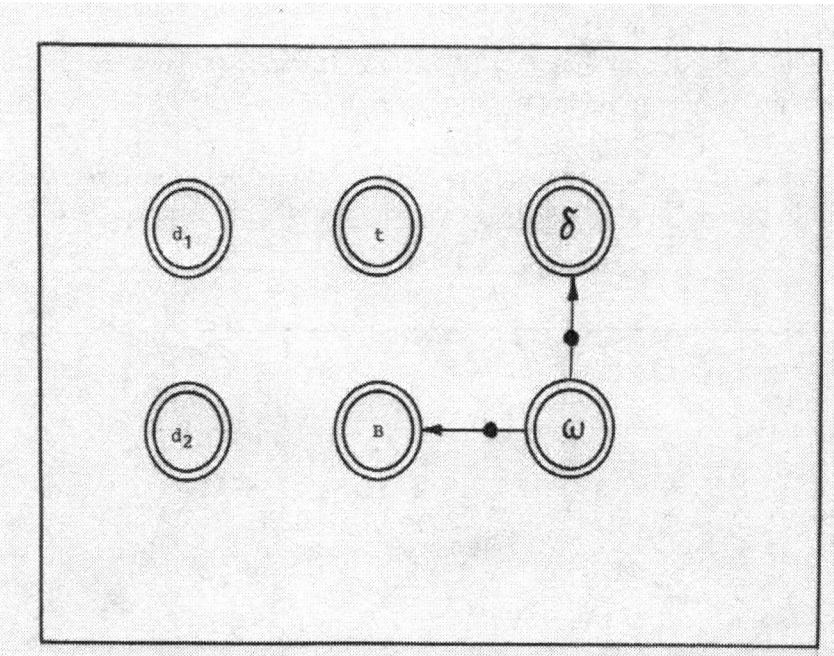

Figure 5 The general functional relations between three variables for the Stainless chip into account of number of chip wraps (ω)

Analysis of Figure 5 shows that we have the following correlations between three variables for the stainless chip having the regression model view of $Y_i=\alpha(X_{i,1};X_{i,2})$ into account:

D. The number (ω) of chip wraps:

1. Dependence of the number (ω) of chip wraps from width (B) and its thickness (δ) ,i.e $\omega=\phi_1(B,\delta)$ ($\bullet$)

38

3.3 Multiple regression analysis for the constructive sizes of the stainless chip having the regression model $Y_i=\gamma(X_{i,1};X_{i,2};X_{i,3})$:

        a) For external diameter (d_2) of chip

        b) For internal diameter (d_1) of chip

        c) For clearance (t) between of chip wraps.

In Figure 6 is shown the functional oriented graphs between the geometrical parameters of the stainless chip into account of its external diameter (d_2) for the regression model $Y_i=\gamma(X_{i,1};X_{i,2};X_{i,3})$.

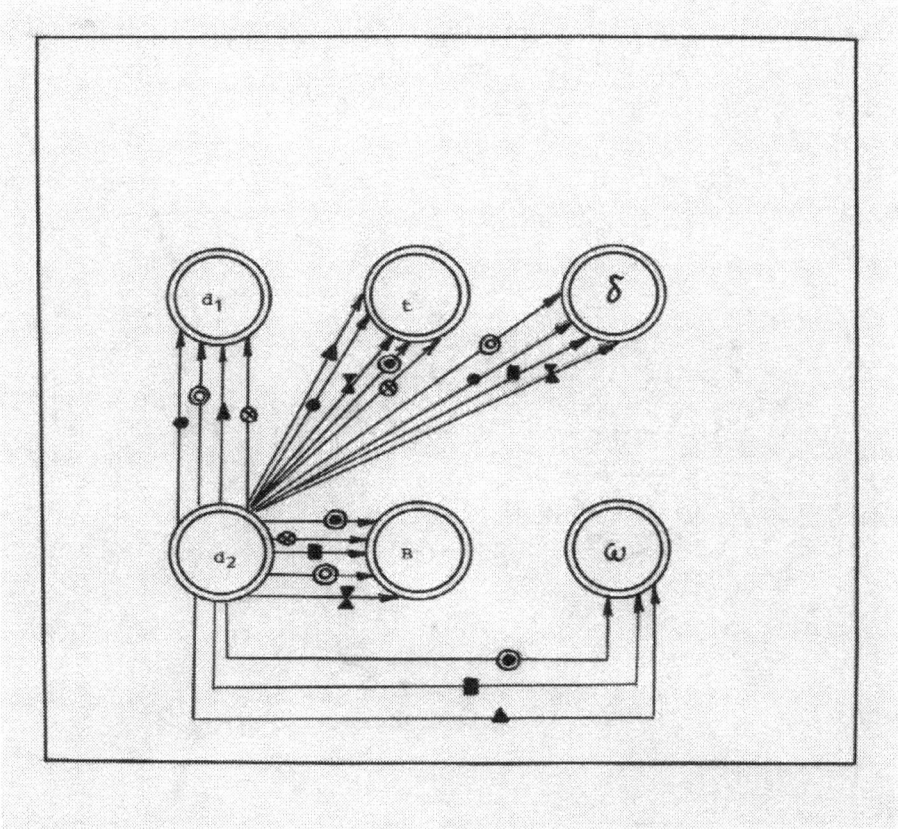

Figure 6 The general functional relations between the geometrical parameters of the stainless chip into account of its external diameter (d_2) for the regression model $Y_i=\gamma(X_{i,1};X_{i,2};X_{i,3})$

Analysis of Figure 6 shows that we have the following functional relations between the geometrical parameters of the stainless chip having the regression model view of $Y_i=\gamma(X_{i,1};X_{i,2};X_{i,3})$ into account:

   A. For external diameter (d_2) of chip:

1.Dependence of external diameter (d_2) from thickness (δ) of chip and also of its internal diameter (d_1) diameter and value of its clearance (t) between of chip wraps ,i.e
$d_2=\alpha_1(\delta,d_1,t)$ (•)

39

2.Dependence of external diameter (d_2) from thickness (δ) of chip and also of its number of chip wraps (ω) and width (B) of this chip ,i.e $d_2= \alpha (\delta ,\omega,B)$ (■)

3.Dependence of external diameter (d_2) from thickness (δ) of chip ,internal diameter (d_1) and width (B) of this stainless chip ,i.e $d_2=\alpha_3(\delta,d_1,B)$ (◎)

4. Dependence of external diameter (d_2) from thickness (δ) of chip ,the value of clearance (t) between of chip wraps and its width (B) ,i.e $d_2=\alpha_4(\delta,t,B)$ (►◄)

5. Dependence of external diameter (d_2) from clearance (t) between of chip wraps ,number (ω) of chip wraps and width (B) of this chip ,i.e $d_2=\alpha_5 (t,\omega,B)$ (◉)

6.Dependence of external diameter (d_2) from internal (d_1) diameter ,clearance (t) between wraps of chip and number (ω) of this chip wraps ,i.e $d_2 =\alpha_6(d_1,t,\omega)$ (▲)

7.Dependence of external diameter (d_2) from width (B) of this chip ,internal diameter (d_1) and clearance (t) between of chip wraps ,i.e $d_2= \alpha_7 (B,d_1,t)$ (⊗) .

In Figure 7 is shown the functional oriented graphs between the geometrical parameters of the stainless chip into account of its internal diameter (d_1) for the regression model $Y_i= \gamma(X_{i,1} ;X_{i,2};X_{i,3})$.

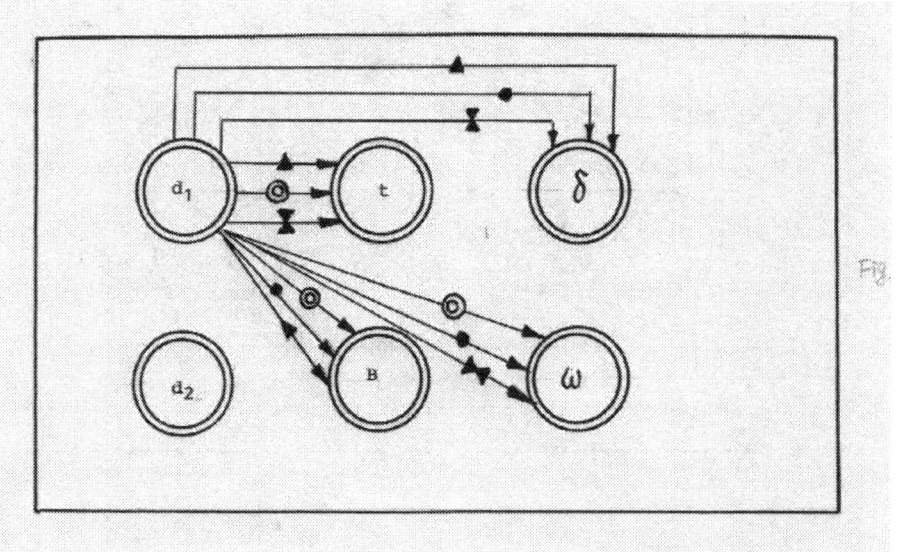

Figure 7 The general functional relations between the geometrical parameters of the stainless chip into account of its internal diameter (d_1) for the regression model $Y_i= \gamma (X_{i,1};X_{i,2};X_{i,3})$.

Analysis of Figure 7 shows that we have the following functional relations between the geometrical parameters of the stainless chip having the regression model view of $Y_i=\gamma (X_{i,1};X_{i,2};X_{i,3})$into account :

   B.   For internal diameter (d_1) of chip:
1.Dependence of internal diameter (d_1) from width (B) of chip, number of chip wraps(ω) and its thickness (δ) ,i.e $d_1=\phi_1(B,\delta,\omega)$ (•)

40

2.Dependence of internal diameter (d_1) from thickness (δ) of chip ,clearance (t) between of chip wraps and width (B) of chip ,i.e $d_1=\phi_2(\delta, t, B)$ ($\blacktriangle$)

3.Dependence of internal diameter (d_1) from width (B) of chip, clearance (t) between of chip wraps and number of chip wraps (ω) ,i.e $d_1=\phi_3(B, t,\omega)$ ($\circledcirc$)

4.Dependence of internal diameter (d_1) from clearance (t) between of chip wraps, number (ω) of chip wraps and thickness (δ) of this stainless chip ,i.e $d_1=\phi_4(t,\omega,\delta)$ ($\blacktriangleright\blacktriangleleft$)

In Figure 8 is shown the functional oriented graphs between the geometrical parameters of stainless chip into account of clearance (t) between of chip wraps for the regression model $Y_i=\gamma (X_{i,1} ;X_{i,2};X_{i,3})$.

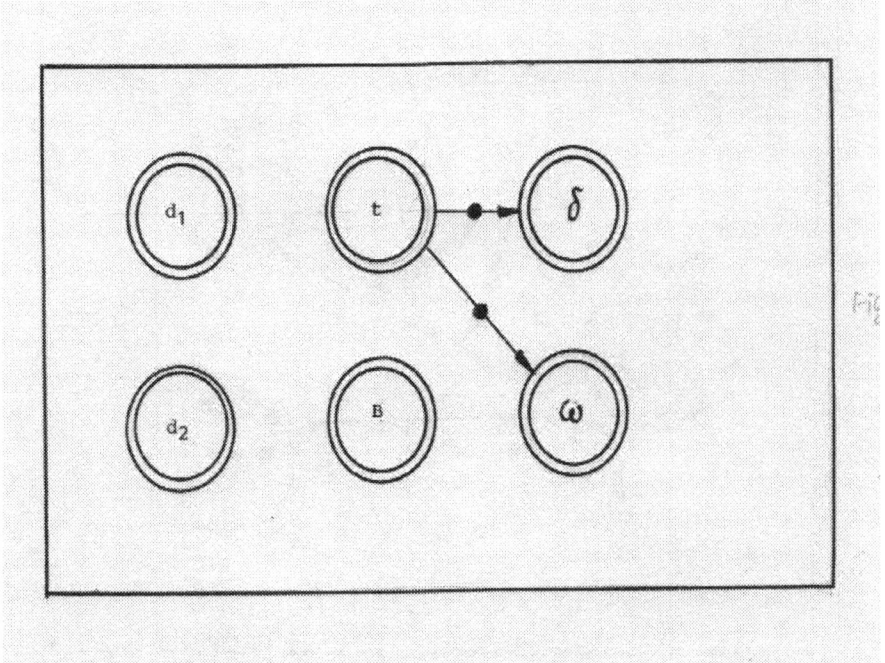

Figure 8 The general functional relations between the geometrical parameters of stainless chip into account of clearance between of chip wraps for the regression model $Y_i=\gamma(X_{i,1} ;X_{i,2} ;X_{i,3})$

Analysis of Figure 8 shows that we have the following functional relations between the geometrical parameters of the stainless chip having the regression model view of $Y_i=\gamma(X_{i,1} ;X_{i,2} ;X_{i,3})$ into account :

   C. *For clearance (t) between of chip wraps* :

1. Dependence of clearance (t) between of chip wraps from thickness (δ) of this chip ,and also width (B) of its and number of chip wraps (ω) ,i.e $t= \theta (\delta,B,\omega)$ ($\bullet$)

41

3.4 Dependence of external and internal diameters of stainless chip from all its constructive parameters having the regression model view of $Y_i=\phi(X_{i,1};X_{i,2};X_{i,3};X_{i,4};X_{i,5})$

In Figure 9 is shown the functional oriented graphs in dependence of external (d_2) and internal (d_1) diameters of stainless chip from all its constructive parameters having the regression model view of $Y_i = \phi (X_{i,1};X_{i,2};X_{i,3};X_{i,4};X_{i,5})$.

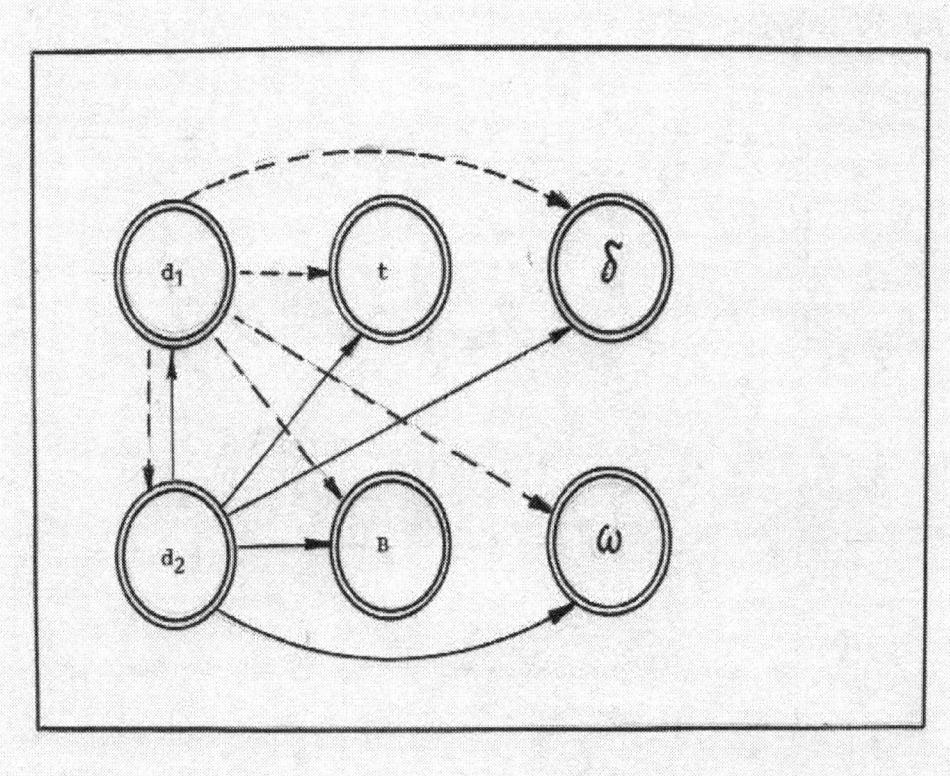

Figure 9   Functional oriented graphs for external and internal diameters of the stainless chip in dependence from all its constructive parameters

Analysis of Figure 9 shows that we have the following functional relations for external (d_2) and internal (d_1) diameters of the stainless chip in dependence from all its constructive parameters.

And besides we see that for independent variable, such as external diameter, depends from internal diameter (d_1) ,clearance (t) between of chip wraps ,width (B) ,thickness (δ) of stainless chip and also from the number (ω) of chip wraps ,i.e we have such functional model view of $d_2 =\mu$ (d_1 ,t, B,δ,ω) and for the independent variable ,such as internal diameter (d_1) this function has view of $d_1=\mu_1$ (d_2 ,t , B ,δ ,ω).

CHAPTER FOUR   APPLIED STATISTICS IN EVALUATION OF CORRELATION BETWEEN PARAMETERS OF THE STAINLESS CHIP

4.1 RELATIONS BETWEEN TWO PARAMETERS OF THE STAINLESS CHIP IN FUNCTION OF $Y_I = \varphi(X_I)$:

A. The main criteria in question of comparing and accepting of hypothesis (linear or non-linear regression model) for given function is taken such statistical factors as:

- Coefficient of determination R^2
- Coefficient of correlation r
- Standard deviation $S_{y/x}$
- Minimization of the mean square error min MSE
- Minimization of the mean absolute deviation min MAD

B. Mathematical symbols and formulas [18] and [19] for calculation of the statistical parameters for the linear regression model such as :

$$R^2 = \frac{[\sum_{i=1}^{n}(Y_i - \bar{Y})^2 - \sum_{i=1}^{n}(Y_i - \hat{Y}_i)^2]}{\sum_{i=1}^{n}(Y_i - \bar{Y})^2} \qquad (1)$$

$$r = \pm(R^2)^{1/2} \qquad (2)$$

$$S = \{[\sum_{i=1}^{n}(Y_i - \hat{Y}_i)^2]/(n-2)\}^{1/2} \qquad (3)$$

and for non-linear regression model:

$$R^2 = [\sum_{i=1}^{n}(\log \hat{Y}_i)^2 - (\log \bar{Y})\sum_{i=1}^{n}\log Y_i] / [\sum_{i=1}^{n}(\log Y_i)^2 - (\log \bar{Y})\sum_{i=1}^{n}\log Y_i] \qquad (4)$$

$$S = \{[\sum_{i=1}^{n}(\log Y_i)^2 - \sum_{i=1}^{n}(\log Y_i)^2]/n\}^{1/2} \qquad (5)$$

and for comparing and accepting of hypothesis for regression model we have:

$$\min MSE = [\sum_{i=1}^{n}(Y_i - \hat{Y}_i)^2/n] \quad (6) \quad \min MAD = [\sum_{i=1}^{n}|Y_i - \hat{Y}_i|/n] \quad (7)$$

4.1.1 Dependence of internal diameter (d_1) chip from its thickness (δ)

In Figure 1 is shown the function $d_1=\varphi(\delta)$ and fitted linear regression equation view of $Y_c=3.34+4.76X$ or $d_1=3.34+4.76\delta$ with such statistical characteristics as: coefficient of determination $R^2=0.871$; coefficient of correlation $r=0.939$; standard deviation $S_{y/x}=0.322$.

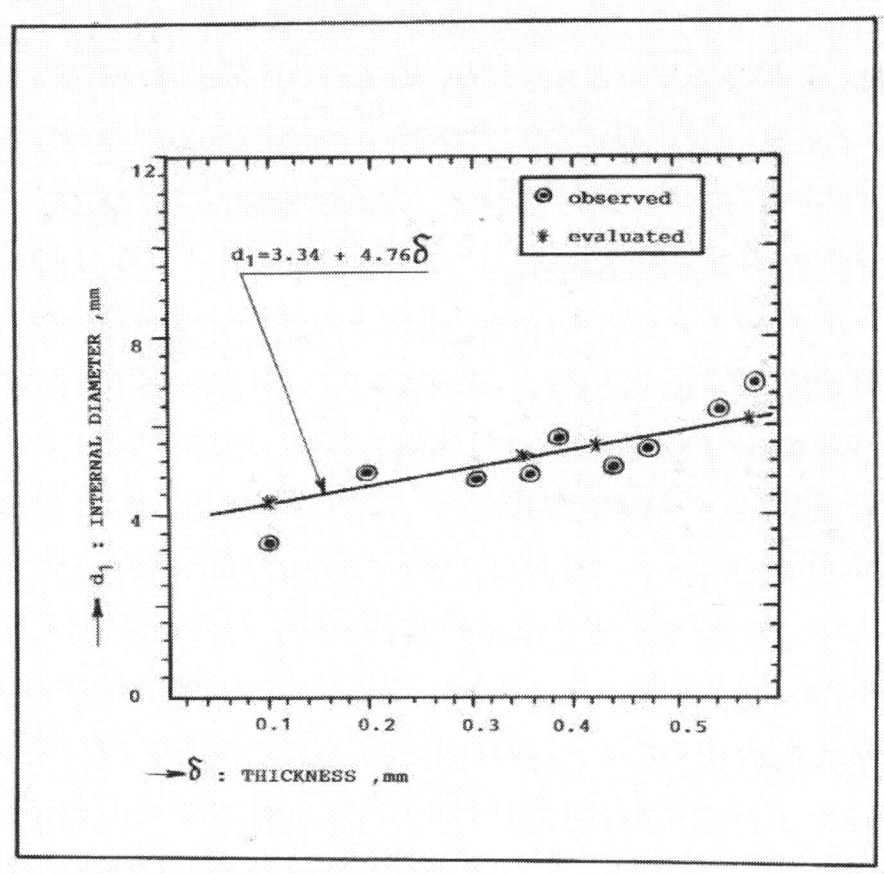

Figure 1 Indicated internal diameter vs. chip thickness

So, we see from Figure 1 that with increasing of chip thickness (δ) ,the value of internal diameter (d_1) of this stainless chip increases considerably and this regression model has the linear character in accordance with the linear equation view of $d_1=3.34+4.76\delta$ (8)

4.1.2 Dependence of external diameter (d_2) of stainless chip from its thickness (δ)

In Figure 2 is shown the fitted linear regression model for function $d_2=\gamma(\delta)$ in view of equation $Y_c=8.81+3.9X$ or $d_2=8.81+3.9\delta$ with such statistical characteristics as: coefficient of determination $R^2=0.796$; coefficient of correlation $r=0.967$; standard deviation $S_{y/x}=0.347$.

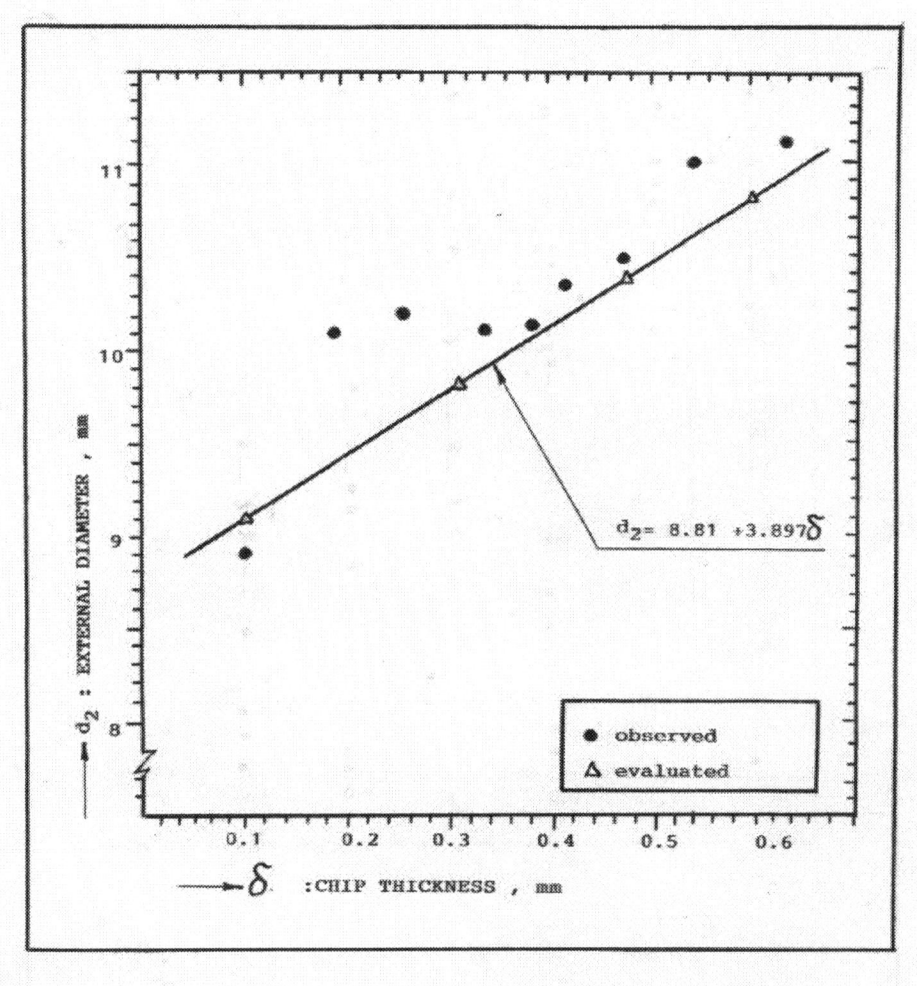

Figure 2 Indicated external diameter vs. chip thickness

So, we see from Figure 2 that with increasing of chip thickness (δ) ,the value of external diameter (d_2) of this stainless chip also increases considerably and this regression model has the linear character in accordance with linear equation view of $\mathbf{d_2 = 8.81 + 3.9\delta}$ (9)

4.1.3 Dependence of clearance (t) between of chip wraps from its thickness (δ)

In Figure 3 are shown both regression models of this function $t = \phi$ (δ). And after of analyzing of the statistical characteristics we can conclude that function $t = \phi$ (δ) better submits to the linear regression model and has the regression equation view of $Y_c = 0.38 - 0.16X$ or $\mathbf{t = 0.38 - 0.16\delta}$ (10) with such statistical characteristics as: coefficient of determination $R^2 = 0.29$;coefficient of correlation $r = -0.788$; standard deviation $S_{y/x} = 0.056$.

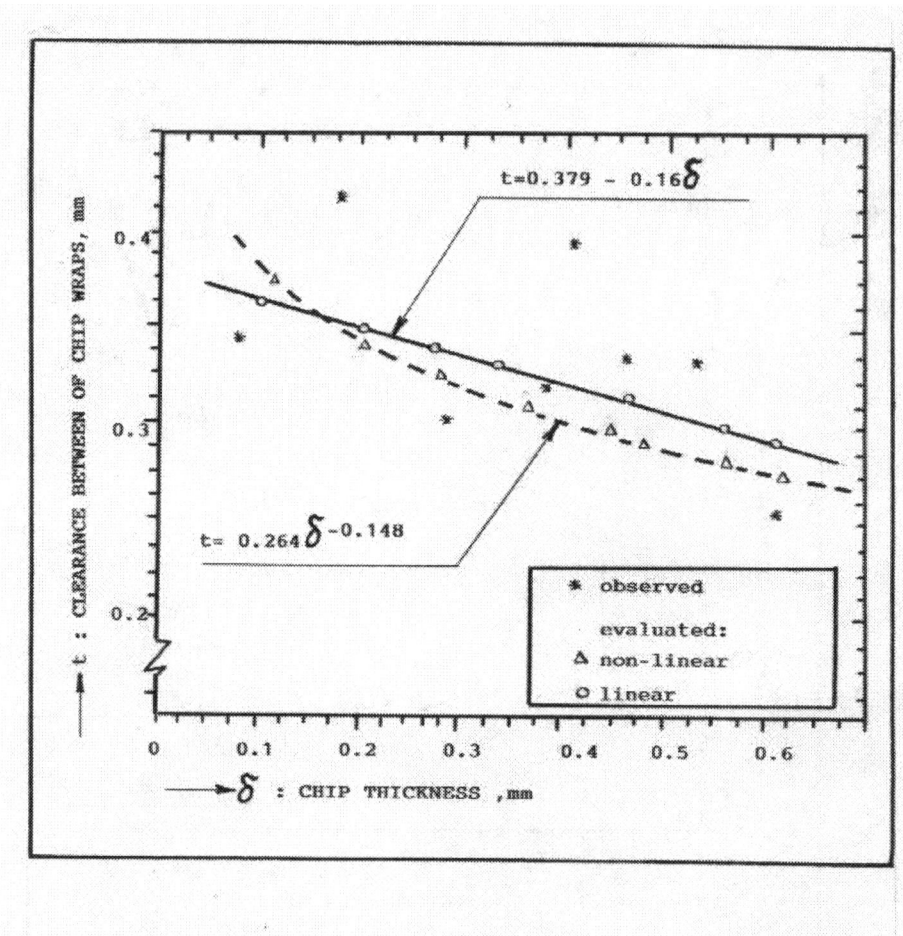

Figure 3 Comparative analysis of functional dependence t= φ (δ)

So, we see from Figure 3 that with increasing of chip thickness (δ) ,the value of clearance (t) between of chip wraps considerably decreases and this regression model better submits to the linear character in accordance with the linear regression equation view of

$$t=0.38-0.16\delta \quad (11).$$

4.1.4 Dependence of number chip wraps (ω) from its thickness (δ)

In Figure 4 are shown both regression models for this function ω =α (δ). And after of analyzing of the statistical characteristics we can conclude that function ω =α (δ) better submits to the non-linear regression model and has the regression equation view of $Y_c=3.09X^{-0.123}$ or ω= $3.09\delta^{-0.123}$ with such statistical characteristics as : coefficient of determination $R^2=0.667$;coefficient of correlation r= 0.816; standard error S=0.026.

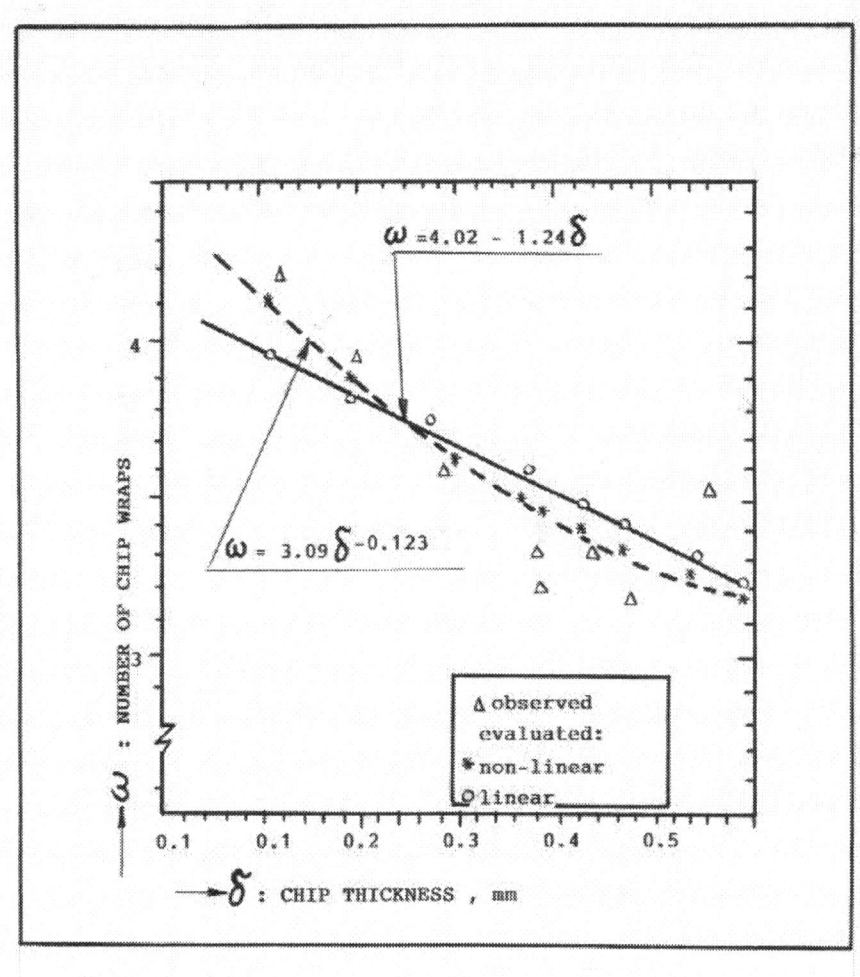

Figure 4 Comparative analysis of functional dependency ω=α(δ) of two regression models (linear and non-linear) between of number chip wraps (ω) and its thickness (δ)

So, we see from Figure 4 that with increasing of chip thickness (δ) , the value of chip wraps (ω) considerably decreases and this regression model better submits to the non-linear function in accordance with the non-linear equation view of $\omega = 3.09\delta^{-0.123}$ (12).

4.1.5 Dependence of chip thickness (δ) from its width (B)

In Figure 5 is given the fitted linear regression model for this function δ =f (B) which has equation view of $Y_c = -1.236 + 0.275X$ or $\delta = -1.236 + 0.275B$ with such statistical characteristics as: coefficient of determination $R^2 = 0.18$; coefficient of correlation r= 0.49 and standard deviation $S_{y/x} = 0.16$.

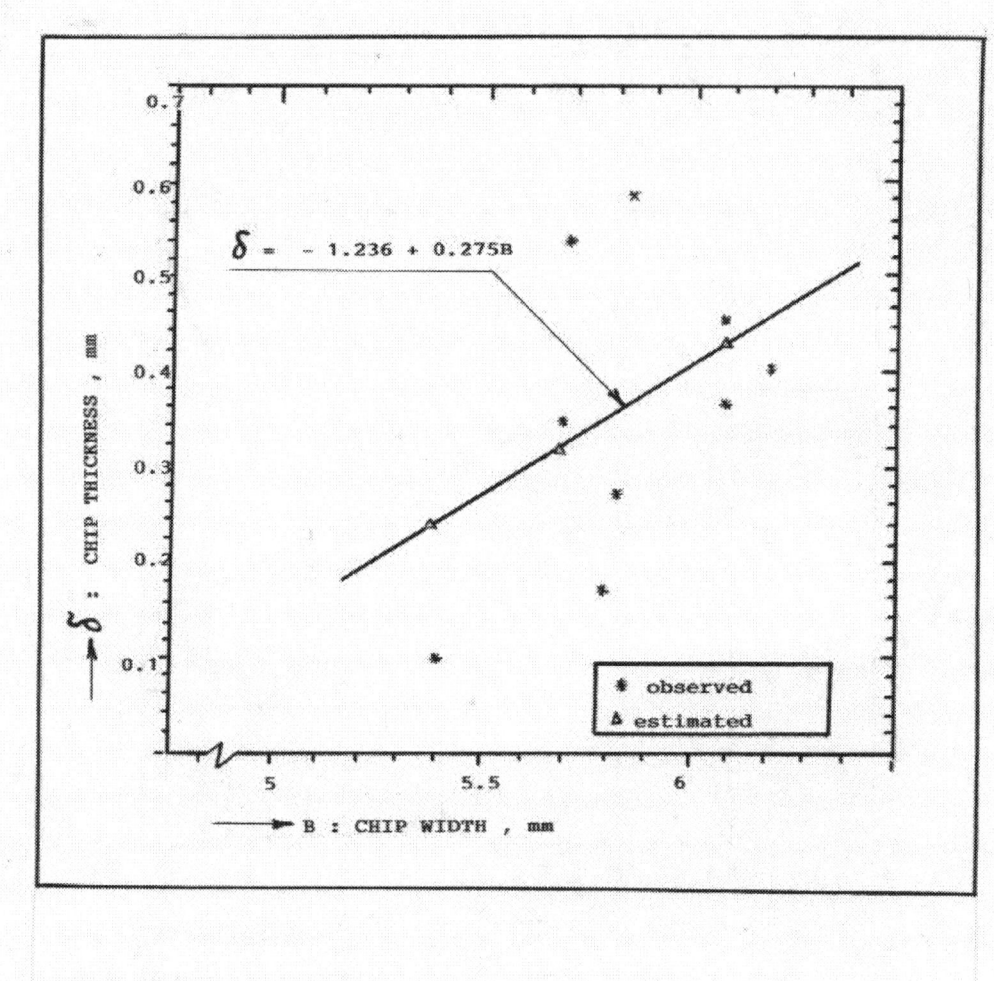

Figure 5 Indicated functional dependency between of chip thickness (δ) and its width (B)

From Figure 5 we see that with increasing of chip width (B).the value of chip thickness (δ) increases considerably and this regression model has the linear character in accordance with the linear equation view of $\delta = -\,1.236 + 0.275\,B$ (13) .

4.1.6 Dependence of internal diameter (d_1) stainless chip from its width (B)

In Figure 6 are shown both regression models of this function $d_1 = \varphi_1(B)$.And after of analyzing of the statistical characteristics we can conclude that function $d_1 = \varphi_1(B)$ better submits to the linear regression model ,although the coefficient of correlation is not so large (r=0.32) and this function has the regression equation view of Y_c=0.06+0.86B or d_1 =0.06+0.863B with such statistical   characteristics as : coefficient of determination R^2=0.10;coefficient of correlation r=o.32; standard deviation $S_{y/x}$=0.864.

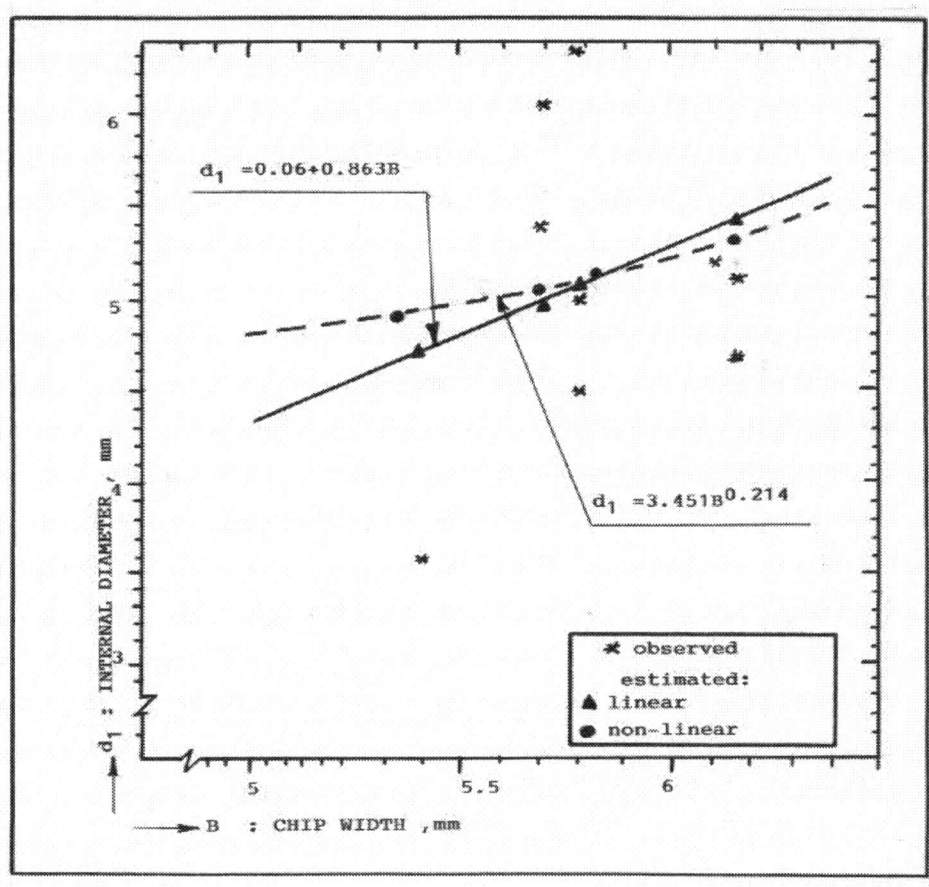

Figure 6 Comparative analysis of functional dependency $d_1 = \varphi_1$ (B) of two regression models (linear and non-linear)between internal diameter (d_1) of stainless chip and its width (B)

So ,we see from Figure 6 that with increasing of chip width (B) ,the value of internal diameter (d_1) of this stainless chip increases considerably and this regression model better submits to the   linear function in accordance with the linear equation view of

$$d_1 = 0.06 + 0.863B \ (\ 14\ ).$$

4.1.7 Dependence of external diameter (d_2) chip from its width (B)

In Figure 7 are shown both regression models for this function $d_2 = \varphi_2$ (B). And after of analyzing of the statistical characteristics we can conclude that function $d_2 = \varphi_2(B)$ better submits to the linear regression model although the coefficient of correlation is not so large (r=0.42) and has the regression equation view of $Y_c = 3.03 + 1.24X$ or $d_2 = 3.03 + 1.24B$ with such statistical characteristics as: coefficient of determination $R^2 = 0.18$;coefficient of correlation r= 0.42;standard deviation $S_{y/x} = 0.694$.

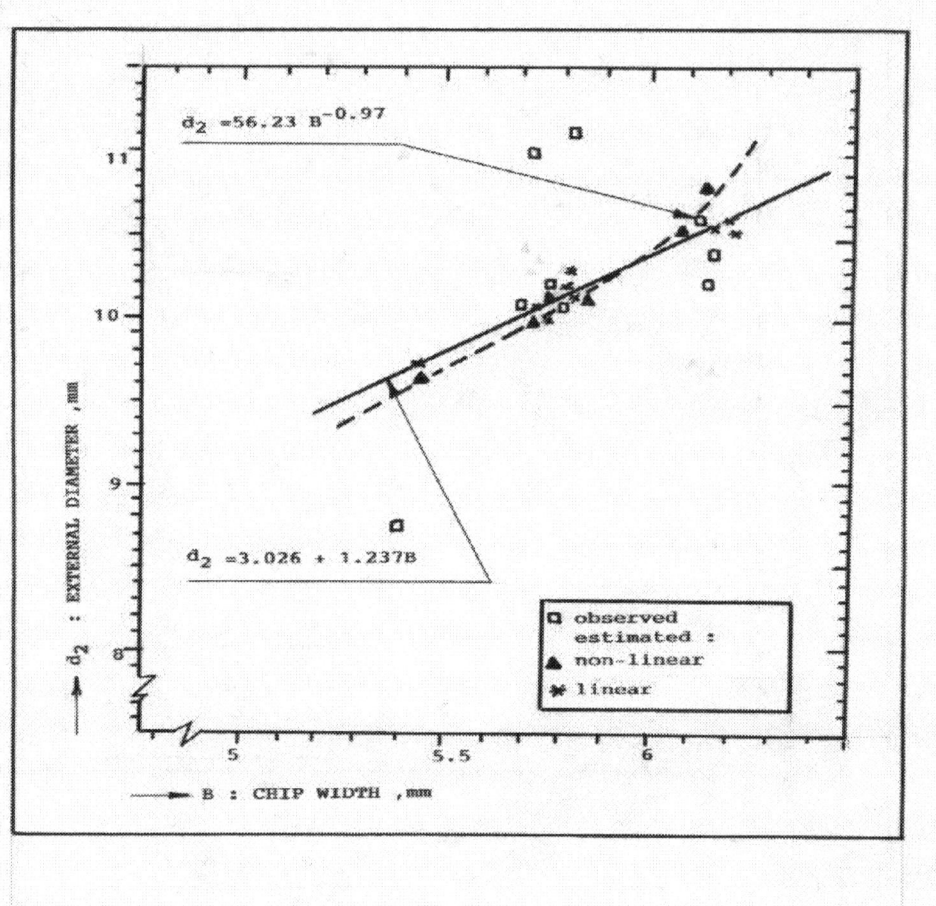

$d_2 = 56.23 \; B^{-0.97}$

$d_2 = 3.026 + 1.237B$

□ observed
estimated :
▲ non-linear
✳ linear

d₂ : EXTERNAL DIAMETER , mm

B : CHIP WIDTH , mm

Figure 7 Comparative analysis of functional dependency $d_2 = \varphi_2$ (B) of two regression models (linear and non-linear) between external diameter (d_2) of stainless chip and its width (B)

So ,we see from Figure 7 that with increasing of chip width (B) ,the value of external diameter (d_2) of this stainless chip increases considerably and this regression model better submits to the linear function in accordance with the linear equation view of   **d_2= 3.03+1.24B** (15).

4.1.8 Dependence of clearance (t) between of chip wraps from its width (B)

In Figure 8 are shown both regression models for this function $t = \gamma_1$ (B). And after of analyzing of the statistical characteristics we can conclude that this function $t = \gamma_1(B)$ better submits to the non-linear regression model although the coefficient of correlation is not so large (r=0.32) and has the regression equation view of $Y_c = 0.032X^{1.296}$ or $t = 0.032B^{1.296}$ with such statistical characteristics as: coefficient of determination R^2=0.10;coefficient of correlation r=0.32; standard error $S_{y/x}$=0.081.

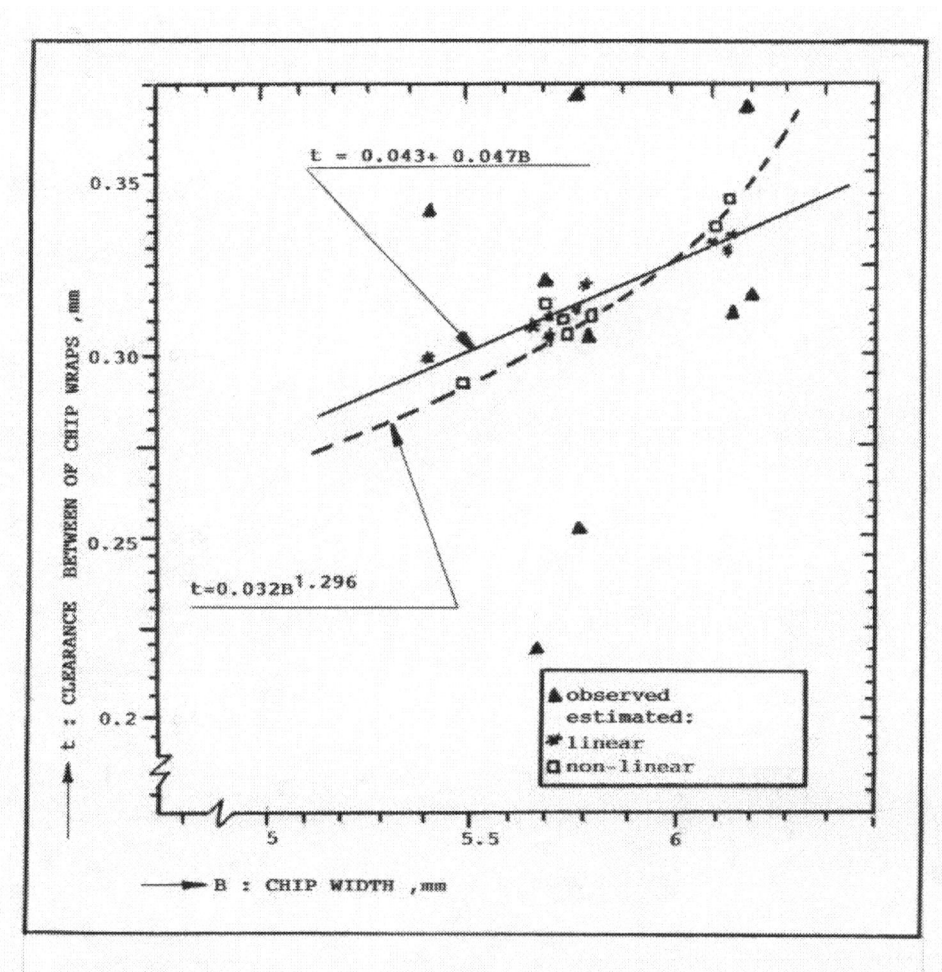

Figure 8 Comparative analysis of functional dependency t=γ₁(B) of two regression models (linear and non-linear) for clearance (t) between of chip wraps and its width (B)

So ,we see from Figure 8 that with increasing of chip width (B) ,the value of clearance (t) between of chip wraps increases considerably and this regression model better submits to the non-linear function in accordance with the non-linear equation view of $\mathbf{t=0.03B^{1.3}}$ (16).

4.1.9 Dependence of the number chip wraps (ω) from its width (B)

In Figure 9 are shown both regression models for this function ω =γ₂(B).And after of analyzing of the statistical characteristics we can conclude that this function ω=γ₂ (B) better submits to the non-linear regression model and has the regression equation view of $Y_c=703.1X^{-3.0}$ or ω =703.1B^{-3.0} with such statistical characteristics as: coefficient of determination $R^2=1.0$;coefficient of correlation r= 1.0;standard error S= 0.

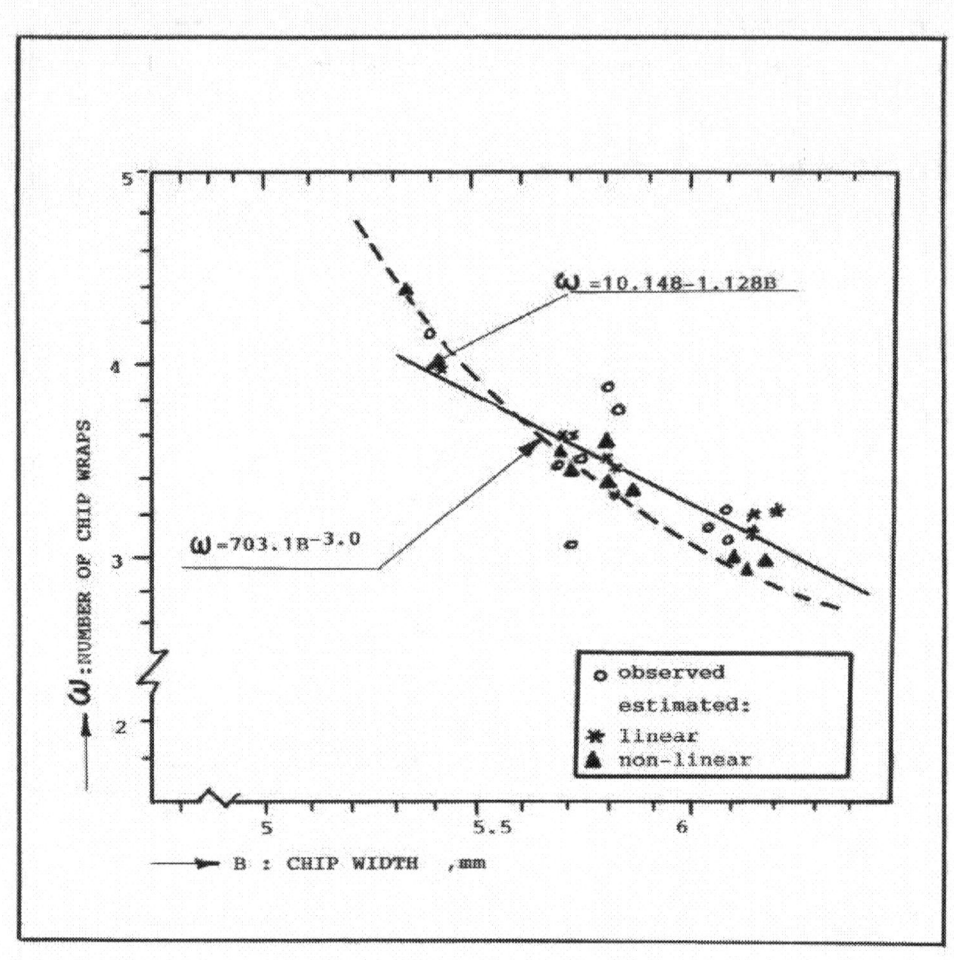

Figure 9 Comparative analysis of functional dependency ω=γ₂(B) of two regression models (linear an non-linear) between of chip wraps (ω) and its width (B)

So, we see from Figure 9 that with increasing of chip width (B) ,the value of chip wraps (ω) decreases considerably and this regression model better submits to the non-linear function in accordance with the non-linear equation view of $\boldsymbol{\omega=703.1B^{-3.0}}$ (17).

4.1.10 Dependence of external diameter (d_2) from the internal diameter (d_1) of this stainless chip

In Figure 10 are shown both regression models for this function $d_2=\varphi_3 (d_1)$. And after of analyzing of the statistical characteristics we can conclude that function $d_2= \varphi_3(d_1)$ better submits to the non-linear regression model and has the regression equation view of $Y_c=5.73X^{0.359}$ or $d_2=5.73d_1^{0.36}$ with such statistical characteristics as : coefficient of determination $R^2=1.0$; coefficient of correlation r=1.0 ; standard error S=0.

52

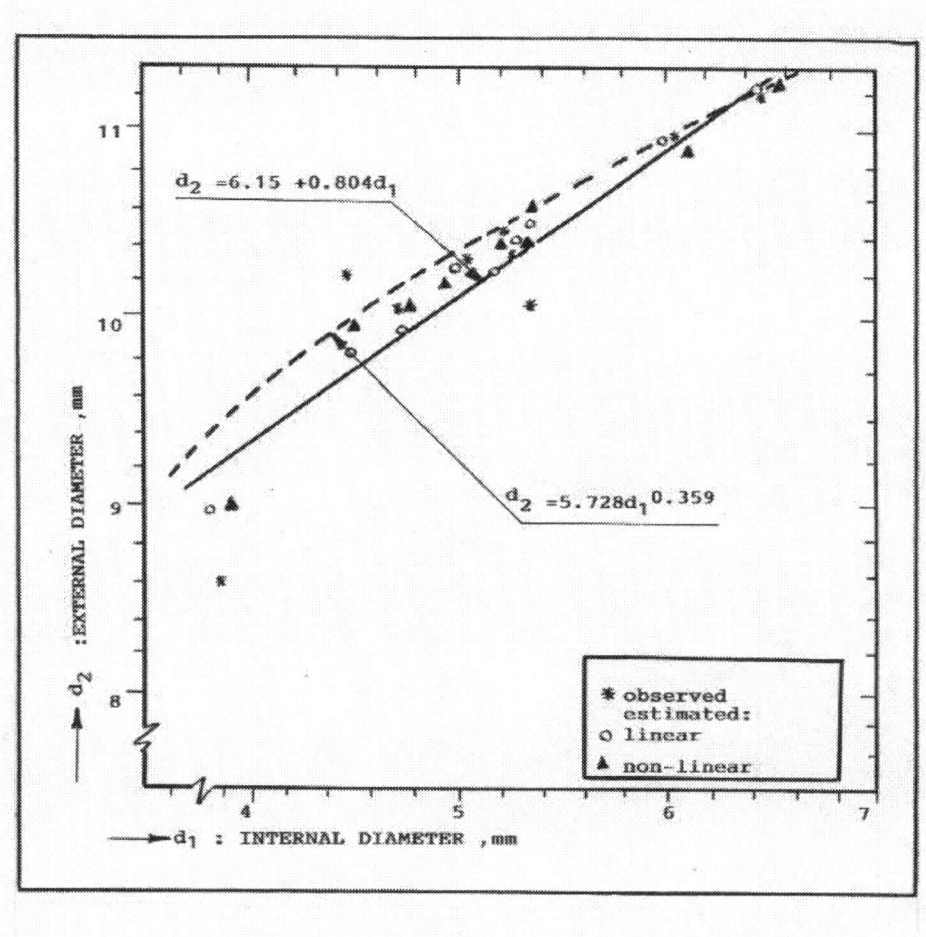

Figure 10 Comparative analysis of functional dependency $d_2 = \varphi_1(d_1)$ of two regression models (linear and non-linear) between external (d_2) and internal (d_1) diameters for stainless chip

So, we see from Figure 10 that with increasing of internal diameter (d_1),the value of external diameter (d_2) increases considerably and this regression model better submits to non-linear function in accordance with the non-linear equation view of $\mathbf{d_2 = 5.73d_1^{0.36}}$ (18).

4.1.11 Dependence of clearance (t) between of chip wraps from its external (d_2) and internal (d_1) diameters

In Figure 11 are shown both regression models for function $t = \gamma_3(d_2)$. After of analyzing of the statistical characteristics we can conclude that function $t = \gamma_3(d_2)$ better submits to the non-linear regression model and has the regression equation view of $Y_c = 1.14X^{-0.56}$ or $t = 1.14d_2^{-0.56}$ with such statistical characteristics as : coefficient of determination $R^2 = 1.0$; coefficient of correlation $r = 1.0$;standard error $S_{y/x} = 0$.

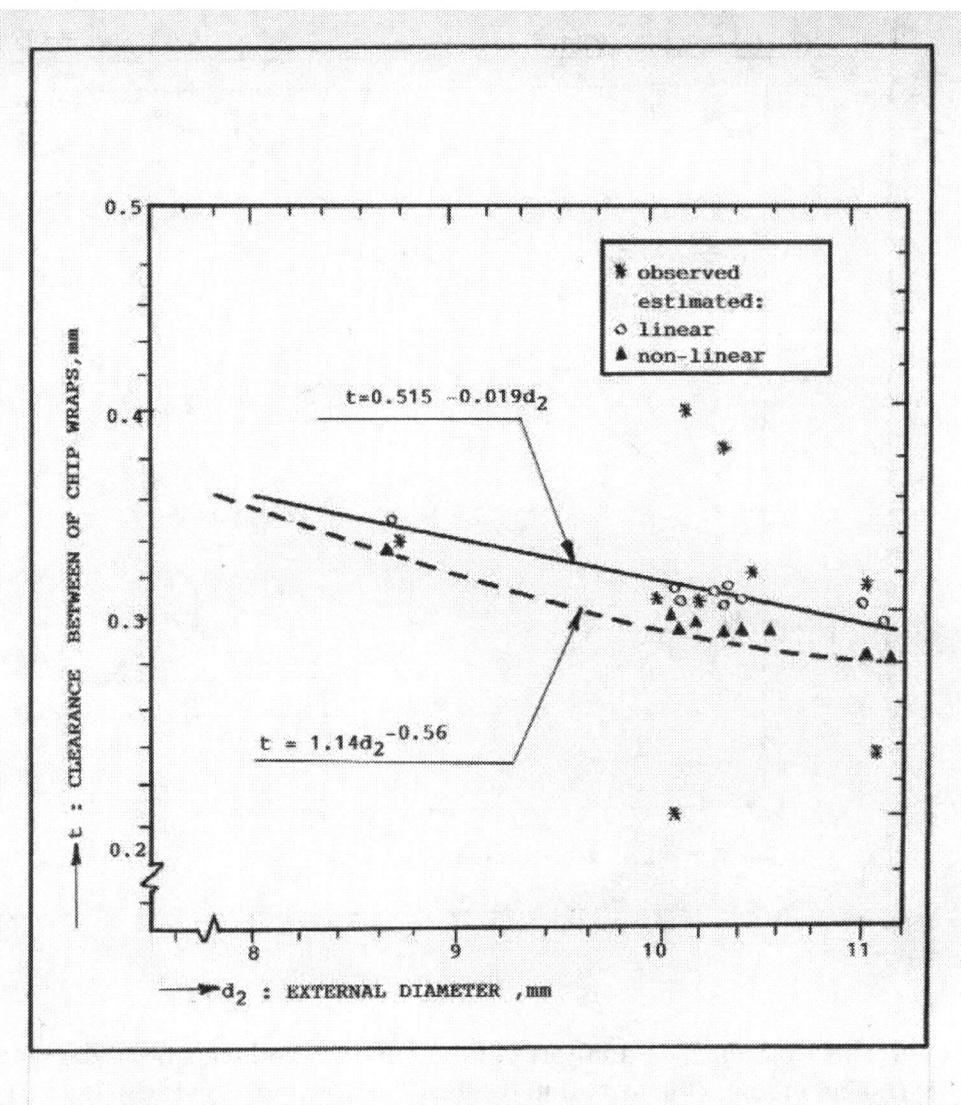

Figure 11 Comparative analysis of functional dependency $t=\gamma_3(d_2)$ of two regression models (linear and non-linear) for clearance (t) between of chip wraps and external diameter (d_2)

So, we see from Figure 11 that with increasing of external diameter (d_2),the value of clearance (t) between of chip wraps decreases considerably and this regression model better submits to the non-linear function in accordance with the non-linear equation view of $t= 1.14d_2^{-0.56}$ (19).

In Figure 12 are shown both regression models for the function $t=\gamma_4(d_1)$.After of analyzing of the statistical characteristics we can conclude that function $t=\gamma_4(d_1)$ better submits to the non-linear regression model and has the regression equation view of $Y_c=0.83X^{-0.61}$ or $t=0.83d_1^{-0.61}$ with such statistical characteristics as : coefficient of determination $R^2=0.43$;coefficient of correlation $r=0.66$; standard error $S_{y/x}=0.06$.

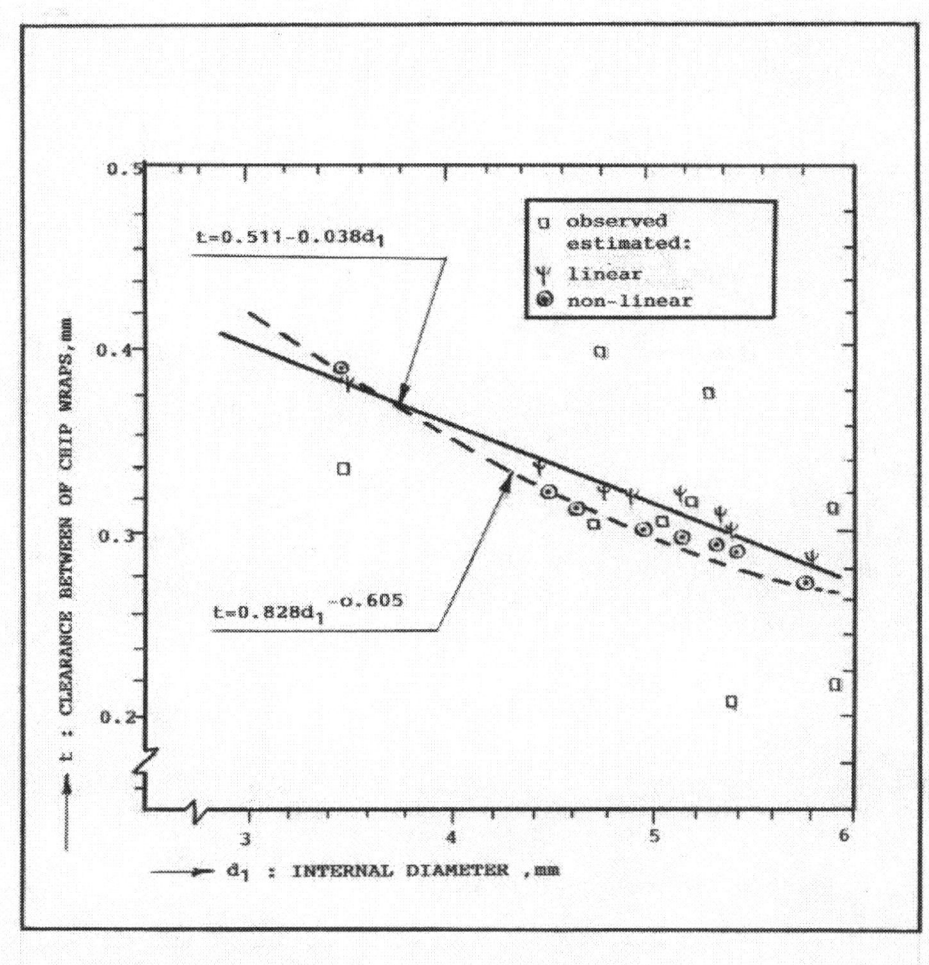

Figure 12   Comparative analysis of functional dependency $t=\gamma_4$ (d_1) of two regression models (linear and non-linear) for clearance (t) between of chip wraps and internal diameter (d_1) of stainless chip

So, we see from Figure 12 that with increasing of internal diameter (d_1) ,the value of clearance (t) between of chip wraps decreases considerably and this regression model better submits to the non-linear function in accordance with the non-linear equation view of $\mathbf{t=0.83d_1^{-0.61}}$ (20).

4.1.12 Dependence of external diameter (d_2) from the number of chip wraps (ω)

In Figure 13 are shown both regression models for this function $d_2=\phi_1(\omega$). And after of analyzing of the statistical characteristics we can conclude that this function $d_2=\phi_1(\omega$) better submits to the linear regression model and has the regression equation view of $Y_c=13.14-0.81X$ or $d_2=13.14-0.81\omega$ with such statistical characteristics as : coefficient of determination $R^2=0.25$;coefficient of correlation r= - 0.44; standard deviation $S_{y/x}=0.699$.

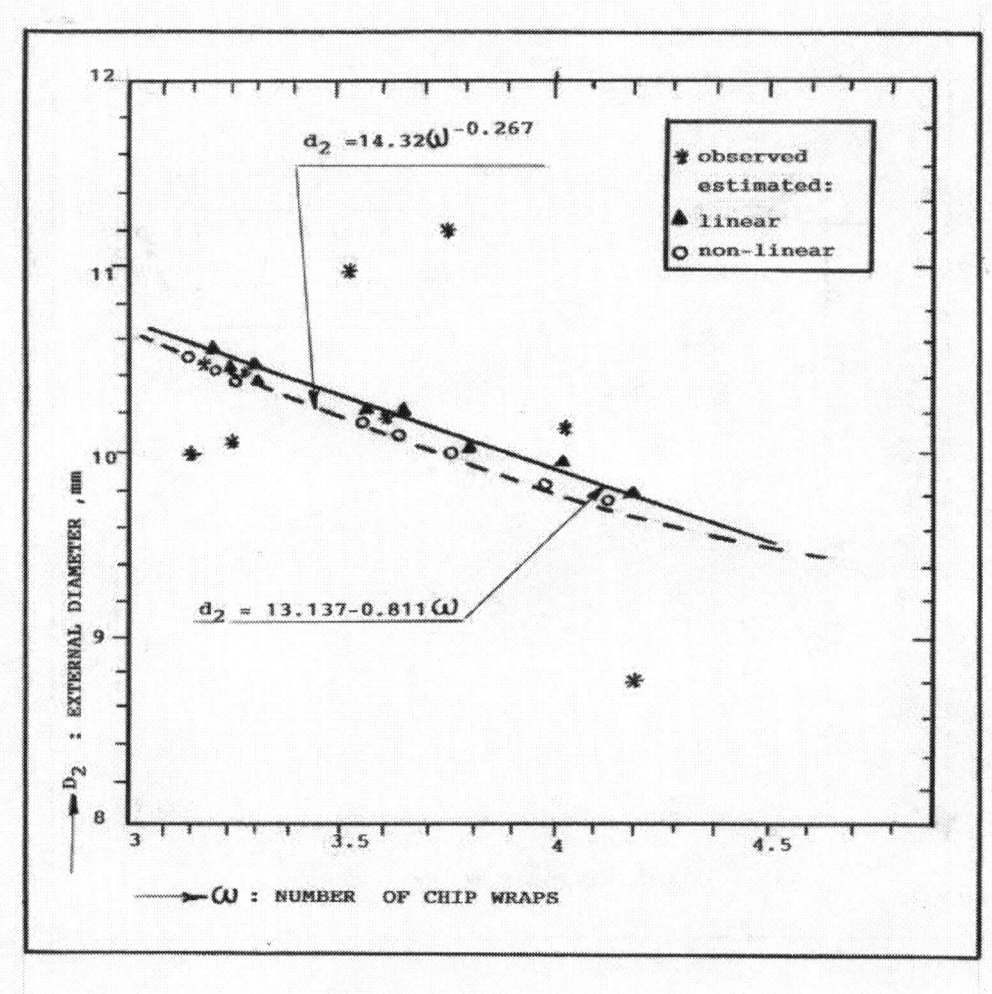

Figure 13 **Comparative analysis of functional dependency $d_2 = \phi_1(\omega)$ of two regression models (linear and non-linear) between of external diameter (d_2) and the number of chip wraps (ω)**

So, we see from Figure 13 that with increasing of the number chip wraps (ω), the value of external diameter (d_2) stainless chip decreases considerably and this regression model better submits to the linear function in accordance with the linear equation view of

$$d_2 = 13.14 - 0.81\omega \ (21).$$

4.1.13 Dependence of internal diameter (d_1) from the number of chip wraps (ω)

In Figure 14 are shown both regression models for this function $d_1 = \phi_3(\omega)$. And after of analyzing of the statistical characteristics we can conclude that function $d_1 = \phi_3(\omega)$ better submits to the non-linear regression model and has the regression equation view of $Y_c = 20.32 X^{-1.1}$ or $d_1 = 20.32\omega^{-1.1}$ with such statistical characteristics as : coefficient of determination $R^2 = 0.58$; coefficient of correlation $r = 0.76$; standard error $S_{y/x} = 0.46$.

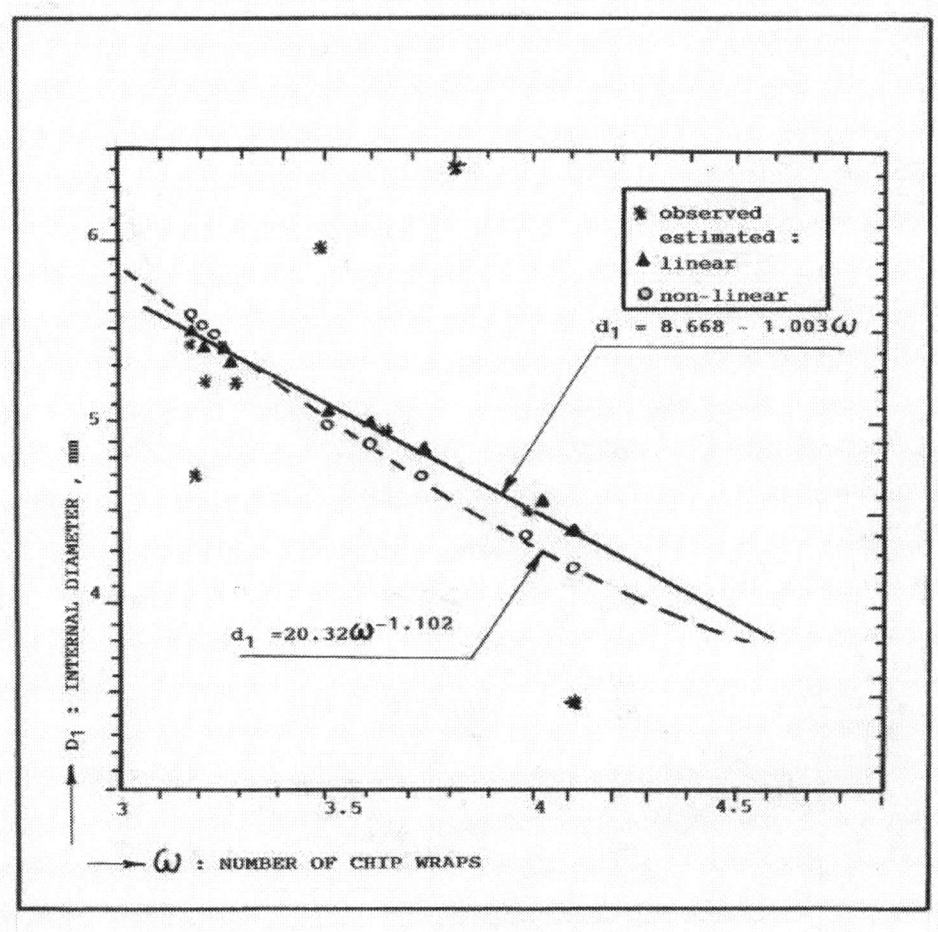

Figure 14 Comparative analysis of functional dependency $d_1=\phi_3(\omega)$ of two regression models (linear and non-linear) between internal diameter (d_1) and the number of chip wraps (ω) for stainless chip

So ,we see from Figure 14 that with increasing of the number chip wraps (ω),the value of internal diameter (d_1) stainless chip decreases considerably and this regression model better submits to the non-linear function in accordance with the non-linear equation view of

$$d_1=20.32\omega^{-1.1} \quad (22).$$

4.1.14 Dependence of the number chip wraps (ω) from the clearance (t) between of chip wraps

In Figure 15 are shown both regression models for this function $\omega=\alpha_1(t)$. And after of analyzing of the statistical characteristics we can conclude that this function $\omega=\alpha_1(t)$ better submits to the non-linear regression model and has the regression equation view of $Y_c=4.23X^{0.152}$ or $\omega=4.23t^{0.152}$ with such statistical characteristics as: coefficient of determination $R^2=0.52$; coefficient of correlation r=0.72; standard error $S_{y/x}=0.05$.

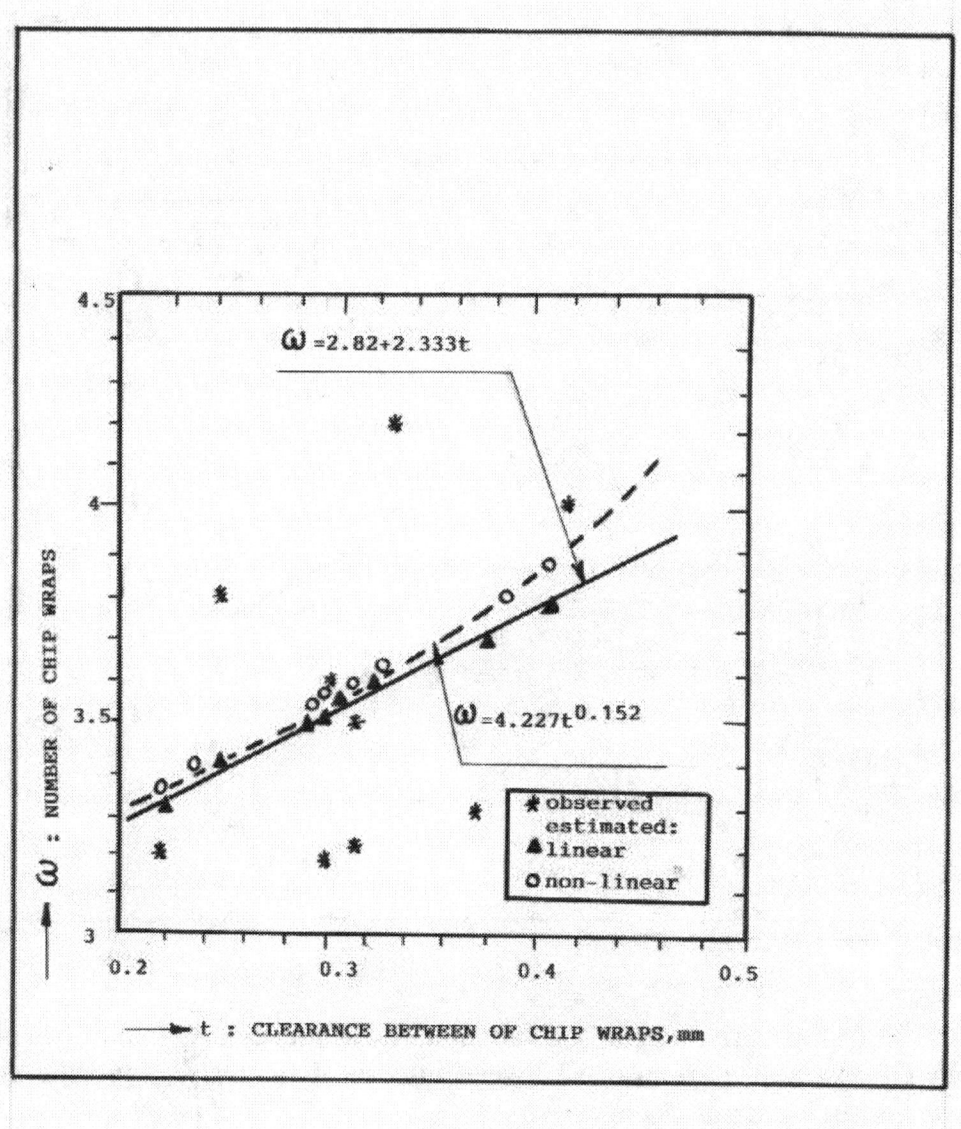

Figure 15 **Comparative analysis of functional dependency $\omega=\alpha_1(t)$ of two regression models (linear and non-linear) for the number chip wraps (ω) and its clearance (t)between of this chip wraps**

So , we see from Figure 15 that with increasing of clearance (t) between of chip wraps ,the value of the number chip wraps (ω) considerably decreases and this regression model better submits to the non-linear function in accordance with the non-linear equation view of $\omega=4.23t^{0.152}$ (23).

In Table 1 is shown summary of relations between of two parameters stainless chip for function view of $Y_i=\varphi(X_i)$ in schematic form.

Table 1 Summary relations between two parameters of stainless chip for function of $Y_i = \varphi(X_i)$

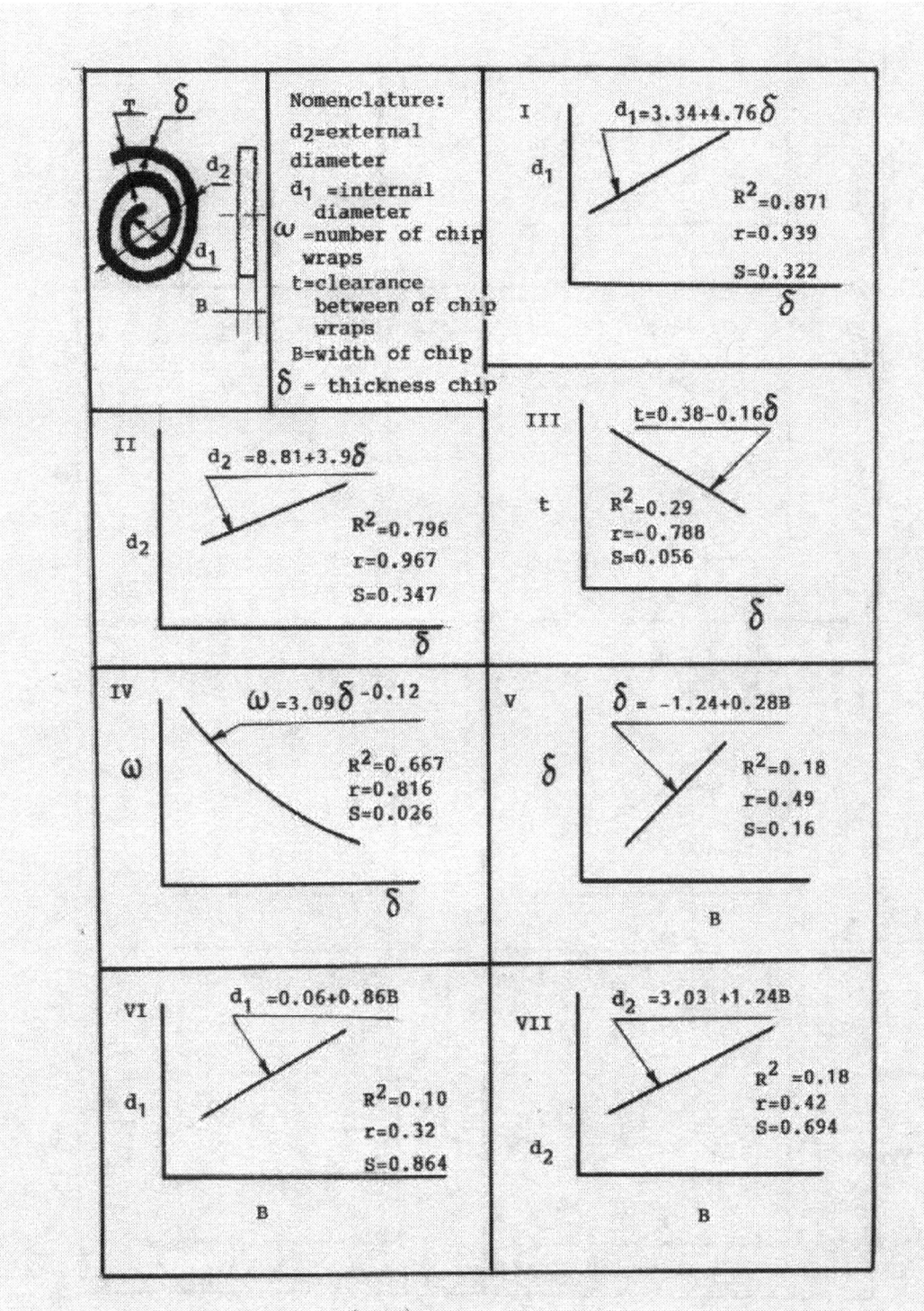

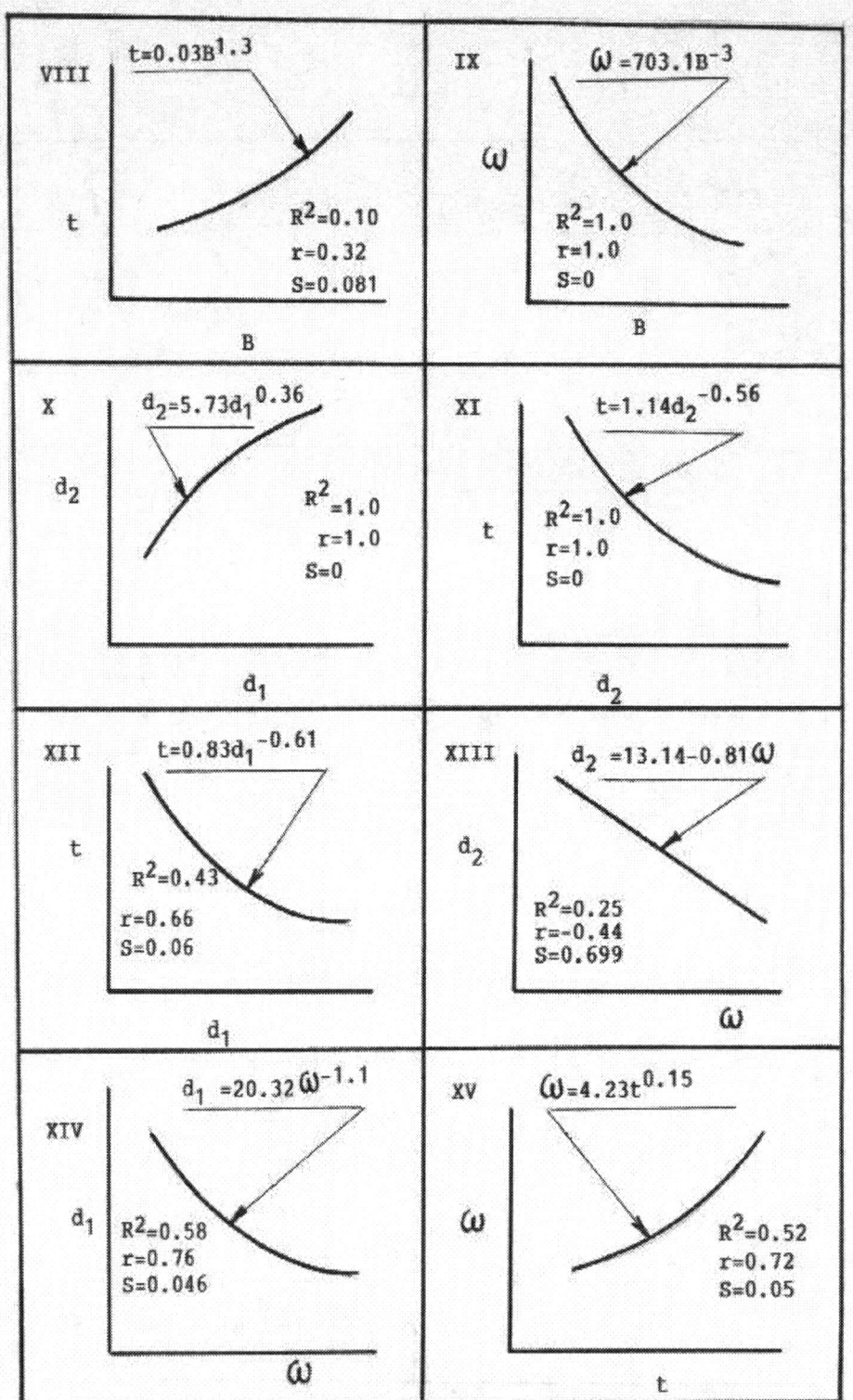

SUMMARY

1. *Functional dependence $d_1 = \varphi(\delta)$ and $d_2 = \gamma(\delta)$ have the good correlation between two parameters of stainless chip with such statistical characteristics, as the coefficient determination(R^2) and correlation (r) and show that parameters of internal (d_1) and external (d_2) diameters considerably depend from the chip thickness for stainless chip:*
   - *With increasing of chip thickness the parameters (d_1) and (d_2) considerably increase and on the contrary-with decreasing of chip thickness (δ) the above-named parameters (d_1) and (d_2) considerably decreases;*
   - *These functional dependencies $d_1 = \varphi(\delta)$ and $d_2 = \gamma(\delta)$ better submit to the linear regression model.*

2. *And besides there is the good correlation and statistical characteristics between internal (d_1) and external (d_2) diameters which better submit to the non-linear regression model:*
   - *With increasing of internal (d_1) diameter stainless chip the value of the external (d_2) considerably increases and on the contrary –with decreasing of internal (d_1) diameter the value of external (d_2) diameter stainless chip also decreases.*

3. *And necessary also to admit that functional dependency $\omega = \alpha(\delta)$ has average statistical characteristics (coefficient determination $R^2 = 0.667$;coefficient of correlation $r = 0.816$ and this function shows that with increasing of chip thickness (δ the number of chip wraps (ω) considerably decreases and besides this function better submits to the non-linear regression model.*

4. *And the same picture has a place for the functional dependency $\omega = \gamma_2(B)$ which shows the good correlation with the statistical characteristics (coefficient determination $R^2 = 1.0$;coefficient correlation $r = 1.0$) and this dependence better submits to the non-linear regression model and indicates that with increasing of width (B) stainless chip ,the number of chip wraps for this considerably decreases and on the contrary –with decreasing of width (B) the number of chip wraps for this stainless chip considerably increases.*

5. *It should also to underline that there is the good correlation in such functional dependencies as $t = \gamma_3(d_2)$ and $t = \gamma_4(d_1)$ which both submit to the non-linear regression model and have the average statistical characteristics .These functions indicate on that fact that with increasing of internal (d_1) and external (d_2) diameters ,the value of clearance (t) between of chip wraps considerably decreases and on the contrary- with decreasing of these diameters (d_1) and (d_2) the value of clearance (t) between of chip wraps increases considerably .*

CHAPTER FIVE   CORRELATION BETWEEN THREE PARAMETERS OF STAINLESS CHIP IN FUNCTION OF $Y_i=\alpha_1(X_{i,1};X_{i,2})$

A. FOR INTERNAL DIAMETER OF STAINLESS CHIP:

1. The main criteria in question of comparing and accepting of hypothesis (linear or non-linear regression model) for given function is taken such statistical factors as:

- Coefficient of determination R^2
- Coefficient of correlation r
- Standard deviation $S_{y/x}$
- Minimization of the mean square error min MSE
- Minimization of the mean absolute deviation min MAD

2. Mathematical symbols and formulas [18] and [19] for calculation of the statistical parameters *for the linear regression model* are such as :

- Multiple linear regression model of $Y_i=b_0+b_1X_{i,1}+b_2X_{i,2}$ (1)

where coefficients b_0, b_1 and b_2 can be determined from the normal equations:

$$\sum Y=nb_0+b_1\sum X_1+b_2\sum X_2$$
$$\sum X_1Y=b_0\sum X_1+b_1\sum X^2_1+b_2\sum X_1X_2\quad(2)$$
$$\sum X_2Y=b_0\sum X_2+b_1\sum X_1X_2+b_2\sum X^2_2\quad,\text{where } X_1,X_2 \text{ –independent variables.}$$

Calculation of the statistical parameters R^2,r ,min MSE and min MAD for the different regression models are given in formulas **(1),(2),(3),(4),(5),(6)** and **(7) of previous chapter 4.**

3. Standard deviation $S_{y/x1,x2}$ for the linear regression model view of $Y_i=\alpha_1(X_{i,1},X_{i,2})$ is equal

$$S_{y/x1,x2}= \left\{ \left[\sum_{i=1}^{n} (Y_i-\hat{Y}_i)^2 \right] / (n-3) \right\}^{1/2}\quad(3)\text{ , where n=number of observation.}$$

- Calculation of given functional surface [20] for the linear regression model is :
  a) The module of vector $|a'b'|=[(X_{12}-X_{11})^2 +(X_{22}-X_{21})^2 +(Y_2-Y_1)^2]^{1/2}$ (4)
  b) The module of vector $|b'c'|=[(X_{13}-X_{12})^2+ (X_{23}-X_{22})^2 +(Y_3-Y_1)^2]^{1/2}$ (5)
  c) The module of vector $|c'd'|=[(X_{14}-X_{13})^2+(X_{24}-X_{23})^2 +(Y_4-Y_3)^2]^{1/2}$ (6)
  d) The module of vector $|d'a'|=[(X_{11}-X_{14})^2+(X_{21}-X_{24})^2+(Y_1-Y_4)^2]^{1/2}$ (7)

e) The area of functional surface is equal : $S_{a'b'c'd'}=1/2\,[(|a'b'|-|c'd'|)\cdot(|b'c'|)]+ +(|b'c'|)\cdot(|c'd'|)$ (8),

where coordinates of point **a:** $(X_{11};X_{21};Y_1)$; **b:** $(X_{12};X_{22};Y_2)$; **c:** $(X_{13};X_{23};Y_3)$.

4. Mathematical symbols and formulas for calculation of the statistical parameters *for the non-linear regression model are such as :*

- Multiple non-linear (curvilinear) model of $Y_i=\alpha_1(X_{i,1};X_{i,2})$ has view

$Y_i=b_0+b_1X_1+b_2X^2_1+b_3X_2$ (9) ,where coefficients b_0,b_1,b_2 and b_3 can be determined from the normal equations:

$$\sum Y =nb_0+b_1\sum X_1+b_2\sum X^2_1+b_3\sum X_2$$
$$\sum X_1Y=b_0\sum X_1+b_1\sum X^2_1+b_2\sum X^3_1+b_3\sum X_1X_2\quad(10)$$
$$\sum X^2_1Y=b_0\sum X^2_1+b_1\sum X^3_1+b_2\sum X^4_1+b_3\sum X^2_1X_2$$
$$\sum X_2Y=b_0\sum X_2+b_1\sum X_1X_2+b_2\sum X^2_1X_2+b_3\sum X^2_2$$

5.1 Dependence of internal diameter (d_1) of stainless chip from its external diameter (d_2) and clearance (t) between of chip wraps

In Figure 1(a) is shown the functional dependence of internal diameter (d_1) stainless chip from of its external diameter (d_2). And as was above –indicated this dependence $d_2 = \varphi_3(d_1)$ has non-linear regression model with equation view of $d_2 = 5.728 d_1^{0.359}$ where **$d_1 = (0.175 d_2)^{2.79}$ (11)** .

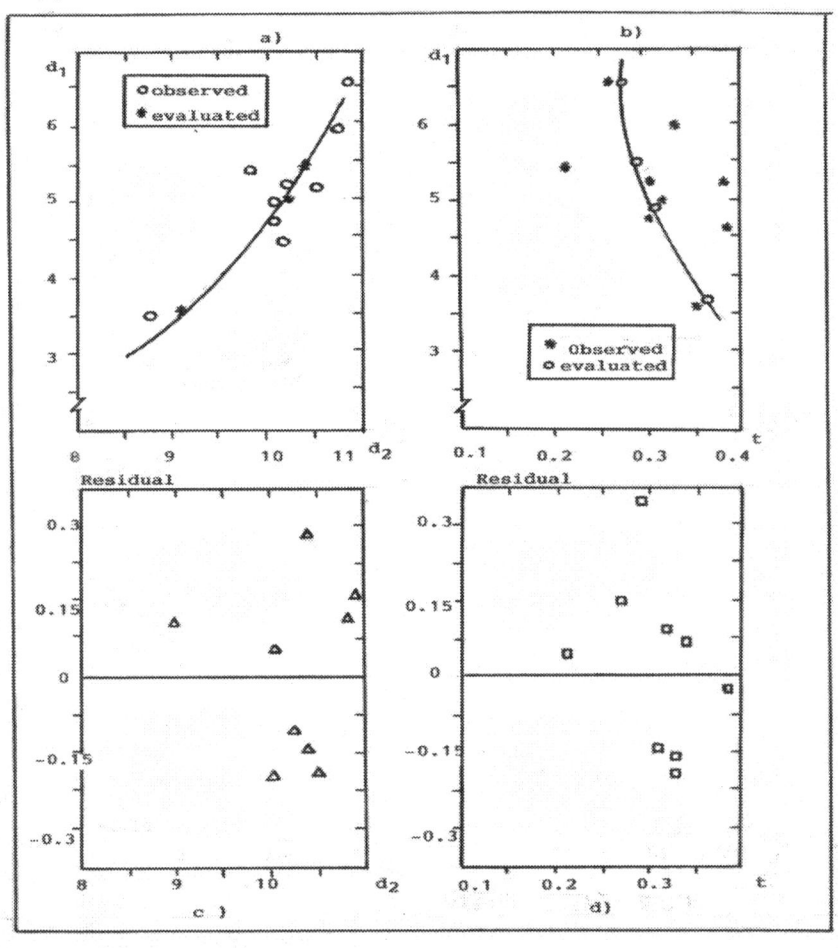

Figure 1

As we see from this Figure 1 (a) that with increasing of external diameter (d_2) stainless chip, the value of its internal diameter (d_1) increases considerably.

In Figure 1 (b) is shown the functional of internal diameter (d_1) from clearance (t) between of chip wraps. And as was above-shown this dependence $d_1 = \gamma_4$ (t) also has non-linear regression model with equation view of $t = 0.828 d_1^{-0.605}$, where **$d_1 = (1.208 t)^{-1.65}$ (12).** And as we see from Figure 1 (b) that with increasing of clearance (t) between of chip wraps, the value of internal diameter (d_1) decreases.

In Figure 1 (c) and 1 (d) are illustrated the residual plots (residual versus d_2 and residual versus t) above-named functional dependencies $d_1=\varphi_3(d_2)$ and $d_1=\gamma_4(t)$ accordingly.

Analyzing the statistical characteristics (coefficient of determination R^2 =0.964, coefficient of correlation r=0.981,standard deviation $S_{y/x}$=0.184) we see that this function better submits to the linear regression equation view of Y_c= −3.712+0.99X_1−4.214X_2 where **d_1= −3.712 +0.99d_2−4.214t   (13)**. In Figure 2 (A) is shown the functional surface $d_1=\varphi_1(d_2,t)$ in three-dimensional drawing for regression linear equation view of d_1= -3.712 +0.99d_2-4.214t. Projection of this function on the plane, as shown in Figure 2(b), has view of trapezium with such coordinate of points (sizes in mm): a (X_{11}=0,X_{21}=0.21, Y_1=5.36) ; b (X_{12}=0,X_{22}=0.21,Y_2=0) ; c (X_{13}=0,X_{23}=0.38,Y_3=0) and d (X_{14}=0,X_{24}=0.38,Y_4=5.22. Using of formulas (4) to (8) we have the value for modules |ab|=5.36,|bc|=0.37,|cd|=5.22 ,|da|=0.53 and area of this surface $S_{a,b,c,d}$=1.9mm^2 .

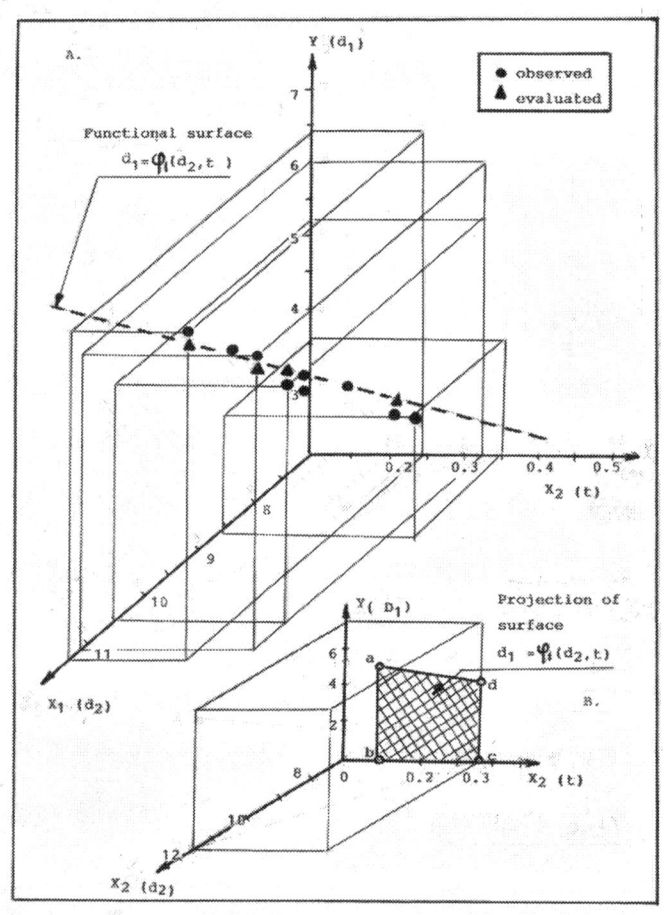

Figure 2

From Figure 2 (A) we see that functional surface $d_1=\varphi_1(d_2,t)$,on which the points are situated of linear regression equation view of d_1=-3.712+0.99d_2- 4.214t ,has the following coordinate of their peaks in three-dimensional drawing (sizes in mm): a'(X_{11}=10.03,X_{21}=0.21,Y_1=5.36);b'(X_{12}=10.03,X_{22}=0.21,Y_2=0);c'(X_{13}=10.36,X_{23}=0.38,Y_3 =0) and d'(X_{14}=10.36,X_{24}=0.38,Y_4=5.22)

64

The result of computations for the linear regression equation $Y_c = -3.712 + 0.99X_1 - 4.24X_2$ are given in Table 1.

Table 1 Evaluation of linear regression equation $Y_c = -3.712 + 0.99X_1 - 4.24X_2$

A. Mean , variance and standard deviation

VARIABLE	MEAN	VARIANCE	STANDARD DEVIATION
X_1	10.25	0.515	0.718
X_2	0.32	0.004	0.062
Y	5.098	0.026	0.159

B. Results of multiple regression of Y on X_1 and X_2

VARIABLE	COEFFICIENTS	STANDARD ERROR	T-VALUE
X_1	0.99	0.866	1.143
X_2	−4.24	1.118	−3.79

C. Analysis of variance results

REGRESSION
- Degrees of freedom 2
- Sum of squares 5.41
- Mean square 2.705

ERROR
- Degrees of freedom 6
- Sum of squares 0.204
- Mean square 0.034
- Standard error of estimate 0.718
- F-value* 79.559 [$F_{0.05;2;6}$=5.14]

As F^*=79.559>$F_{0.05;2;6}$ we reject the hypothesis that all regression coefficients are zero

D. Determination of residuals

NUMBER	OBSERVED	ESTIMATED	RESIDUAL
1	5.36	5.33	0.03
2	5.22	4.94	0.28
3	3.52	3.45	0.07
4	4.51	4.54	- 0.03
5	4.98	5.13	-0.15
6	4.74	4.94	-0.20
7	5.15	5.32	-0.17
8	5.96	5.88	0.08
9	6.44	6.29	0.15

** *all sizes in mm*

5.2 Dependence of internal diameter (d_1) stainless chip from width (B) and thickness (δ) of this chip

Analyzing of the data of this function $d_1=\varphi_2(B,\delta)$, we see that it better submits to the non-linear regression model than for the linear regression model[1] for some reasons: coefficients of determination ($R^2=0.87$) and correlation ($r=0.932$) are larger and have the good values; standard deviation $S_{y/x1,x2}=0.348$ is smaller; minimization of the mean square error (min MSE=0.08) is smaller. And this function has the non-regression equation view $Y_c= -2.60 +2.67X_1 -0.28X^2_1+ 5X_2$, where $d_1= -2.60+2.67B-0.28B^2+5\delta$ (14).

In Figure 3(a) is shown the functional dependence of internal diameter (d_1) from width(B) of stainless chip. And as was above-shown this functional dependence $d_1=\varphi_1(B)$ has the linear regression model with equation view of $Y_c=0.06+0.863X$, where $d_1=0.06+0.863B$.

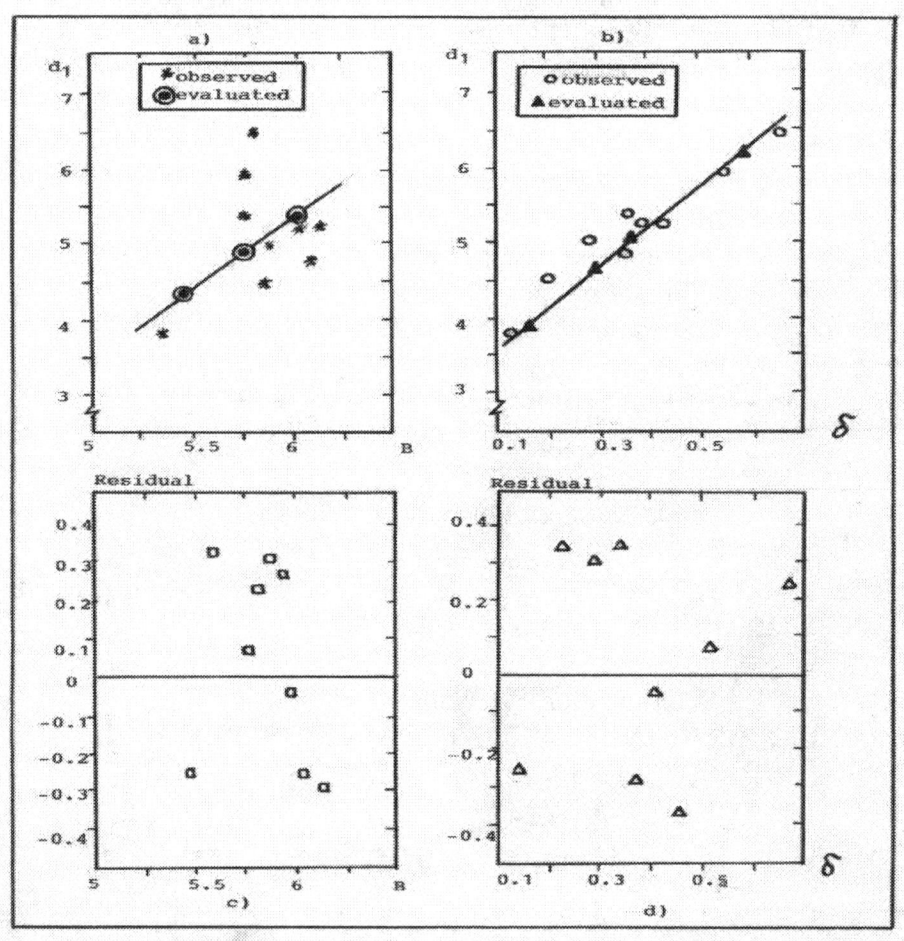

Figure 3

---

1. Linear regression equation has view of $d_1= -0.05+0.89B -0.08\,\delta$

66

From Figure 3(a) we see that with increasing of width (B) stainless chip, the value of internal diameter (d_1) increases considerably. In Figure 3(b)is shown the functional dependence of internal diameter (d_1) from thickness (δ) of this chip. And as was above-shown this functional dependence $d_1 = \varphi(\delta)$ has the linear regression model with equation view of $Y_c = 3.338 + 4.762X$, where $d_1 = 3.338 + 4.762\delta$.

As we see from this Figure 3 (b) that with increasing of chip thickness (δ), the value of internal diameter (d_1) increases considerably. In Figure 3 (c) and 3(d) are illustrated the residual plots (residual versus X_1 and residual versus X_2) of above-named functional dependencies $d_1 = \varphi_1(B)$ and $d_1 = \varphi(\delta)$ accordingly. In Figure 4 we see that functional surface $d_2 = \varphi_2(B, \delta)$ is shown in view of parabola in three-dimensional drawing for regression non-linear equation $Y_c = -2.6 + 2.67X_1 - 0.28X^2_1 + 5X_2$.

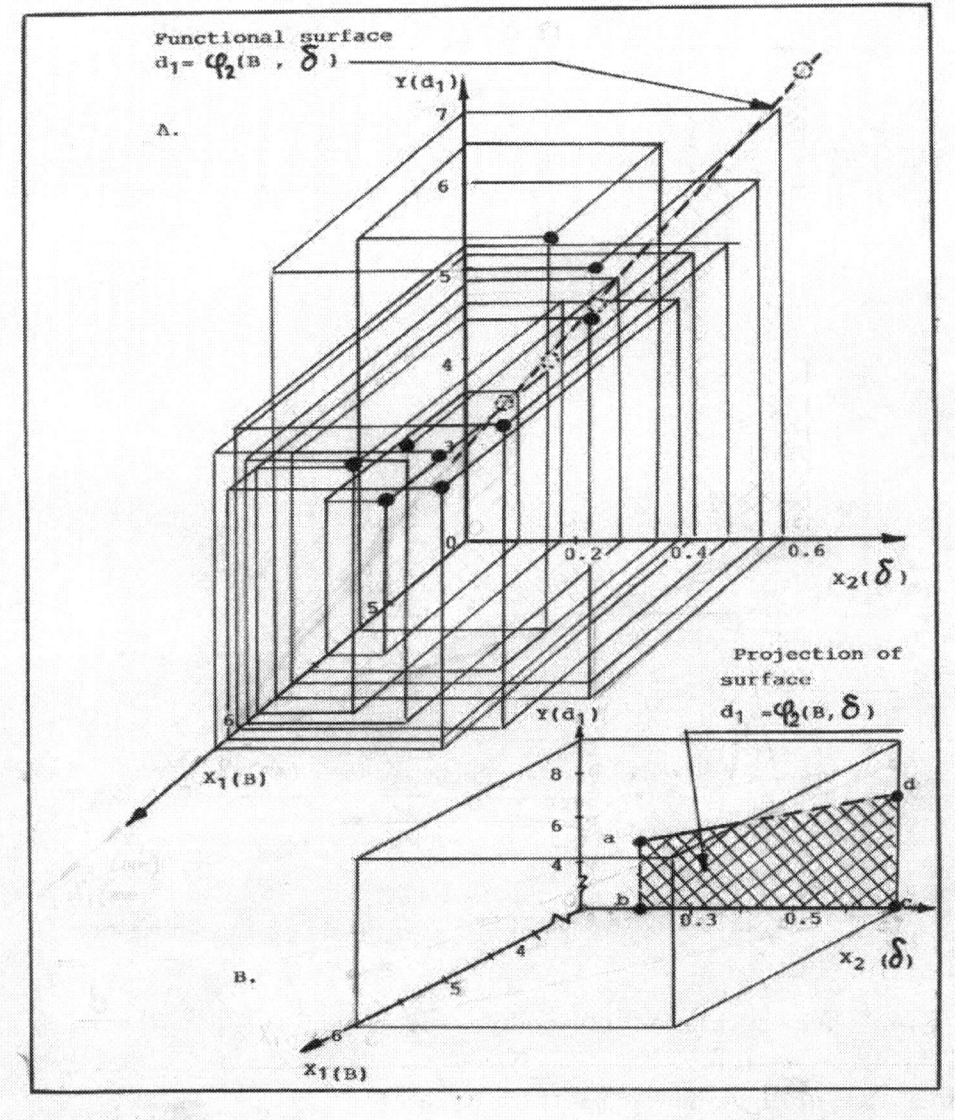

Figure 4

Projection of function $d_1 = \varphi_2(B, \delta)$ on the plane YOX_2 has view of trapezium with such coordinates as for points (sizes in mm): a $(X_{11}=0, X_{21}=0.18, Y_1=4.32)$; b$(X_{12}=0, X_{22}=0.18, Y_2=0)$; c$(X_{13}=0, X_{23}=0.6, Y_3=0)$; d $(X_{14}=0, X_{24}=0.6, Y_4=6.4)$.

From Figure 4 (A) we see that functional surface $d_1 = \varphi_2(B, \delta)$ on which are located the points of non-linear regression equation view of $d_1 = -2.6 + 2.67B - 0.28B^2 + 5\delta$ has a curvilinear character. In Figure 5 schematically is shown this functional surface and graph for calculation of module for vectors $|b'c'|$ and $|a'd'|$.

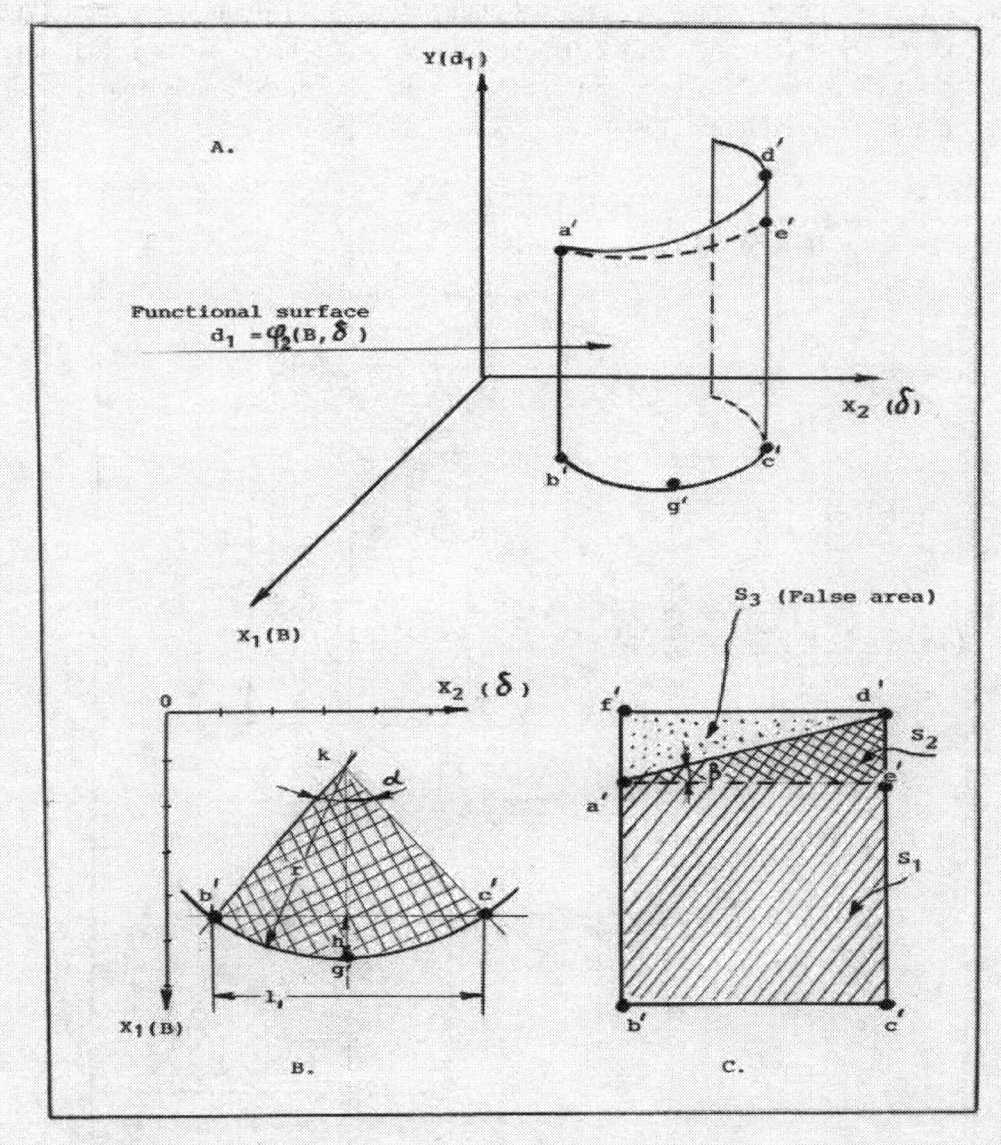

Figure 5 Graph for calculation of module vectors $|b'c'|$, $|a'd'|$ and area of functional surface $d_1 = \varphi_2(B, \delta)$:

 A-total view of this functional surface

 B-graph for calculation of module vector $|b'c'|$

 C – graph for calculation of module vector $|a'd'|$ and area of this surface

68

a)The module of vectors $\left|a'b'\right|$ and $\left|c'd'\right|$ can be calculated from the projection of this functional surface $d_1=\varphi_2(B,\delta)$ on the plane YOX_2 (Figure 4B). So, the module of vectors are equal: $\left|a'b'\right|=4.32$ and $\left|c'd'\right|=6.4$.

b) And module of vector $\left|b'c'\right|$ or it length can be calculated by formula **$L=$ $\left|b'c'\right|=0.01745\cdot r\cdot\alpha$ [21]** , where L= length of arc (b'c');r=radius of sector; α=angle of sector.

Coordinates of points b', c' and g' for this sector (k b'c') are equal: $k(X_{11}=0,X_{21}=0);b'(X_{12}=5.79,X_{22}=0.18);c'(X_{13}=5.75,X_{23}=0.6)$ and $g'(X_{14}=6.12,X_{24}=0.4)$. So, the module of vector $\left|kg'\right|$ is equal: $r=\left|kg'\right|=[(X_{14}-X_{11})^2+(X_{24}-X_{21})^2]^{1/2}=6.14$. The angle of $\alpha=82°$ (evaluated from drawing) and then module of vector $\left|b'c'\right|=8.79$ and other parameters of circular segment which are equal: **$h=r[1-\cos(\alpha/2)]$ and $l_1=2[h(2r-h)]^{1/2}$** , where h=1.5 and $l_1=8.04$.

The graph for calculation of module vectors $\left|b'c'\right|,\left|a'd'\right|$ and area of functional surface $d_1=\varphi_2(B,\delta)$ is shown in Figure 5.

c) The module of vector $\left|a'd'\right|$ and area of this functional surface, as shown in Figure 5c, can be calculated by the following way: coordinates of points $a'(X_{11}=5.79,X_{21}=0.18,Y_1=4.32);$ $d'(X_{12}=5.75,X_{22}=0.6,Y_2=6.4)$,where module of length for curve $\left|a'e'\right|=\left|b'c'\right|$,i.e $\left|a'e'\right|=8.79$ and module of length is equal for $\left|d'e'\right|=\left|c'd'\right|-\left|a'b'\right|=2.08$.

d) For calculation of area S_1,S_2 and S_3 we find the angle β which is equal $\tan\beta=\left|d'e'\right|/\left|b'c'\right|$,where $\beta=13°20'$ and then the module of length for curve $\left|a'd'\right|=\left|d'e'\right|/\sin13°20'=9.25$. Coordinate of point f' for this condition is equal $(X_1=5.79,X_2=0.18,Y_1=6.4)$ and module of length $\left|f'd'\right|=\left|b'c'\right|=8.79$ and area of $S_2=S_3=S/2=(\left|d'e'\right|\cdot\left|b'c'\right|)/2=9.14mm^2$.For these conditions, we have the area $S_1=\left|b'c'\right|\cdot\left|a'b'\right|=38mm^2$ and total area of this functional surface $d_1=\varphi_2(B,\delta)$ is equal $\sum S=47.14mm^2$.

So ,the parameters of this functional surface $d_1=\varphi_2(B,\delta)$ are the following:
- *This function better submits to the non-linear regression model in view of curvilinear surface, as parabola, and describes by the equation* **$d_1=-2.6+2.67B-0.28B^2+5\delta$**
- *The total area of functional surface $d_1=\varphi_2(B,\delta)$ is equal $\sum S=47.14\ mm^2$ with such coordinates in three-dimensional drawing for this surface(sizes in mm):*
  $a'(X_{11}=5.79,X_{21}=0.18,Y_1=4.32)$
  $b'(X_{12}=5.79,X_{22}=0.18,Y_2=0)$
  $c'(X_{13}=5.75,X_{23}=0.60,Y_3=0)$
  $d'(X_{14}=5.75,X_{24}=0.60,Y_4=6.40)$

The results of these computations for the non-linear regression equation are given in Table 2.

Table 2 Evaluation of the non-linear regression equation $Y_c = -2.6 + 2.67X_1 - 0.28X^2_1 + 5X_2$

A. MEAN, VARIANCE AND STANDARD DEVIATION *

Variable	Mean	Variance	Standard deviation
X_1	5.836	0.515	0.718
X_2	0.37	0.21	0.458
Y	5.098	5.608	2.368

B. RESULTS OF MULTIPLE REGRESSION OF Y ON X_1 AND X_2

Variable	Coefficients	Standard error	T-value
X_1	2.67	0.239	11.172
X_2	5	0.153	32.679

C.   ANALYSIS OF VARIANCE RESULTS

Regression

Degrees of freedom	2
Sum of squares	4.881
Mean square	2.441

Error

Degrees of freedom	6
Sum of squares	0.728
Mean square	0.121
Standard error of estimate	0.718
F-value	20.17 [$F_{0.05,2,6}$=5.14]

As the value F>[$F_{0.05,2,\ 6}$] we reject the hypothesis that all regression coefficients are zero.

D. DETERMINATION OF RESIDUALS

Number	Observed	Estimated	Residual
1	5.36	5.27	0.09
2	5.22	5.28	- 0.06
3	3.52	4.13	-0.61
4	4.51	4.32	0.19
5	4.98	4.81	-0.17
6	4.74	5.05	-0.31
7	5.15	5.52	-0.37
8	5.96	6.17	-0.21
9	6.44	6.40	0.04

* all sizes in mm

5.3 Dependence of internal diameter (d_1) from number of chip wraps (ω) and thickness (δ) of stainless chip

Analyzing of the data and statistical characteristics, we see that function $d_1=\varphi_3(\omega,\delta)$ better submits to linear than for the non-linear regression model[2] for some reasons: minimization of the mean square error (min MSE=0.075) and absolute deviation (min MAD=0) are smaller; coefficients of determination ($R^2=0.88$) and correlation($r=0.938$) are larger; standard deviation ($S_{y/x1,x2}= 0.335$) is smaller. And this function has the regression equation view of **$Y_c=2.292+0.258X_1+5.131X_2$ or $d_1=2.292+0.258\omega+5.131\delta$ (15).**

In Figure 6(a) is shown the functional dependence of internal diameter (d_1) from the number of chip wraps (ω). And as was above-identified this dependence $d_1=\phi_3(\omega)$ has the non-linear regression model with equation $Y_c=20.32X^{-1.102}$ or $d_1=20.32\omega^{-1.102}$ with such statistical characteristics as: coefficient of determination ($R^2=0.58$), coefficient correlation ($r=0.76$), standard error ($S_{y/x}=0.046$).

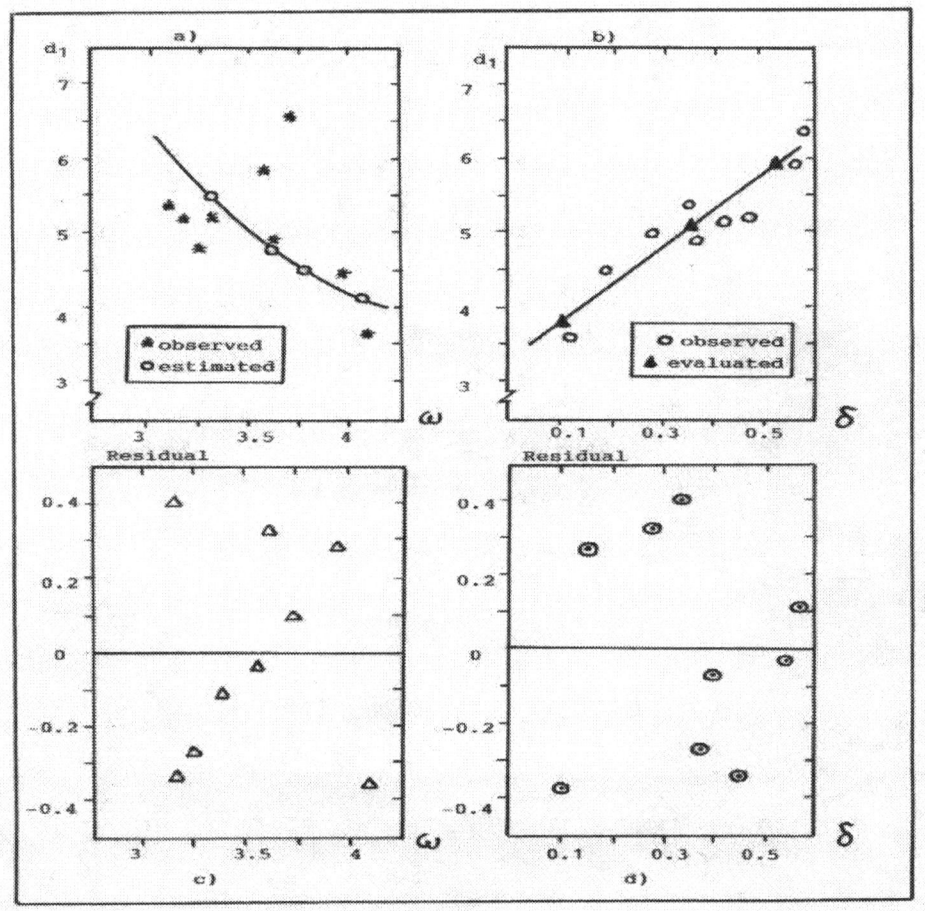

Figure 6

----

2   *non-linear regression equation has view of $d_1=11.032-4.714\omega+0.689\omega^2+5.568\delta$*

From Figure 6(a) we see that with increasing of number chip wraps (ω), the value of internal diameter (d_1) stainless chip decreases considerably. In Figure 6(b) is shown the functional dependence of internal diameter (d_1) from the thickness (δ) of stainless chip and as was above –shown this dependence $d_1=\varphi(\delta)$ has the linear regression model with equation $Y_c=3.338+4.762X$ or $d_1=3.338+4.762\delta$ with such statistical characteristics as : coefficient of determination ($R^2=0.871$),coefficient of correlation ($r=0.939$),standard deviation ($S_{y/x}=0.322$).

So, we see from Figure 6(b) that with increasing of chip thickness (δ) ,the value of internal diameter (d_1) of this stainless chip increases considerably. In Figure 6(c) and 6(d) are illustrated residual plots (residual versus ω and residual versus δ) the above-named functional dependencies of $d_1=\phi_3(\omega)$ and $d_1=\varphi(\delta)$accordingly. Analyzing of Figure 7(A),we see that functional surface $d_1=\varphi_3(\omega,\delta)$ is shown in three-dimensional drawing and has the linear regression model with equation view of $Y_c=2.292+0.258X_1+5.131X_2$ or $d_1=2.292+0.258\omega+5.131\delta$.

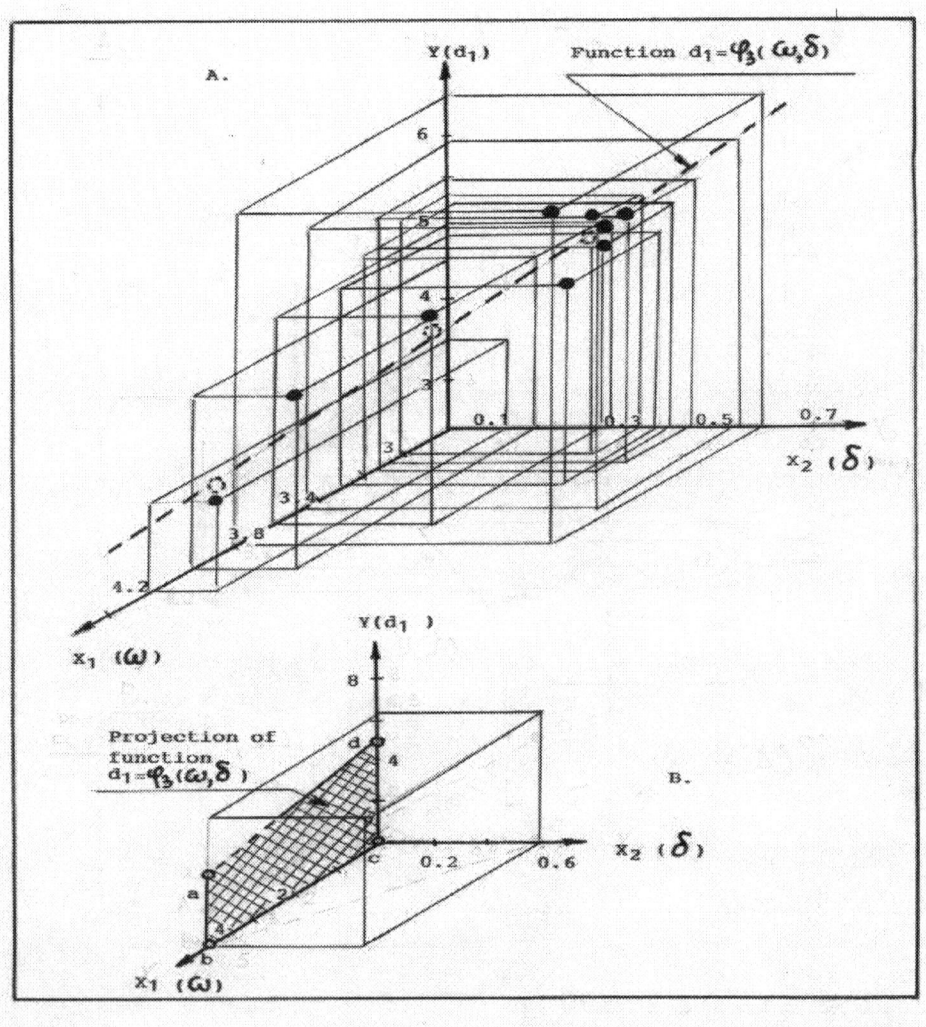

Figure 7

72

In Figure 7(B) is shown the projection of function $d_1=\varphi_3(\omega,\delta)$ in view of trapezium (abcd) with such coordinates of their points: $a(X_{11}=4.22, X_{21}=0, Y_1=3.89)$; $b(X_{12}=4.22, X_{22}=0, Y_2=0)$; $c(X_{13}=3.21, X_{23}=0, Y_3=0)$; $d(X_{14}=3.21, X_{24}=0, Y_4=5.48)$.

From Figure 7 (A) we see that functional surface $d_1=\varphi_3(\omega,\delta)$, on which are situated the points of the linear regression equation view of $d_1=2.292+0.258\omega+5.131\delta$ has the following coordinate of its peaks in three-dimensional drawing (sizes in mm): $a'(X_{11}=4.22, X_{21}=0.1, Y_1=3.89)$; $b'(X_{12}=4.22, X_{22}=0.1 , Y_2=0)$; $c'(X_{13}=3.21, X_{23}=0.46, Y_3=0)$; $d'(X_{14}=3.21, X_{24}=0.46, Y_4=5.48)$. The modulus of vectors are equal : $|a'b'|=3.89, |b'c'|=1.07, |c'd'|=5.48, |d'a'|=1.92$ and total area of this functional surface $S_{a'b'c'd'}=5.01 mm^2$.

The result of computations for linear regression equation $Y_c=2.292+0.258X_1+5.131X_2$ are given in Table 3.

Table 3 Evaluation of the linear regression equation $Y_c=2.292+0.258X_1+5.131X_2$

A. MEAN, VARIANCE AND STANDARD DEVIATION			
Variable	Mean	Variance	Standard deviation
X_1	3.56	1.114	1.055
X_2	0.37	0.21	0.458
Y	5.10	5.607	2.368

B. RESULTS OF MULTIPLE REGRESSION OF Y ON X_1 AND X_2			
Variable	Coefficients	Standard error	T-value
X_1	0.258	0.352	0.245
X_2	5.131	0.153	11.203

C. ANALYSIS OF VARIANCE RESULTS

Regression

Degrees of freedom	2	
Sum of squares	4.932	
Mean square	2.221	

Error

Degrees of freedom	6	
Sum of squares	0.675	
Mean square	0.113	
Standard error of estimate	1.055	
F-value	19.655	$[F_{0.05,2,6}=5.14]$

As the value $F>F_{0.05,2,6}$ we reject the hypothesis that all regression coefficients are zero

D. DETERMINATION OF RESIDUALS			
Number	Observed	Estimated	Residual
1	5.36	4.96	0.40
2	5.22	5.29	-0.07
3	3.52	3.89	-0.37
4	4.51	4.25	0.26
5	4.98	4.66	0.32
6	4.74	5.03	-0.29
7	5.15	5.48	-0.33
8	5.96	5.97	-0.01
9	6.44	6.34	0.10

So, the parameters of this functional surface $d_1 = \varphi_3(\omega, \delta)$ are the following:

1. This function better submits to the linear regression model, as shown in Figure 7(A) and describes by equation $d_1 = 2.292 + 0.258\omega + 5.131\delta$

2. The total area of this functional surface is equal $S = 5.01$ mm^2 with such coordinate of its peaks in three-dimensional drawing (sizes in mm): a'($X_{11}=4.22, X_{21}=0.10, Y_1=3.89$); b'($X_{12}=4.22, X_{22}=0.10, Y_2=0$); c'($X_{13}=3.21, X_{23}=0.46, Y_3=0$); d'($X_{14}=3.21, X_{24}=0.46, Y_4=5.48$).

5.4 Dependence of internal diameter (d_1) from external diameter (d_2) of stainless chip and its width(B)

Analyzing of data and statistical characteristics we see that function $d_1 = \varphi_4(d_2, B)$ better submits to the non-linear regression model than for the linear regression model[3] for some reasons: coefficients of determination ($R^2 = 0.90$) and correlation ($r = 0.949$) are larger; minimization of the mean square error (min MSE = 0.062) and absolute deviation (min MAD = 0.098) are smaller; standard deviation ($S_{y/x1,x2} = 0.305$)is smaller. And this function has the non-linear regression equation $Y_c = -3.132 + 1.192X_1 - 2.52 \cdot 10^{-6}X^2_1 - 0.683X_2$ where

$$d_1 = -3.132 + 1.192d_2 - 2.52 \cdot 10^{-6} d^2_1 - 0.683B \quad (16).$$

In Figure 8(a) is shown the functional dependency of internal diameter (d_1) from external diameter (d_2),i.e we have $d_1 = \varphi_3(d_2)$.

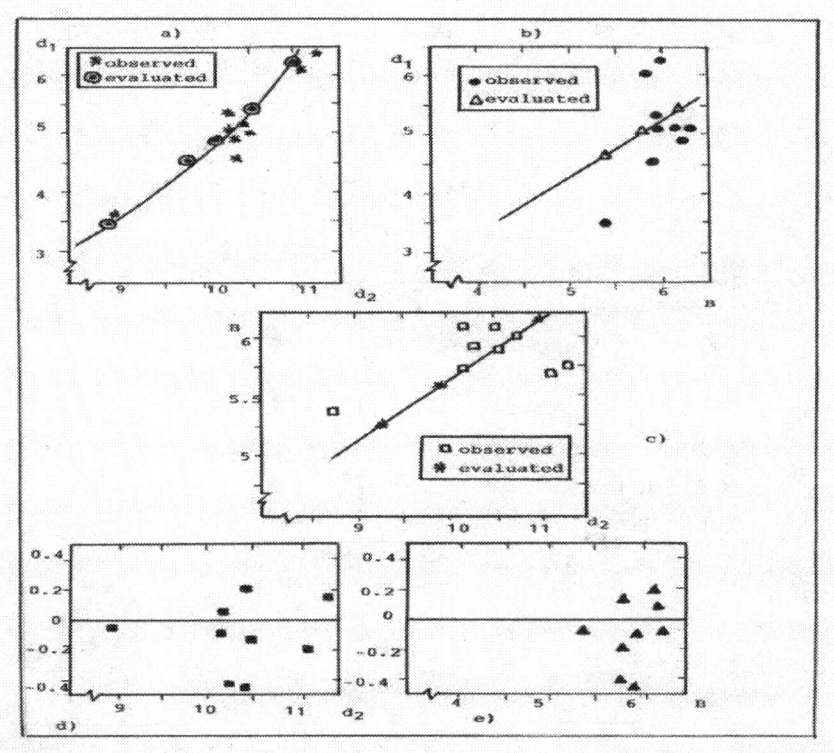

Figure 8

---

3   *Linear regression equation has view of $d_1 = -6.129 + 1.096d_2 - 0.001B$*

And as was above-shown this dependence $d_1 = \varphi_3(d_2)$ has non-linear regression model with equation $Y_c = 5.728X^{0.359}$ or $d_2 = 5.728d_1^{0.359}$ where $\mathbf{d_1 = (0.175d_2)^{2.79}}$ **(17).**

So, from this equation we can conclude that with increasing of external diameter (d_2), the value of internal diameter (d_1) increases considerably.

In Figure 8(b)is shown the functional dependency of internal diameter $(d_1)$from width (B) of stainless chip. This dependence, as was above-shown, better submits to the linear regression model with equation $Y_c = 0.06 + 0.863X$ or $d_1 = 0.06 + 0.863B$. As we see from Figure 8(b) that with increasing of width (B) for stainless chip, the value of internal diameter (d_1) increases accordingly.

In Figure 8 (c) is shown the functional dependence of stainless chip width(B) from external diameter (d_2) and this dependence $B = \varphi_2(d_2)$ has the linear regression model with equation $Y_c = 3.026 + 1.237X$ or $d_2 = 3.026 + 1.237B$, where $\mathbf{B = 0.808d_2 - 2.446}$ **(18).**

And besides from Figure 8 (c) we see that with increasing of external diameter (d_2), the value of width (B) of stainless chip increases accordingly.

In Figure 8 (d) and 8(e) are illustrated residual plots (residual versus d_2 and residual versus B) the above-named functional dependencies $d_1 = \varphi_3(d_2)$ and $B = \varphi_2(d_2)$ accordingly.

Analyzing the Figure 9(A) we see that functional surface $d_1 = \varphi_4(d_2, B)$ is shown in view of three-dimensional drawing for non-linear regression model with equation $Y_c = -3.132 + 1.192X_1 - 2.52 \cdot 10^{-6}X^2_1 - 0.683X_2$ or $d_1 = -3.132 + 1.192d_2 - 2.52 \cdot 10^{-6}d^2_2 - 0.683B$.

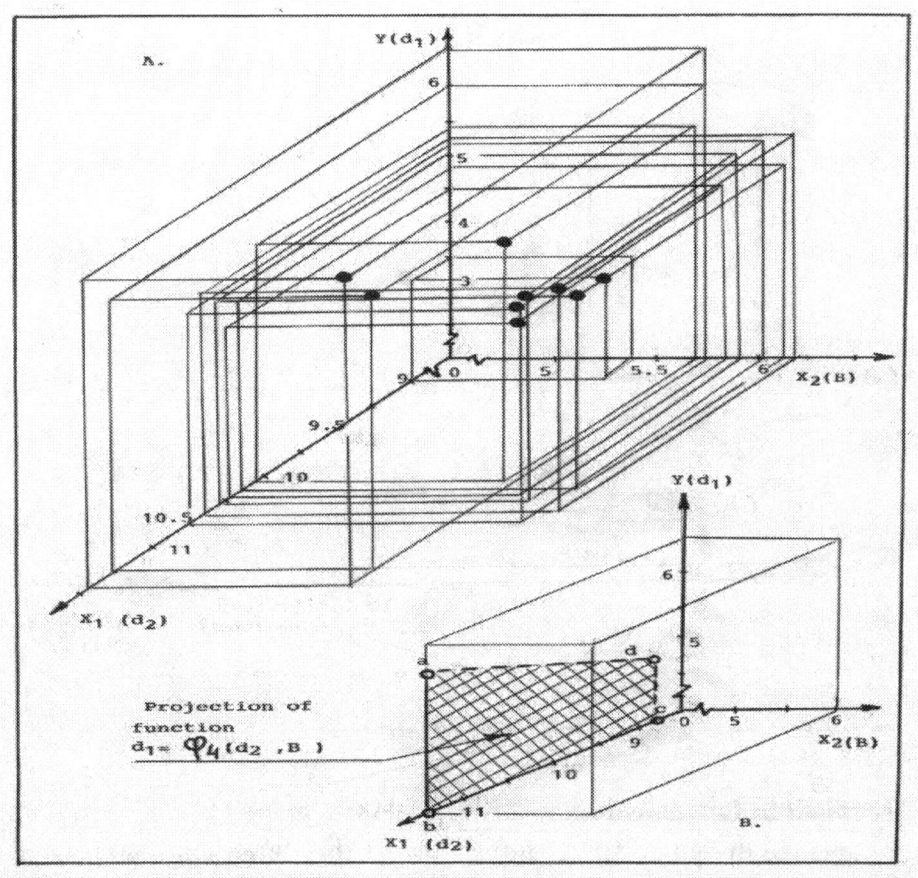

Figure 9

75

In Figure 9(B) is shown that projection of this functional surface $d_1=\varphi_4(d_2,B)$ on the plane YOX_1 has view of trapezium with coordinate for such points as (sizes in mm): $a(X_{11}=8.69,X_{21}=0,Y_1=3.56)$; $b(X_{12}=8.69,X_{22}=0,Y_2=0)$; $c(X_{13}=11.18,X_{23}=0,Y_3=0)$ and $d (X_{14}=11.18, X_{24}=0, Y_4=6.27)$.

Analysis of functional surface $d_1= \varphi_4(d_2,B)$

From Figure 9(A) we see that this functional surface, on which located the points of non-linear regression equation view of $d_1= - 3.132+1.192d_2 -2.52\cdot10^{-6}d^2_2 - 0.683B$ has a curvilinear character with the following coordinate of their peaks in three-dimensional drawing (sizes in mm): $d'(X_{11}=8.69,X_{21}=5.37,Y_1=3.56)$; $c'(X_{12}=8.69,X_{22}=5.37,Y_2=0)$; $b'(X_{13}=11.18,X_{23}=5.75,Y_3=0)$; $a'(X_{14}=11.18,X_{24}=5.75,Y_4=6.27)$.

In Figure 10 schematically is shown the functional surface $d_1=\varphi_4(d_2,B)$ and graph for calculation of module for vectors $|b'c'|$ and $|a'd'|$.

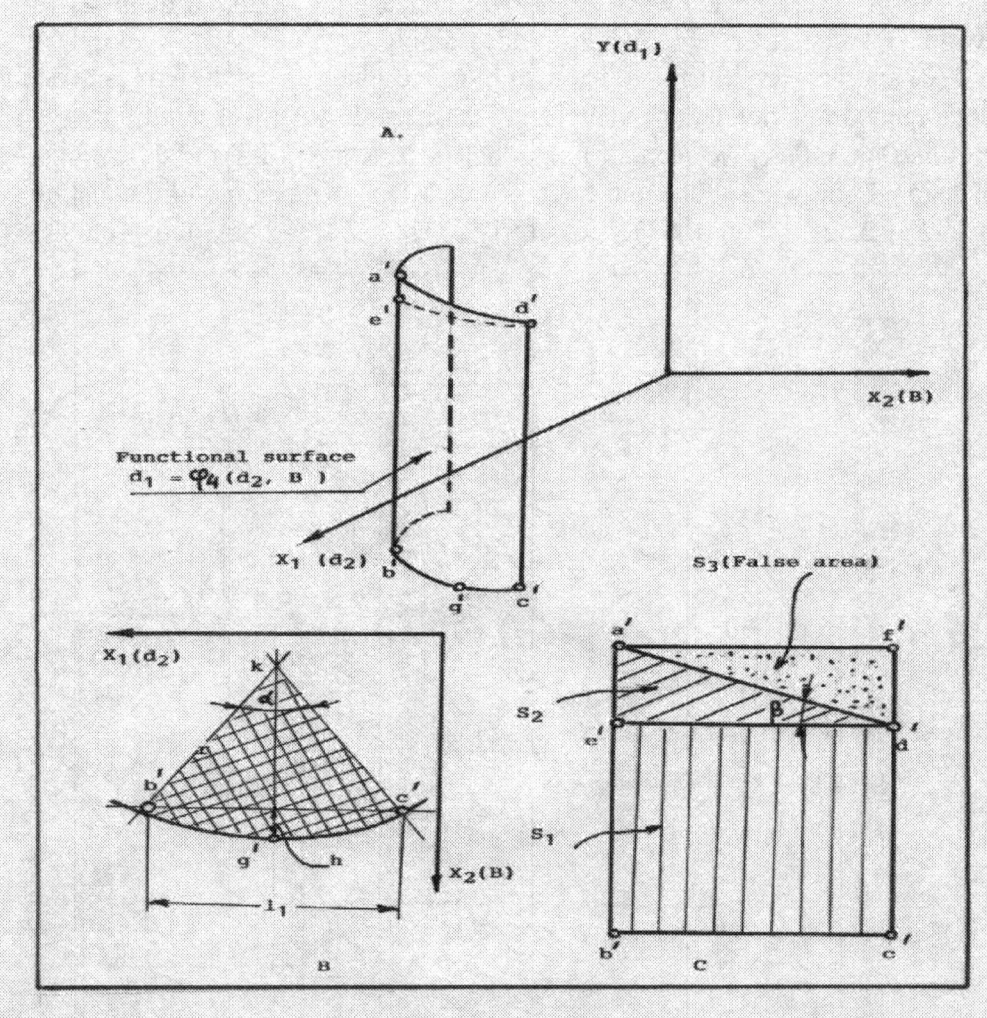

Figure 10 Graph for calculation of module vectors $|b'c'|$, $|a'd'|$ and area of functional surface $d_1=\varphi_4(d_2,B)$: A- total view of this functional surface; B-graph for calculation of module vector $|b'c'|$; C-graph for calculation of module vector $|a'd'|$ and area of this surface

76

The module of vector $|a'b'|$ and $|c'd'|$ can be calculated from the projection of this functional surface $d_1=\varphi_4(d_2,B)$ on the plane YOX_1 (Figure 9B and 10A).So, the module of vectors $|c'd'|=3.56$ and $|a'b'|=6.27$.

The module of vector $|b'c'|$ or it length can be calculated by formula $L=|b'c'|=0.01745r\alpha$, where L=length of arc (b'c');r=radius of sector; α=angle of sector. So, the module of vector $|kg'|$ is equal: $r=|kg'|=[(X_{16}-X_{15})^2+(X_{26}-X_{25})^2]^{1/2}=11.54$,where the coordinates of points g' and k: g'($X_{16}=10.03,X_{26}=5.71$) and k($X_{15}=0,X_{25}=0$). At data $\alpha=82°$,r=11.54 we have the module of vector $|b'c'|=16.51$ and other parameters of circular segment which are equal h= $r[1-\cos(\alpha/2)]=2.83$ and $l_1=2[h(2r-h)]^{1/2}=15.14$.

The graph for calculation of module vectors $|b'c'|$, $|a'd'|$ and area of functional surface $d_1=\varphi_4(d_2,B)$ is shown in Figure 10.

The module of vector $|a'd'|$ and area of this functional surface $d_1=\varphi_4(d_2,B)$,as shown in Figure 10 (c) ,can be calculated the following way: the module of length for curve $|d'e'|$ is equal $|d'e'|=|b'c'|$,i.e $|d'e'|=16.51$ and module of length for $|a'e'|$ is equal $|a'e'|=|a'b'|-|c'd'|=2.71$.

For calculation of area S_1,S_2 and S_3 we find the angle β which is equal $\tan\beta=(|a'e'|/(|b'c'|)$,where $\beta=9°20'$ and then module of length for curve $|a'd'|$ is equal $|a'd'|=(|a'e'|)/\sin9°20'=16.73$.Coordinate of point f '($X_{11}=8.69,X_{21}=5.37,Y_4=6.27$) and module of length $|f'a'|=|b'c'|=16.51$ and area of $S_2=S_3=S/2=(|f'a'|\cdot|a'e'|)/2$.

For these conditions, we have are $S_1=|c'd'|\cdot|b'c'|=58.78mm^2$ and total area ΣS of this functional surface $d_1=\varphi_4(d_2,B)$ is equal $\Sigma S=S_1+S_2=81.15mm^2$.

So, the parameters of this functional surface $d_1=\varphi_4 (d_2,B)$ are the following:

1.This function better submits to the non-linear regression model in view of curvilinear surface, as parabola, and schematically is shown in Figure 10.And besides this function surface describes by the equation view of $Y_c= -3.132+1.192X_1-2.52\cdot10^{-6}X^2_1-0.683X_2$ where $d_1= -3.132+1.192d_2 -2.52\cdot10^{-6}d^2_2-0.683B$.

2.The total area of functional surface $d_1=\varphi_4(d_2,B)$ is equal $\Sigma S=81.15mm^2$ with such coordinate of its points in three-dimensional drawing for this surface(sizes in mm):

    d' ($X_{11}=8.69,X_{21}=5.37,Y_1=3.56$)
    c' ($X_{12}=8.69,X_{22}=5.37,Y_2=0$)
    b' ($X_{13}=11.18,X_{23}=5.75,Y_3=0$)
    a' ($X_{14}=11.18,X_{24}=5.75,Y_4=6.27$)

The results of computations for the non-linear regression equation $Y_c= -3.132+1.192X_1-2.52\cdot10^{-6}X^2_1-0.683X_2$ is given in Table 4.

Table4 Evaluation of non-regression equation $Y_c = -3.132 + 1.192X_1 - 2.52 \cdot 10^{-6}X^2_1 - 0.683X_2$

A. MEAN, VARIANCE AND STANDARD DEVIATION

Variable	Mean	Variance	Standard deviation
X_1	10.25	2.263	1.504
X_2	5.836	0.064	0.253
Y	5.098	5.609	2.368

B. RESULTS OF MULTIPLE REGRESSION OF Y ON X_1 AND X_2

Variable	Coefficients	Standard error	T-value
X_1	1.192	0.473	2.52
X_2	− 0.683	0.027	−25.296

C. ANALYSIS OF VARIANCE RESULTS

Regression
- Degrees of freedom 2
- Sum of squares 5.048
- Mean square 2.524

Error
- Degrees of freedom 6
- Sum of squares 0.561
- Mean square 0.094

Standard error of estimate 1.504

F-value* 26.851 [$F_{0.05,2,6}$]=5.147

Since F^* >[$F_{0.05,2,6}$] we reject the hypothesis that all regression coefficients are zero.

D. DETERMINATION OF RESIDUALS

Number	Observed	Estimated	Residual
1	5.36	4.93	− 0.43
2	5.22	5.01	0.21
3	3.52	3.56	−0.04
4	4.51	4.99	− 0.48
5	4.98	5.08	−0.10
6	4.74	4.69	0.05
7	5.15	5.21	−0.06
8	5.96	6.16	−0.20
9	6.44	6.27	0.17

5.5 Dependence of internal diameter (d_1) from clearance (t) between of chip wraps and its thickness (δ)

Analyzing of the data and statistical characteristics, we see that this function $d_1=\varphi_5(t,\delta)$ better submits to the non-linear regression model than for the linear regression model [4] for some reasons:

- Coefficients of determination (R^2=0.882) and correlation (r=0.939) are larger
- Standard deviation ($S_{y/x1,x2}$) is smaller
- Minimization of the mean square error (min MSE= 0.074) and absolute deviation (min MAD=0) are smaller

And besides this function $d_1=\varphi_5(t,\delta)$ has non-linear regression equation view of **$Y_c=3.935-2.281X_1+1.657X^2_1+4.657X_2$ or $d_1=3.935-2.281t+1.657t^2+4.657\delta$ (19)** .

In Figure 11(a) is shown the functional dependency of internal diameter (d_1) from clearance (t) between of chip wraps ,i.e $d_1=\gamma_4(t)$.

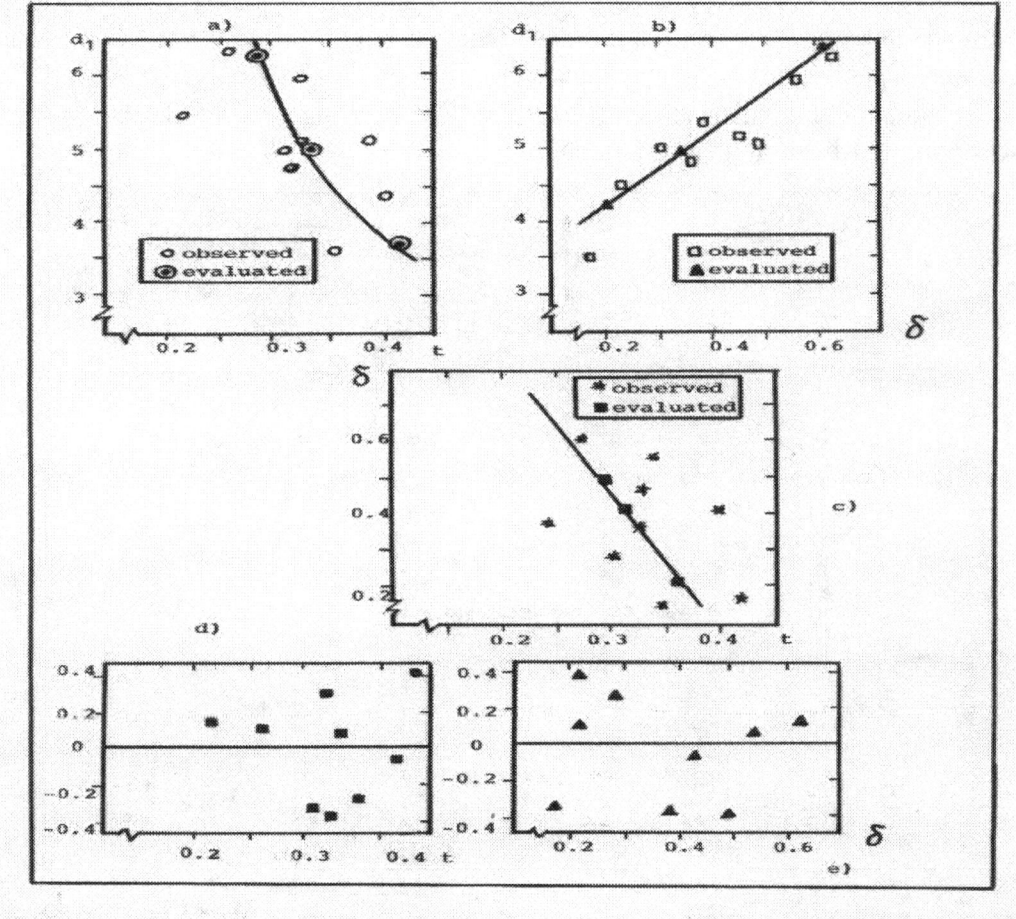

Figure 11

----

4   *Linear regression model has the equation view of d_1= 3.778 −1.237 t +4.653δ*

And as was above-identified this dependency $d_1=\gamma_4(t)$ has the non-linear regression model with equation view of $t=0.828d_1^{-0.605}$,where $d_1=(1.207t)^{-1.653}$ with such statistical characteristics as :coefficients of determination ($R^2=0.43$) and correlation ($r=0.66$),standard error ($S_{y/x1}=0.06$).So , we see from Figure 11(a) that with increasing of clearance (t) between of chip wraps ,the value of internal diameter (d_1) decreases accordingly.

In Figure 11(b) is shown the functional dependency of internal diameter (d_1) from thickness (δ) of stainless chip, i.e $d_1=\varphi(\delta)$. This dependency has the linear regression model with equation view of $Y_c=3.338+4.762X_2$ or $d_1=3.338+4.762\delta$. From Figure 11(b) we see that with increasing of thickness (δ) stainless chip , the value of internal diameter (d_1) increases considerably. In Figure 11 (c) is shown the functional dependency of thickness (δ) stainless chip from clearance (t) between of chip wraps ,i.e $\delta =\phi(t)$. And as we saw in chapter 4 this dependency better submits to the linear regression equation view of $t=0.379-0.16\delta$, where $\delta = (2.369- 6.25t)$ (20) with such statistical characteristics as : coefficients of determination ($R^2=0.29$) and correlation ($r= -0.788$),standard deviation ($S_{y/x}=0.056$). In Figure 1 (d) and 11(e) are illustrated residual plots (residual versus t and residual versus δ) the above-named functional dependencies of $d_1=\gamma_4(t)$ and $d_1=\varphi(\delta)$ accordingly. Analyzing Figure 12 (A) ,we see that functional surface $d_1=\varphi_5(t,\delta)$ is shown in view of three-dimensional drawing for non-linear regression model.

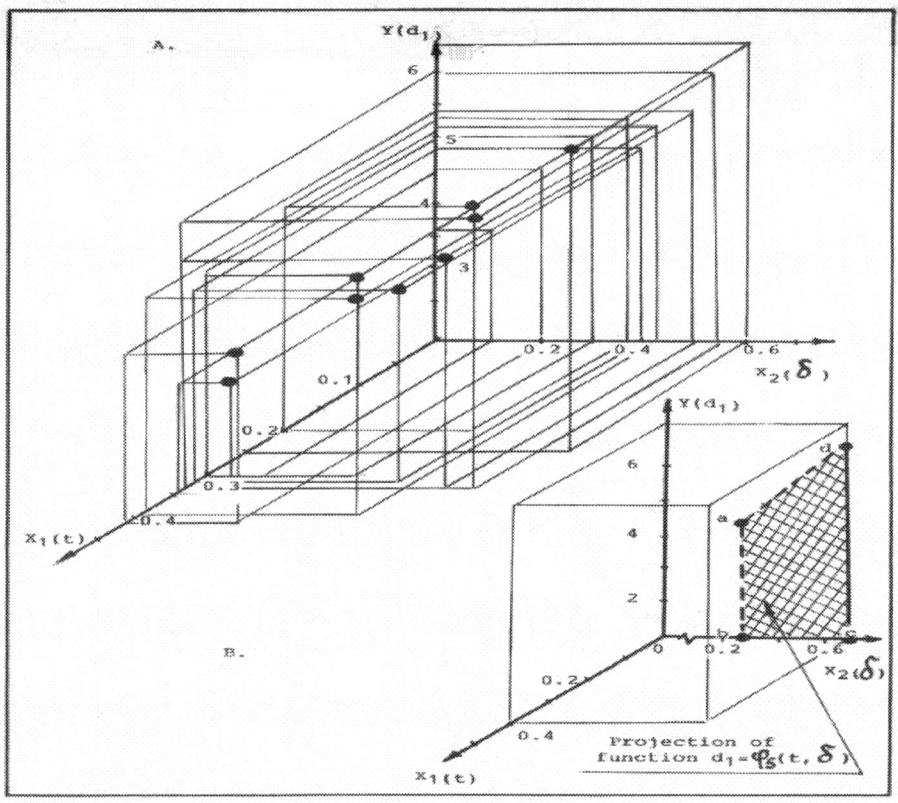

Figure 12

This function has the equation view of $Y_c=3.935-2.281X_1+1.657X^2_1+4.657X_2$ or $d_1=3.935$ $-2.281t+1.657t^2+4.657\delta$ .In Figure 12(B) is shown the projection of this functional surface $d_1=\varphi_5(t,\delta)$ which has view of trapezium with such coordinates for points (sizes in mm): a $(X_{11}=0,X_{21}=0.28,Y_1=4.70)$; b $(X_{12}=0,X_{22}=0.28,Y_2=0)$;c $(X_{13}=0,X_{23}=0.60,Y_3=0)$; d $(X_{14}=0,X_{24}=0.60,Y_4=6.26)$.

Analysis of functional surface $d_1=\varphi_5(t,\delta)$

From Figure 12 (A) we see that functional surface $d_1=\varphi_5(t,\delta)$,on which the points are located, of the non-linear regression equation view of $d_1=3.935 -2.281t+1.657t^2+4.657\delta$, has a curvilinear character with the following coordinate of their peaks in three-dimensional drawing (sizes in mm):a' $(X_{11}=0.3,X_{21}=0.28,Y_1=4.7)$; b'$(X_{12}=0.3,X_{22}=0.28,Y_2=0)$; c' $(X_{13}=0.25,X_{23}=0.6,Y_3=0)$;d'$(X_{14}=0.25,X_{24}=0.6,Y_4=6.26)$.

In Figure 13 schematically is shown the functional surface $d_1=\varphi_5(t,\delta)$ and graph for calculation of module for vectors $|a'b'|,|b'c'|,|c'd'|$ and $|a'd'|$.

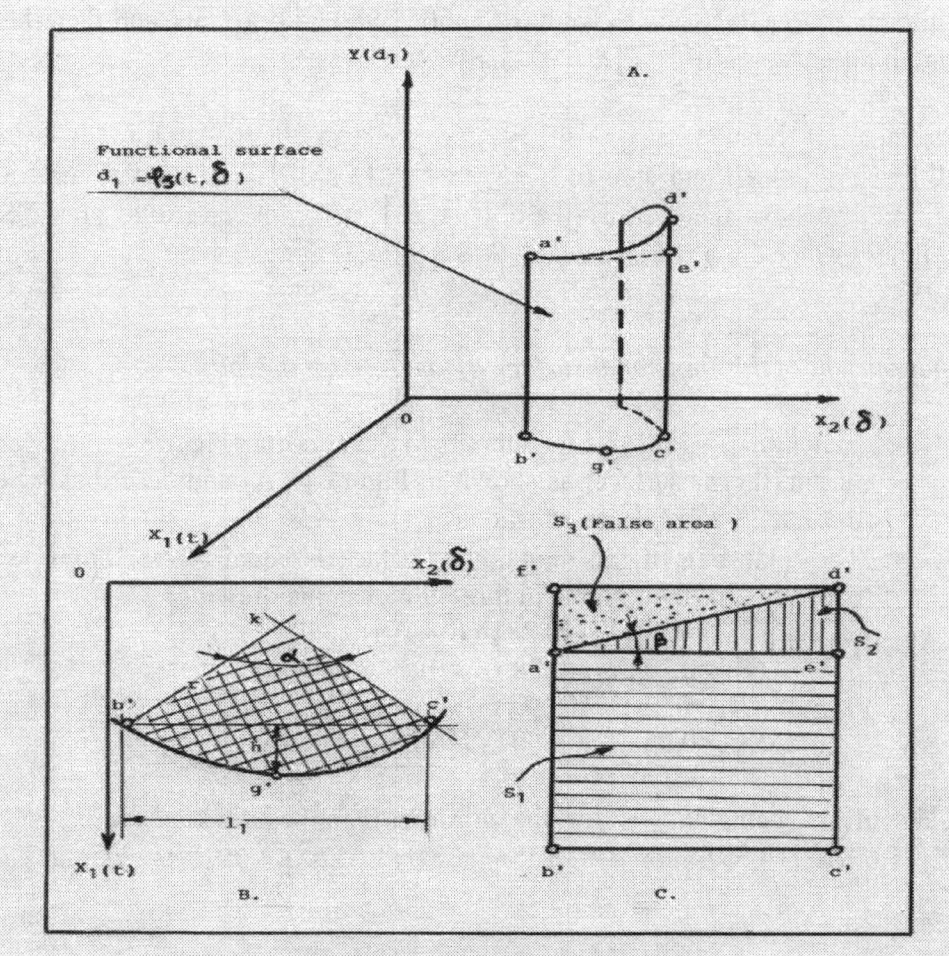

Figure 13 Graph for calculation of module vectors $|a'b'|,|b'c'|,|c'd'|,|a'd'|$ and area of functional surface $d_1=\varphi_5(t,\delta)$: A-total view of this functional surface; B-graph for calculation of module vector $|b'c'|$;C-graph for calculation of module vector $|a'd'|$ and area of this surface.

- The module of vector $|a'b'|$ and $|c'd'|$ can be calculated from the projection of this functional surface $d_1=\varphi_5(t,\delta)$ on the plane YOX_2 (Figure 12B).So, the module of vectors are equal $|a'b'|=4.7$ and $|c'd'|=6.26$. And module of vector $|b'c'|=0.01745\cdot r\cdot\alpha$, where $r=|kg'|=[(X_{16}-X_{15})^2+(X_{26}-X_{25})^2]^{1/2}=0.56$ and coordinate of points g' and k: $k(X_{15}=0, X_{25}=0)$;$g'(X_{16}=0.32, X_{26}=0.46)$. At data $\alpha=82°$,$r=0.56$ we have the module of vector $|b'c'|=0.80$ and other parameters of circular segment h= $r[1-\cos(\alpha/2)]=0.14$,$l_1=2[h(2r-h)]^{1/2}=0.74$.

- The graph for calculation of module vectors $|b'c'|$,$|a'd'|$ and area of functional surface $d_1=\varphi_5(t,\delta)$ is shown in Figure 13.

- The module of vector $|a'd'|$ and area of this functional surface ,as shown in Figure 13(c) ,can be calculated by the following way: module of length for curve $|a'e'|=|b'c'|=0.80$. And module of length for $|d'e'|=|c'd'|-|a'b'|=1.56$. For calculation of area S_1,S_2 and S_3 we have $\tan\beta=1.95$ and $\beta=62°50'$ and then the module of length for curve $|a'd'|=(|d'e'|/\sin62°50'=1.75$.

- Coordinate of point f ' ($X_{11}=0.30$,$X_{21}=0.28$,$Y_4=6.26$)and module of length $|f'd'|=|b'c'|=0.80$ and area of $S_2=S_3=S/2=(|f'd'|\cdot|d'e'|)/2=0.62mm^2$.For these conditions we have the area $S_1=|a'b'|\cdot|b'c'|=3.76mm^2$ and total area $\sum S$ of this functional surface $d_1=\varphi_5(t,\delta)$ is equal $\sum S=S_1+S_2=4.38mm^2$.

So ,the parameters of this functional surface $d_1=\varphi_5(t,\delta)$ are the following:

- Function $d_1=\varphi_5(t,\delta)$ better submits to the non-linear regression model in view of curvilinear surface, as shown in Figure 13(A) and describes by equation $d_1=3.935-2.281t+1.657t^2+4.657\delta$.
- The total area of this functional surface is equal $\sum S=4.38mm^2$ with such coordinate of their points in three-dimensional drawing:
    a' ($X_{11}=0.30$, $X_{21}=0.28$,$Y_1=4.70$)
    b' ($X_{12}=0.30$,$X_{22}=0.28$,$Y_2=0$)
    c' ($X_{13}=0.25$,$X_{23}0.60$,$Y_3=0$)
    d' ($X_{14}=0.25$,$X_{24}=0.60$,$Y_4=6.26$)

The results of the computations for the non-regression equation view of $Y_c=3.935-2.281X_1+1.657X^2_1+4.657X_2$ or $d_1=3.935-2.281t+1.657t^2+4.657\delta$ is given in Table 5.

Table 5 Evaluation of non- regression equation $Y_c=3.935-2.281X_1+1.657X^2_1+4.657X_2$

A. MEAN, VARIANCE AND STANDARD DEVIATION

Variable	Mean	Variance	Standard deviation
X_1	0.32	0.031	0.176
X_2	0.37	0.21	0.458
Y	5.098	5.608	2.368

B.RESULTS OF MULTIPLE REGRESSION OF Y ON X_1 AND X_2

Variable	Coefficients	Standard error	T-value
X_1	-2.281	0.059	-38.661
X_2	4.657	0.153	30.438

C. ANALYSIS OF VARIANCE RESULTS

Regression
- Degrees of freedom   2
- Sum of squares   4.945
- Mean square   2.473

Error
- Degrees of freedom   6
- Sum of squares   0.664
- Mean squares   0.111
- Standard error of estimate   0.176
- F-value[*]   14.051   [$F_{0.05,2,6}$]=5.14

Since $F^* > F_{0.05,2,6}$ we reject the hypothesis that all regression coefficients are zero.

E. DETERMINATION OF RESIDUALS

Number	Observed	Estimated	Residual
1	5.36	5.21	0.15
2	5.22	5.26	−0.04
3	3.52	3.82	−0.30
4	4.51	4.11	0.40
5	4.98	4.70	0.28
6	4.74	5.11	−0.37
7	5.15	5.52	−0.37
8	5.96	5.89	0.07
9	6.44	6.26	0.18

5.6 Dependence of internal diameter (d_1) from width (B) of stainless chip and number of chip wraps (ω)

Analyzing of the data and statistical characteristics, we see that function $d_1=\varphi_6(B,\omega)$ better submits to the non- linear regression model than for the linear regression model [5] for some reasons:

- The coefficients of determination ($R^2=0.216$) and correlation ($r=0.465$) are larger
- Standard deviation ($S_{y/x1,x2} =0.856$)is smaller
- Minimization of the mean square error (min MSE=0.489) and absolute deviation (min MAD=0) are smaller

And for this reason we can conclude that this function $d_1=\varphi_6(B,\omega)$ better submits to the non-linear regression equation view of $\mathbf{Y_c=6.3+1.241X_1-0.133X^2_1-1.098X_2}$ or
$\mathbf{d_1=6.3+1.241B -0.133B^2 -1.098\omega}$ $\mathbf{(21)}$.

In Figure 14(a) is shown the functional dependence of internal diameter (d_1) from width (B) stainless chip, i.e we have $d_1=\varphi_1(B)$.

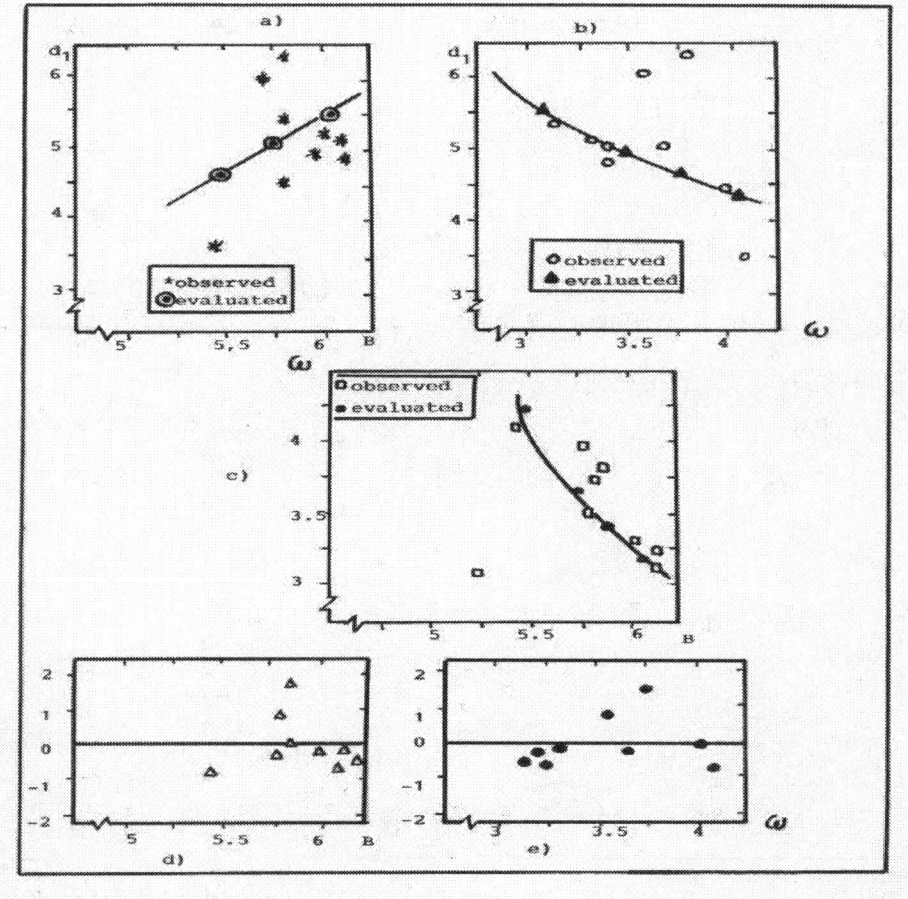

Figure 14

----
5   *Linear regression equation has view of $d_1=14.225-0.721B-1.382\omega$*

84

This dependence $d_1=\varphi_1(B)$ has the linear regression model with equation view of $Y_c=0.06+0.863X$ or $d_1=0.06+0.863B$. So, we see from Figure 14(a) that with increasing of chip width (B), the value of internal diameter (d_1) increases considerably. The statistical characteristics of this linear regression model are: coefficient of determination ($R^2=0.10$), coefficient correlation ($r=0.32$), standard deviation ($S_{y/x}=0.864$).

In Figure 14(b) is shown the functional dependence of internal diameter (d_1) from the number of chip wraps (ω) ,i.e we have $d_1=\phi_3(\omega)$. This dependence has the non-linear regression model with equation view of $Y_c=20.32X^{-1.102}$ or $d_1=20.32\omega^{-1.102}$ with such statistical characteristics :coefficient of determination ($R^2=0.58$),coefficient of correlation ($r=0.76$),standard error ($S_{y/x}=0.046$). And as we see from Figure 14(b0 that with increasing of chip wraps (ω) ,the value of internal diameter (d_1) decreases accordingly.

In Figure 14(c) is shown functional dependence of number chip wraps (ω) from width (B) of this stainless chip, i.e we have $\omega=\gamma_2(B)$. This dependence better submits to the non-linear regression model with equation $Y_c=703.1X^{-3}$ or $\omega=703.1B^{-3}$. And besides from Figure 14(c) we see that with increasing of width (B) stainless chip ,the value of chip wraps(ω) decreases considerably in accordance with regression equation $\omega=703.1B^{-3}$. In Figure 14(d) and 14(e) are illustrated residual plots (residual versus B and residual versus ω) of the above-named functional dependencies $d_1=\varphi_1(B)$ and $d_1=\phi_3(\omega)$ accordingly.

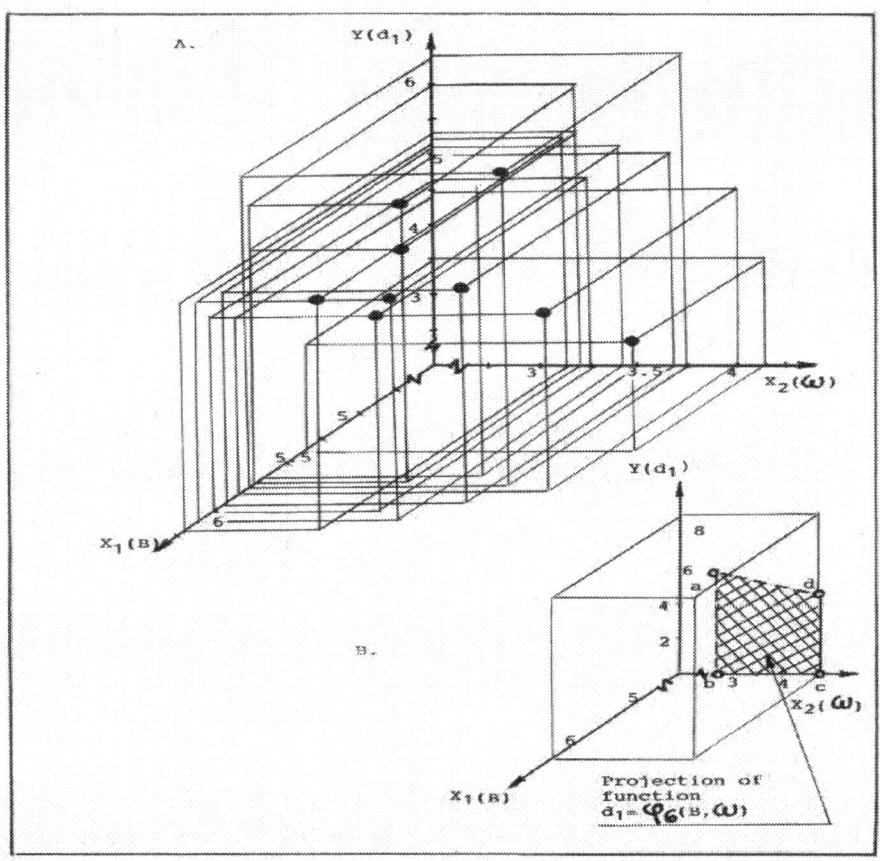

Figure 15

85

Analyzing the Figure 15(A) ,we see that functional surface $d_1=\varphi_6(B,\omega)$ is shown in view of three-dimensional drawing for the non-linear regression model with equation $Y_c=6.3+1.241X_1-0.133X^2_1-1.098X_2$ or $d_1=6.3+1.241B-0.133B^2-1.098\omega$. And Figure 15(B) shows that projection of functional surface $d_1=\varphi_6(B,\omega)$ has view of trapezium with such coordinate of its points (sizes in mm): a $(X_{11}=0,X_{21}=3.25,Y_1=5.35)$; b$(X_{12}=0,X_{22}=3.25,Y_2=0)$;c$(X_{13}=0,X_{23}=4.22,Y_3=0)$; d$(X_{14}=0,X_{24}=4.22,Y_4=4.49)$.

Analysis of functional surface $d_1=\varphi_6(B,\omega)$

From Figure 15(A) we see that functional surface on which the points are located of non-linear regression equation view of $d_1=6.3+1.241B-0.133B^2-1.098\omega$ has a curvilinear character with the following coordinate of their peaks in three-dimensional drawing: a'$(X_{11}=6.12,X_{21}=3.25,Y_1=5.35)$;b'$(X_{12}=6.12,X_{22}=3.25,Y_2=0)$; c'$(X_{13}=5.37,X_{23}=4.22,Y_3=0)$; d' $(X_{14}=5.37,X_{24}=4.22,Y_4=4.49)$. In Figure 16 schematically is shown the functional surface $d_1=\varphi_6(B,\omega)$ and graph for calculation of module for vectors $|a'b'|$, $|b'c'|$, $|c'd'|$ and $|a'd'|$.

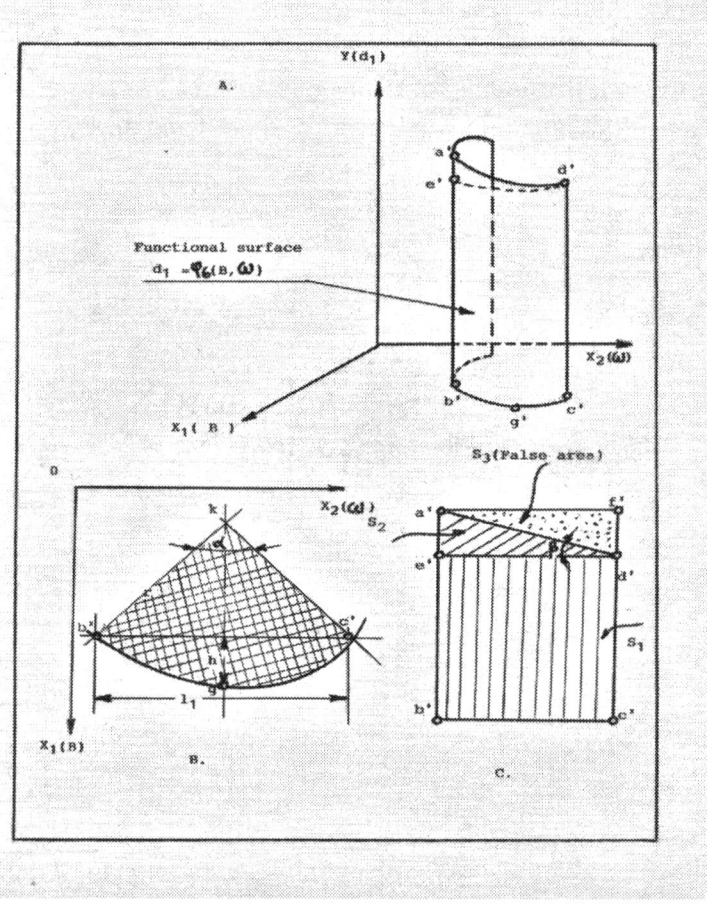

Figure 16 **Graph for calculation of module vectors $|a'b'|$,$|b'c'|$,$|c'd'|$ and $|a'd'|$ and area of functional surface $d_1=\varphi_6(B,\omega)$:A-total view of this functional surface; B-graph for calculation of module vector $|b'c'|$; C-graph for calculation of module vector $|a'd'|$ and area of this functional surface.**

- The module of vector $|a'b'|$ and $|c'd'|$ can be calculated from projection of this functional surface $d_1=\varphi_6(B,\omega)$ on the plane YOX_2(Figure 15B) ,where $|a'b'|=5.35$ and $|c'd'|=4.49$.
- The module of vector $|b'c'|=0.01745r\alpha$,where $r=|kg'|=[(X_{16}-X_{15})+(X_{26}-X_{25})]^{1/2}$ and coordinate of points g' ,k are equal: g'$(X_{16}=5.75,X_{26}=3.76)$;k$(X_{15}=0,X_{25}=0)$. At data $\alpha=82°$,$r=6.87$ we have $|b'c'|=9.83$ and other parameters of circular segment h=r$[1-\cos(\alpha/2)]=1.68$,$l_1=2[$ h(2r-h)$]^{1/2}=9.0$.
  The graph for calculation of module vectors $|b'c'|$,$|a'd'|$ and area of this functional surface is shown in Figure 16.
- The module of vector $|a'd'|$ and area of this functional surface $d_1=\varphi_6(B,\omega)$,as shown in Figure 16(c) ,can be calculated by the following way: the module of length for curve $|d'e'|=|b'c'|=9.83$ and module of length $|a'e'|=|a'b'|-|c'd'|=0.86$.Then the module of length $|a'd'|=(|a'e'|)/\sin\beta$,where $\tan\beta=(|a'e'|)/(|d'e'|)=0.087$ and angle $\beta=5°$.
- For calculation of area S_1,S_2 and S_3 we have $S_2=S_3=S/2 =(|f'd'|\cdot|d'e'|)/2$,where $|f'd'|=|a'e'|=0.86$ and $S_2=S_3=4.23mm^2$,$S_1=|c'd'|\cdot|b'c'|=44.14mm^2$. The total area ΣS of functional surface $d_1=\varphi_6(B,\omega)$ is equal $\Sigma S=S_1+S_2=48.37mm^2$.

So, the parameters of this functional surface $d_1 =\varphi_6 (B,\omega)$ are the following:

1.This function $d_1=\varphi_6(B,\omega)$ better submits to the non-linear regression model in view of curvilinear surface , parabola as shown in Figure 16(A) and   describes by equation $Y_c=6.3+1.241X_1-0.133X^2_1-1.098X_2$ or $d_1=6.3+1.241B-0.133B^2-1.098\omega$.

2.The total area of functional surface $d_1=\varphi_6(B,\omega)$ is equal $\Sigma S =48.37$ mm^2 with such coordinate of its points in three-dimensional drawing for this surface (sizes in mm):

   a' $(X_{11}=6.12$,$X_{21}=3.25$,$Y_1=5.35)$
   b' ($X_{12}=6.12$,$X_{22}=3.25$,$Y_2=0$ (
   c' $(X_{13}=5.37$, $X_{23}=4.22$,$Y_3=0)$
   d' ($X_{14}=5.37$,$X_{24}=4.22$,$Y_4=4.49)$

The results of computations for the regression equation $Y_c=6.3+1.241X_1-0.133X^2_1-1.098X_2$ is given in Table 6.

Table 6 Evaluation of regression equation $Y_c=6.3+1.241X_1-0.133X^2_1-1.098X_2$

A.MEAN,VARIANCE AND STANDARD DEVIATION

Variable	Mean	Variance	Standard deviation
X_1	5.836	0.582	0.763
X_2	3.559	1.113	1.055
Y	5.098	5.609	2.368

B.RESULTS OF MULTIPLE REGRESSION OF Y ON X_1 AND X_2

Parameter	Variable	Coefficients	Standard error	T-value
β_1	X_1	1.241	0.254	4.886
β_2	X_2	-1.098	0.352	-3.119

C.ANALYSIS OF VARIANCE RESULTS

Regression
- Degrees of freedom 2
- Sum of squares   1,211
- Mean square   0.606

Error
- Degrees of freedom 6
- Sum of squares   4.398
- Mean square   0.733

Standard error of estimate 0.763

F-value* 0.827 [$F_{0.05,2,6}=5.14$]

Since $F^* < F_{0.05,2,6}$ we can not reject the hypothesis that both β_1 and β_2 are zero. Notice that our t-test (confidence interval) lead us not to conclude that $\beta_1=0$. So, we have for β_1:

$$b_1 - t_{\alpha/2;n-3} S_{b1} \leq \beta_1 + t_{\alpha/2;n-3} S_{b1} \quad \text{where}$$

$1.241-(2.571)(0.254) \leq \beta_1 \leq 1.241+(2.571)(0.254)$ and $0.588 \leq \beta_1 \leq 1.894$

Thus ,we 95 percent confident that β_1 lies between 0.588 and 1.894. Since this interval does not contain zero, we would further conclude at the 0.05 significance level that β_1 is not zero.

And for β_2 we have :

$$b_2 - t_{\alpha/2;n-3} S_{b2} \leq b_2 + t_{\alpha/2;n-3} S_{b2}$$

$-1.098 -(2.571)(0.352) \leq \beta_2 \leq -1.098+(2.571)(0.352)$ and $-2.003 \leq \beta_2 \leq -0.193$

Thus ,we are 95 percent confident that β_2 lies between -2.003 and -0.193. Since this interval does not contain zero, we would further conclude at the 0.05 significance level that β_2 is not zero.

D.DETERMINATION OF RESIDUALS

Number	Observed	Estimated	Residual
1	5.36	5.57	−0.21
2	5.22	5.30	−0.08
3	3.52	4.49	−0.97
4	4.51	4.65	−0.14
5	4.98	5.04	−0.06
6	4.74	5.35	−0.61
7	5.15	5.40	−0.25
8	5.96	5.18	0.78
9	6.44	4.91	1.53

5.7 Dependence of internal diameter (d₁) from clearance (t) between of chip wraps and number of its wraps (ω)

Analysis of data and statistical characteristics (coefficient of determination R^2=0.308 , coefficient of correlation r=0.554,standard deviation $S_{y/x}$=0.805) and also the minimization of mean square error (min MSE=0.43) and minimization of the mean absolute deviation (min MAD=0) show that this functional dependence $d_1=\varphi_7(t,\omega)$ better submits to the linear than to the non-linear [6] regression model and has the equation view of $Y_c= 9.074-5.067X_1 -0.666X_2$ or $d_1=9.074-5.067t-0.666\omega$ (22).

In Figure 17(a) is shown the functional dependence of internal diameter (d₁) from clearance (t) between of chip wraps for stainless chip, i.e we have $d_1=\gamma_4(t)$.

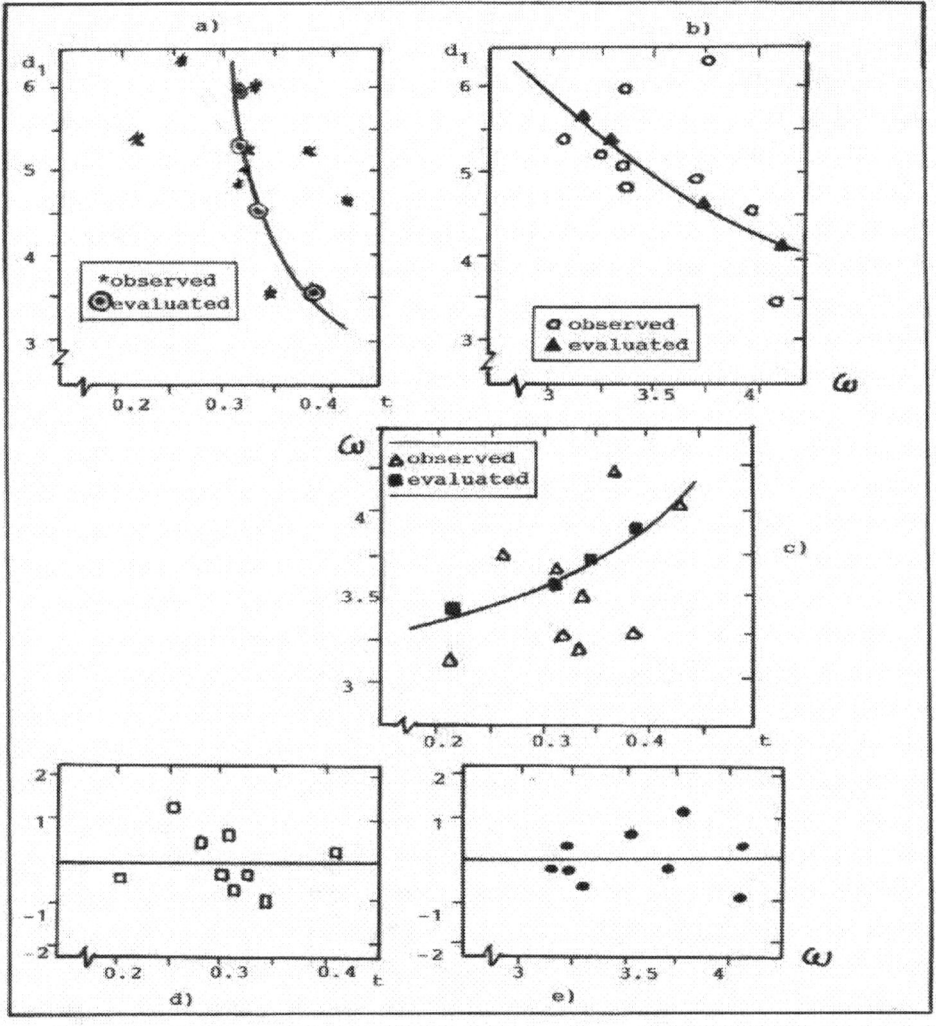

Figure 17

----

6   *This non-linear regression equation has view of d_1=9.374 –9.356t +7.184t² –0.578ω*

89

This dependence $d_1=\varphi_4(t)$ has the non-linear regression model with equation view of $Y_c=0.828X^{-0.605}$ or $t=0.828d_1^{-0.605}$ where $d_1=(1.208t)^{-1.653}$. So ,we see from Figure 17(a) that with increasing of clearance (t) between of chip wraps ,the value of external diameter (d_1) decreases.

In Figure 17(b) is shown the functional dependence of internal diameter (d_1) from number (ω) of chip wraps, i.e we have $d_1=\phi_3(\omega)$.This dependence $d_1=\phi_3(\omega)$ has the non-linear regression model with equation view of $Y_c= 20.32\omega^{-1.102}$ or $d_1=20.32\omega^{-1.102}$. As we see from Figure 17(b) that with increasing of number (ω) wraps for stainless chip ,the value of internal diameter (d_1) decreases considerably.

In Figure 17(c) is shown the functional dependence of number (ω) chip wraps from the clearance (t) between of these chip wraps and as was above-shown this dependence can be expressed in view of function $\omega=\alpha_1(t)$ and has the non- linear regression model with equation $Y_c=4.227X^{0.152}$ or $\omega=4.227t^{0.152}$. And besides from Figure 17(c) we see that with increasing of clearance (t) between of chip wraps, the value of number (ω) chip wraps increases considerably in accordance with the non-regression equation $\omega=4.227t^{0.152}$.In Figure 17(d) and 17(e) are illustrated residual plots (residual versus t and residual versus ω) the above-named functional dependencies of $d_1=\varphi_4(t)$ and $d_1=\phi_3(\omega)$accordingly. Analyzing the Figure 18(A),we see that functional surface $d_1=\varphi_7(t,\omega)$ is shown in view of three-dimensional drawing for linear regression model with equation $Y_c=9.074-5.067X_1-0.666X_2$ or $d_1=9.074-5.067t-0.666\omega$.

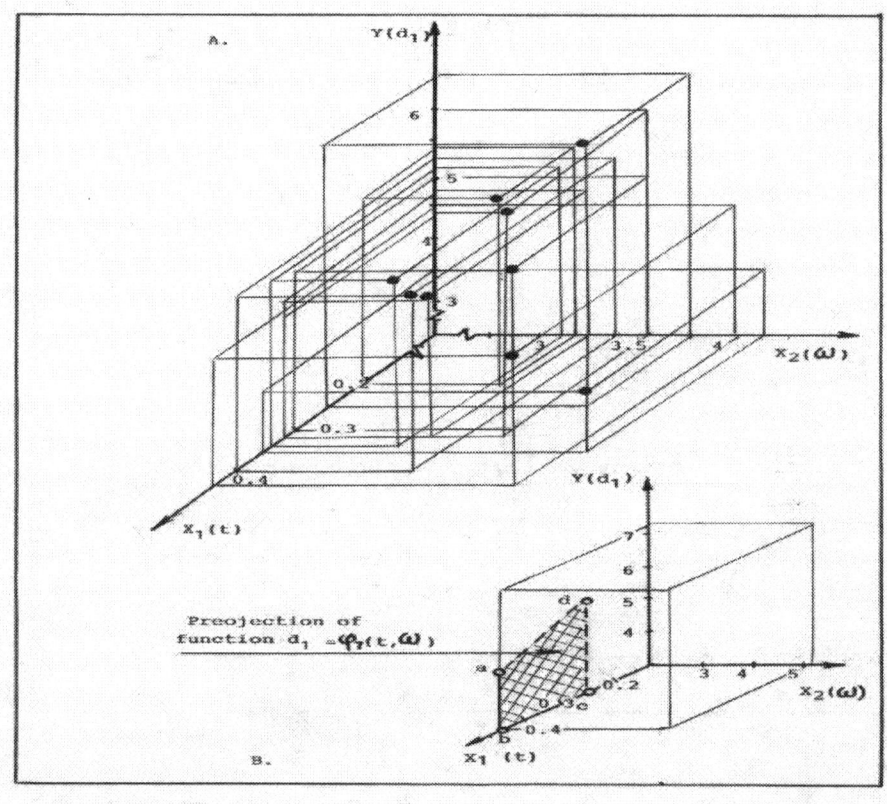

Figure 18

90

And Figure 18(B) shows that this functional surface $d_1=\varphi_7(t,\omega)$ has view of trapezium on the face YOX_1 with such coordinate of their points (sizes in mm): $a(X_{11}=0.38,X_{21}=0,Y_1=4.96)$; $b(X_{12}=0.38,X_{22}=0,Y_2=0)$; $c(X_{13}=0.25,X_{23}=0,Y_3=0)$ and $d(X_{14}=0.25,X_{24}=0,Y_4=5.32)$.

Analysis of functional surface $d_1=\varphi_7(t,\omega)$

From Figure 18(A) we see that this functional surface $d_1=\varphi_7(t,\omega)$, on which are situated the points of the linear regression equation view of $Y_c=9.074 -5.067X_1-0.666X_2$ or $d_1=9.074-5.067t-0.666\omega$ has the following coordinate of their peaks in three-dimensional drawing (sizes in mm): $a'(X_{11}=0.38,X_{21}=3.28,Y_1=4.96)$; $b'(X_{12}=0.38,X_{22}=3.28,Y_2=0)$;$c'(X_{13}=0.25,X_{23}=3.76,Y_3=0)$;$d'(X_{14}=0.25,X_{24}=3.76,Y_4=5.32)$. The module of vectors for this functional surface $d_1=\varphi_7(t,\omega)$ are equal: $|a'b'|=4.96$; $|b'c'|=0.50$; $|c'd'|=5.32$; $|d'a'|=0.62$ and area $S_{a'b'c'd'}$ can be calculated by formula $S_{a'b'c'd'}=1/2\,[(|c'd'|-|a'b'|)\cdot|b'c'|]+(|b'c'|\cdot|a'b'|)=2.57mm^2$.

The results of the computations for this linear regression equation view of $Y_c=9.074-5.067X_1-0.666X_2$ is given in Table 7.

Table 7 Evaluation of regression equation view of $Y_c=9.074-5.067X_1-0.666X_2$

A.MEAN, VARIANCE AND STANDARD DEVIATION

Variable	Mean	Variance(σ^2)	Standard deviation(σ)
X_1	0.317	0.032	0.179
X_2	3.559	1.114	1.055
Y	5.098	5.609	2.368

B. RESULTS OF MULTIPLE REGRESSION OF Y ON X_1 AND X_2

Parameter	Variable	Coefficient	Standard error	T-value
β_1	X_1	$b_1=-5.067$	$S_{b1}=\sigma/(n)^{1/2}=0.059$	$t_1=b_1/S_{b1}=-85.881$
β_2	X_2	$b_2=-0.666$	$S_{b2}=0.352$	$t_2=-1.892$

C. ANALYSIS OF VARIANCE RESULTS

Regression
Degrees of freedom 2
Sum of squares 1.724
Mean square 0.862
Error
Degrees of freedom 6
Sum of squares 3.885
Mean square 0.648
Standard error of estimate
F-value* 1.330 [$F_{0.05,2,6}=5.14$]

Since $F=1.330< F_{0.05,2,6}$ we can not reject the hypothesis that both β_1 and β_2 are zero.

D. DETERMINATION OF RESIDUALS

Number	Observed	Estimated	Residual
1	5.36	5.89	-0.53
2	5.22	4.97	0.25
3	3.52	4.54	-1.02
4	4.51	4.29	0.22
5	4.98	5.14	-0.16
6	4.74	5.34	-0.60
7	5.15	5.32	-0.17
8	5.96	5.10	0.86
9	6.44	5.30	1.14

So , the parameters of this functional surface $d_1=\varphi_7(t,\omega)$ are the following:

1. Function $d_1=\varphi_7(t,\omega)$ better submits to the linear regression model with equation view of $Y_c=9.074-5.067X_1-0.666X_2$ or $d_1=9.074-5.067t-0.666\omega$ with such statistical characteristics:

2. Coefficient of determination $R^2=0.308$;coefficient of correlation r=0.554; standard deviation $S_{y/x1,x2}=0.805$;minimization of the mean square error (min MSE= 0);minimization of the mean absolute deviation (min MAD=0)

3. The total area of functional surface $d_1=\varphi_7(t,\omega)$ is equal $\sum S=2.57$ mm^2 with such coordinates of its points in three-dimensional drawing for this surface (sizes in mm): a' ($X_{11}=0.38$,$X_{21}=3.28$,$Y_1=4.96$);b' ($X_{12}=0.38$,$X_{22}=3.28$,$Y_2=0$); c' ($X_{13}=0.25$,$X_{23}=3.76$,$Y_3=0$); d' ($X_{14}=0.25$, $X_{24}=3.76$,$Y_4=5.32$).

5.8 Dependence of internal diameter (d_1) from external diameter (d_2) and thickness (δ) of stainless chip

Analysis of data and statistical characteristics of linear regression model (coefficient of determination $R^2=0.925$,coefficient of correlation r=0.962, standard deviation $S_{y/x1,x2}=0.264$)and also the minimization of the mean square error (min MSE =0.046) and minimization of the mean absolute deviation (min MAD=0) show that this functional dependence $d_1=\varphi_8(d_2,\delta)$ better submits to the linear than to the non-linear regression model [7] with equation view of $Y_c=-2.762+0.693X_1+2.061X_2$ or $d_1=-2.762+0.693d_2+2.062\delta$ (23) .In Figure 19 (a) is shown the functional dependence of internal diameter (d_1)from the external diameter(d_2), i.e we have $d_1=\varphi_3(d_2)$.

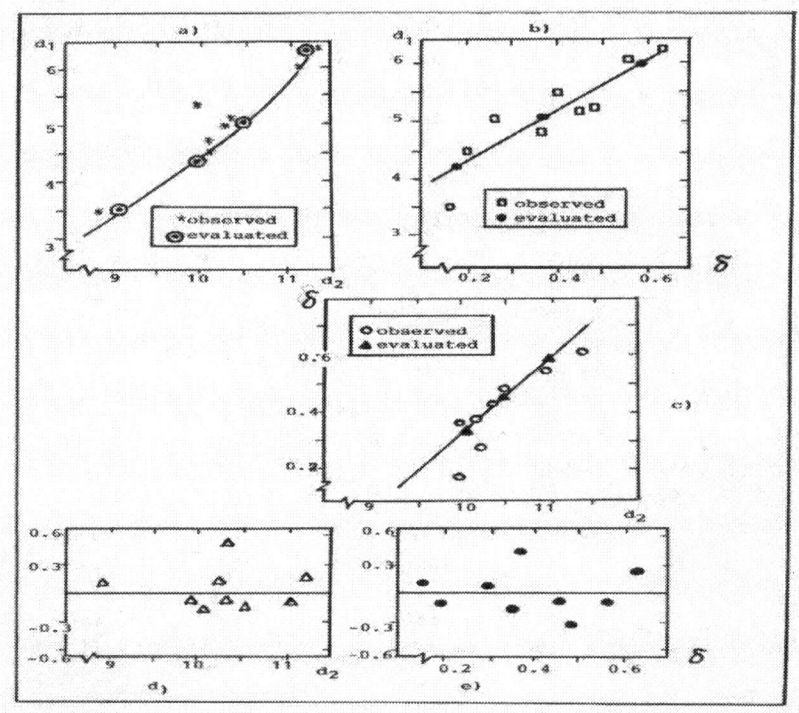

Figure 19

----

7   *The non-linear regression equation has view of $d_1=-1.254+0.513d_2-3.043\cdot10^{-6}d^2_2+2.976\delta$*

This functional dependence $d_1=\varphi_3(d_2)$ has the non-regression model with equation view of $Y_c=5.728X^{0.359}$ or $d_2=5.728d_1^{0.359}$,where $d_1=(0.175d_2)^{2.786}$. So ,we see from Figure 19(a) that with increasing of external diameter (d_2) , the value of internal diameter(d_1) increases accordingly with equation $d_1=(0.175d_2)^{2.786}$.

In Figure 19(b) is shown functional dependency of internal diameter (d_1) from thickness (δ) stainless chip ,i.e we have $d_1=\varphi(\delta)$ and as was above-shown this dependence has the linear regression model with equation view of $Y_c=3.338+4.762X$ or $d_1=3.338+4.762\delta$. And as we see from Figure 19(b) that with increasing of thickness (δ) stainless chip, the value of internal diameter (d_1) increases in accordance with equation $d_1=3.338+4.762\delta$. In Figure 19(c) is shown the functional dependence of thickness (δ) stainless chip from external diameter (d_2) of this chip. This dependency can be expressed in view of $\delta=\gamma(d_2)$ and has the linear regression model with equation $Y_c=8.808+3.897X$ or $d_2=8.808+3.897\delta$, where $\delta=0.26d_2-2.26$. And besides from Figure 19(c) ,we see that with increasing of external diameter (d_2) stainless chip ,the value of thickness (δ) this chip increases considerably with the linear regression equation view of $\delta=0.26d_2-2.26$.

In Figure 19(d) and 19(e) are illustrated residual plots (residual versus d_2 and residual versus δ the above-named functional dependencies of $d_1=\varphi_3(d_2)$ and $d_1=\varphi(\delta)$ accordingly.

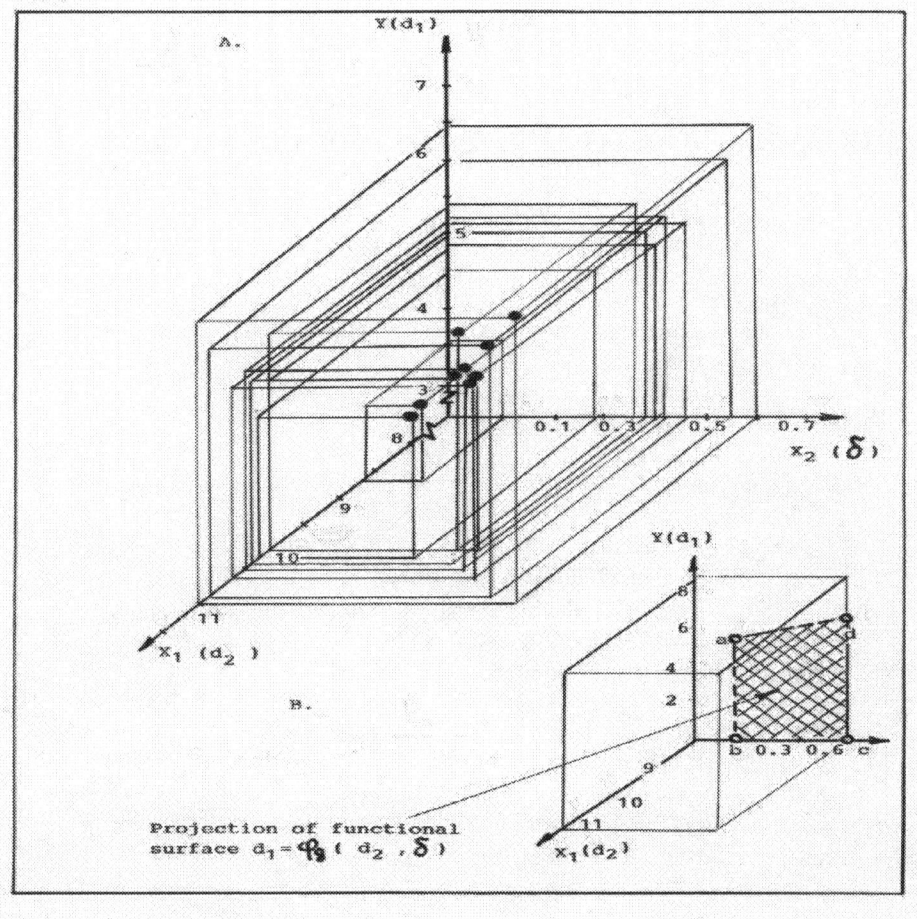

Figure 20

93

Analyzing the Figure 20(A) ,we see that functional surface $d_1=\varphi_8(d_2,\delta)$ is shown in view of three-dimensional drawing for the linear regression model with equation $Y_c=-2.762+0.693X_1+2.061X_2$ or $d_1=-2.762+0.693d_2+2.061\delta$. And Figure 20(B) shows that projection of functional surface $d_1=\varphi_8(d_2,\delta)$ has view of trapezium on the plane YOX_2 with the following coordinate of their points(sizes in mm): $a(X_{11}=0,X_{21}=0.36,Y_1=4.93)$; $b(X_{12}=0,X_{22}=0.36,Y_2=0)$; $c(X_{13}=0,X_{23}=0.60,Y_3=0)$; $d(X_{14}=0,X_{24}=0.60,\ Y_4=6.22)$.

From Figure 20(A) we see that functional surface $d_1=\varphi_8(d_2,\delta)$,on which are situated the points of the linear regression equation view of $Y_c=-2.762+0.693X_1+2.061X_2$ or $d_1=-2.762+0.693d_2+2.061\delta$,has the following coordinate of their peaks in three-dimensional drawing (sizes in mm):$a'(X_{11}=10.03,X_{21}=0.36,Y_1=4.93)$; $b'(X_{12}=10.03,X_{22}=0.36,Y_2=0)$; $c'(X_{13}=11.18,X_{23}=0.60,Y_3=0)$;$d'(X_{14}=11.18,X_{24}=0.60,Y_4=6.22)$.

The module of vectors and area of this functional surface are equal: $|a'b'|=4.93$, $|b'c'|=0.24$, $|c'd'|=6.22$, $|d'a'|=1.74$ and $S_{a'b'c'd'}=1/2[(|c'd'|-|a'b'|)\cdot|b'c'|]+$ $+(|b'c'|\cdot|a'b'|)=1.34mm^2$.All results of the computations for the linear regression equation view of $Y_c=-2.762+0.693X_1+2.061X_2$ are given in Table 8.

Table 8 Evaluation of regression equation view of $Y_c=-2.762+0.693X_1+2.061X_2$

A. MEAN, VARIANCE AND STANDARD DEVIATION

Variable	Mean	Variance	Standard deviation
X_1	10.248	4.120	2.029
X_2	0.368	0.209	0.457
Y	5.098	5.609	2.368

B.RESULTS OF MULTIPLE REGRESSION OF Y ON X_1 AND X_2

Parameter	Variable	Coefficients	Standard error	T-value
β_1	X_1	0.693	0.676	1.025
β_2	X_2	2.061	0.152	13. 559

C. ANALYSIS OF VARIANCE RESULTS

Regression
    Degrees of freedom 2
    Sum of squares   5.191
    Mean square   2.596
Error
    Degrees of freedom   6
    Sum of squares   0.418
    Mean square   0.069
Standard error of estimate   2.029
F-value[*]   37.623   [$F_{0.05,2,6}=5.14$]

Since F =37.623 > $F_{0.05,2,6}$ we can reject the hypothesis that both coefficients β_1 and β_2 are zero

D. DETERMINATION OF RESIDUALS

Number	Observed	Estimated	Residual
1	5.36	4.93	0.43
2	5.22	5.28	−0.06
3	3.52	3.47	0.05
4	4.51	4.63	−0.12
5	4.98	4.90	0.08
6	4.74	4.98	− 0.24
7	5.15	5.46	−0.31
8	5.96	6.02	− 0.06
9	6.44	6.22	0.22

So , the parameters of this functional surface $d_1=\varphi_8(d_2,\delta)$ are the following:

1.Function $d_1=\varphi_8(d_2,\delta)$ better submits to the linear regression model with equation view of $Y_c = -2.762 +0.693X_1+2.061X_2$ or $d_1= -2.762 +0.693d_2+2.061\delta$ with such statistical characteristics : coefficient of determination $R^2=0.925$,coefficient of correlation r=0.962, standard deviation $S_{y/x1,x2} =0.264$,minimization of the mean square error (min MSE=0.046) and minimization of the mean absolute deviation (min MAD=0).

2.The total area of functional surface $d_1=\varphi_8 (d_2,\delta)$ is equal $\sum S=1.34$ mm^2 with such coordinates of its points in three-dimensional drawing for this surface (sizes in mm):
a'($X_{11}=10.03,X_{21}=0.36,Y_1=4.93$);b'($X_{12}=10.03$, $X_{22}=0.36,Y_2=0$); c'($X_{13}=11.18,X_{23}=0.60$,$Y_3=0$) ;d'($X_{14}=11.18,X_{24}=0.60$,$Y_4=6.22$).

5.9 Dependence of internal diameter (d_1) from clearance (t) between of chip wraps and width (B) of stainless chip

Analysis of data and also the statistical characteristics showed that coefficients of determination ($R^2=0.341$) and correlation(r=0.584), standard deviation ($S_{y/x1,x2} =0.785$) are larger for the non-linear than for linear [8] regression model and besides the minimization of the mean square error (min MSE=0.41) ,and minimization of the mean absolute deviation (min MAD=0) are smaller for this regression model. And for this reasons we can conclude that function $d_1=\varphi_9(t,B)$ better submits to the non-linear regression equation **$Y_c=1.603-4.903X_1-3.868X^2_1+0.934X_2$ or $d_1=1.603-4.903t-3.868t^2+ 0.934B$ (24).**

In Figure 21 (a) is shown the functional dependence of internal diameter (d_1) from clearance (t) between of chip wraps ,i.e we have function $d_1=\gamma_4(t)$.

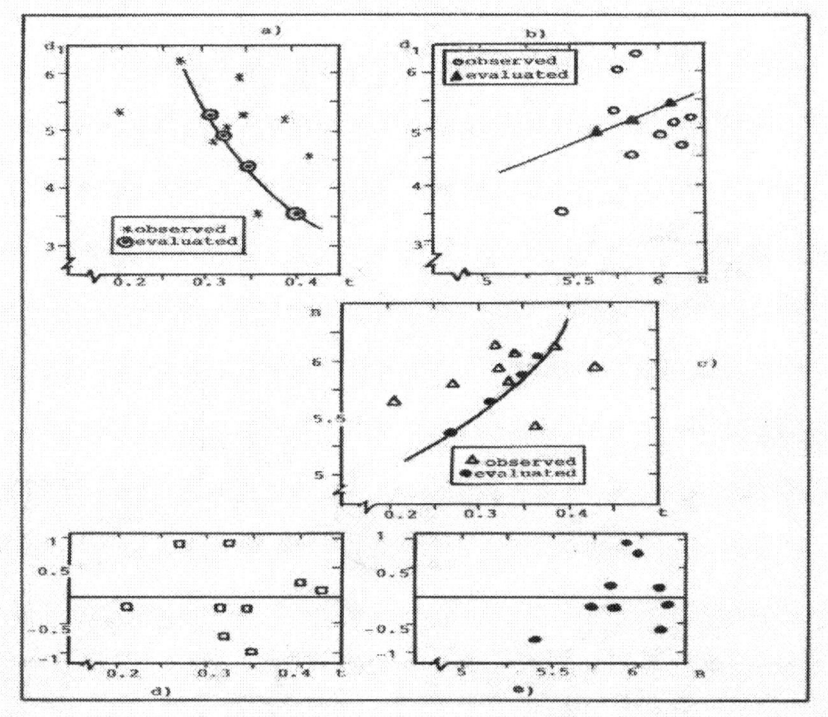

Figure 21

----

8   The linear regression equation has view of $d_1=7.216 -6.681t-3.295 \cdot 10^{-5}B$

This dependence $d_1=\gamma_4(t)$ has the non-linear regression model with equation view of $Y_c=0.828X^{-0.605}$ or $t=0.828d_1^{-0.605}$, where $d_1=(1.208t)^{-1.653}$. So , we see from Figure 21(a) that with increasing of clearance (t) between of chip wraps, the value of internal diameter(d_1) decreases considerably with non-linear regression equation $d_1=(1.208t)^{-1.653}$.

In Figure 21(b) is shown the functional dependence of internal diameter (d_1) from width (B) of stainless chip ,i.e we have function $d_1=\varphi_1(B)$. And as was above-identified this dependence $d_1=\varphi_1(B)$ has the linear regression model with equation view of $Y_c=0.06+0.863X$ or $d_1=0.06+0.863B$.As we see from Figure 21(b) that with increasing of width (B) stainless chip, the value of internal diameter (d_1) increases in accordance with regression equation $d_1=0.06+0.863B$.

In Figure 21 (c) is shown the functional dependence of width (B) stainless chip from the clearance (t) between of chip wraps and this dependence can be expressed in view of $B=\gamma_1(t)$ and has the non-linear regression model with equation $Y_c=0.032X^{1.296}$ or $t=0.032B^{1.296}$,where $\mathbf{B=(31.25t)^{0.772}}$ (25). And besides from Figure 21(c) we see that with increasing of clearance (t) between of chip wraps ,the value of width (B) for this stainless chip increases considerably with regression equation view of $B=(31.25t)^{0.772}$.

In Figure 21(d) and 21 (e) are illustrated residual plots (residual versus t and residual versus B) the above-named functional dependencies of $d_1=\gamma_4(t)$ and $d_1=\varphi_1(B)$ accordingly.

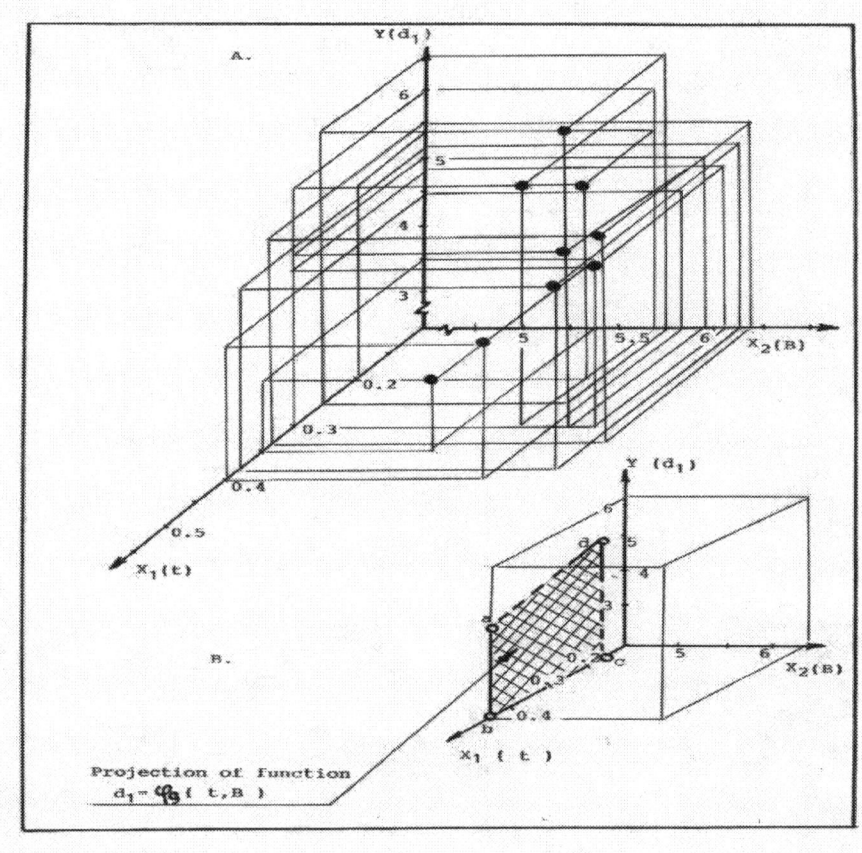

Figure 22

96

Analyzing the Figure 22(A) we see that functional surface $d_1=\varphi_9(t,B)$ is shown in view of three-dimensional drawing for non-linear regression model with equation view of $Y_c=1.603-4.903X_1-3.868X^2_1+0.934X_2$ or $d_1=1.603-4.903t-3.868t^2+0.934B$. And Figure 22(B) shows that functional surface $d_1=\varphi_9(t,B)$ has view of trapezium on the plane YOX_1 with such coordinate of its points (sizes in mm): a $(X_{11}=0.21,X_{21}=0,Y_1=5.74)$; b$(X_{12}=0.21,X_{22}=0,Y_2=0)$; c$(X_{13}=0.42,X_{23}=0,Y_3=0$) ;d $(X_{14}=0.42,X_{24}=0,Y_4=4.27)$.

Analysis of functional surface $d_1=\varphi_9(t,B)$

From Figure 22(A) we see that functional surface $d_1=\varphi_9(t,B)$,on which located the points of non-linear regression equation view of $Y_c1.603-4.903X_1-3.868X^2_1+0.934X_2$ or $d_1=1.603-4.903t-3.868t^2+0.934B$,has a curvilinear character with the following coordinate of their peaks in three-dimensional drawing(sizes in mm): d'$(X_{11}=0.21,X_{21}=5.71,Y_1=5.74)$; c'$(X_{12}=0.21,X_{22}=5.71,Y_2=0)$;b'$(X_{13}=0.42$, $X_{23}=5.79$, $Y_3=0)$; a' $(X_{14}=0.42,X_{24}=5.79,Y_4=4.27)$. In Figure 23 schematically is shown the functional surface $d_1=\varphi_9(t,B)$ and graph for calculation of module for vectors $|b'c'|$ and $|a'd'|$.

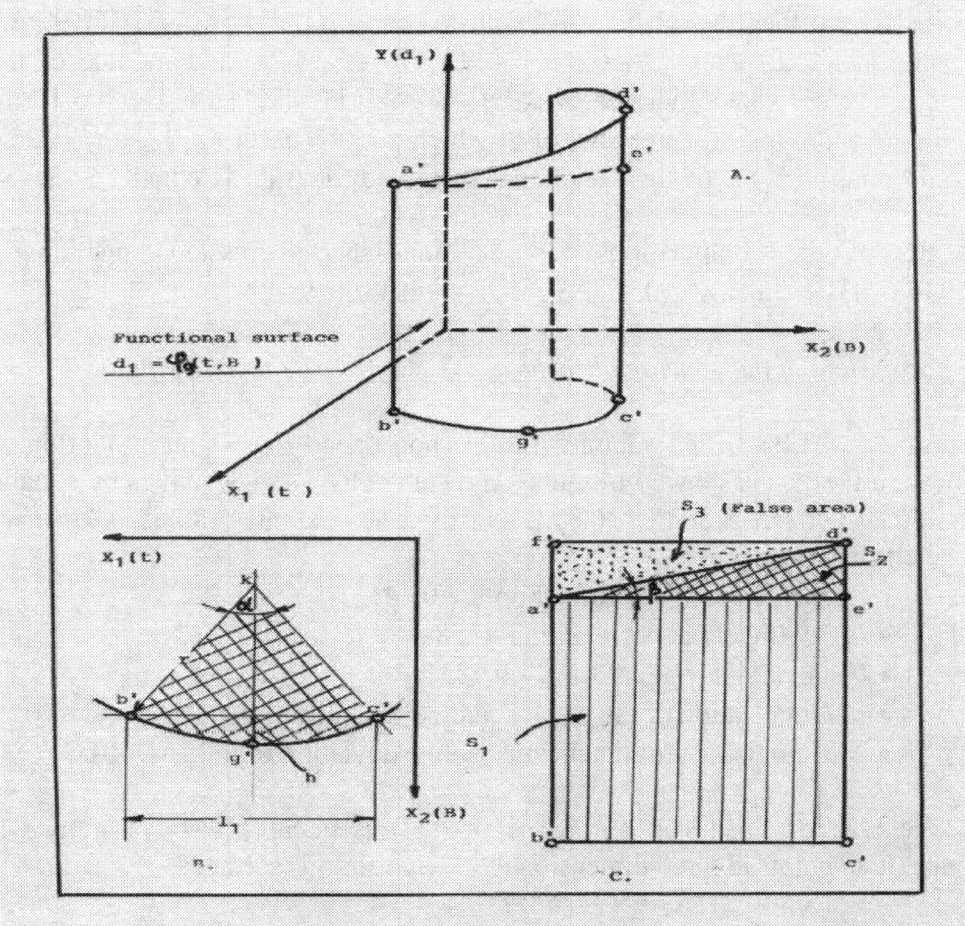

Figure 23   Graph for calculation of module vectors $|b'c'|,|a'd'|$ and area of functional surface $d_1=\varphi_9(t,B)$: A-total view of this functional surface ;B- graph for calculation of module vector $|b'c'|$; C-graph for calculation of module vector $|a'd'|$ and area of this surface.

97

- The module of vector $|a'b'|$ and $|c'd'|$ can be calculated from the projection of this functional surface $d_1=\varphi_9(t,B)$ on plane YOX_1 (Figure 22A and 23A).So ,the module of vectors $|c'd'|=5.74$ and $|a'b'|=4.27$.
- The module of vector $|b'c'|$ or it length can be calculated by formula $L=|b'c'|=0.01745 \cdot r \cdot \alpha$,where L=length of arc(b'c');r=radius of sector;α=angle of sector. So ,the module of vector $r=|kg'|=[(X_{16}-X_{15})^2+(X_{26}-X_{25})^2]^{1/2}=5.83$, where coordinate of points g' and k: g'($X_{16}=0.30$,$X_{26}=5.82$) and k($X_{15}=0,X_{25}=0$). At data $\alpha=82°$,r=5.83 we have the module of vector $|b'c'|=8.34$ and other parameters of circular segment which are equal: $h=r[1-\cos(\alpha/2)]=1.43$ and $l_1=2[h(2r-h)]^{1/2}=6.4$ .The graph for calculation of module vectors $|b'c'|$, $|a'd'|$ and area of functional surface $d_1=\varphi_9(t,B)$ is shown in Figure 23.
- The module of vector $|a'd'|$ and area of this functional surface, as shown in Figure 23 (c) can be calculated by the following way: the modules of length for curve $|a'e'|=|b'c'|=8.34$ and vector $|d'e'|=|c'd'|-|a'b'|=1.47$.For calculation of area S_1,S_2 and S_3 we find the angle β which is equal: $\tan\beta=|d'e'|/|a'e'|=0.176$,$\beta=10°$ and then the module of length for curve $|a'd'|=|d'e'|/\sin10°=8.45$. The coordinates of point f ' ($X_{14}=0.42,X_{24}=5.79$,$Y_4=5.74$) and module of length $|f'd'|=|b'c'|=8.34$ and area of $S_2=S_3=S/2==(|f'd'| \cdot |d'e'|)/2=6.13$ mm^2. For these conditions, we have area of S_1 which is equal $S_1=|a'b'| \cdot |b'c'|=35.61$ mm^2 and total area ΣS of this functional surface $d_1=\varphi_9(t,B)$ is equal $\Sigma S=S_1+S_2=41.74$ mm^2.

All results of the computations for the non-linear regression equation view of $Y_c=1.603-4.903X_1-3.868X^2_1+0.934X_2$ are given in Table 9.

So, the parameters of this functional surface $d_1=\varphi_9(t,B)$ are the following:

1.Function $d_1=\varphi_9(t,B)$ better submits to the non-linear regression model in view of curvilinear surface ,as is shown in Figure 22(A) and 23(A) and describes by equation view of $Y_c=1.603 -4.903X_1-3.868X^2_1+0.934X_2$ or $d_1=1.603-4.903t-3.868t^2+0.934B$ with such statistical characteristics:
- Coefficient of determination $R^2=0.341$
- Coefficient of correlation r=0.584
- Standard deviation $S_{y/x1,x2}=0.785$
- Minimization of the mean square error (min MSE=0.41)
- Minimization of the mean absolute deviation (min MAD=0)

2.The total area of functional surface $d_1=\varphi_9(t,B)$ is equal $\Sigma S=41.74$ mm^2 with such coordinate of its points in three-dimensional drawing for this surface:
   - d' ($X_{11}=0.21$,$X_{21}=5.71$,$Y_1=5.74$)
   - c' ($X_{12}=0.21$,$X_{22}=5.71$,$Y_2=0$)
   - b' ($X_{13}=0.42$,$X_{23}=5.79$,$Y_3=0$)
   - a' ($X_{14}=0.42$,$X_{24}=5.79$, $Y_4=4.27$)

Table 9 Evaluation of regression equation $Y_c=1.603-4.903X_1-3.868X^2_1+0.934X_2$

A. Mean, variance and standard deviation

Variance	Mean	Variance	Standard deviation
X_1	0.317	0.030	0.173
X_2	5.836	0.514	0.717
Y	5.098	5.609	2.368

B. Results of multiple regression of Y on X_1 and X_2

Parameter	Variable	Coefficients	Standard error	T-value
β_1	X_1	-4.903	0.058	-84.534
β_2	X_2	0.934	0.239	3/908

C. Analysis of variance results

Regression
- Degrees of freedom 2
- Sum of squares 1.915
- Mean square 0.958

Error
- Degrees of freedom 6
- Sum of squares 3.694
- Mean square 0.616
- Standard error of estimate 0.173
- F-value * 1.555   [$F_{0.05,2,6}=5.14$]

Since $F^*=1.555 < 5.14$ we can not reject the hypothesis that both β_1 and β_2 are zero

D. Determination of residuals

Number	Observed	Estimated	Residual
1	5.36	5.74	-0.38
2	5.22	4.94	0.28
3	3.52	4.50	-0.98
4	4.51	4.27	0.24
5	4.98	5.22	-0.24
6	4.74	5.43	-0.69
7	5.15	5.34	-0.19
8	5.96	4.96	1.0
9	6.44	5.50	0.94

5.10 Dependence of internal diameter (d_1) from external diameter (d_2) and number (ω) of chip wraps for stainless chip

Analysis of data and statistical characteristics (coefficient of determination $R^2=0.885$, coefficient of correlation $r=0.941$, standard deviation $S_{y/x1,x2}=0.327$) and also the minimization of the mean square error (min MSE=0.071) and minimization of the mean absolute deviation (min MAD=0) show that this functional dependence $d_1=\varphi_{10}(d_2,\omega)$

better submits to the linear than to non-linear [9] regression model with equation view of
$Y_c= -5.329+1.064X_1-0.134X_2$ or $d_1= -5.329+1.064d_2-0.134\omega$ (26).

In Figure 24 (a) is shown the functional dependency of internal diameter (d_1) from external diameter (d_2), i.e we have function $d_1=\varphi_3(d_2)$.

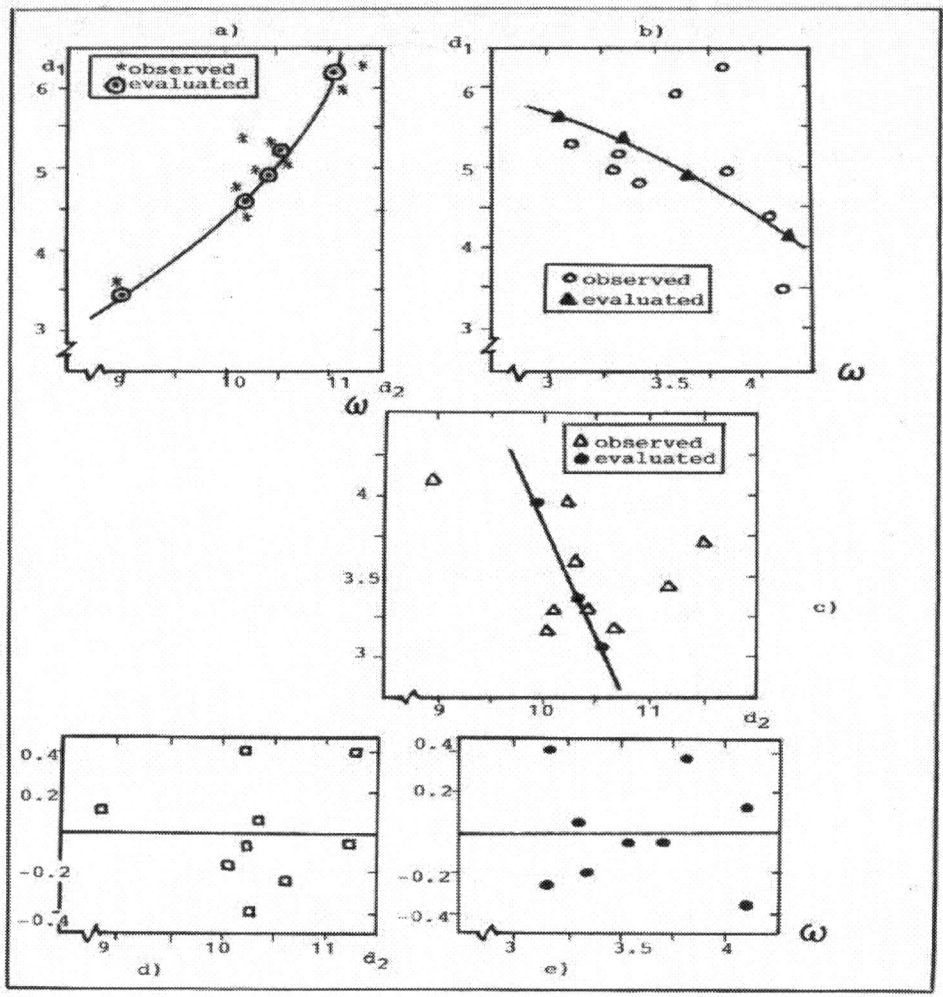

Figure 24

This dependence has the non-linear regression model with equation view of $Y_c=5.728X^{0.359}$ or $d_2=5.728d_1^{0.359}$, where $d_1=(0.175d_2)^{2.786}$. So, we see from Figure24(a) that with increasing of external diameter (d_2)stainless chip ,the value of internal diameter (d_1) of this chip increases considerably in accordance with non-linear regression equation $d_1=(0.175d_2)^{2.786}$.In Figure 24(b) is shown the functional dependence of internal diameter(d_1) from number (ω) of chip wraps for stainless chip, i.e we have function $d_1=\phi_3(\omega)$.

----

9   *The non-linear regression equation has view of $d_1= -27.593 + 5.174d_2 -0.201d^2{}_2+ 0.244\omega$*

This dependence $d_1=\phi_3(\omega)$ has non-linear regression model with equation view of $Y_c=20.32X^{-1.102}$,where $d_1=20.32\omega^{-1.102}$.So ,we see from Figure 24(b) that with increasing of number (ω) chip wraps for stainless chip ,the value of internal diameter (d_1) decreases in accordance with regression equation $d_1=20.32\omega^{-1.102}$.

In Figure 24(c) is shown the functional dependence of number (ω) of chip wraps for stainless chip from the external diameter (d_2) and this dependence can be expressed in view of function of $\omega=\phi_1(d_2)$ and has the linear regression model with equation $Y_c=13.137-0.811X$ or $d_2=13.137-0.811\omega$,where $\boldsymbol{\omega=16.199-1.233d_2}$ **(27)**.

So, we see from Figure 24 (c) that with increasing of external diameter (d_2) stainless chip ,the value of number (ω) chip wraps considerably decreases in accordance with regression equation $\omega=16.199-1.233d_2$. In Figure 24(d) and 24(e) are illustrated residual plots (residual versus d_2 and residual versus ω) above-named functional dependencies $d_1=\varphi_3(d_2)$ and $d_1=\phi_3(\omega)$ accordingly.

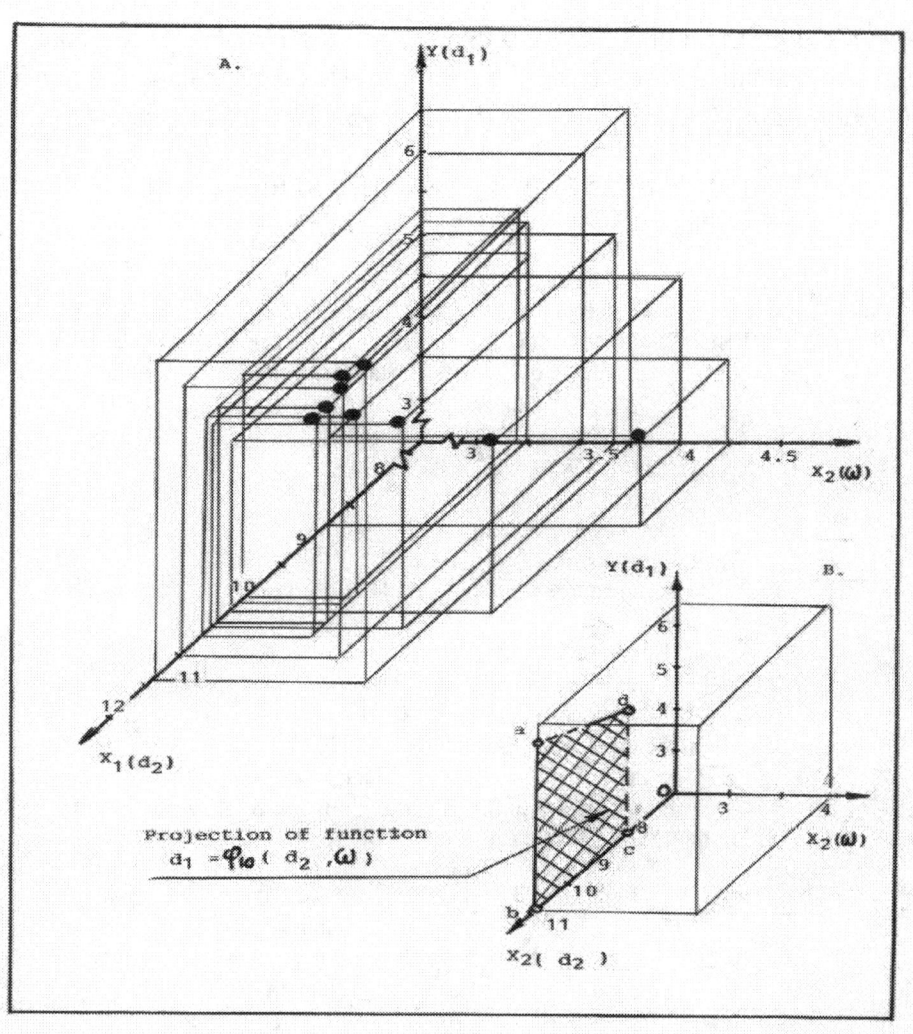

Figure 25

101

Analyzing the Figure 25(A), we see that functional surface $d_1=\varphi_{10}(d_2,\omega)$ is shown in view of three-dimensional drawing for linear regression model with equation view of $Y_c=-5.329+1.064X_1-0.134X_2$ or $d_1=-5.329+1.064d_2-0.134\omega$. And Figure 25(B) shows that functional surface $d_1=\varphi_{10}(d_2,\omega)$ as view of trapezium on the plane YOX_2 with such coordinate of its points (sizes in mm): a ($X_{11}=11.18$,$X_{21}=0,Y_1=6.06$) ; b($X_{12}=11.18,X_{22}=0,Y_2=0$); c($X_{13}=10.03,X_{23}=0,Y_3=0$) ;d($X_{14}=10.03,X_{24}=0$,$Y_4=4.92$).

From Figure 25(A) we see that functional surface $d_1=\varphi_{10}(d_2$,$\omega)$, on which are situated the points of the linear regression equation view of $Y_c=-5.329+1.064X_1-0.134X_2$ or $d_1=-5.329+1.064d_2-0.134\omega$,has the following coordinate of its peaks in three-dimensional drawing(sizes in mm): a'($X_{11}=11.18$,$X_{21}=3.76$,$Y_1=6.06$) ;b' ($X_{12}=11.18$, $X_{22}=3.36$,$Y_2=0$) ;c'($X_{13}=10.03$,$X_{23}=3.17,Y_3=0$) ;d'($X_{14}=10.03,X_{24}=3.17,Y_4=4.92$).The module of vectors and area of functional surface $d_1=\varphi_{10}(d_2,\omega)$ are equal (sizes in mm): $|a'b'|=6.06, |b'c'|=1.29, |c'd'|=4.92, |d'a'|=1.72$ and $S_{a'b'c'd'}=1/2\cdot[(|a'b'|--|c'd'|)\cdot|b'c'|]+(|b'c'|\cdot|c'd'|)=7.09mm^2$.All results of the computations for the linear regression equation view of $Y_c=-5.329+1.064X_1-0.134X_2$ are given in Table 10.

Table 10 Evaluation of regression equation $Y_c=-5.329+1.064X_1-0.134X_2$

A.MEAN ,VARIANCE AND STANDARD DERIVIATION

Variable	Mean	Variance	Standard deviation
X_1	10.248	4.122	2.030
X_2	3.559	1.114	1.055
Y	5.098	5.609	2.368

B. RESULTS OF MULTIPLE REGRESSION OF Y ON X_1 AND X_2

Parameter	Variable	Coefficients	Standard error	T-value
β_1	X_1	1.064	0.677	1.572
β_2	X_2	−0.134	0.352	− 0.381

C.ANALYSIS OF VARIANCE RESULTS

Regression
Degrees of freedom   2
Sum of squares   4.97
Mean square   2.484
Error
Degrees of freedom   6
Sum of squares   0.642
Mean square   0.107
Standard error of estimate   2.030
F-value[*]   23.21   [$F_{0.05,2,6}$]=5.14

Since F^*=23.21>5.14 we can reject the hypothesis that both β_1 and β_2 are zero

D.DETERMINATION OF RESIDUALS

Number	Observed	Estimated	Residual
1	5.36	4.92	0.44
2	5.22	5.26	0.04
3	3.52	3.35	0.17
4	4.51	4.91	−0.40
5	4.98	5.06	−0.08
6	4.74	4.95	−0.21
7	5.15	5.40	−0.25
8	5.96	5.97	−0.01
9	6.44	6.06	0.38

So, the parameters of this functional surface $d_1 = \varphi_{10}(d_2,, \omega)$ are the following:

1.Function $d_1 = \varphi_{10}(d_2, \omega)$ better submits to the linear regression model with equation view of $Y_c = -5.329 + 1.064X_1 - 0.134X_2$ or $d_1 = -5.329 + 1.064d_2 - 0.134\omega$ with such statistical characteristics: coefficient of determination $R^2 = 0.885$, coefficient of correlation $r = 0.941$, standard deviation $S_{y/x1,x2} = 0.327$, minimization of the mean square error (min MSE = 0.071) and minimization of the mean absolute deviation (min MAD = 0).

2.The total area of functional surface $d_1 = \varphi_{10}(d_2, \omega)$ is equal $\sum S = 7.09 \text{mm}^2$ with such coordinate of its points in three- dimensional drawing for this surface (sizes in mm): a'($X_{11} = 11.18$,$X_{21} = 3.76$,$Y_1 = 6.06$) ;b'($X_{12} = 11.18$,$X_{22} = 3.76$,$Y_2 = 0$) ; c'($X_{13} = 10.03$, $X_{23} = 3.17$, $Y_3 = 0$) , d' ($X_{14} = 10.03$, $X_{24} = 3.17$, $Y_4 = 4.92$).

In Table 11 is given the summary of _Multiple regression analysis between two independent variable parameters in function of $Y_i = \gamma(X_{i,1}, X_{i,2})$_ for internal diameter($d_1$) of stainless chip
.

Table 11 Summary characteristics for internal diameter (d_1) of stainless chip

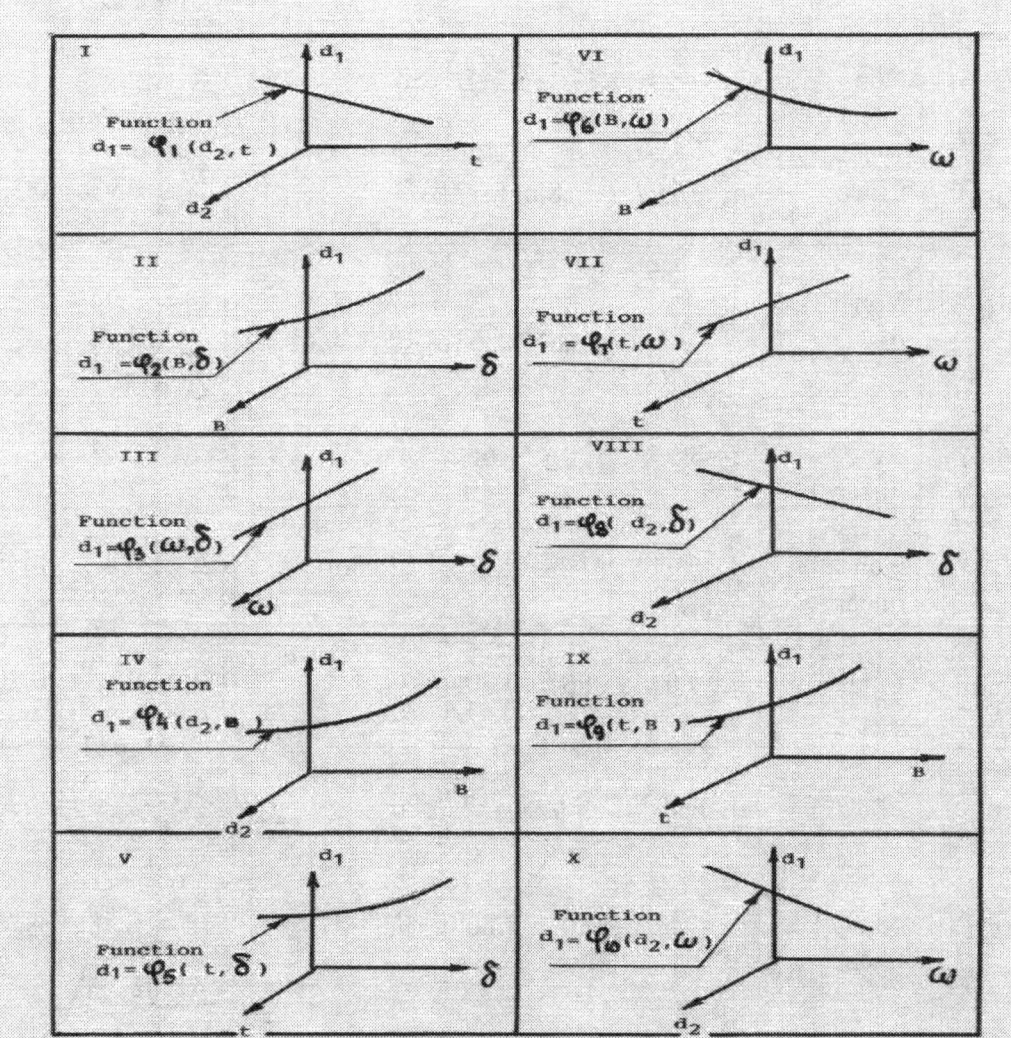

And in Table 12 is shown the summary statistical characteristics and general empirical equations for this function $Y_i = \gamma(X_{i,1}, X_{i,2})$.

Table 12 Summary statistical characteristics and general empirical equations in function of $Y_i = \gamma(X_{i,1}, X_{i,2})$ for INTERNAL DIAMETER(d_1) OF STAINLESS CHIP

View of function	Regression model	Empirical equation	Statistical characteristics		
			R^2	r	$S_{y/x1,x2}$
$d_1 = \varphi_1(d_2, t)$	Linear	$d_1 = -3.712 + 0.99d_2 - 4.214t$	0.964	0.981	0.184
$d_1 = \varphi_2(B, \delta)$	Non-linear	$d_1 = -2.6 + 2.67B - 0.28B^2 + 5\delta$	0.870	0.932	0.348
$d_1 = \varphi_3(\omega, \delta)$	Linear	$d_1 = 2.292 + 0.258\omega + 5.131\delta$	0.880	0.938	0.335
$d_1 = \varphi_4(d_2, B)$	Non-linear	$d_1 = -3.132 + 1.192d_2 - 2.52 \cdot 10^{-6}(d^2{}_2) - 0.683B$	0.900	0.949	0.305
$d_1 = \varphi_5(t, \delta)$	Non-linear	$d_1 = 3.935 - 2.281t + 1.657t^2 + 4.657\delta$	0.882	0.939	0.333
$d_1 = \varphi_6(B, \omega)$	Non-linear	$d_1 = 6.3 + 1.241B - 0.133B^2 - 1.098\omega$	0.216	0.465	0.856
$d_1 = \varphi_7(t, \omega)$	Linear	$d_1 = 9.074 - 5.067t - 0.666\omega$	0.368	0.554	0.805
$d_1 = \varphi_8(d_2, \delta)$	Linear	$d_1 = -2.762 + 0.693d_2 + 2.062\delta$	0.925	0.962	0.264
$d_1 = \varphi_9(t, B)$	Non-linear	$d_1 = 1.603 - 4.903t - 3.868t^2 + 0.934B$	0.341	0.584	0.785
$d_1 = \varphi_{10}(d_2, \omega)$	Linear	$d_1 = -5.329 + 1.064d_2 - 0.134\omega$	0.885	0.941	0.327

SUMMARY

Analysis of correlation between three parameters of stainless chip in function of $Y_i=\gamma(X_{i,1},X_{i,2})$ advantageously for internal diameter (d_1) of this chip in dependence from the different its parameters have showed the following results:

1. The functions of $d_1=\varphi_1(d_2,t)$, $d_1=\varphi_3(\omega,\delta)$,$d_1=\varphi_7(t,\omega)$, $d_1=\varphi_8(d_2,\delta)$ and $d_1=\varphi_{10}(d_2,\omega)$ better submit to the linear regression models;

2. The functions of $d_1=\varphi_2(B,\delta)$,$d_1=\varphi_4(d_2,B)$,$d_1=\varphi_5(t,\delta)$,$d_1=\varphi_6(B,\omega)$ and $d_1=\varphi_9(t,B)$ better submit to the non-linear regression models;

3. And besides we see that between parameters of stainless chip there are good correlation in such functions as $d_1=\varphi_1(d_2,t)$,$d_1=\varphi_2(B,\delta)$,$d_1=\varphi_3(\omega,\delta)$, $d_1=\varphi_4(d_2,B)$, $d_1=\varphi_5(t,\delta)$,$d_1=\varphi_8(d_2,\delta)$ and $d_1=\varphi_{10}(d_2,\omega)$,where the statistical characteristics (coefficients of determination R^2 and correlation r) have the larger values;

4. Analyzing of each function for these relations between parameters of stainless chip ,we see that internal diameter (d_1) could be evaluated approximately by the different empirical formulas ,as shown in Table 12.

5. And besides we can also to conclude that internal diameter (d_1) of stainless chip is the function of such multiple independent variables as : external diameter (d_2),clearance (t) between chip wraps, width (B) of stainless chip, thickness (δ) of this chip and number (ω) of chip wraps ,i.e we have function $d_1=\mu(d_2,t,B,\delta,\omega)$;

6. The most influence on increasing of size for internal diameter (d_1) of stainless chip plays such parameters as thickness (δ),width (B) and external diameter (d_2) and as we see this parameter (d_1) increases considerably with increasing of above-named parameters (δ,B,d_2);

7. The most influence on decreasing of size for internal diameter (d_1) of stainless chip plays such parameters as clearance (t) between of chip wraps and number (ω) of chip wraps. As we see this parameter (d_1) decreases considerably with increasing of above-named parameters (t,ω).

B. FOR EXTERNAL DIAMETER OF STAINLESS CHIP

5.11 Dependence of external diameter (d_2) of stainless chip from clearance (t) between of chip wraps and its width (B)

Analysis of data and statistical characteristics (coefficient of determination $R^2=0.27$ and coefficient of correlation r=0.52 ,standard deviation $S_{y/x1,x2}=0.7100$ and also the minimization of the mean square error (min MSE=0.34) and minimization of the mean absolute deviation (min MAD=0)) show that this functional dependence $d_2=\gamma(t,B)$ better submits to the linear than to non-linear [10] regression model with equation view of **$Y_c=3.761-3.452X_1+1.299X_2$ or $d_2=3.761-3.452t+1.299B$ (28)** .

----

10   *The non-linear regression equation has view of $d_2=3.886-1.139t-3.982t^2+1.223B$*

In Figure 26(a) is shown functional dependence of external diameter (d_2) stainless chip from clearance (t) between of chip wraps, i.e we have the function $d_2=\gamma_3(t)$.

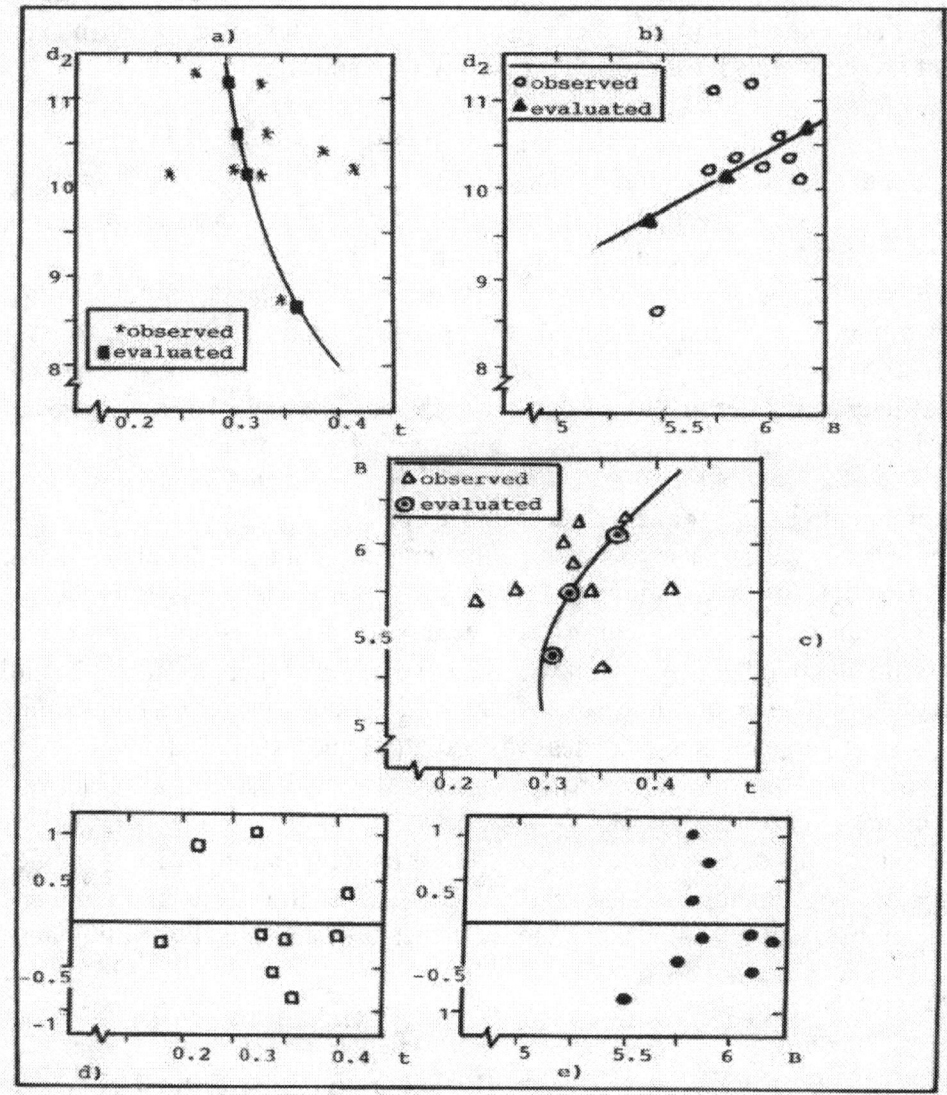

Figure 26

This functional dependence $d_2=\gamma_3(t)$ has the non-linear regression model with equation view of $Y_c=1.14X^{-0.56}$ or $t=1.14d_2^{-0.56}$,where $\mathbf{d_2=(0.878t)^{-1.786}}$ (29) . So ,we see from Figure 26(a) that with increasing of clearance (t) between of chip wraps, the value of external diameter (d_2) stainless chip decreases in accordance with the non- regression equation $d_2=(0.878t)^{-1.786}$.

In Figure 26(b) is shown the functional dependence of external diameter (d_2) from width (B) stainless chip ,i.e we have the function $d_2=\varphi_2(B)$. And as was above-identified ,this functional dependence $d_2=\varphi_2(B)$ has the linear regression model with equation $Y_c=3.026+1.237X$ or $d_2=3.026+1.237B$.

So ,we see from Figure 26(b) that with increasing of width (B) stainless chip ,the value of external diameter (d_2) increases considerably in accordance with the linear regression equation $d_2 =3.026+1.237B$.

In Figure 26 (c) is shown the functional dependence of width (B) stainless chip from the clearance (t) between of chip wraps. This dependence can be expressed in view of function $B=\gamma_1(t)$ and has the non- linear model with equation $Y_c=0.032X^{1.296}$ or $t=0.032B^{1.296}$,where $B=(31.25t)^{0.772}$. So , we see from Figure 26(c) that with increasing of clearance (t) between of chip wraps , the value of width (B) stainless chip increases considerably with the non-linear regression equation view of $B=(31.25t)^{0.772}$.

In Figure 26(d) and 26(e) are illustrated residual plots (residual versus t and residual versus B) the above-named functional dependencies of $d_2=\gamma_3(t)$ and $d_2=\varphi_2(B)$ accordingly.

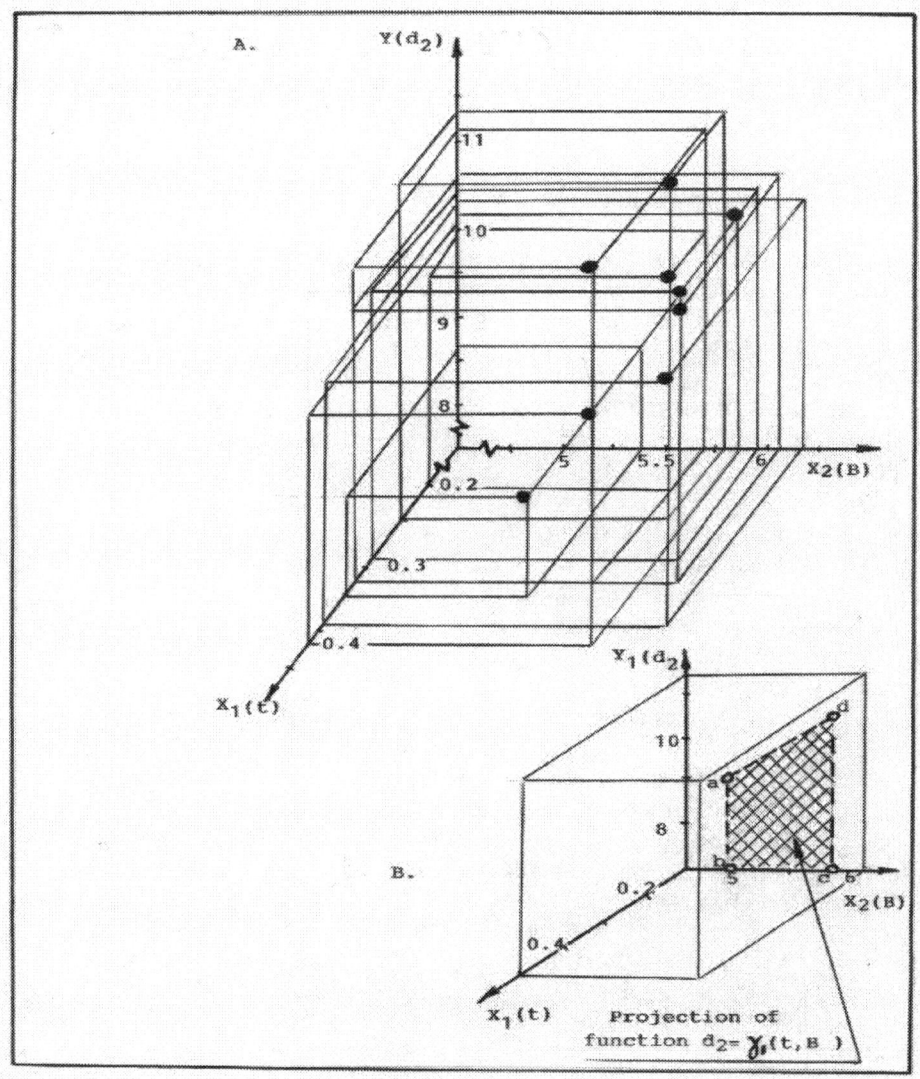

Figure 27

Analyzing the Figure 27(A) we see that functional surface $d_2=\gamma_1(t,B)$ is shown in view of three-dimensional drawing for the linear regression model with equation view of **$Y_c=3.761-3.452X_1+1.299X_2$** or **$d_2=3.761-3.452t+1.299B$** (**30**) . And Figure 27(B) shows that functional surface $d_2=\gamma_1(t,B)$ has view of trapezium on the plane YOX_2 with such coordinate of its points (sizes in mm): a($X_{11}=0$, $X_{21}=5.70$,$Y_1=10.06$) ; b($X_{12}=0,X_{22}=5.70,Y_2=0$) ;c ($X_{13}=0$,$X_{23}=6.12$,$Y_3=0$) ; d($X_{14}=0$,$X_{24}=10.64$) .

From Figure 27(A) we see that functional surface $d_2=\gamma(t,B)$,on which are situated the points of the linear regression equation view of $Y_c=3.761-3.452X_1+1.299X_2$ or $d_2=3.761-3.452t+1.299B$ has the following coordinate of its peaks in three-dimensional drawing (sizes in mm): a' ($X_{11}=0.32$,$X_{21}=5.70$,$Y_1=10.06$) ; b' ($X_{12}=0.32$, $X_{22}=5.70$, $Y_2=0$) ; c' ($X_{13}=0.31$,$X_{23}=6.12$,$Y_3=0$) ; d'($X_{14}=0.31$,$X_{24}=6.12$,$Y_4=10.64$).The module of vectors and area of this functional surface $d_2=\gamma_1(t,B)$ are equal (sizes in mm): $|a'b'|=10.06$, $|b'c'|=0.42$, $|c'd'|=10.64$, $|d'a'|=0.72$ and area $S_{a'b'c'd'}$ = $=1/2\{[|c'd'|-|a'b'|]\cdot|b'c'|\}+(|a'b'|\cdot|b'c'|)$ =4.35mm^2. All results of the computations for the linear regression equation view of $Y_c=3.761-3.452X_1+1.299X_2$ are given in Table 13.

Table 13 Evaluation of linear regression equation $Y_c=3.761-3.452X_1+1.299X_2$

A. MEAN, VARIANCE AND STANDARD DEVIATION

Variable	Mean	Variance	Standard deviation
X_1	0.320	0.031	0.176
X_2	5.836	0.515	0.718
Y	10.250	4.119	2.029

B. RESULTS OF MULTIPLE REGRESSION OF Y ON X_1 AND X_2

Parameter	Variable	Coefficients	Standard error	T-value
β_1	X_1	−3.452	0.059	− 58.508
β_2	X_2	1.299	0.239	5.435

C. ANALYSIS OF VARIANCE RESULTS

Regression

Degrees of freedom 2

Sum of squares 1.094

Mean square 0.547

Error

Degrees of freedom 6

Sum of squares 3.025

Mean square 0.504

Standard error of estimate 0.176

F-value* 1.085 [$F_{0.05,2,6}=5.14$]

Since $F^* < F_{0.05,2,6}$ we can not reject the hypothesis that both β_1 and β_2 are zero

DETERMINATION OF RESIDUALS

Number	Observed	Estimated	Residual
1	10.03	10.45	−0.42
2	10.36	10.45	− 0.09
3	8.69	9.56	− 0.87
4	10.13	9.83	0.30
5	10.22	10.29	− 0.07
6	10.07	10.64	−0.57
7	10.49	10.58	−0.09
8	11.06	10.06	1.0
9	11.18	10.37	0.81

So , the parameters of this functional surface $d_2 = \gamma_1(t,B)$ are the following:

1. 1.Function $d_2 = \gamma_1$ (t,B) better submits to the linear regression model with equation view of $Y_c = 3.761 - 3.452X_1 + 1.299X_2$ or $d_2 = 3.761 - 3.452t + 1.299B$ with such statistical characteristics: coefficient of determination $R^2 = 0.27$,coefficient of correlation r=0.52,standard deviation $S_{y/x1,x2} = 0.710$,minimization of the mean square error (min MSE=0.34),minimization of the mean absolute deviation (min MAD=0);

2. The total area of functional surface $d_2 = \gamma_1$ (t,B) is equal $\sum S = 4.35$ mm^2 with such coordinate of its points in three-dimensional drawing for this surface (sizes in mm): a'$(X_{11}=0.32 ,X_{21}=5.70,Y_1=10.06)$;b' $(X_{12}=0.32 ,X_{22}=5.70 ,Y_2=0)$; c' $(X_{13}=0.31 ,X_{23}=6.12 ,Y_3=0)$;d'$(X_{14}=0.31 ,X_{24}=6.12 ,Y_4=10.64)$.

5.12 Dependence of external diameter (d$_2$) stainless chip from number (ω) of chip wraps and thickness (δ) of this chip

Analysis of data and statistical characteristics (coefficient of determination $R^2 = 0.801$, coefficient of correlation r=0.895 ,standard deviation $S_{y/x1,x2} = 0.369$ and also the minimization of the mean square error (min MSE=0.091) and minimization of the mean absolute deviation (min MAD=0)) show that this functional dependence $d_2 = \gamma_2(\omega,\delta)$ better submits to the linear than non-linear [11] regression model with equation view of $Y_c = 7.985 + 0.207X_1 + 4.149X_2$ or $d_2 = 7.985 + 0.207\omega + 4.149\delta$ (31) .

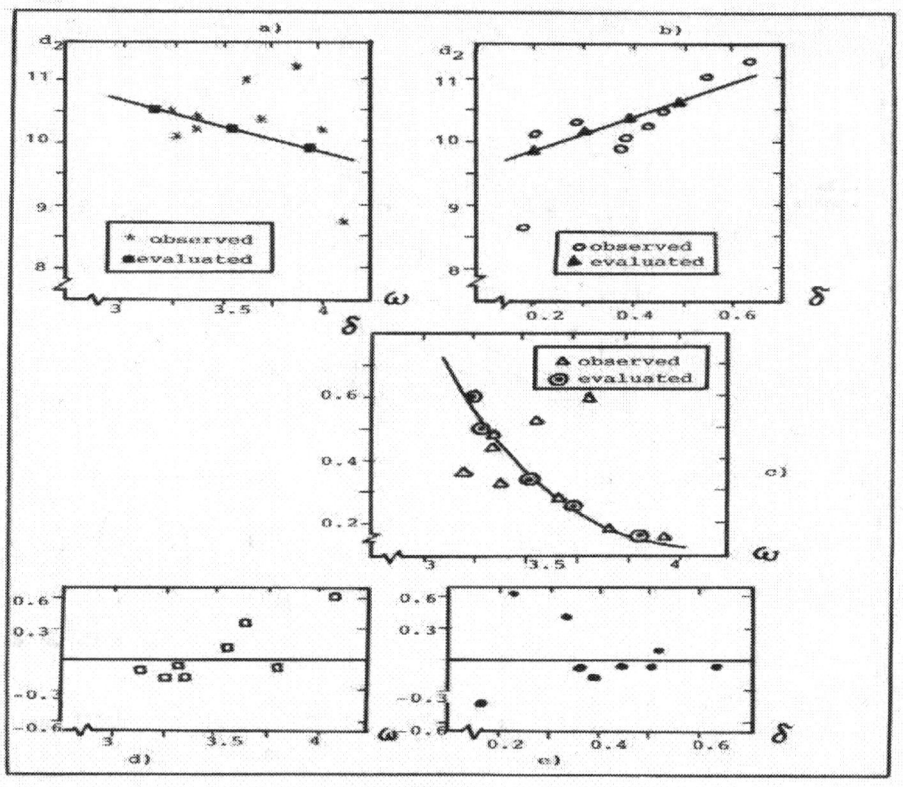

Figure 28

---

[11] *The non-linear regression equation has view $d_2 = -0.304 + 1.991\omega - 0.03\omega^2 + 10.463\delta$*

In Figure 28(a) is shown the functional dependence of external diameter (d_2) stainless chip from number (ω) of chip wraps ,i.e we have the function view $d_2=\phi_1(\omega)$.This dependence $d_2=\phi_1(\omega)$ has the linear regression model with equation view of $Y_c=13.137- -0.811X$ or $d_2=13.137-0.811\omega$. So ,we see from Figure 28(a) that with increasing of number (ω) chip wraps stainless chip ,the value of external diameter (d_2) of this chip decreases considerably with regression equation $d_2=13.137-0.811\omega$.

In Figure 28(b) is shown the functional dependence of external diameter (d_2) from thickness (δ) stainless chip ,i.e we have the function $d_2=\gamma(\delta)$ which has the linear regression model with equation view of $Y_c=8.808 +3.897X$ or $d_2=8.808+3.897\delta$. So ,we see from Figure 28(b) that with increasing of thickness (δ) stainless chip, the value of external diameter (d_2) increases considerably in accordance with the linear regression equation $d_2=8.808+3.897\delta$.

In Figure 28 (c) is shown the functional dependence of thickness (δ) stainless chip from the number (ω) of chip wraps and this dependence can be expressed in view of function $\delta=\alpha(\omega)$ and has the non-linear model with equation $Y_c=3.09X^{-0.123}$ or $\omega=3.09\delta^{-0.123}$,where $\boldsymbol{\delta=(0.324\omega)^{-8.13}}$ (32). So ,we see from Figure 28(c0 that with increasing of number (ω) chip wraps ,the value of thickness (δ) decreases considerably with regression equation $\delta =(0.324\omega)^{-8.13}$ .In Figure 28(d) and 28(e) are illustrated residual plots (residual versus ω and residual versus δ) the above-named functional dependencies of $d_2=\phi_1(\omega)$ and $d_2=\gamma(\delta)$ accordingly.

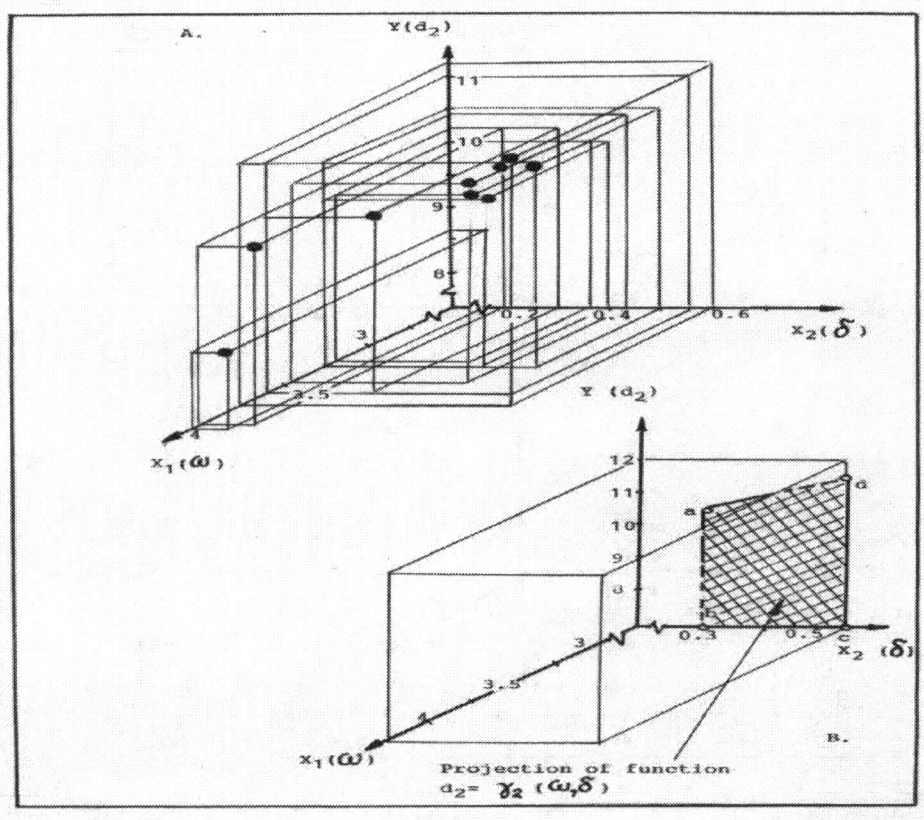

Figure 29

110

Analyzing the Figure 29(A) ,we see that functional surface $d_2=\gamma_2(\omega,\delta)$ is shown in view of three-dimensional drawing for the linear regression model with equation view of $Y_c=7.985+0.207X_1+4.149X_2$ or $d_2=7.985+0.207\omega+4.149\delta$. And Figure 29(B) shows that functional surface $d_2=\gamma_2(\omega,\delta)$ has view of trapezium on the plane YOX_2 with such coordinate of its points (sizes in mm): $a(X_{11}=0,X_{21}=0.36$,$Y_1=10.14)$; $b(X_{12}=0$, ,$X_{22}=0.36,Y_2=0)$;$c(X_{13}=0,X_{23}=0.60$,$Y_3=0)$;d ($X_{14}=0$, $X_{24}=0.60$,$Y_4=11.25)$.

From Figure 29(A) we see that functional surface $d_2=\gamma_2(\omega,\delta)$,on which are situated the points of the linear regression equation view of $Y_c =7.985+0.207X_1+4.149X_2$ or $d_2=7.985+0.207\omega+4.149\delta$ has the following coordinate of its peaks in three-dimensional drawing (sizes in mm): a' $(X_{11}=3.17$,$X_{21}=0.36$,$Y_1=10.14)$; b'$(X_{12}=3.17$,$X_{22}=0.36,Y_2=0)$;c'$(X_{13}=3.76$,$X_{23}=0.60$,$Y_3=0)$; d' ($X_{14}=3.76,X_{24}=0.60$,$Y_4=11.25)$. The value of modules and area of functional surface $d_2=\gamma_2(\omega,\delta)$ are equal: $|a'b'|=10.14$, $|b'c'|=0.64$, $|c'd'|=11.25$, $|d'a'|=1.28$ and area $S_{a'b'c'd'}=1/2\{[|c'd'|-|a'b'|]\cdot|b'c'|\}+$ $+(|b'c'|\cdot|a'b'|=6.85$ mm^2.All results of the computations for the linear regression equation view of $Y_c=7.985+0.207X_1+4.149X_2$ are given in Table 14.

Table 14 Evaluation of linear regression equation $Y_c=7.985+0.207X_1+4.149X_2$

A. MEAN, VARIANCE AND STANDARD DEVIATION			
Variable	Mean	Variance	Standard deviation
X_1	3.559	1.113	1.055
X_2	0.368	0.209	0.457
Y	10.248	4.121	2.030

B. RESULTS OF MULTIPLE REGRESSION OF Y ON X_1 AND X_2				
Parameter	Variable	Coefficients	Standard error	T-value
β_1	X_1	0.207	0.352	0.588
β_2	X_2	4.149	0.152	27.296

C. ANALYSIS OF VARIANCE RESULTS

Regression
- Degrees of freedom 2
- Sum of squares 3.301
- Mean square 1.651

Error
- Degrees of freedom 6
- Sum of squares 0.819
- Mean square 0.137

Standard error of estimate 1.055
F-value* 12.051 $[F_{0.05,2,6}=5.14]$

Since $F^*=12.051 >5.14$,we can reject the hypothesis that both β_1 and β_2 are zero

D. DETERMINATION OF RESIDUALS

Number	Observed	Estimated	Residual
1	10.03	10.14	−0.11
2	10.36	10.41	−0.05
3	8.69	9.27	−0.58
4	10.13	9.56	0.57
5	10.22	9.89	0.33
6	10.07	10.19	− 0.12
7	10.49	10.56	−0.07
8	11.06	10.96	0.10
9	11.18	11.25	−0.07

So , the parameters of this functional surface $d_2=\gamma_2(\omega,\delta)$ are the following:

- Function $d_2=\gamma_2(\omega,\delta)$ better submits to the linear regression model with equation view of $Y_c=7.985+0.207X_1+4.149X_2$or $d_2=7.985+0.207\omega+4.149\delta$ with such statistical characteristics: coefficient of determination $R^2=0.801$,coefficient of correlation r= 0.895,standard deviation $S_{y/x1,x2}=0.369$,minimization of the mean square error (min MSE=0.091),minimization of the mean absolute deviation (min MAD =0));

- The total area of functional surface $d_2=\gamma_2(\omega,\delta)$ is equal $\sum S=6.85$ mm^2 with such coordinate of its points in three-dimensional drawing for this functional surface (sizes in mm): a'($X_{11}=3.17,X_{21}=0.36$,$Y_1=10.14$);b'($X_{12}=3.17,X_{22}=0.36$,$Y_2=0$); c'($X_{13}=3.76,X_{23}=0.60$,$Y_3=0$) ; d'($X_{14}=3.76$,$X_{24}=0.60$,$Y_4=11.25$).

5.13 Dependence of external diameter (d_2) stainless chip from width (B) and thickness (δ) of this chip

Analysis of data and also the statistical characteristics show that coefficients of determination ($R^2=0.787$) and correlation (r=0.887),standard deviation $S_{y/x1,x2}=0.382$ are larger for non-linear regression model and besides the minimization of the mean square error (min MSE=0.097) and minimization of the mean absolute deviation (min MAD=0) are smaller for this regression model than for the linear regression model. And for this reason we can conclude that this function $d_2=\gamma_3(B,\delta)$ better submits to the non-linear than linear[12] regression model and has the equation view of $Y_c=12.466-1.432X_1+0.139X^2_1+3.799X_2$ or $d_2=12.466-1.432B+0.139B^2+3.799\delta$ (33).

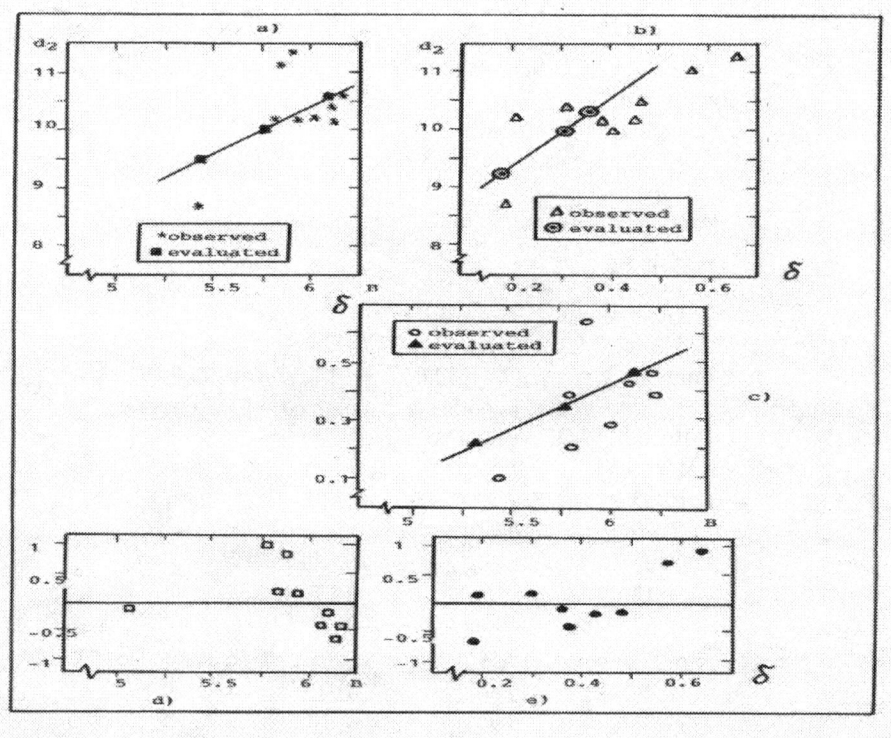

Figure 30

---

12 The linear regression equation has view of $d_2=3.541+1.114B+0.559\delta$

In Figure 30 (a) is shown the functional dependence of external diameter (d_2) stainless chip from its width (B) ,i.e we have the function $d_2=\varphi_2(B)$ and this dependence   has the linear regression model with equation view of $Y_c=3.026+1.237X$ or $d_2=3.026+1.237B$. So ,we see from Figure 30(a) that with increasing of width (B) stainless chip ,the value of external diameter   (d_2) of this chip increases   considerably with regression equation $d_2=3.026+1.237B$.

In Figure 30(b) is shown the functional dependence of external diameter (d_2) from thickness (δ) stainless chip ,i.e we have   the function $d_2=\gamma(\delta)$ which has the linear regression model with equation view of $Y_c=8.808+3.897X$ or $d_2=8.808+3.897\delta$. So ,we from Figure 30(b) that with increasing of thickness (δ) stainless chip ,the value of external diameter (d_2)increases in accordance with regression equation $d_2=8.808+3.897\delta$.

In Figure 30(c) is shown the functional dependence of thickness (δ) stainless chip from width (B) of this   stainless chip. This dependency can be expressed in view of function $\delta=f(B)$ and besides has the linear model with equation view of $Y_c= -1.236+0.275X$, where $\delta= -1.236+ 0.275B$. So , we see from Figure 30(c)   that with increasing of width (B) stainless chip ,the value of thickness (δ) of this stainless chip increases considerably with regression equation $\delta= -1.236+0.275B$.

In Figure 30 (d) and 30 (e) are illustrated residual plots (residual versus B and residual versus δ) of the above-named functional dependencies $d_2=\varphi_2(B)$ and $d_2=\gamma(\delta)$ accordingly.

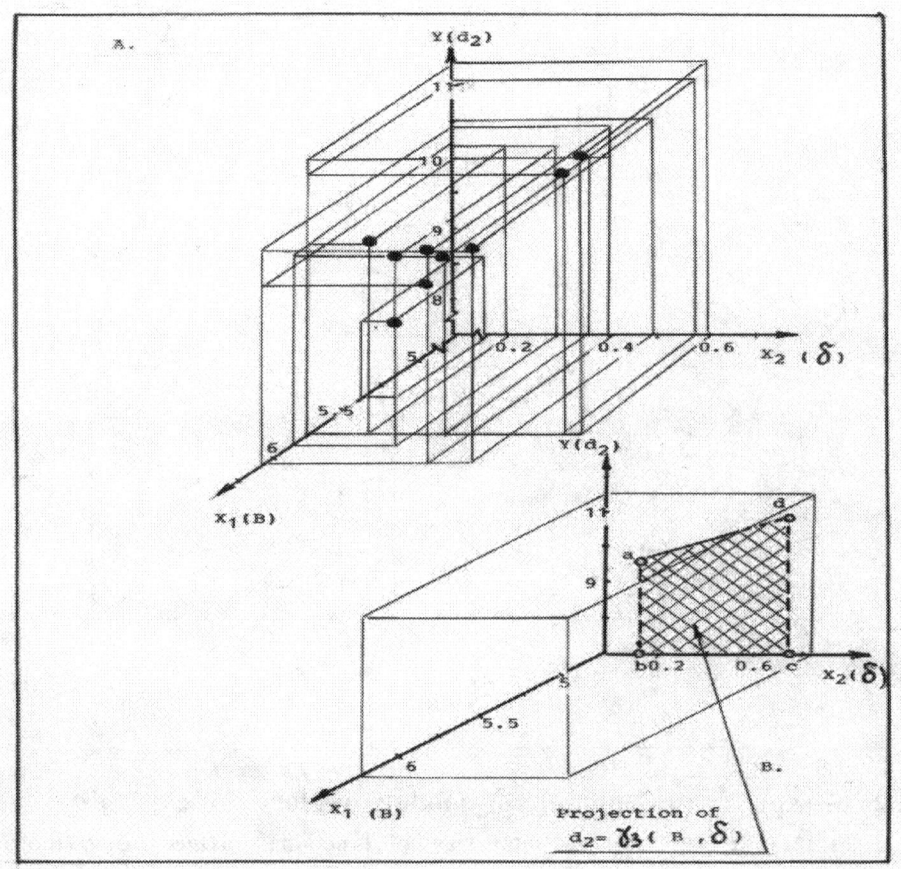

Figure 31

113

Analyzing the Figure 31 (A) we see that functional surface $d_2=\gamma_3(B,\delta)$ is given in view of three-dimensional drawing for the non- linear regression model with equation view of $Y_c=12.466-1.432X_1+0.139X^2_1+3.799X_2$ or $d_2=12.466-1.432B+0.139B^2+3.799\delta$. And Figure 31(B) shows that functional surface $d_2=\gamma_3(B,\delta)$ has view of trapezium on the plane YOX_2 with such coordinate for points (sizes in mm): a $(X_{11}=0,X_{21}=0.10,Y_1=9.16)$;b($X_{12}=0, X_{22}=0.10,Y_2=0)$; c$(X_{13}=0,X_{23}=0.60,Y_3=0)$; d$(X_{14}=0,X_{24}=0.60,Y_4=11.11)$.

Analysis of functional surface $d_2=\gamma_3(B,\delta)$

 From Figure 31(A) we see that functional surface $d_2=\gamma_3(B,\delta)$, on which located the points of non-linear regression equation view of $Y_c=12.466-1.432X_1+0.139X^2_1+3.799X_2$ or $d_2=12.466-1.432B+0.139B^2+3.799\delta$ has a curvilinear character with following coordinate of their peaks in three-dimensional drawing: a'$(X_{11}=5.37,X_{21}=0.10,Y_1=9.16)$; b'$(X_{12}=5.37,X_{22}=0.10,Y_2=0)$;c'$(X_{13}=5.75,X_{23}=0.60,Y_3=0)$;d'$(X_{14}=5.75,X_{24}=0.6,Y_4=11.11$). In Figure 32 schematically is shown the functional surface $d_2=\gamma_3(B,\delta)$ and graph for calculation of module for vectors $|b'c'|$ and $|a'd'|$.

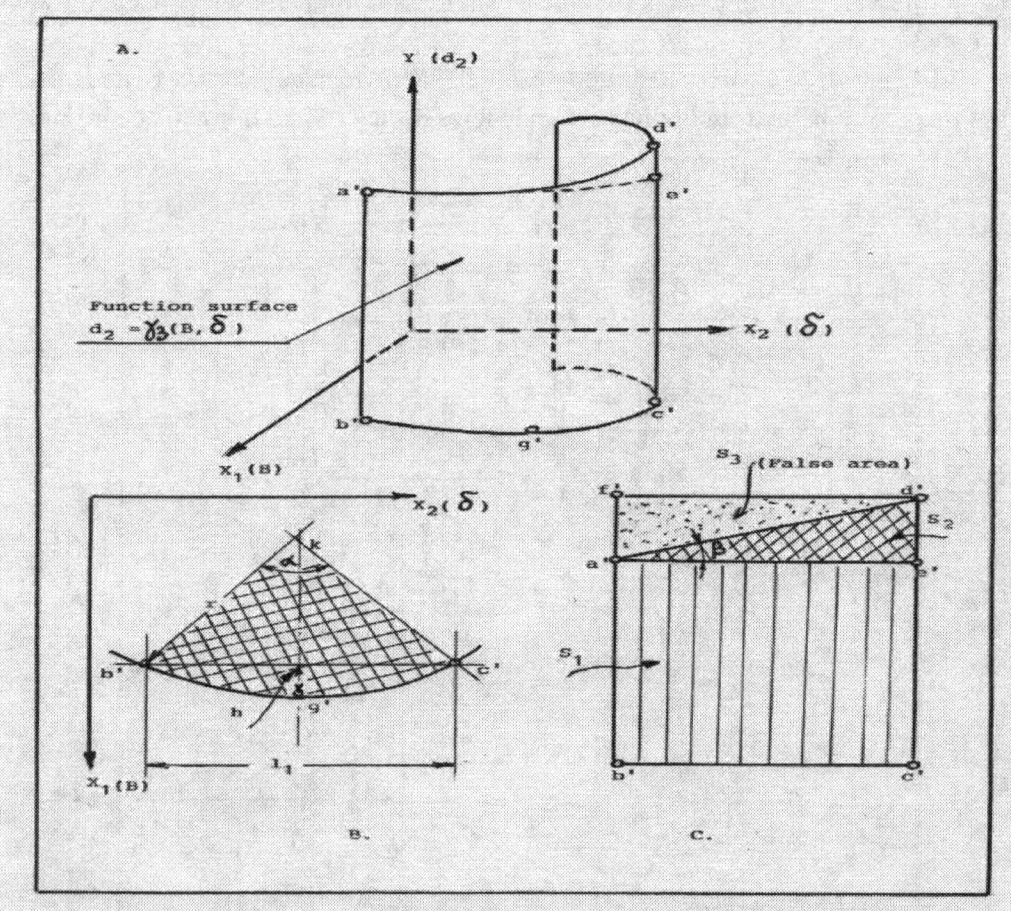

Figure 32 Graph for calculation of module vectors $|b'c'|$, $|a'd'|$ and area of functional surface $d_2=\gamma_3(B,\delta)$: A- total view of functional surface $d_2=\gamma_3(B,\delta)$;B-graph for calculation of module vector $|b'c'|$; C-graph for calculation of module vector $|a'd'|$ and area of this surface.

- The module of vector $|a'b'|$ and $|c'd'|$ can be calculated from the projection of this functional surface $d_2=\gamma_3(B,\delta)$ on the plane YOX_2 (Figure 31A and 32A). So ,the module of vectors $|a'b'|=9.16$ and $|c'd'|=11.11$.
- The module of vector $L=|b'c'|=0.01745r\alpha$, where L= length of arc(b'c'), r=radius of sector,α=angle of sector. And module of vector $r=|kg'|=$ $[(X_{16}-X_{15})^2+(X_{26}-X_{25})^2]^{1/2}=6.13$,where coordinate of points :g' $(X_{16}=6.12, X_{26}=0.37$) and $k(X_{15}=0,X_{25}=0)$. At data r=6.13,α=82° we have $|b'c'|=8.77$ and other parameters of circular segment $h=r[1-\cos(\alpha/2)]=1.5$ and $l_1=2[h(2r--h)]^{1/2}=8.03$. The graph for calculation of module vectors $|b'c'|$, $|a'd'|$ and area of functional surface $d_2=\gamma_3(B,\delta)$ is shown in Figure 32.
- The module of vector $|a'd'|$ and area of this functional surface $d_2=\gamma_3(B,\delta)$,as shown in Figure 32 (c) ,can be calculated by the following way: the module of length for curve $|a'e'|$ $=|b'c'|=8.77$ and module of length for $|d'e'|=|c'd'|-|a'b'|$ $=1.95$.For calculation the module of vector $|a'd'|$ and area S_1,S_2 and S_3 we find angle β which is equal : $\tan\beta=|d'e'|/|a'e'|=0.222$ and then $\beta=12°5'$.The module of length for curve $|a'd'|=|d'e'|/\sin12°5'=9.03$. The coordinates of point f ' is equal $(X_{11}=5.37 ,X_{21}=0.10 ,Y_4=11.11)$ and module of length $|f'd'|=|b'c'|=8.77$, area of $S_2=S_3=S/2 = 0.5 [|f'd'|\cdot|d'e'|]=8.55$ mm^2. For these conditions ,we have the area $S_1=|a'b'|\cdot|b'c'|=80.33$mm^2 and total area $\sum S$ of this functional surface $d_2=\gamma_3(B,\delta)$ is equal $\sum S=S_1+S_2=88.9$ mm^2.

All results of computations for the non-linear equation view of $Y_c=12.466-1.432X_1+0.139X^2_1+3.799X_2$ are given in Table 15.

So ,the parameters of this functional surface $d_2=\gamma_3(B,\delta)$ are the following:

1. Function $d_2=\gamma_3(B,\delta)$ better submits to the non-linear regression model in view of curvilinear surface, as shown in Figure 31 A and 32 A and describes by equation view of $Y_c=12.466-1.432X_1+0.139X^2_1+3.799X_2$ or $d_2=12.466-1.432B+0.139B^2+3.799\delta$ with such statistical characteristics :
   - Coefficient of determination $R^2=0.787$
   - Coefficient of correlation r=0.887
   - Standard deviation $S_{y/x1,x2}=0.382$
   - Minimization of the mean square error (min MSE=0.097)
   - Minimization of the mean absolute deviation (min MAD=0)
2. The total area of this functional surface $d_2=\gamma_3(B,\delta)$ is equal $\sum S=88.9$mm^2 with such coordinates of its points in three- dimensional drawing for this surface:
   a' $(X_{11}=5.37 , X_{21}=0.10 ,Y_1=9.16)$
   b' $(X_{12} =5.37 ,X_{22}=0.10 ,Y_2=0)$
   c' $(X_{13}=5.75 ,X_{23}=0.60 , Y_3=0)$
   d' $(X_{14}=5.75 ,X_{24}=0.60 ,Y_4=11.11)$

Table 15 Evaluation of non-regression equation $Y_c=12.466-1.432X_1+0.139X^2_1+3.8X_2$

A. MEAN, VARIANCE AND STANDARD DEVIATION

Variable	Mean	Variance	Standard deviation
X_1	5.836	0.515	0.718
X_2	0.368	0.209	0.457
Y	10.248	4.121	2.030

B. RESULTS OF MULTIPLE REGRESSION OF Y ON X_1 AND X_2

Parameter	Variable	Coefficients	Standard error	T-value
β_1	X_1	−1.432	0.239	−5.992
β_2	X_2	3.799	0.152	24.993

C. ANALYSIS OF VARIANCE RESULTS

Regression
- Degrees of freedom   2
- Sum of squares   3.241
- Mean square   1.621

Error
- Degrees of freedom   6
- Sum of squares   0.877
- Mean square   0.146
- Standard error of estimate   0.718
- F-value*   11.102   [$F_{0.05,2,6}$=5.14]

Since F^*=11.102 >$F_{0.05,2,6}$ we can reject the hypothesis that both coefficients β_1 and β_2 are zero

D. DETERMINATION OF RESIDUALS

Number	Observed	Estimated	Residual
1	10.03	10.19	−0.16
2	10.36	10.52	−0.16
3	8.69	9.16	−0.47
4	10.13	9.52	0.61
5	10.22	9.90	0.32
6	10.07	10.32	−0.25
7	10.49	10.65	−0.16
8	11.06	10.87	0.19
9	11.18	11.11	0.07

5.14 Dependence of external diameter (d_2) stainless chip from clearance (t) between of chip wraps and its thickness (δ)

Analysis of data and statistical characteristics (coefficient of determination R^2=0.827, coefficient of correlation r=0.909, standard deviation $S_{y/x1,x2}$=0.344) and also the minimization of the mean square error (min MSE=0.079) and minimization of the mean absolute deviation (min MAD=0) show that this functional dependence $d_2=\gamma_4(t,\delta)$ better submits to the linear than the non-linear [13] regression model with equation view of $Y_c=7.593+3.167X_1+4.473X_2$ or $d_2=7.593+3.167t+4.473\delta$ (34)

----

[13]   *The non-linear regression equation has view of $d_2=7.97+1.851t+1.051t^2+4.298\delta$*

In Figure 33 (a) is shown the functional dependence of external diameter (d_2) of stainless chip from clearance (t) between of chip wraps ,i.e we have the function $d_2=\gamma_3(t)$ and as was above-shown this dependence has the non-linear regression model with equation view of $Y_c=1.14X^{-0.56}$ or $t=1.14d_2^{-0.56}$,where $d_2=(0.878t)^{-1.785}$.

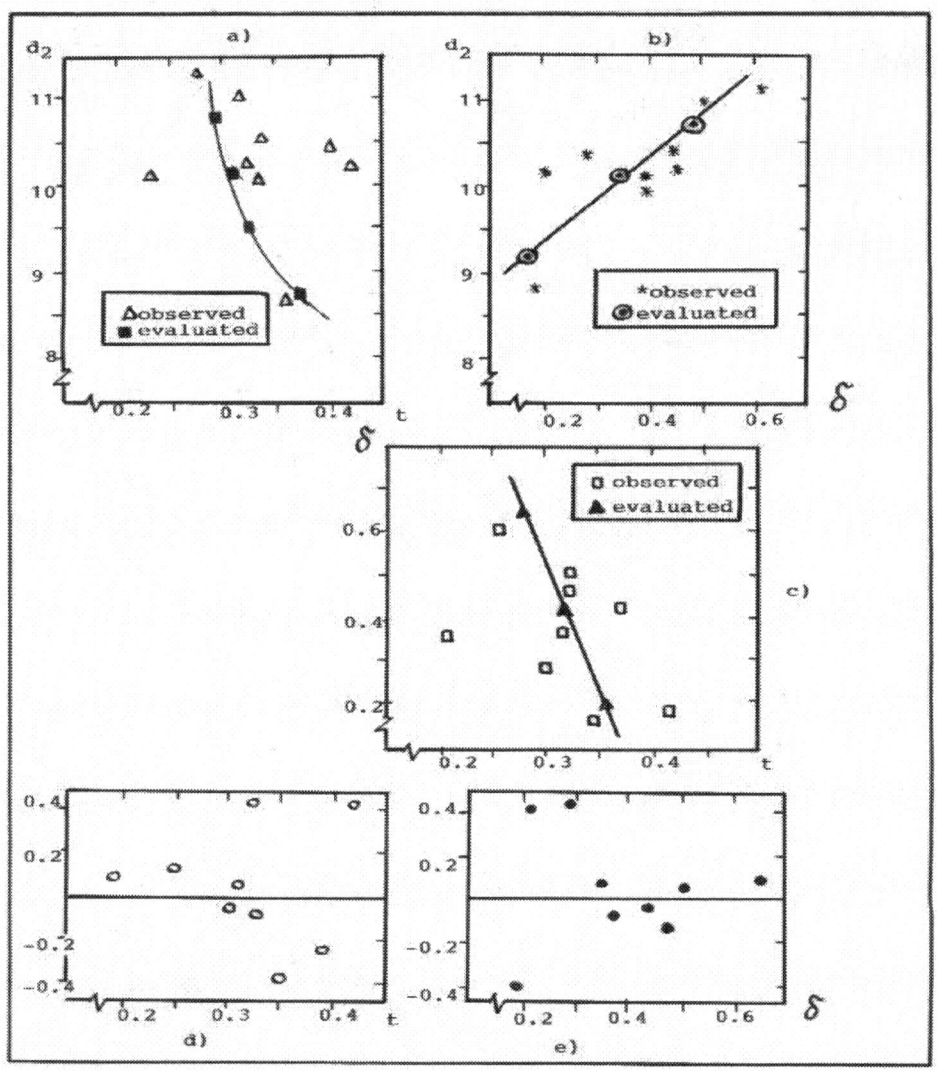

Figure 33

So ,we see from Figure 33 (a) that with increasing of clearance (t) between of chip wraps ,the value of external diameter (d_2) stainless chip decreases considerably in accordance with the regression equation $d_2=(0.878t)^{-1.785}$.

In Figure 33 (b) is shown the functional dependence of external diameter (d_2) from thickness (δ) stainless chip ,i.e we have the function $d_2=\gamma(\delta)$.This dependence has the linear regression model with equation view of $Y_c=8.808+3.897X$ or $d_2=8.808+3.897\delta$. So ,we see from Figure 33 (b) that with increasing of thickness (δ) stainless chip ,the value of external   diameter (d_2) increases in accordance with the regression equation $d_2=8.808+3.897\delta$.

In Figure 33(c) is shown the functional dependency of thickness (δ) stainless chip from clearance (t) between of chip wraps of this chip. And this dependence can be expressed in view of $\delta=\phi(t)$ and has the linear regression model with equation $Y_c=0.379-0.16X$ or $t=0.379-0.160\delta$, where $\delta=2.369-6.25t$. So , we see from Figure 33(c) that with increasing of clearance (t) between of chip wraps for stainless chip ,the value of thickness (δ) of this chip considerably decreases.

In Figure 33(d) are illustrated the residual plots (residual versus t and residual versus δ the above-named functional dependencies of $d_2=\varphi_2(t)$ and $d_2=\gamma(\delta)$ accordingly.

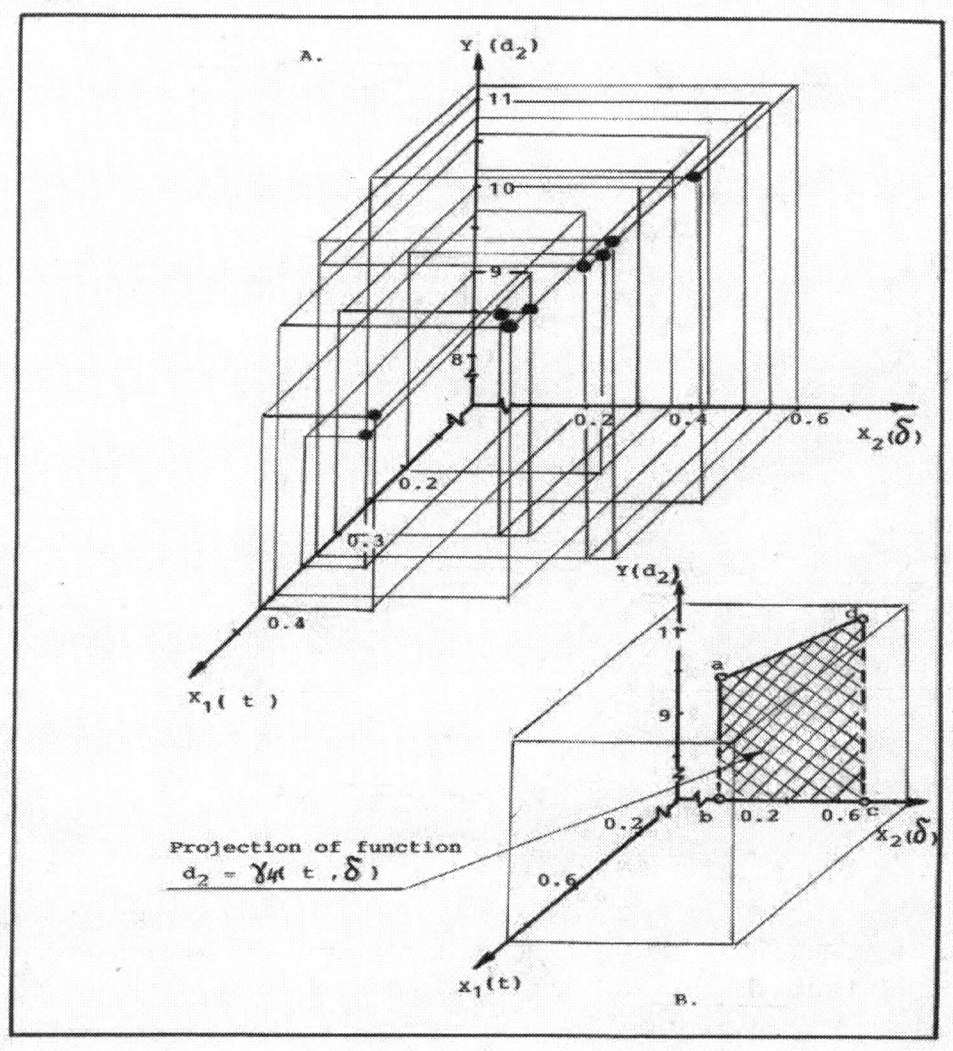

Figure 34

Analyzing the Figure 34(A) we see that functional surface $d_2=\gamma_4(t,\delta)$ is shown in view of three- dimensional drawing for the linear regression model with equation view of $Y_c=7.593+3.167X_1+4.473X_2$ or $d_2=7.593+3.167t+4.473\delta$.

And Figure 34(B) shows that this functional surface $d_2=\gamma_4(t,\delta)$ has view of trapezium on the plane YOX_2 with such coordinate of their points (sizes in mm): a $(X_{11}=0, X_{21}=0.18, Y_1=9.73)$; b$(X_{12}=0 \quad, X_{22}=0.18, Y_2=0)$, c$(X_{13}=0, X_{23}=0.60 \quad, Y_3=0)$; d$(X_{14}=0, \quad X_{24}=0.60, Y_4=11.07)$.

Analysis of functional surface $d_2=\gamma(t,\delta)$

From Figure 34(A) we see that functional surface $d_2=\gamma_4(t,\delta)$,on which are situated the points of the linear regression equation view of $Y_c=7.593+3.167X_1+4.473X_2$ or $d_2=7.593+3.167t+4.473\delta$, has the following coordinate on their peaks in three-dimensional drawing (sizes in mm): a'$(X_{11}=0.42, X_{21}=0.18, Y_1=9.73)$; b'$(X_{12}=0.42, X_{22}=0.18, Y_2=0)$; c'$(X_{13}=0.25, X_{23}=0.60, Y_3=0)$; d' $(X_{14}=0.75, X_{24}=0.60, Y_4=11.07)$. The value of modules and area of functional surface $d_2=\gamma_4(t,\delta)$ are equal : $|a'b'|=9.73$, $|b'c'|=0.54$, $|c'd'|=11.07$, $|d'a'|=1.44$ and area $S_{a'b'c'd'}=1/2 \ [(|c'd'|-|a'b'|)\cdot |b'c'|] + |a'b'| \cdot |b'c'| = 5.61$ mm^2. All results of computation for the linear equation view of $Y_c=7.593+3.167X_1+4.473X_2$ are given in Table 16.

Table 16 Evaluation of linear regression equation $Y_c=7.593+3.167X_1+4.473X_2$

A. MEAN , VARIANCE AND STANDARD DEVIATION			
Variable	Mean	Variance	Standard deviation
X_1	0.31	0.031	0.176
X_2	0.368	0.209	0.457
Y	10.248	4.121	2.030

B. RESULTS OF MULTIPLE REGRESSION OF Y ON X_1 AND X_2				
Parameter	Variable	Coefficients	Standard error	T-value
β_1	X_1	3.167	0.059	53.678
β_2	X_2	4.473	0.152	29.428

C. ANALYSIS OF VARIANCE RESULTS

Regression
  Degrees of freedom 2
  Sum of squares   3.409
  Mean squares   1.705
Error
  Degrees of freedom   6
  Sum of squares   0.712
  Mean square   0.119
Standard error of estimate   0.176
F-value* 14.327 $[F_{0.05,2,6}]=5.14$

Since $F^*=14.327>5.14$,we can reject the hypothesis that both β_1 and β_2 are zero.

D. DETERMINATION OF RESIDUALS			
Number	Observed	Estimated	Residual
1	10.03	9.87	0.16
2	10.36	10.67	−0.31
3	8.69	9.12	−0.43
4	10.13	9.73	0.40
5	10.22	9.80	0.42
6	10.07	10.23	−0.16
7	10.49	10.66	−0.17
8	11.06	11.02	0.04
9	11.18	11.07	0.11

So ,the parameters of this functional surface $d_2=\gamma_4(t,\delta)$ are the following:

1.Function $d_2=\gamma_4(t,\delta)$ better submits to the linear regression model with equation view of $Y_c=7.593+3.167X_1+4.473X_2$ or $d_2=7.593+3.167t+4.473\delta$ with such statistical characteristics : (coefficient of determination $R^2=0.827$;coefficient of correlation $r=0.909$;standard deviation $S_{y/x1,x2}=0.344$;minimization of the mean square error (min MSE=0.079) and minimization of the mean absolute deviation (min MAD=0);

2. The total area of functional surface $d_2=\gamma_4(t,\delta)$ is equal $\sum S=5.61$ mm^2 with such coordinate of their points in three- dimensional drawing for this surface (sizes in mm):

a' ($X_{11}=0.42$,$X_{21}=0.18$,$Y_1=9.73$);   b' ($X_{12}=0.42$,$X_{22}=0.18$,$Y_2=0$)

c' ($X_{13}=0.75$,$X_{23}=0.60$,$Y_3=0$);   d' ($X_{14}=0.75$,$X_{24}=0.60$,$Y_4=11.07$)

5.15 Dependence of external diameter (d_2) stainless chip from width (B) and number (ω) of chip wraps

Analysis of data and statistical characteristics (coefficient of determination $R^2=0.20$, coefficient of correlation $r=0.448$,standard deviation $S_{y/x1,x2}=0.741$) and also the minimization of the mean square error (min MSE=0.366) and minimization of the mean absolute deviation (min MAD=0) show that this functional dependence $d_2=\gamma_5(B,\omega)$ better submits to the linear than non-linear [14] regression model with equation view of $Y_c=7.245+0.767X_1-0.414X_2$ or $d_2=7.245+0.767B-0.414\omega$ (35) .

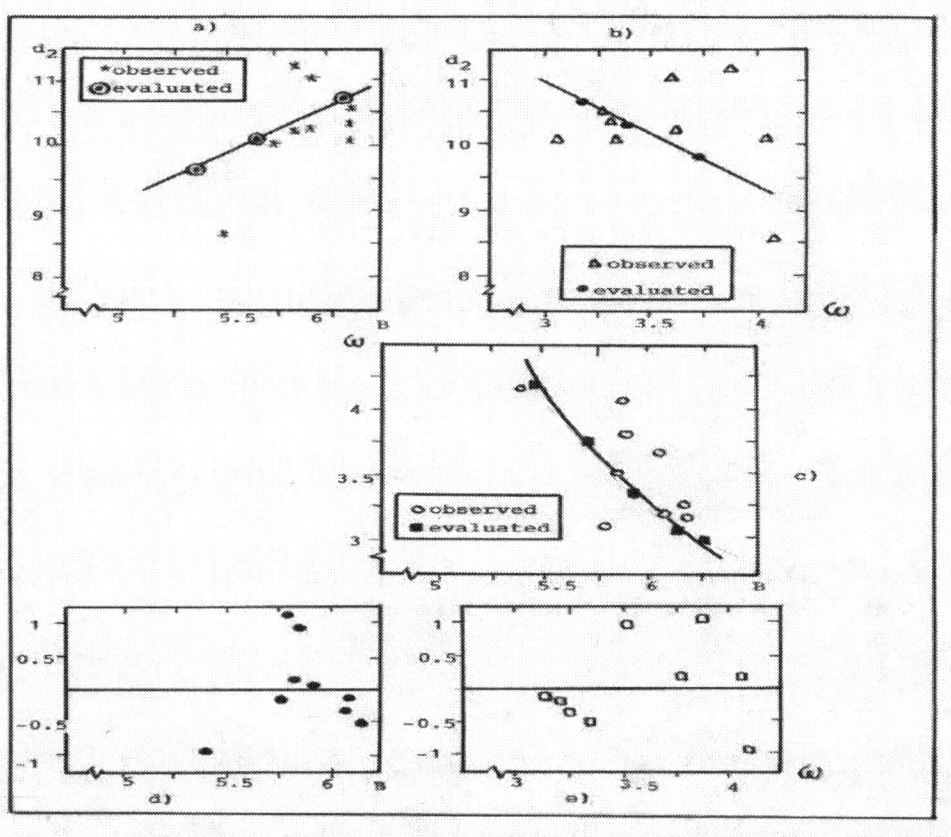

Figure 35

---

14 The non-linear regression equation has view of $d_2=12.726-0.668B+0.101B^2-0.569\omega$

120

In Figure 35(a) is shown the functional dependence of external diameter (d_2) stainless chip from its width (B) ,i.e we have function $d_2=\varphi_2(B)$ which has the linear regression model with equation view of $Y_c=3.026+1.237X$ or $d_2=3.026+1.237B$. So, we see from Figure 35(a) that with increasing of width (B) stainless chip, the value of external diameter (d_2) increases in accordance with regression linear equation $d_2=3.026+1.237B$.

In Figure 35(b) is shown the functional dependence of external diameter (d_2) from number of chip wraps (ω) for stainless chip, i.e we have the function $d_2=\phi_1(\omega)$ which has the linear regression model with equation view of $Y_c=13.137-0.811X$ or $d_2=13.137-0.811\omega$. So ,we see from Figure 35(b) that with increasing of number of chip wraps (ω) ,the value of external diameter (d_2) decreases considerably with regression equation $d_2=13.137-0.811\omega$ .In Figure 35(c) is shown the functional dependence of number chip wraps (ω) stainless chip from the width(B) of this chip, i.e we have the function $\omega=\gamma_2(B)$ which can be expressed in view of the non-linear regression equation view of $\omega=703.1B^{-3.0}$. And From Figure 35 (c) we see that with increasing of width (B) stainless chip, the value of number chip wraps (ω) decreases considerably with the non-linear regression equation $\omega=703.1B^{-3.0}$. In Figure 35 (d) and 35(e) are illustrated residual plots (residual versus B and residual versus ω) the above-named functional dependencies $d_2=\varphi_2(B)$ and $d_2=\phi_1(\omega)$ accordingly.

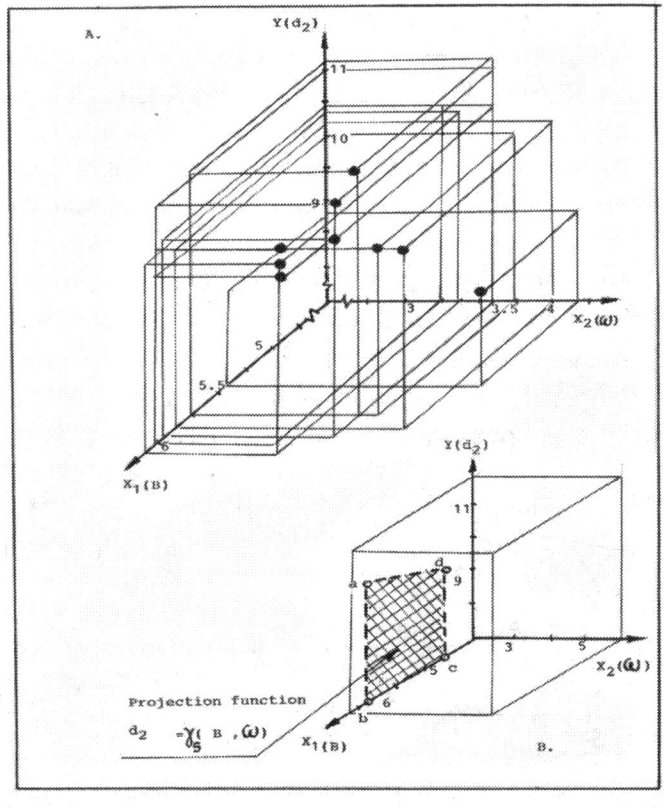

Figure 36

121

Analyzing the Figure 36(A) we see that functional dependence $d_2=\gamma_5(B,\omega)$ is shown in view of three-dimensional drawing for the linear regression model with equation view of $Y_c=7.245+0.767X_1-0.414X_2$ or $d_2=7.245+0.767B-0.414\omega$.

And Figure 36(B) shows that this functional surface $d_2=\gamma_5(B,\omega)$ has view of trapezium on the plane YOX_1 with such coordinate of their points (sizes in mm):a $(X_{11}=6.12,X_{21}=0$,$Y_1=10.59)$;b$(X_{12}=6.12,X_{22}=0,\ Y_2=0)$; c $(X_{13}=5.37,X_{23}=0,Y_3=0)$; d $(X_{14}=5.37,\ X_{24}=0,$ $Y_4=9.62)$.

Analysis of functional surface $d_2=\gamma_5(B,\omega)$

From Figure 36(A) we see that functional surface $d_2=\gamma_5(B,\omega)$, on which are situated the points of the linear regression equation view of $Y_c=7.245+0.767X_1-0.414X_2$ or $d_2=7.245+0.767B-0.414\omega$,has the following coordinate of their peaks in three-dimensional drawing (sizes in mm): a'$(X_{11}=6.12,X_{21}=3.25,Y_1=10.59)$; b'$(X_{12}=6.12,$,$X_{22}=3.25,Y_2=0)$;c'$(X_{13}=5.37,X_{23}=4.22,Y_3=0)$;d'$(X_{14}=5.37,X_{24}=4.22,Y_4=9.62)$.

The module of vectors and area of functional surface $d_2=\gamma_5(B,\omega)$ are equal: $|a'b'|=10.59, |b'c'|=1.23, |c'd'|=9.62, |d'a'|$ and area $S_{a'b'c'd'}=1/2\cdot[(|a'b'|-$ $-|c'd'|)\cdot|b'c'|]+(|b'c'|\cdot|c'd'|=12.43\text{mm}^2$.

All results of computations for the linear regression equation view of $Y_c=7.245+0.767X_1-0.414X_2$ are given in Table 17.

So ,the parameters of this functional surface $d_2=\gamma_5(B,\omega)$ are the following:

1.Function $d_{2=}\gamma_5(B,\omega)$ better submits to the linear regression model with equation view of $Y_c=7.245+0.767X_1-0.414X_2$ or $d_2=7.245+0.767B-0.414\omega$ with   such   statistical characteristics:

- Coefficient of determination $R^2=0.20$
- Coefficient of correlation $r=0.448$
- Standard deviation $S_{y/x1,x2}=0.741$
- Minimization of the mean square error (min MSE=0.366)
- Minimization of the mean absolute deviation (min MAD =0)

2.The total area of functional surface $d_2=\gamma_5(B,\omega)$ is equal $\sum S=12.43\text{mm}^2$ with such coordinate of their points in three- dimensional drawing for this surface (sizes in mm):

a' ($X_{11}=6.12,X_{21}=3.25$,$Y_1=10.59)$
b' ($X_{12}=6.12$,$X_{22}=3.25$,$Y_2=0)$
c' ($X_{13}=5.37$,$X_{23}=4.22$, $Y_3=0)$
d' ($X_{14}=5.37$,$X_{24}=4.22$, $Y_4=9.62)$

Table 17 Evaluation of linear regression equation $Y_c=7.245+0.767X_1-0.414X_2$

A. MEAN, VARIANCE AND STANDARD DEVIATION

Variable	Mean	Variance	Standard deviation
X_1	5.836	0.515	0.718
X_2	3.559	1.113	1.055
Y	10.248	4.121	2.030

B. RESULTS OF MULTIPLE REGRESSION OF Y ON X_1 AND X_2

Parameter	Variable	Coefficient	Standard error	T-value
β_1	X_1	0.767	0.239	3.209
β_2	X_2	−0.414	0.352	−1.176

C. ANALYSIS OF VARIANCE RESULTS

Regression
  Degrees of freedom 2
  Sum of squares   0.827
  Mean squares   0.414
Error
  Degrees of freedom 6
  Sum of squares   3.294
  Mean square   0.549
Standard error of estimate   0.718
F-value[*]   0.754 [$F_{0/05,2,6}=5.14$]

Since $F^*=0.754<5.14$ we can not reject the hypothesis that both β_1 and β_2 are zero.

D. DETERMINATION OF RESIDUALS

Number	Observed	Estimated	Residual
1	10.03	10.31	−0.28
2	10.36	10.61	−0.25
3	8.69	9.62	−0.93
4	10.13	10.03	0.10
5	10.22	10.21	0.01
6	10.07	10.59	−0.52
7	10.49	10.59	−0.10
8	11.06	10.16	0.90
9	11.18	10.09	1.09

5.16 Dependence of external diameter (d_2) of stainless chip from clearance (t) between of chip wraps and number(ω) of chip wraps

Analysis of data and statistical characteristics (coefficient of determination $R^2=0.205$, coefficient of correlation r= 0.453, standard deviation $S_{y/x1,x2}=0.739$) and also the minimization of the mean square error(min MSE=0.364) and minimization of the mean absolute deviation (min MAD=0)show that this functional dependence $d_2=\gamma_6(t,\omega)$ better submits to the linear than non-linear [15] regression model with equation view of $Y_c=13.234-0.685X_1-0.778X_2$ or $d_2=13.234-0.685t-0.778\omega$ (36) .

---

15 *The non-linear regression equation has view of $d_2=13.467-2.464t+2.766\ t^2-0.766\ \omega$*

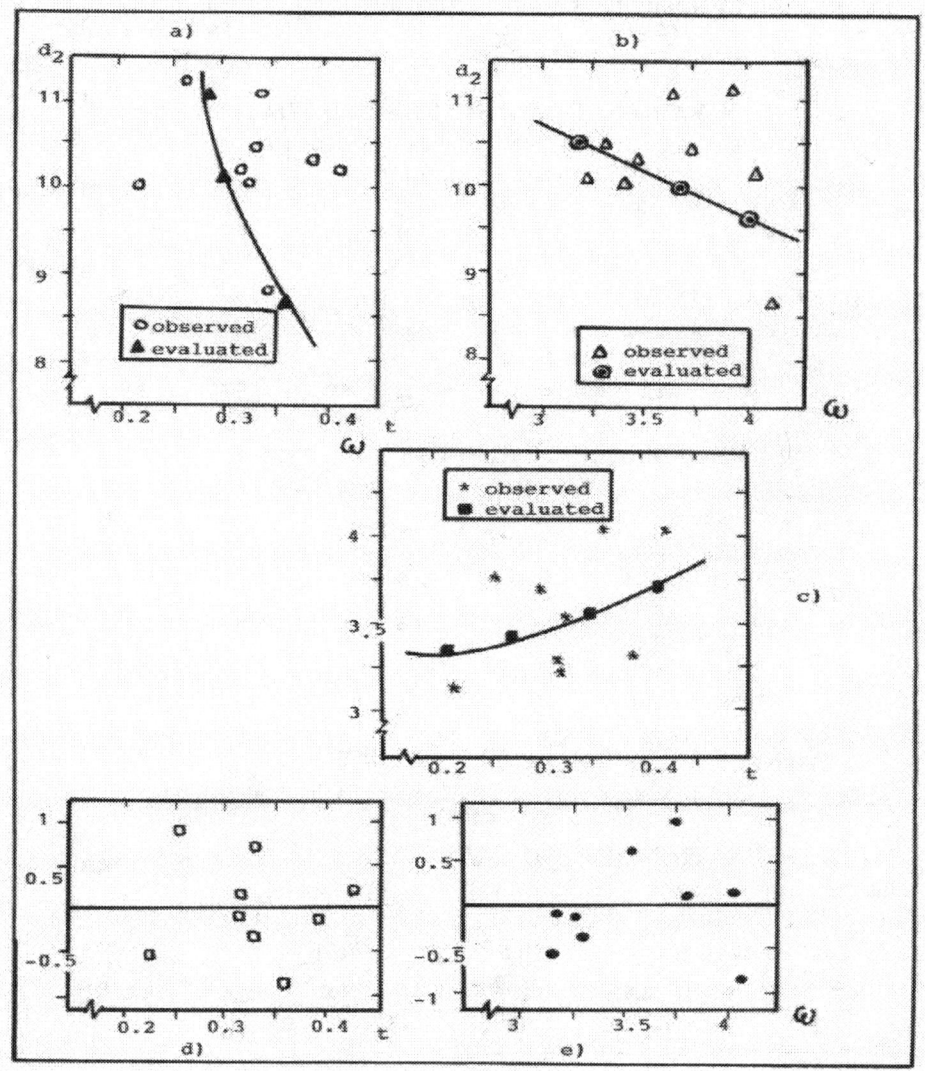

Figure 37

In Figure 37(a) is shown the functional dependency of external diameter (d_2) stainless chip from clearance (t) between of chip wraps, i.e we have the function $d_2=\varphi_2(t)$ which has the non-linear regression model with equation view of $Y_c=1.14X^{-0.56}$ or $t=1.14d_2^{-0.56}$,where $\mathbf{d_2=(0.878t)^{-1.786}}$.So ,we see from Figure 37(a) that with increasing of clearance (t) between of chip wraps ,the value of external diameter (d_2) stainless chip decreases considerably with regression equation $d_2=(0.878t)^{-1.786}$.

In Figure 37(b) is shown the functional dependence of external diameter (d_2) from number (ω) of chip wraps for stainless chip ,i.e we have the function $d_2=\varphi_3(\omega)$ which has the linear regression model with equation view of $Y_c=13.137-0.811X$ or $d_2=13.137-0.811\omega$. So ,we see from Figure 37(b) that with increasing of number of chip wraps (ω)for stainless chip ,the value of external diameter (d_2) decreases in accordance with linear regression equation $d_2=13.137-0.811\omega$.

In Figure 37(c) is shown the functional dependence of number chip wraps (ω) from clearance (t) between of chip wraps for stainless chip. This dependency can be expressed in view of function view of $\omega=\varphi_6(t)$ and has the non-linear regression model with equation view of $Y_c=4.227X$ or $\omega=4.227t$. From Figure 37(c) we see that with increasing of clearance (t) between of chip wraps ,the value of number chip wraps (ω) increases considerably in accordance with regression equation $\omega=4.227t$.

In Figure 37(d) and 37(e) are illustrated the residual plots (residual versus t and residual versus ω) of above-named functional dependencies $d_2=\varphi_2(t)$ and $d_2=\varphi_3(\omega)$ accordingly.

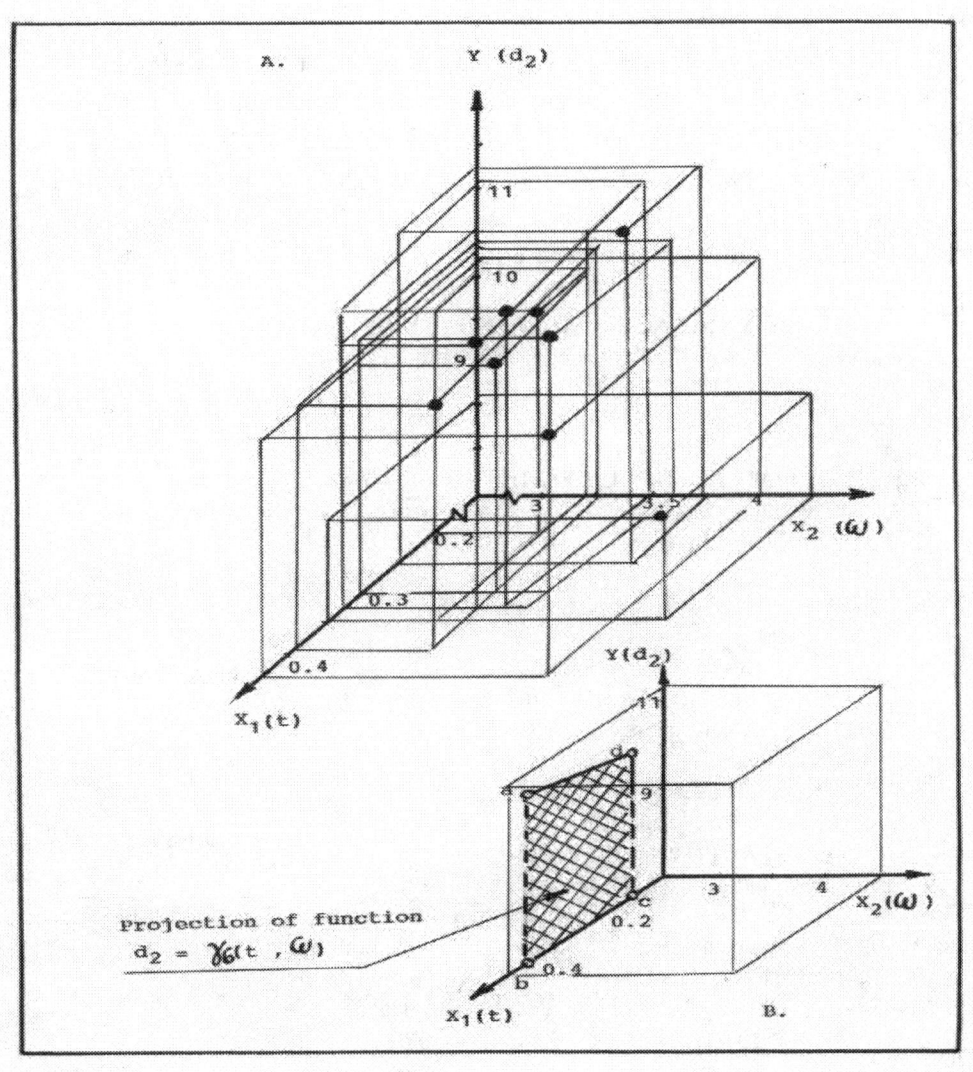

Figure 38

Analyzing the Figure 38(A) ,we see that functional surface $d_2=\gamma_6(t,\omega)$ is shown in view of three-dimensional drawing for the linear regression model with equation view of $Y_c=13.324-0.685X_1-0.778X_2$ or $d_2=13.324-0.685t-0.778\omega$.

And Figure 38(B) shows that this functional surface $d_2=\gamma_6(t,\omega)$ has view of trapezium on the plane YOX_1 with such coordinate of their points (sizes in mm): $a(X_{11}=0.38\ ,X_{21}=0,$ $,Y_1=10.42)$; $b(X_{12}=0.38,X_{22}=0,Y_2=0)$; $c(X_{13}=0.25,X_{23}=0,Y_3=0)$; $d(X_{14}=0.25,X_{24}=0,$ $,Y_4=10.19)$.

Analysis of functional surface $d_2=\gamma_6(t,\omega)$

From Figure 38(A) we see that functional surface $d_2=\gamma_6(t,\omega)$,on which are situated the points of the linear regression equation view of $Y_c=13.234-0.685X_1-0.778X_2$ or $d_2=13.234-0.685t-0.778\omega$, has the following coordinate of their peaks in three-dimensional drawing : a' $(X_{11}=0.38,X_{21}=3.28\ ,Y_1=10.42)$;b'$(X_{12}=0.38,X_{22}=3.28,Y_2=0)$; c'$(X_{13}=0.25,X_{23}=3.76,Y_3=0)$;d'$(X_{14}=0.25\ ,X_{24}=3.76,Y_4=10.19)$. The module of vectors and area of functional surface $d_2=\gamma_6(t,\omega)$are equal: $|$a'b'$|=10.42$, $|$b'c'$|=0.50$, $|$c'd'$|=10.19$, $|$d'a'$|=0.6$ and area of this functional surface $S_{a'b'c'd'}=$ $=1/2[(|$a'b'$|-|$c'd'$|)\cdot|$b'c'$|]+$ $|$b'c'$|\cdot|$c'd'$|=5.16$ mm^2. All results of computation for the linear equation view of $Y_c=13.234-0.685X_1-0.778X_2$ are given in Table 18.

Table 18 Evaluation of linear regression equation $Y_c=13.234-0.685X_1-0.778X_2$

A. MEAN ,VARIANCE AND STANDARD DEVIATION			
Variable	Mean	Variance	Standard deviation
X_1	0.317	0.031	0.176
X_2	3.559	1.113	1.055
Y	10.248	4.121	2.030

B.RESULTS OF MULTIPLE REGRESSION OF Y ON X_1 AND X_2				
Parameter	Variable	Coefficients	Standard error	T-value
β_1	X_1	−0.685	0.059	−11.61
β_2	X_2	−0.778	0.352	−2.21

C. ANALYSIS OF VARIANCE RESULTS

Regression

Degrees of freedom	2
Sum of squares	0.742
Mean squares	0.371

Error

Degrees of freedom	6	
Sum of squares	3.379	
Mean square	0.563	
Standard error estimate	0.176	
F-value*	0.658	[$F_{0.05,2,6}=5.14$]

Since $F^*=0.658<5.14$ we can not reject the hypothesis that both $\beta_1$and β_2 are zero.

D. DETERMINATION OF RESIDUALS

Number	Observed	Estimated	Residual
1	10.03	10.64	−0.61
2	10.36	10.42	−0.06
3	8.69	9.72	−1.03
4	10.13	9.86	0.27
5	10.22	10.20	0.02
6	10.07	10.48	−0.41
7	10.49	10.50	−0.01
8	11.06	10.26	0.80
9	11.18	10.15	1.03

1.Function $d_2=\gamma_6$ (t,ω) better submits to the linear regression model with equation view of $Y_c=13.234-0.685X_1-0.778X_2$ or $d_2=13.234-0.685t-0.778\omega$ with such statistical characteristics: coefficient of determination $R^2=0.21$,coefficient of correlation r=0.453, standard deviation $S_{y/x1,x2}=0.739$,minimization of the mean square error (min MSE=0.364) and minimization of the mean absolute (min MAD=0);

2.The total area of functional surface $d_2=\gamma_6(t,\omega)$ is equal $\sum S=5.16$ mm^2 with such coordinate of their points in three- dimensional drawing for this surface: a'($X_{11}=0.38$,$X_{21}=3.28$,$Y_1=10.42$); b'($X_{12}=0.38$,$X_{22}=3.28$,$Y_2=0$);c'($X_{13}=0.25$,$X_{23}=3.76$, ,$Y_3=0$); d'($X_{14}=0.25$,$X_{24}=3.76$,$Y_4=10.19$).

In Table 19 is given the *Summary Multiple Analysis* between two independent variable parameters in function of $Y_i=\gamma(X_{i,1};X_{i,2})$ for external diameter(d_2) of stainless chip.

Table 19 Summary Multiple Analysis

B.FOR EXTERNAL DIAMETER OF STAINLESS CHIP

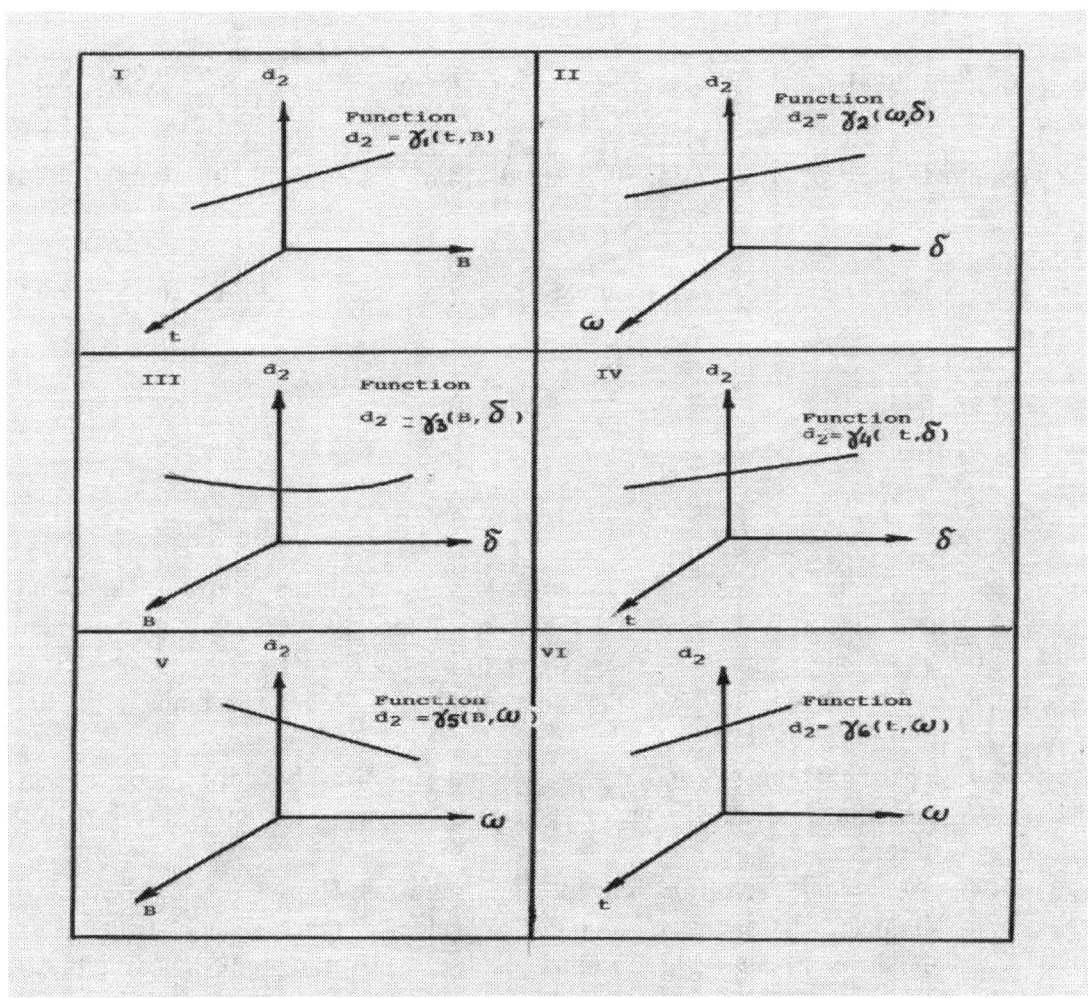

In Table 20 is shown the summary statistical characteristics and empirical equations for function $Y_i=\gamma(X_{i,1},X_{i,2})$.

Table 20 Summary statistical characteristics and empirical equations for function $Y_c=\gamma(X_{i,1},X_{i,2})$

B. For external diameter (d_2) of stainless chip

View of function	Regression model	Empirical equation	Statistical characteristics		
			R^2	r	$S_{y/x1,x2}$
$d_2=\gamma_1(t,B)$	Linear	$d_2=3.761-3.452t+1.299B$	0.27	0.52	0.710
$d_2=\gamma_2(\omega,\delta)$	Linear	$d_2=7.985+0.207\omega+4.149\delta$	0.801	0.895	0.369
$d_2=\gamma_3(B,\delta)$	Non-linear	$D_2=12.466-1.432B+0.139B^2+3.799\delta$	0.787	0.887	0.382
$d_2=\gamma_4(t,\delta)$	Linear	$d_2=7.593+3.167t+4.473\delta$	0.827	0.909	0.344
$d_2=\gamma_5(B,\omega)$	Linear	$D_2=7.245+0.767B-0.414\omega$	0.20	0.448	0.741
$d_2=\gamma_6(t,\omega)$	Linear	$d_2=13.234-0.685t-0.778\omega$	0.21	0.453	0.739

SUMMARY

Analysis of correlation between three parameters of stainless chip in function of $Y_i=\gamma(X_{i,1},X_{i,2})$advantageously for external diameter of this stainless chip in dependence from the different of their parameters showed the following results:

- All functions, besides of the function $d_2=\gamma_3(B,\delta)$, better submits to the linear regression models;
- The function $d_2=\gamma_2(\omega,\delta),d_2=\gamma_3(B,\delta)$ and $d_2=\gamma_4(t,\delta)$ have the good correlation and other statistical characteristics between the parameters of this stainless chip;
- Analyzing each function for this stainless chip, we see that external diameter (d_2) of this chip could be evaluated approximately with using of the different empirical formulas which are shown in above-known tables;

- And besides we can conclude that external diameter (d_2) of stainless chip is the function such independent variable as : clearance (t) between of chip wraps, width(B),thickness (δ)and number of chip wraps (ω) ,i.e we have the function $d_2=\rho(t,B,\delta,\omega)$;
- The most influence on increasing of size for external diameter (d_2) of stainless chip plays such parameters as width (B), thickness (δ). And as we see that this parameter (d_2) increases considerably with increasing of above-named parameters (δ,B);
- The most influence on decreasing of size for external diameter (d_2) of stainless chip plays such parameters as clearance (t) between of chip wraps and number of chip wraps (ω). And as we this parameter (d_2) decreases considerably with increasing of above-named parameters (t,ω).

C. FOR OTHER PARAMETERS OF STAINLESS CHIP

5.17 Dependence of clearance (t) between of chip wraps from width (B) and thickness (δ)of this stainless chip

Analysis of data and statistical characteristics (coefficient of determination $R^2=0.452$,coefficient of correlation r=0.672,standard deviation $S_{y/x1,x2}=0.53$) and also the minimization of the mean square error (min MSE=0.002) and minimization of the mean absolute deviation (min MAD=0) show that this functional dependence $t=\alpha_1(B,\delta)$ better submits to the linear than non-linear [16] regression model with equation view of $Y_c=0.021+0.064X_1-0.212X_2$ or $t=0.021+0.064B-0.212\delta$ (37) .

In Figure 39(a) is shown the functional dependence of clearance (t) between of chip wraps from its width (B) ,i.e we have the function $t=\gamma_1(B)$ and this dependence has the non-linear regression model with equation view of $Y_c=0.032X^{1.296}$ or $t=0.032B^{1.296}$. So , we see from Figure 39 (a) that with increasing of width (B) stainless chip ,the value of clearance (t) between of chip wraps increases considerably with regression equation $t=0.032B^{1.296}$.

In Figure 39(b) is shown the functional dependence   of clearance (t) between of chip wraps stainless chip from its thickness ,i.e we have the function view of $t=\phi(\delta)$. This dependence has the linear regression model with equation view of $Y_c=0.379-0.160X$ or $t=0.379-0.160\delta$. So ,we see from Figure 39(b) that with increasing of thickness (δ) stainless chip ,the value of clearance (t) between of chip wraps decreases considerably in accordance with regression equation $t=0.0379-0.160\delta$.

In Figure 39 (c) is shown the functional dependence of thickness(δ) stainless chip from its width(B) and this dependence can be expressed in view of $\delta= f$ (B) and has the linear regression model with equation view of $Y_c= -1.236+0.275$ or $\delta= -1.236+0.275B$.So ,we see from 39(c) that with increasing of width (B) stainless chip ,the value of thickness (δ) for this chip increases considerably with regression equation $\delta= -1.236+0.275B$.

----

16 The non-linear regression model has equation view of $t= -2.297+0.825B-0.062B^2-0.234\delta$

129

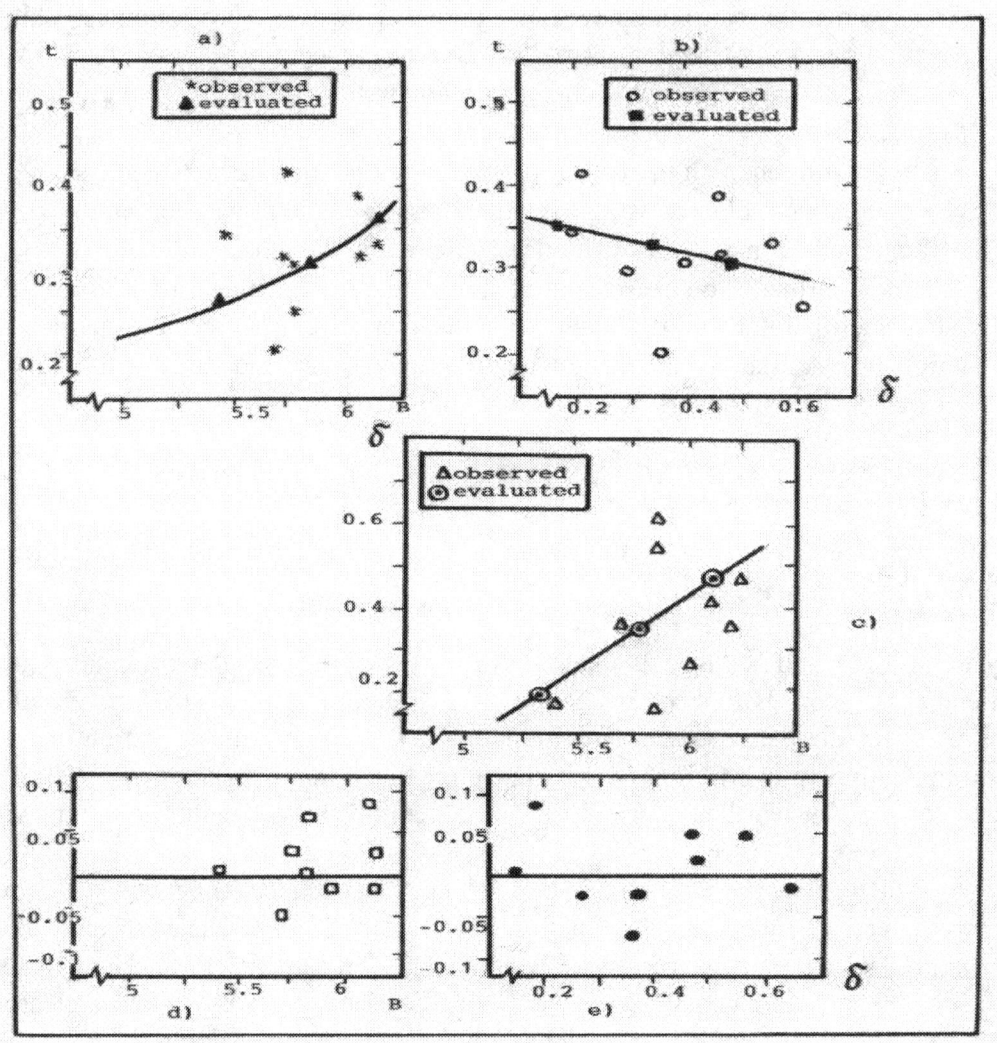

Figure 39

In Figure 39(d) and 39(e) are illustrated residual plots (residual versus B and residual versus δ) of the above-named functional dependencies $t=\gamma_1(B)$ and $t=\phi(\delta)$ accordingly.

Analyzing the Figure 40 (A) ,we see that functional surface $t=\alpha_1(B,\delta)$ is shown in view of three-dimensional drawing for the linear regression model with equation view of $Y_c=0.021+0.064X -0.212X_2$ where $t=0.021+0.064B -0.212\delta$.

And Figure 40(B) shows that this functional surface $t=\alpha_1(B,\delta)$ has view of trapezium on the plane YOX_2 with such coordinate of their points (sizes in mm): a $(X_{11}=0,X_{21}=0.10,Y_1=0.34)$; b $(X_{12}=0,X_{22}=0.10 , Y_2=0)$; c $(X_{13}=0,X_{23}=0.46,Y_3=0)$; d $(X_{14}=0 ,X_{24}=0.24 ,Y_4=0.31)$.

From Figure 40(A) we see also that functional surface $t=\alpha_1(B,\delta)$,on which are situated the points of the linear regression equation view of $Y_c=0.021 +0.064X_1-0.212X_2$ or $t=0.021+0.064B-0.212\delta$ has the following coordinate of their peaks in three-dimensional drawing (sizes in mm) :

a' (X_{11}=5.37 ,X_{21}=0.10,Y_1=0.34) ;b'(X_{12}=5.37 ,X_{22}=0.10,Y_2=0); c' (X_{13}=6.10,X_{23}=0.46 ,Y_3=0) ; d' (X_{14}=6.10 ,X_{24}=0.46 ,Y_4=0.31).

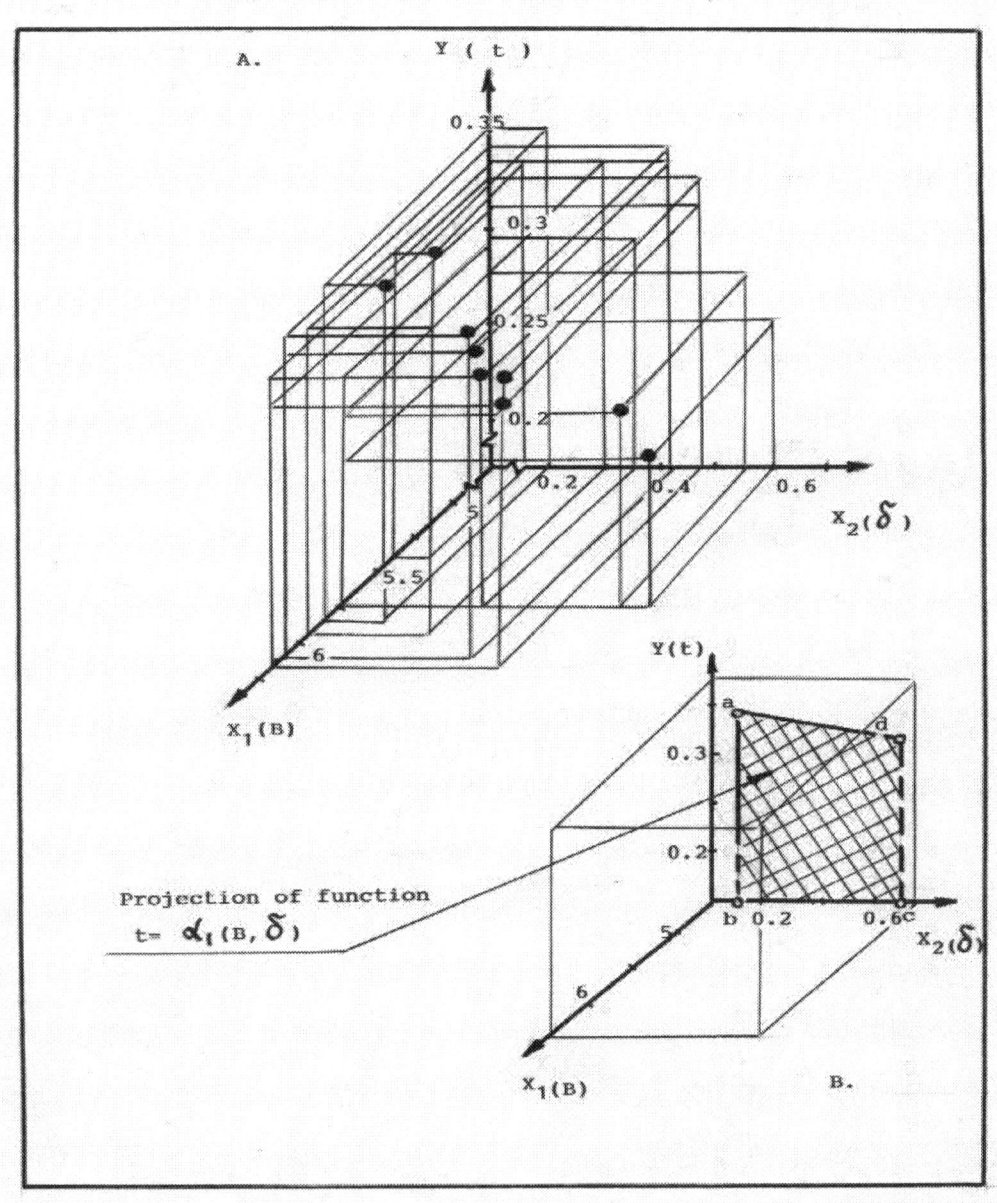

Figure 40

The value of modules and area of functional surface t=α_1(B,δ) are equal : $\left| a'b' \right|$=0.34 , $\left| b'c' \right|$=0.81 , $\left| c'd' \right|$=0.31 , $\left| d'a' \right|$=0.82 and area $S_{a'b'c'd'}$=1/2· ·[($\left| a'b' \right|$ − $\left| c'd' \right|$)· $\left| b'c' \right|$] + $\left| b'c' \right|$ · $\left| c'd' \right|$ =0.26 mm^2 .

All results of computation for the linear regression equation view of Y_c=0.021+0.064X_1−0.212X_2 are given in Table 21.

Table 21 Evaluation of linear regression equation $Y_c=0.021+0.064X_1-0.212X_2$

A. MEAN , VARIANCE AND STANDARD DEVIATION

Variable	Mean	Variance	Standard deviation
X_1	5.836	0.514	0.717
X_2	0.368	0.209	0.457
Y	0.317	0.031	0.

B . RESULTS OF MULTIPLE REGRESSION OF Y ON X_1 AND X_2

Parameter	Variable	Coefficients	Standard error	T-value
β_1	X_1	0.064	0.239	0.268
β_2	X_2	−0.212	0.152	−1.395

C.ANALYSIS OF VARIANCE RESULTS

Regression

Degrees of freedom	2
Sum of squares	0.014
Mean squares	0.007

Error

Degrees of freedom	6
Sum of squares	0.017
Mean square	0.003

Standard error of estimate 0.717

F-value* 2.333   [$F_{0.05,2,6}$=5.14]

Since F^*=2.333< 5.14 we can not reject the hypothesis that both β_1 and β_2 are zero

D. DETERMINATION OF RESIDUALS

Number	Observed	Estimated	Residual
1	0.21	0.29	−0.08
2	0.38	0.33	0.05
3	0.34	0.34	0
4	0.42	0.35	0.07
5	0.30	0.33	−0.03
6	0.31	0.34	−0.03
7	0.32	0.31	0.01
8	0.32	0.27	0.05
9	0.25	0.26	−0.01

So ,the parameters of this functional surface $t=\alpha_1(B,\delta)$ are the following:

1.Function $t=\alpha_1(B,\delta)$ better submits to the linear regression model with equation view of $Y_c=0.021+0.064X_1-0.212X_2$ or $t=0.021+0.064B-0.212\delta$ with such statistical characteristics: coefficient of determination R^2=0.452,coefficient of correlation r=0.672,standard deviation $S_{y/x1,x2}$=0.053,minimization of the mean square error (min MSE=0.002)and minimization of the mean absolute (min MAD =0);

2.The total area of functional surface $t=\alpha_1(B,\delta)$ is equal $\sum S$=0.26mm^2 with such coordinate of their points in three-dimensional drawing (sizes in mm): a' $(X_{11}=5.37,X_{21}=0.10,Y_1=0.34)$;b'$(X_{12}=5.37, X_{22}=0.10 ,Y_2=0)$;c'$(X_{13}=6.10 ,X_{23}=0.46, ,Y_3=0)$; d'$(X_{14}=6.10 ,X_{24}=0.46 ,Y_4=0.31)$.

5.18 Dependence of clearance (t)between of chip wraps from the number of wraps (ω) and thickness (δ) of this stainless chip

Analysis of data and statistical characteristics (coefficient of determination $R^2=0.30$,coefficient of correlation r=0.55,standard deviation $S_{y/x1,x2}$) and also the minimization of the mean square error (min MSE=0.002) ,minimization of the mean absolute deviation (min MAD=0) show that this functional dependence $t=\alpha_2(\omega,\delta)$ better submits to the linear regression model than non-linear[17] model with equation view of **$Y_c=0.31+0.017X_1-0.147X_2$ or $t=0.31+0.017\omega-0.147\delta$ (38).**

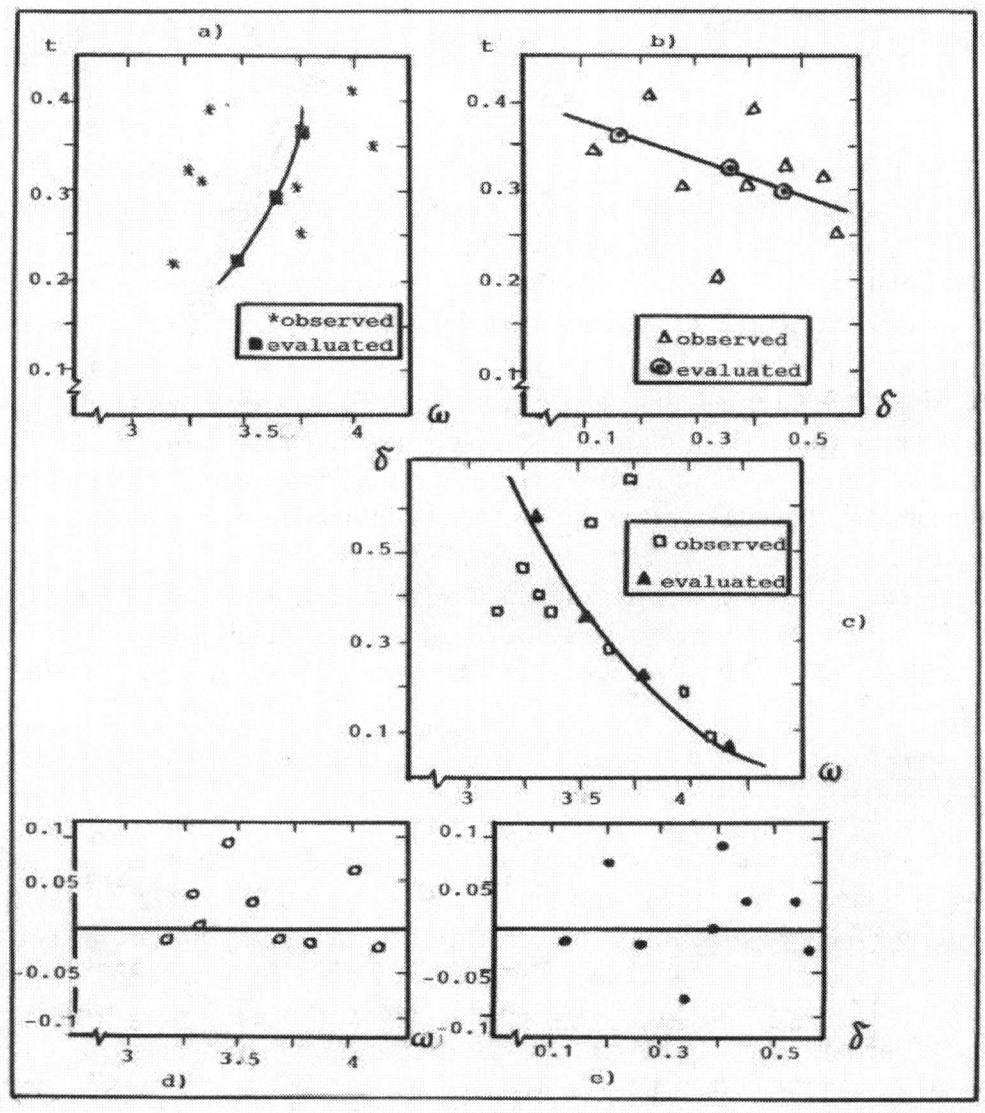

Figure 41

----

17   The non-linear regression model has equation view of $t=-6.501+3.833\omega-0.535\omega^2-0.431\delta$

In Figure 41 (a) is shown the functional dependence of clearance (t) between of chip wraps from the number of wraps (ω) stainless chip ,i.e we have the function $t=\alpha_1(\omega)$ and this dependence has the non-linear regression model with equation view of $Y_c=4.227X^{0.152}$ or $\omega=4.227t^{0.152}$ where $\mathbf{t=(0.237\omega)^{6.579}}$ (**39**).

So ,we see from Figure 41(a) that with increasing of number wraps(ω) stainless chip ,the value of clearance (t) between of chip wraps increases considerably with regression equation $t=(0.237\omega)^{6.579}$.

In Figure 41(b) is shown the functional dependence of clearance (t) between of chip wraps from thickness (δ) of this stainless chip, i.e we have function view of $t=\phi(\delta)$ and this dependence has the linear regression model with equation view of $Y_c=0.379-0.160\delta$.

So ,we see from Figure 41 (b) that with increasing of thickness (δ) stainless chip ,the value of clearance (t) between of chip wraps decreases considerably with regression equation $t=0.379-0.160\delta$.

In Figure 41(c) is shown the functional dependence of thickness (δ) stainless chip from number of wraps (ω) for this chip and as was admitted this dependence can be expressed in view of $\delta=\alpha$ (ω) and has the non-linear regression model with equation view of $Y_c=3.09X^{-0.123}$ or $\omega=3.09\delta^{-0.123}$,where $\boldsymbol{\delta=(0.324\omega)^{-8.13}}$.

So, we see from Figure 41 (c) that with increasing of number wraps (ω) for stainless chip, the value of thickness(δ) of this stainless chip decreases considerably with regression equation $\delta=(0.324\omega)^{-8.13}$.

In Figure 41(d) and 41(e) are illustrated residual plots (residual versus ω and residual versus δ) of the above-named functional dependencies $t=\alpha_1(\omega)$ and $t=\phi(\delta)$ accordingly.

Analyzing the Figure 42 (A), we see that functional surface $t=\alpha_2(\omega,\delta)$ is shown in view of three- dimensional drawing for the linear regression model with equation view of $Y_c=0.31+0.017X_1-0.147X_2$ or $t=0.31+0.017\omega-0.147\delta$.

And Figure 42(B) shows that this functional surface $t=\alpha_2(\omega,\delta)$ has view of trapezium on the plane YOX_2 with such coordinate of their points (sizes in mm): $a(X_{11}=0,X_{21}=0.10,Y_1=0.37)$;b $(X_{12}=0,X_{22}=0.10,Y_2=0)$;c $(X_{13}=0,X_{23}=0.46,Y_3=0)$; d $(X_{14}=0,X_{24}=0.46,Y_4=0.29)$.

Analysis of functional surface $t=\alpha_2(\omega,\delta)$

From Figure 42(A) we see that functional surface $t=\alpha_2(\omega,\delta)$ on which are situated the points of the linear regression equation view of $Y_c=0.31+0.017X_1-0.147X_2$ or $t=0.31+0.017\omega-0.147\delta$ has the following coordinate of their peaks in three-dimensional drawing (sizes in mm): a' $(X_{11}=4.22,X_{21}=0.10,Y_1=0.37)$; b'($X_{12}=4.22$, $X_{22}=0,Y_2=0$); c' $(X_{13}=3.21,X_{23}=0.46,Y_3=0)$;d'$(X_{14}=3.21,X_{24}=0.46,Y_4=0.29)$.

The value of modules and area of functional surface $t=\alpha_2(\omega,\delta)$ are equal: $|a'b'|=0.37$, $|b'c'|=1.07$, $|c'd'|=0.29$, $|d'a'|=1.08$ and area $S_{a'b'c'd'}$ = $=1/2[(|a'b'|-|c'd'|)\cdot|b'c'|]+(|b'c'|\cdot|c'd'|=0.35$ mm^2.

All results of computation for the linear regression equation view of $Y_c=0.31+0.017X_1-0.147X_2$ are given in Table 22.

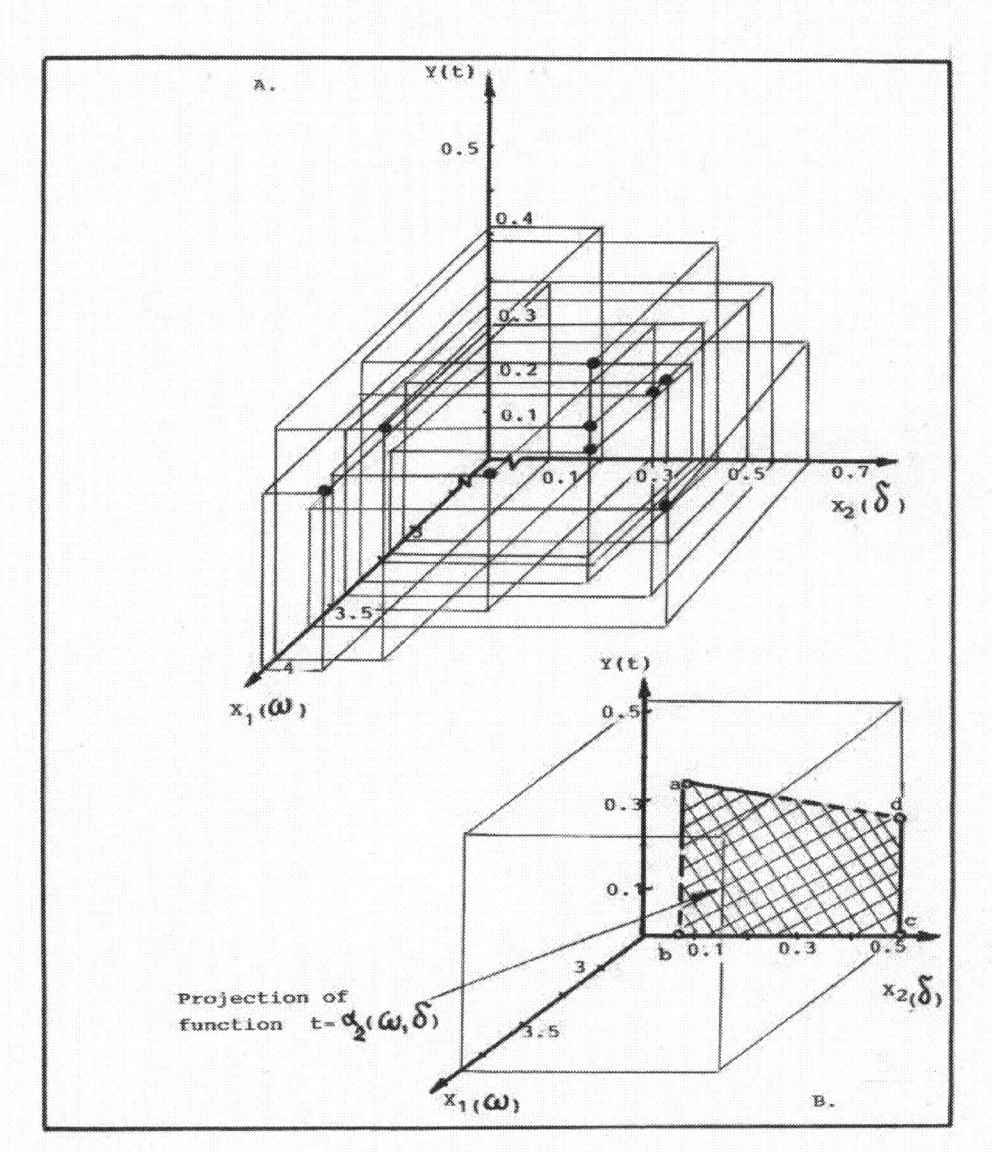

Figure 42

So, the parameters of this functional surface $t = \alpha_2(\omega, \delta)$ are the following:

1. Function $t = \alpha_2(\omega, \delta)$ better submits to the linear regression model with equation view of $Y_c = 0.31 + 0.017X_1 - 0.147X_2$ or $t = 0.31 + 0.017\omega - 0.147\delta$ with such statistical characteristics: (coefficient of determination $R^2 = 0.30$, coefficient of correlation $r = 0.55$, standard deviation $S_{y/x1,x2} = 0.059$, minimization of the mean square error (min MSE = 0.002), minimization of the mean absolute deviation (min MAD = 0);

2. The total area of functional surface $t = \alpha_2(\omega, \delta)$ is equal $\sum S = 0.35$ mm^2 with such coordinate of their points in three-dimensional drawing for this surface (sizes in mm): a'($X_{11} = 4.22, X_{21} = 0.10, Y_1 = 0.37$); b'($X_{12} = 4.22, X_{22} = 0.10, Y_2 = 0$); c'($X_{13} = 3.21, X_{23} = 0.46$, ,$Y_3 = 0$); d'($X_{14} = 3.21, X_{24} = 0.46, Y_4 = 0.29$).

Table 22 Evaluation of linear regression equation $Y_c=0.31+0.017X_1-0.147X_2$

A. MEAN, VARIANCE AND STANDARD DEVIATION

Variable	Mean	Variance	Standard deviation
X_1	3.559	1.113	1.055
X_2	0.368	0.209	0.457
Y	0.317	0.030	0.173

B. RESULTS OF MULTIPLE REGRESSION OF Y ON X_1 AND X_2

Parameter	Variable	Coefficients	Standard error	T-value
β_1	X_1	0.017	0.352	0.048
β_2	X_2	-0.147	0.152	-0.967

C. ANALYSIS OF VARIANCE RESULTS

Regression
- Degrees of freedom   2
- Sum of squares   0.009
- Mean squares   0.005

Error
- Degrees of freedom   6
- Sum of squares   0.021
- Mean square   0.004

Standard error of estimate   1.055

F-value*   0.80   [$F_{0.05,2,6}$]=5.147

Since $F^*=0.80 < 5.147$,we can not reject the hypothesis that both β_1 and β_2 are zero

D. DETERMINATION OF RESIDUALS

Number	Observed	Estimated	Residual
1	0.21	0.31	-0.10
2	0.38	0.30	0.08
3	0.34	0.34	-0.03
4	0.42	0.35	0.07
5	0.30	0.33	-0.03
6	0.31	0.31	0
7	0.32	0.29	0.03
8	0.32	0.29	0.03
9	0.25	0.28	-0.03

5.19 Dependence of clearance (t) between of chip wraps from width (B) and number of wraps (ω) for this stainless chip

Analysis of data and statistical characteristics (coefficient of determination $R^2=0.16$,coefficient of correlation r=0.40,standard   deviation $S_{y/x1,x2}=0.066$ 0 and also the minimization of the mean square error (min MSE=0.003) and minimization of the mean absolute deviation (min MAD=0) show that this functional dependence $t=\alpha_3(B,\omega)$ better submits to the linear than non-linear [18] regression model with equation view of $Y_c= -0.02+0.002X_1+0.09X_2$ or $t= -0.02+0.002B+0.09\omega$ (40).

---

18   *The non-linear regression equation view of* $t= 0.275+0.001B+0.443\cdot10^{-3}B^2+0.01\omega$

In Figure 43 (a) is shown the functional dependence of clearance (t) between of chip wraps from width (B) of this stainless chip ,i.e we have the function t=γ_1(B) and this dependence has the non-linear regression model with equation view of $Y_c=0.032X^{1.296}$ or t=$0.032B^{1.296}$.

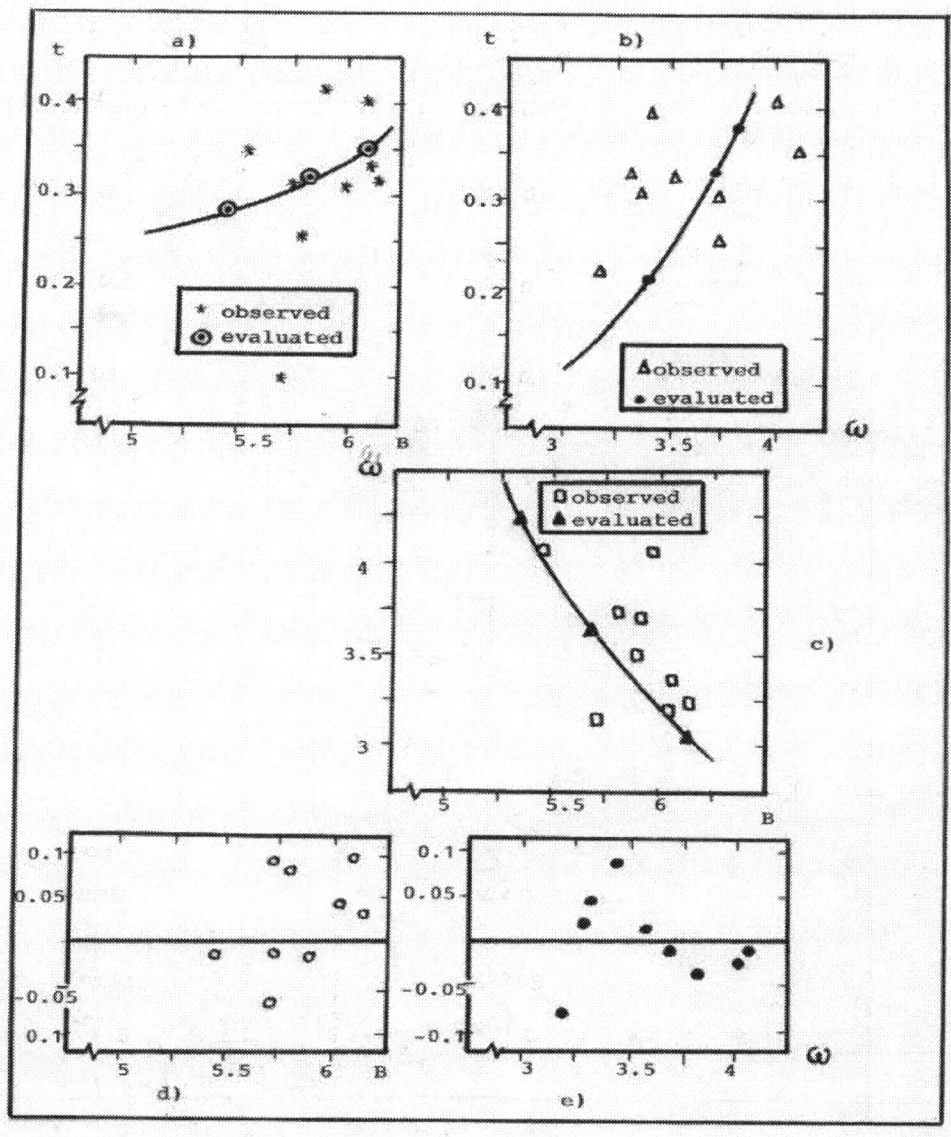

Figure 43

So ,we see from Figure 43(a) that with increasing of width (B) stainless chip ,the value of clearance (t) between of chip wraps increases considerably with regression equation t=$0.032B^{1.296}$.

In Figure 43(b) is shown the functional dependence of clearance (t) between of chip wraps from the number of wraps (ω),i.e we have the function view of t=α_1(ω).

This dependence has the non-linear regression model with equation view of $Y_c=4.227X^{0.152}$ or $\omega=4.227t^{0.152}$ where $\mathbf{t=(0.237\omega)^{6.579}}$ (**41**). So ,we see from Figure 43(b) that with increasing of number chip wraps (ω) ,the value of clearance (t) between stainless chip increases considerably with regression equation $t=(0.237\omega)^{6.579}$.

In Figure 43 (c) is shown the functional dependence of number chip wraps (ω) stainless chip from width (B) of this chip and this dependence can be expressed in view of function view of $\omega=\gamma_2(B)$ and has the non-linear regression model with equation view of $Y_c=703.1$ $X^{-3.0}$ or $\omega=703.1B^{-3.0}$. So ,we see from Figure 43(c) that with increasing of width (B) stainless chip ,the value of number chip wraps (ω) of this chip decreases considerably with regression equation view of $\omega=703.1B^{-3.0}$.

In Figure 43(d) and 43(e) are illustrated residual plots (residual versus B and residual versus ω) of the above-named functional dependencies $t=\alpha_1(\omega)$ accordingly.

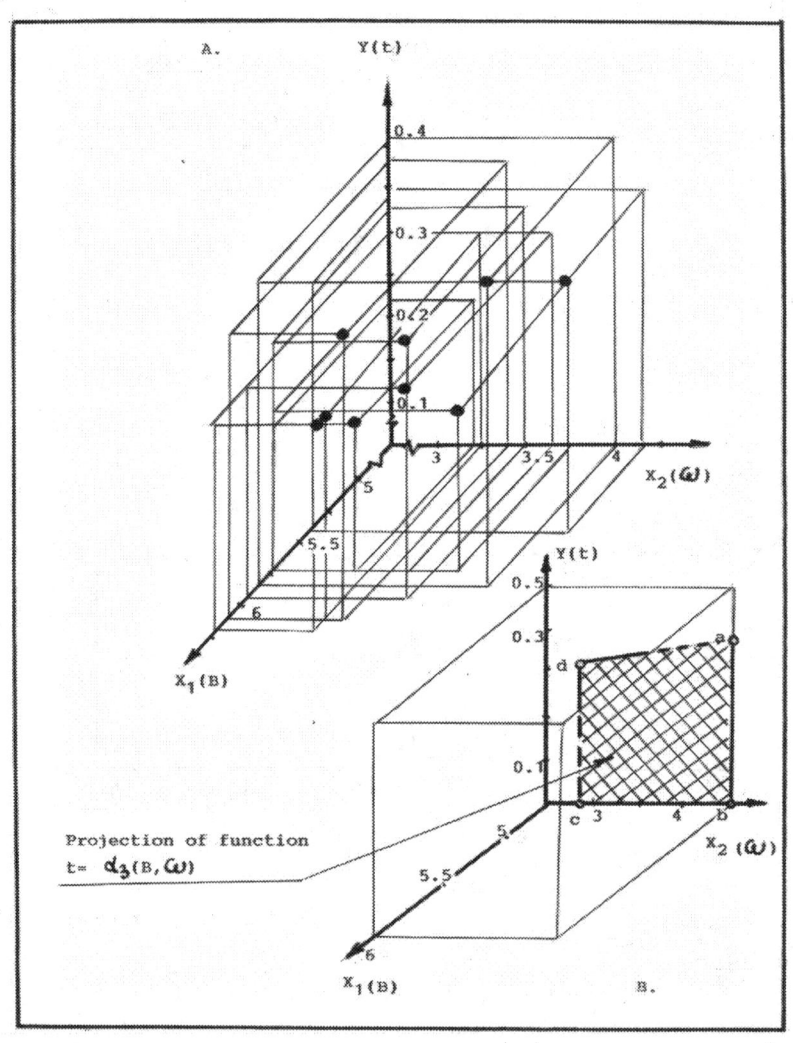

Figure 44

138

Analyzing the Figure 44 (A) we see that functional surface $t=\alpha_3(B,\omega)$ is shown in view of three-dimensional drawing for the linear regression model with equation view of $Y_c= = -0.02+0.002X_1+0.09X_2$ or $t= -0.02+0.002B+0.09\omega$. And Figure 44(B) shows that this functional surface $t=\alpha_3(B,\omega)$ has view of trapezium on the face YOX_2 with such coordinate of their points (sizes in mm): a($X_{11}=0$,$X_{21}=4.22,Y_1=0.36$); b($X_{12}=0$, ,$X_{22}=4.22,Y_2=0$);c($X_{13}=0,X_{23}=3.25,Y_3=0$); d($X_{14}=0,X_{24}=3.25,Y_4=0.29$).

Analysis of functional surface $t=\alpha_3(B,\omega)$

From Figure 44(A) we see that functional surface $t=\alpha_3(B,\omega)$,on which are situated the points of the linear regression equation view of $Y_c= -0.02+0.002X_1+0.09X_2$ or $t= -0.02+0.002B+0.09\omega$,has the following coordinate of their peaks in three-dimensional drawing (sizes in mm): a'($X_{11}=5.37,X_{21}=4.22,Y_1=0.36$) ;b'($X_{12}=5.37,X_{22}=4.22,Y_2=0$), ,c'($X_{13}=6.12,X_{23}=3.25,Y_3=0$);d' ($X_{14}=6.12,X_{24}=6.12$,$X_{24}=3.25$,$Y_4=0.29$). The value of modules and area of functional surface $t=\alpha_3(B,\omega)$ are equal: $|a'b'|=0.36$, $|b'c'|=1.23$,$|c'd'|=0.29, |d'a'|=1.24$ and area $S_{a'b'c'd'}=1/2[(|a'b'|-|c'd'|)\cdot|b'c'|]+$ $+(|b'c'|\cdot|c'd'|)=0.40$ mm^2.All results of computations for the linear equation view of $Y_c= -0.02+0.002X_1+0.09X_2$ are given in Table 23.

Table 23 Evaluation of linear regression equation $Y_c= -0.02+0.002X_1+0.09X_2$

A. MEAN ,VARIANCE AND STANDARD DEVIATION			
Variable	Mean	Variance	Standard deviation
X_1	5.836	0.514	0.717
X_2	3.559	1.113	1.055
Y	0.317	0.031	0.176

B.RESULTS OF MULTIPLE REGRESSION OF Y ON X_1 AND X_2				
Parameter	Variable	Coefficients	Standard error	T-value
β_1	X_1	0.002	0.239	0.008
β_2	X_2	0.09	0.352	0.256

C. ANALYSIS OF VARIANCE RESULTS

   Regression

      Degrees of freedom   2

      Sum of squares   0.005

      Mean squares   0.003

   Error

      Degrees of freedom 6

      Sum of squares   0.026

      Mean squares   0.004

   Sum error of estimate 0.717

    F-value* 0.75   [$F_{0.05.2,6}$]=5.147

Since $F^*=0.75<5.147$ we can not reject the hypothesis that both β_1 and β_2 are zero.

D. DETERMINATION OF RESIDUALS			
Number	Observed	Estimated	Residual
1	0.21	0.28	−0.07
2	0.38	0.29	0.09
3	0.34	0.36	−0.02
4	0.42	0.35	0.07
5	0.30	0.32	−0.02
6	0.31	0.29	0.02
7	0.32	0.28	0.04
8	0.32	0.31	0.01
9	0.25	0.33	−0.08

So, the parameters of this functional surface t=α₃(B,ω) are the following:

1. Function $t=\alpha_3(B,\omega)$ better submits to the linear regression model with equation view of $Y_c= -0.02+0.002X_1+0.09X_2$ or $t= -0.02 +0.002B+0.09\omega$ with such statistical characteristics: coefficient of determination $R^2=0.16$, coefficient of correlation $r=0.40$, standard deviation $S_{y/x1,x2}=0.066$, minimization of the mean square error (min MSE=0.003), minimization of the mean absolute deviation (min MAD=0).

2. The total area of functional surface $t=\alpha_3(B,\omega)$ is equal $\sum S =0.40$ mm² with such coordinate of their points in three-dimensional drawing for this surface (sizes in mm): a' $(X_{11}=5.37, X_{21}=4.22, Y_1=0.36)$; b'$(X_{12}=5.37, X_{22}=4.22, Y_2=0)$; c'$(X_{13}=6.12, X_{23}=3.25, , Y_3=0$); d' $(X_{14}=6.12, X_{24}=3.25, Y_4=0.29)$.

5.20 Dependence of number chip wraps (ω) from width (B) and thickness (δ) of stainless chip

Analysis of data and also statistical characteristics show that coefficients of determination ($R^2=0.644$) and correlation ($r=0.803$), standard deviation $S_{y/x1,x2}=0.257$ are larger for the non-linear regression model than for linear regression model[19] and besides the minimization of the mean square error (min MSE=0.044) and minimization of the mean absolute deviation (min MAD=0) are smaller for this regression model. And for this reason we can conclude that this function $\omega=\phi_1(B,\delta)$ better submits to the non-linear regression equation and has view of $\mathbf{Y_c=4.021+0.837X_1-0.149X^2_1-0.718X_2}$ or $\mathbf{\omega=4.021+0.837B-0.149B^2-0.718\,\delta}$ (42) .

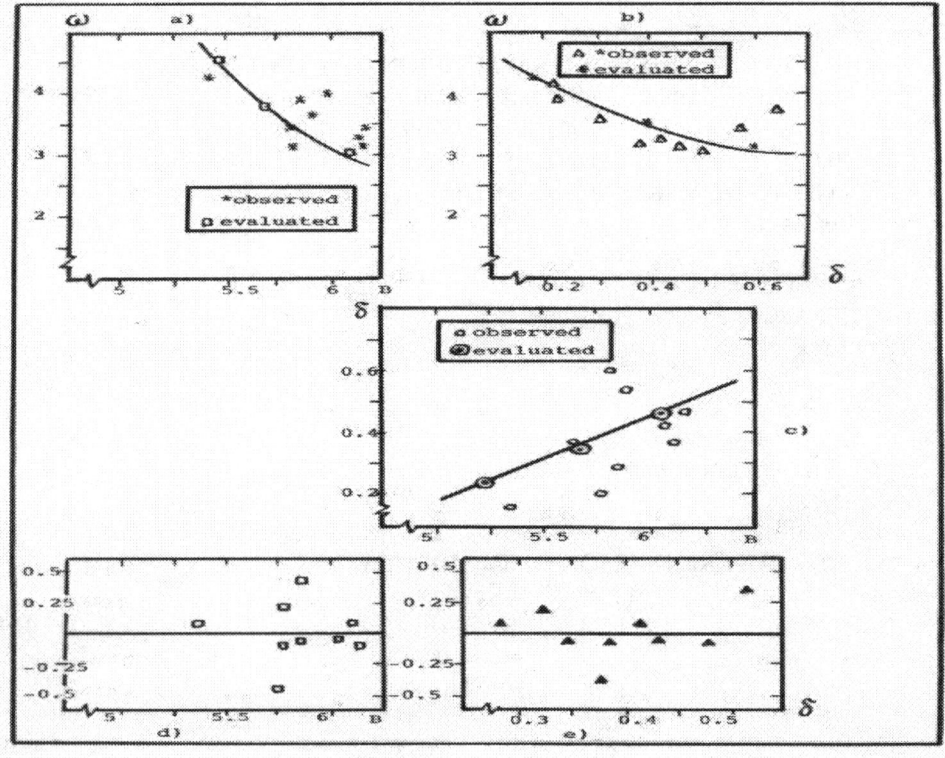

Figure 45

---

19   The linear regression equation has view of $\omega=8.51-0.802B-0.739\delta$

140

In Figure 45(a) is shown the functional dependence of number chip wraps (ω) stainless chip from width (B) of this chip, i.e we have the function $\omega=\gamma_2(B)$ and this function has the non-linear regression model with equation view of $Y_c=703.1X^{-3.0}$ or $\omega=703.1B^{-3.0}$. So ,we see from Figure 45(a) that with increasing of width (B) stainless chip ,the value of number (ω) chip wraps decreases accordingly.

In Figure 45(b) is shown the functional dependence of number (ω) chip wraps stainless chip from its thickness (δ) ,i.e we have the function view of $\omega=\alpha(\delta)$ which has the non-linear regression model with equation $Y_c=3.09X^{-0.123}$ where $\omega=3.09\delta^{-0.123}$. So ,we see from Figure 45(b) that with increasing of thickness (δ) stainless chip ,the value of number (ω) chip wraps decreases accordingly.

In Figure 45 (c) is shown the functional dependence of thickness (δ) stainless chip from its width (B) and this dependence can be expressed in view of function $\delta=f(B)$ and has the linear regression model with equation $Y_c= -1.236+0.275X$ or $\delta= -1.236+0.275B$. From Figure 45 (c) we see that with increasing of width (B) stainless chip ,the value of thickness (δ) of this chip increases considerably.

In Figure 45(d) and 45(e) are illustrated the residual plots (residual versus B and residual (δ) of the above-named functional dependencies $\omega=\gamma_2(B)$ and $\omega=\alpha(\delta)$ accordingly.

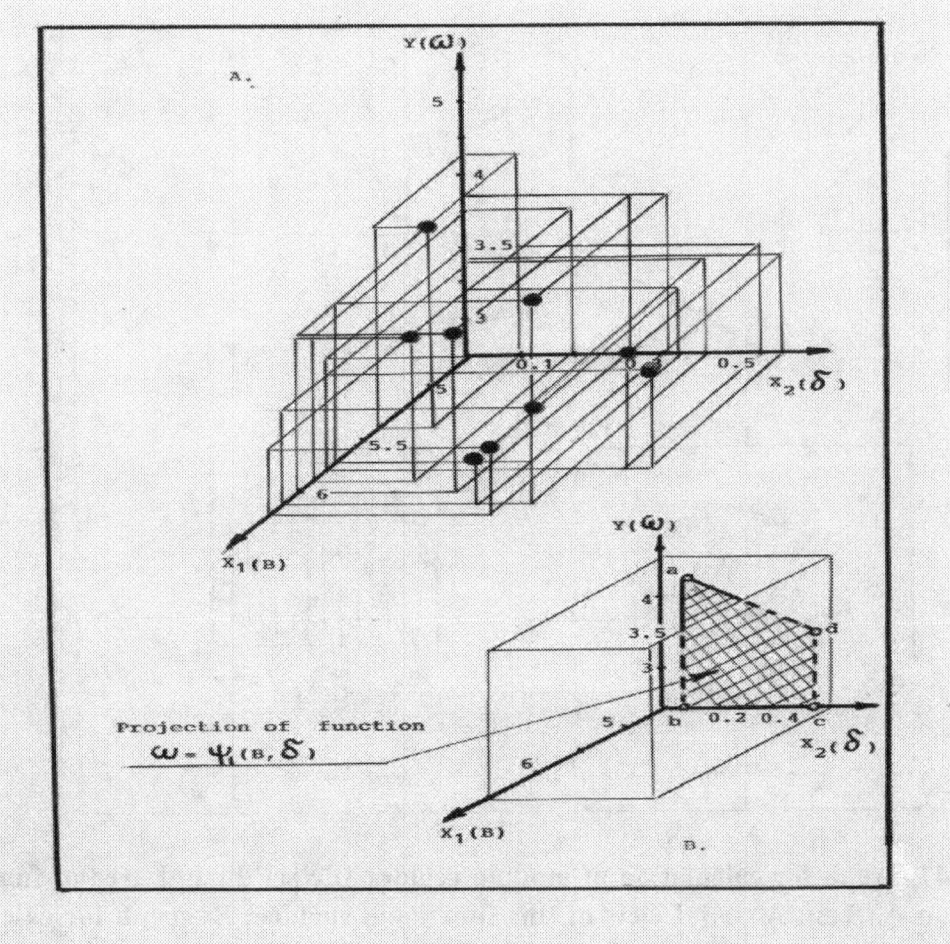

Figure 46

141

Analyzing the Figure 46(A) we see that this functional surface $\omega=\phi_1(B,\delta)$ is shown in view of three-dimensional drawing for the non-linear regression model with equation view of $Y_c=4.021+0.837X_1-0.149X^2{}_1-0.718X_2$ or $\omega=4.021+0.837B-0.149B^2-0.718\delta$.

And Figure 46 (B) shows that this functional surface $\omega=\phi_1(B,\delta)$ has view of trapezium on the face YOX_2 with such coordinate of their points (sizes in mm): $a(X_{11}=0,X_{21}=0.10,Y_1=4.15\quad)$; $\quad b(X_{12}=0,X_{21}=0.10,Y_2=0); c(X_{13}=0,X_{23}=0.60\quad,Y_3=0)\quad$; $d(X_{14}=0,X_{24}=0.60,Y_4=3.48)$.

Analysis of functional surface $\omega=\phi_1(B,\delta)$

From Figure 46(A) we see that functional surface $\omega=\phi_1(B,\delta)$,on which located the points of non-linear regression equation view of $Y_c=4.021+0.837X_1-0.149X^2{}_1-0.718X_2$ or $\omega=4.021+0.837B-0.149B^2-0.718\delta$ has a curvilinear character with the following coordinate of their pears in three-dimensional drawing (sizes in mm): $a'(X_{11}=5.37,X_{21}=0.10,Y_1=4.15)$; $\quad b'(X_{12}=5.37\quad,X_{22}=0.10\quad,Y_2=0); c'(X_{13}=5.75,X_{23}=0.60,$ $,Y_3=0\ )$; $\quad d'\ (X_{14}=5.75\ ,X_{24}=0.60\ ,Y_4=3.48)$.

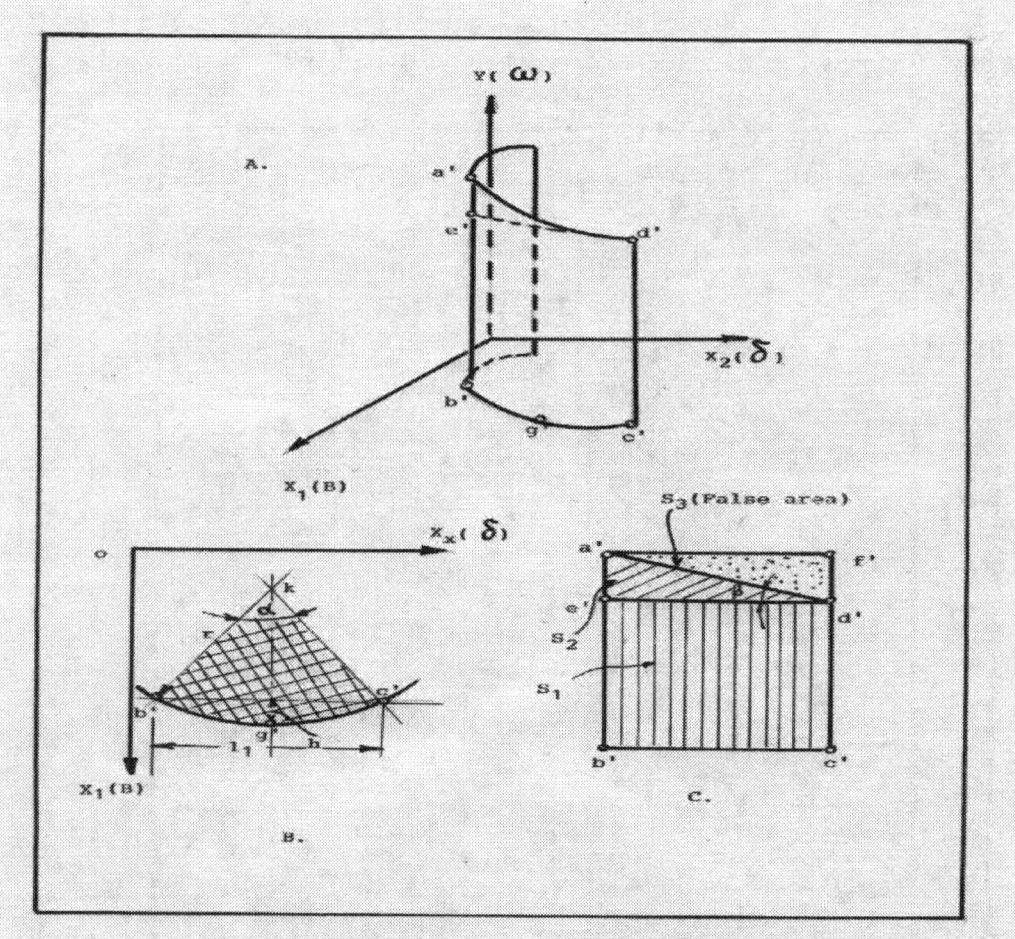

Figure 47 Graph for calculation of module vectors |b'c'|,|a'd'| and area of functional surface $\omega=\phi_1(B,\omega)$: A =total view of this functional surface; B=graph for calculation of module vector |b'c'|;C= graph for calculation of module vector |a'd'| and area of this surface.

In Figure 47 schematically is shown the functional surface $\omega=\phi_1(B,\delta)$ and graph for calculation of module for vectors $|b'c'|$ and $|a'd'|$.

- The module of vector $|a'b'|$ and $|c'd'|$ can be calculated from the projection of this functional surface $\omega=\phi_1(B,\delta)$ on the plane YOX$_2$ (Figure 46 A and 47A). So, the module of vectors $|a'b'|=4.15$, $|c'd'|=3.48$.

- The module of vector $|b'c'|$ or it length can be calculated by formula $L=|b'c'|=0.01745r\alpha$, where L= length of arc (b'c'),r=radius of sector,α=angle of sector. So, the module of vector r is equal $r=|k'g'|=[(X_{16}-X_{15})^2+(X_{26}-X_{25})^2]^{1/2}=5.83$, where the coordinate of points g' and k are equal $g'(X_{16}=5.82,X_{26}=0.28);k(X_{15}=0,X_{25}=0)$. At data $\alpha=82°$,r=5.83 we have the module of vector $|b'c'|=8.34$ and other parameters of circular segment which are equal: $h=r[1-\cos(\alpha/2)]=1.43$ and $l_1=2[h(2r-h)]^{1/2}=7.65$.

- The graph for calculation of module vectors $|b'c'|$,$|a'd'|$ and area of functional surface $\omega=\phi_1(B,\delta)$ is shown in Figure 47. The module of vector $|a'd'|$ and area of this functional surface ,as shown in Figure 47(c) ,can be calculated by the following way: the module of length for curve $|d'e'|$ is equal $|d'e'|=|b'c'|$,i.e $|d'e'|=8.34$ and module of length for $|a'e'|=|a'b'|-|c'd'|=0.67$.

- For calculation of area S_1,S_2 and S_3 we find the angle β which is equal $\tan\beta=|a'e'|/|d'e'|=0.080$,where $\beta=4°30'$ and then the module of length for curve $|a'd'|=(|a'e'|)/\sin4°30'=8.54$. The coordinates of points f' is equal f' $(X_{13}=5.75,X_{23}=0.60,Y_1=4.15)$ and module of length $|f'a'|=|b'c'|=8.34$ and area $S_2=S_3=S/2=(|a'e'|\cdot|d'e'|)/2=2.79$ mm^2.

- For these conditions, we have the area $S_1=(|c'd'|\cdot|b'c'|=29.02$ mm^2 and total area $\sum S$ of this functional surface $\omega=\phi_1(B,\delta)$ is equal $\sum S=S_1+S_2=31.81$mm^2.

All results of computations for the linear equation view of $Y_c=4.021+0.837X_1-0.149X^2_1-0.718X_2$ are given in Table 24.

So ,the parameters of the functional surface $\omega=\phi_1(B,\delta)$ are the following:

1.Function $\omega=\phi_1(B,\delta)$ better submits to the non-linear regression model in view of curvilinear surface as is shown in Figure 46 and 47(A) and describes by equation view of $Y_c=4.021+0.837X_1-0.149X^2_1-0.718X_2$ or $\omega=4.021+0.837B-0.149B^2-0.718\delta$ with such statistical characteristics:

- Coefficient of determination R^2=0.644
- Coefficient of correlation r=0.803
- Standard deviation S$_{y/x1,x2}$=0.257
- Minimization of the mean square error (min MSE=0.044)
- Minimization of the mean absolute deviation (min MAD=0)

2.The total area of functional surface $\omega=\phi_1(B,\delta)$ is equal $\sum S=31.81$ mm^2 with such coordinate of their points in three- dimensional drawing for this surface(sizes in mm): a'$(X_{11}=5.37,X_{21}=0.10,Y_1=4.15)$;b'$(X_{12}=5.37,X_{22}=0.10,Y_2=0)$;c'$(X_{13}=5.75,X_{23}=0.60,Y_3=0)$;d' $(X_{14}=5.75,X_{24}=0.60,Y_4=3.48)$.

Table 24 Evaluation of regression equation $Y_c=4.021+0.837X_1-0.149X^2_1-0.718X_2$

A. MEAN, VARIANCE AND STANDARD DEVIATION

Variable	Mean	Variance	Standard deviation
X_1	5.836	0.515	0.718
X_2	0.368	0.209	0.457
Y	3.559	1.113	1.055

B. RESULTS OF MULTIPLE REGRESSION OF Y ON X_1 AND X_2

Parameter	Variable	Coefficients	Standard error	T-value
β_1	X_1	0.837	0.239	3.502
β_2	X_2	-0.718	0.152	-4.724

C. ANALYSIS OF VARIANCE RESULTS

Regression
- Degrees of freedom 2
- Sum of squares 0.717
- Mean squares 0.359

Error
- Degrees of freedom 6
- Sum of squares 0.396
- Mean squares 0.066

Standard error of estimate 0.718

F-value * 5.439 $[F_{0.05,2,6}]=5.147$

Since $F^*=5.439 > 5.147$ we can reject the hypothesis that both β_1 and β_2 are zero.

D. DETERMINATION OF RESULTS

Number	Observed	Estimated	Residual
1	3.17	3.68	-0.51
2	3.28	3.22	0.06
3	4.22	4.15	0.07
4	3.99	3.78	0.21
5	3.62	3.64	-0.02
6	3.25	3.30	-0.05
7	3.21	3.25	-0.04
8	3.53	3.56	-0.03
9	3.76	3.48	0.28

In Table 25 is shown the *Summary Multiple Regression Analysis* between two independent variable parameters in function of $Y_i=\gamma(X_{i,1};X_{i,2})$.

In Table 26 is shown the *Summary Statistical Characteristics* and empirical regression equations for this function view of $Y_i=\gamma(X_{i,1};X_{i,2})$.

C. FOR OTHER PARAMETERS OF STAINLESS CHIP

Table 25 Summary Multiple Regression Analysis

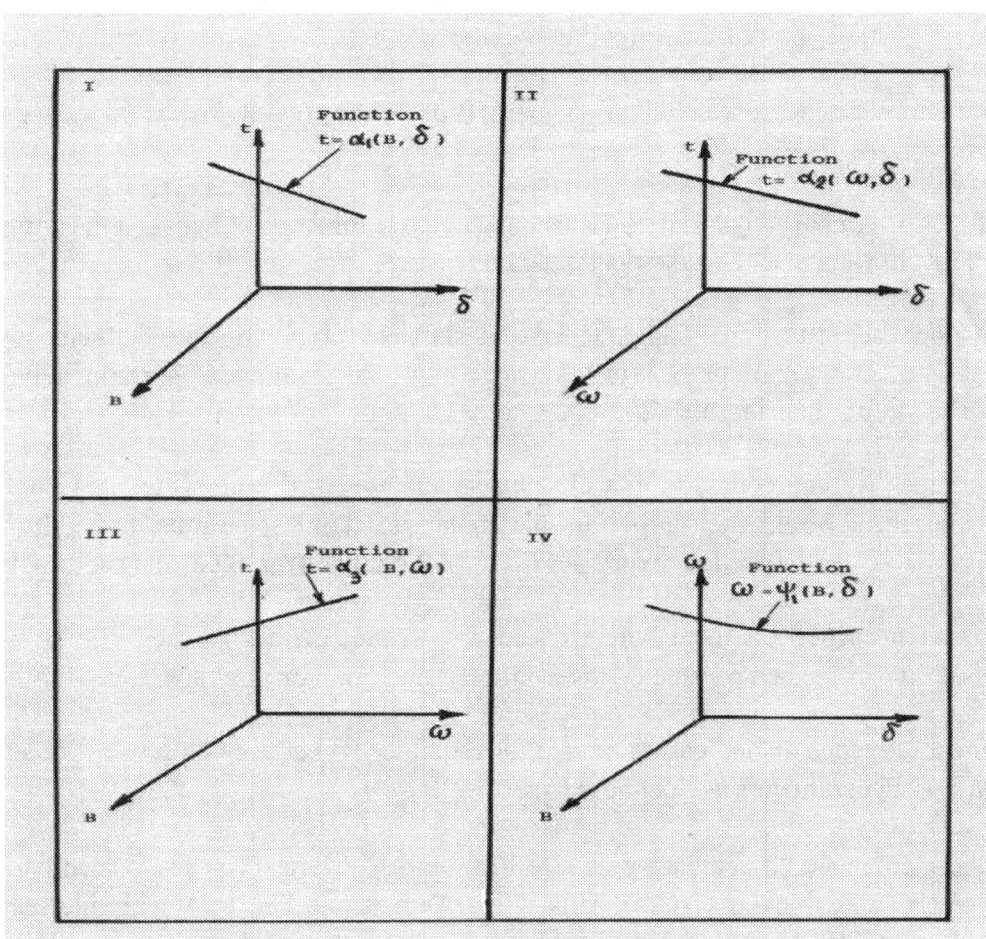

Table 26   Summary statistical characteristics and empirical regression equations for function $Y_c = \gamma(X_{i,1}, X_{i,2})$

View of function	Regression model	Empirical equation	Statistical characteristics		
			R^2	r	$S_{y/x1,x2}$
$t = \alpha_1(B,\delta)$	Linear	$T = 0.021 + 0.064\,B - 0.212\delta$	0.452	0.672	0.053
$t = \alpha_2(\omega,\delta)$	Linear	$t = 0.31 + 0.017\,\omega - 0.147\delta$	0.30	0.55	0.059
$t = \alpha_3(B,\omega)$	Linear	$t = -0.02 + 0.002\,B + 0.09\omega$	0.16	0.40	0.066
$\omega = \phi_1(B,\delta)$	Non-linear	$\omega = 4.021 + 0.837\,B - 0.149B^2 - 0.718\delta$	0.644	0.803	0.257

SUMMARY

Analysis of correlation between three parameters for stainless chip in function $Y_i = \gamma(X_{i,1}; X_{i,2})$ have showed the following results :

1. All functions besides of the function $\omega = \phi_1(B, \delta)$ better submit to the linear regression model;

2. The functions $t = \alpha_1(B, \delta)$ and $\omega = \phi_1(B, \delta)$ have average correlation between the parameters of stainless chip and also the statistical characteristics;

3. Analyzing each functional dependence for stainless chip, we see that clearance (t) between chip wraps and number (ω) of wraps could be evaluated approximately by the above –shown formulas in Table 26;

4. And besides we can conclude that clearance (t) between of chip wraps and number of wraps (ω) depends from such important parameters of stainless chip as its width (B) and thickness (δ), i.e each function expresses as $t = \rho(B, \delta, \omega)$ and $\omega = \theta(B, \delta)$ accordingly ;

5. The most influence on increasing of size for clearance (t) between chip wraps for stainless chip plays such parameters as width (B) and number of wraps (ω) of this chip ;

6. The most influence on decreasing of size for clearance (t) between chip wraps of stainless chip plays such parameters as thickness (δ);

7. The most influence on decreasing of size for number of wraps (ω) plays such parameter as width (B) and thickness (δ) of this chip. And as we see that with increasing of parameters width (B) and thickness (δ), the number of wraps (ω) decreases considerably.

CHAPTER SIX   MULTIPLE REGRESSION ANALYSIS FOR EXTERNAL DIAMETER OF STAINLESS CHIP IN DEPENDENCE FROM SOME GENERAL ITS PARAMETERS IN FUNCTION OF $Y_I=\gamma(X_{i,1};X_{i,2};X_{i,3})$

6.1 Dependence of external diameter (d_2) from thickness (δ), internal diameter (d_1) and clearance (t) between of chip wraps, and also their modification

A. The main formulas for calculation are the following:

- The multiple linear model has view $Y_c=b_0+b_1X_{i,1}+b_2X_{i,2}+b_3X_{i,3}$, where coefficients b_0,b_1,b_2 and b_3 can be determined from four normal equations:

$$\sum Y=nb_0+b_1\sum X_1+b_2\sum X_2+b_3\sum X_3$$
$$\sum X_1Y=b_0\sum X_1+b_1\sum X^2_1+b_2\sum X_1X_2+b_3\sum X_1X_3 \quad (1)$$
$$\sum X_2Y=b_0\sum X_2+b_1\sum X_1X_2+b_2\sum X^2_2+b_3\sum X_2X_3$$
$$\sum X_3Y=b_0\sum X_3+b_1\sum X_1X_3+b_2\sum X_2X_3+b_3\sum X^2_3$$

- And besides we can hypothecate the multiple curvilinear correlation for any given functional dependency, for example view of $d_2=\alpha_1(\delta,d_1,t)$ in view of $Y=b_0+b_1X_1+b_2X^2_1+b_3X_2+b_4X_3$ where b_0,b_1,b_2,b_3 and b_4 coefficients of regression equation which could be defined from the following normal equations:

$$\sum Y=nb_0+b_1\sum X_1+b_2\sum X^2_1+b_3\sum X_2+b_4\sum X_3$$
$$\sum X_1Y=b_0\sum X_1+b_1\sum X^2_1+b_2\sum X^3_1+b_3\sum X_1X_2+b_4\sum X_1X_3 \quad (2)$$
$$\sum X^2_1Y=b_0\sum X^2_1+b_1\sum X^3_1+b_2\sum X^4_1+b_3\sum X^2_1X_2+b_4\sum X_1X_3$$
$$\sum X_2Y=b_0\sum X_2+b_1\sum X_1X_2+b_2\sum X^2_1X_2+b_3\sum X^2_2+b_4\sum X_2X_3$$
$$\sum X_3Y=b_0\sum X_3+b_1\sum X_1X_3+b_2\sum X^2_1X_3+b_3\sum X_2X_3+b_4\sum X^2_3$$

where X_1, X_2 and X_3 -independent variables.

B. Analysis of function $d_2=\alpha_1(\delta,d_1,t)$

Analysis of the data and statistical characteristics (coefficient of determination $R^2=0.95$, coefficient of correlation r=0.974 , standard deviation $S_{y/x1, x2}=0.203$) and also the minimization of the mean square error (MSE=0.023)and minimization of the mean absolute deviation (min MAD=0) show this functional dependence **$d_2=\alpha_1(\delta,d_1,t)$ better** submits to the non-linear than to linear [1] regression model with equation view of **$Y_c=3.564 -0.211X_1-0.905X^2_1 +1.124X_2 +3.708X_3$ or $d_2 =3.564 -0.211\delta--0.905\delta^2 +1.124d_1+3.708t$ (3)** .

In Figure 1(a) is shown the functional dependence of external diameter (d_2) stainless chip from its thickness (δ), i.e $d_2=\gamma(\delta)$. And as was above-shown , this dependence $d_2=\gamma(\delta)$ has

the linear regression model with equation view of $Y_c=8.808 +3.897X$ or $d_2=8.808+3.897\delta$ .So, we see from Figure 1(a) that with increasing of thickness (δ) stainless chip, the value of external diameter (d_2) increases accordingly with regression equation $d_2=8.808+3.897\delta$.
In Figure 1(b) is shown the functional dependence of external diameter (d_2) stainless chip from its internal diameter (d_1) ,i.e we have the function $d_2=\varphi_3(d_1)$. This dependence has the non-linear model with equation view of $Y_c=5.728X^{0.359}$ or $d_2=5.728d_1^{0.359}$.
So ,we see from Figure 1 (b) that with increasing of internal diameter (d_1) stainless chip ,the value of external diameter (d_2) increases accordingly with regression equation $d_2=5.728d_1^{0.359}$.

1- Linear regression model has view $d_2 =3.564 -0.211\delta -0.905\delta^2 +1.124d_1 +3.708t$

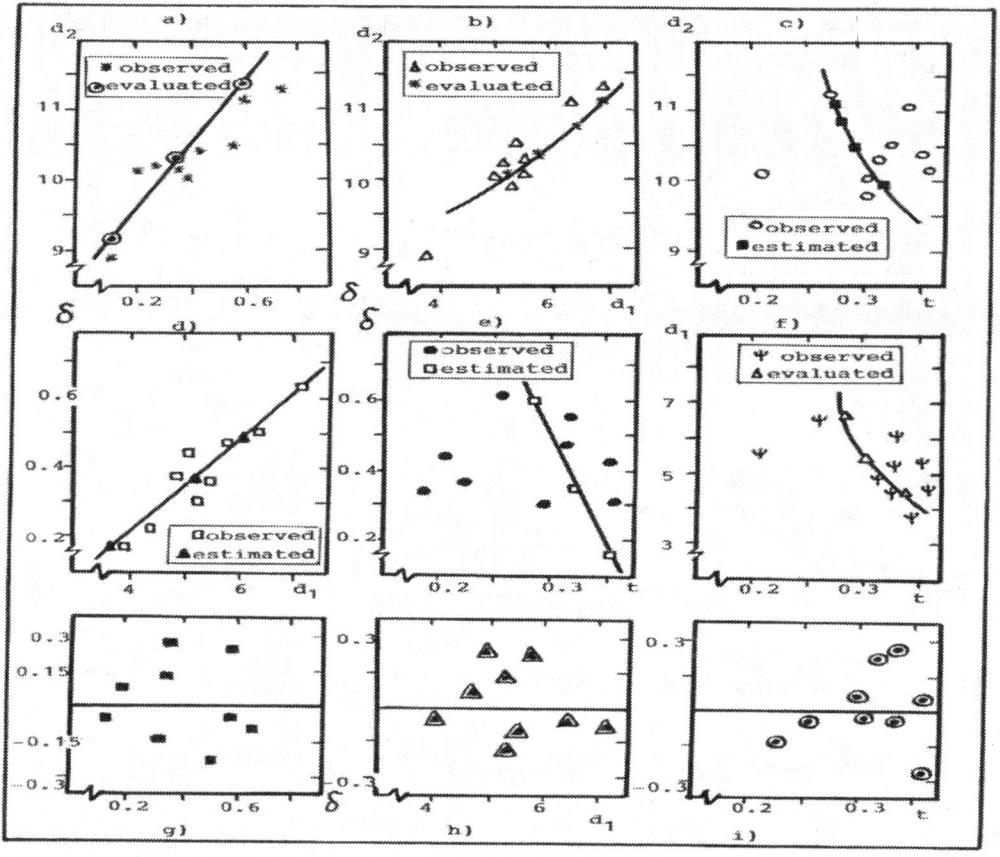

Figure 1
In Figure 1(c) is shown the functional dependence of external diameter (d_2) stainless chip from clearance (t) between of chip wraps of this stainless chip, i.e we have the function view of $d_2=\gamma_3(t)$. This dependence, as was above-shown, has   the non-linear regression model with equation view of $Y_c=1.14X^{-0.56}$ or $t=1.14d_2^{-0.56}$ where $\mathbf{d_2=(0.877t)^{-1.786}}$ (**4**) .So , we see from Figure 1(c) that with increasing of clearance (t) between of chip

wraps stainless chip ,the value of its external diameter (d_2) decreases considerably with regression equation view of $d_2 = (0.877t)^{-1.786}$.

In Figure 1 (d) is shown the functional dependence of thickness (δ) stainless chip from its internal diameter (d_1) ,i.e we have the function view of $\delta = \varphi(d_1)$. This function has the linear regression model with equation view of $Y_c = 3.338 + 4.762X$ or $d_1 = 3.338 + 4.762\delta$ where $\boldsymbol{\delta = 0.21d_1 - 0.70}$ **(5)** . So, we see from Figure 1(d) that with increasing of internal diameter (d_1) stainless chip, the value of its thickness (δ) increases accordingly with regression equation $\delta = 0.21d_1 - 0.70$. In Figure 1 (e) is shown the functional dependence of thickness (δ) stainless chip from clearance (t) between of chip wraps for this stainless chip ,i.e we have the function view of $\delta = \phi(t)$.

This dependence has the linear regression model with equation view of $Y_c = 0.379 - 0.16X$ or $t = 0.379 - 0.16\delta$ where $\boldsymbol{\delta = 2.37 - 6.25t}$ **(6)**.

So ,we see from Figure 1(e) that with increasing of clearance (t) between of chip wraps stainless chip ,the value of its thickness (δ) decreases considerably with regression equation $\delta = 2.37 - 6.25t$. In Figure 1(f) is shown the functional dependence of internal diameter (d_1) stainless chip from clearance (t) between of chip wraps stainless chip ,i.e we have the function view of $t = \gamma_4(d_1)$. And as was above-shown this dependence $t = \gamma_4(d_1)$ has the non-linear regression model with equation view of $Y_c = 0.828X^{-0.605}$ or $t = 0.828d_1^{-0.605}$, where $\boldsymbol{d_1 = (1.208t)^{-1.653}}$ **(7)** . So , we see from Figure 1 (f) that with increasing of clearance (t) between of chip wraps stainless chip, the value of its internal diameter (d_1) decreases considerably in accordance with regression equation $d_1 = (1.208t)^{-1.653}$ .In Figures 1(g) and 1(h),1(i) are illustrated residual plots (residual versus δ ,d_1 and t) of the above-named functional dependencies $d_2 = \gamma(\delta), d_2 = \varphi_3(d_1)$ and $d_2 = \gamma_3(t)$ accordingly. All results of computation for the non-linear equation view of $Y_c = 3.564 - 0.211X_1 - 0.905X^2_1 + 1.124X_2 + 3.708X_3$ are given in Table 1.

Table 1 Evaluation of regression equation $Y_c=3.56-0.21X_1-0.91X^2_1+1.12X_2+3.71X_3$

A. Mean ,variance and standard deviation			
Variable	Mean	Variance	Standard deviation
X_1	0.368	0.209	0.457
X_2	5.098	5.609	2.368
X_3	0.317	0.031	0.176
Y	10.248	4.120	2.030

B. Results of multiple regression of Y on X_1,X_2 and X_3				
Parameter	Variable	Coefficients	Standard error	T-value
β_1	X_1	-0.211	0.152	-1.388
β_2	X_2	1.124	0.789	1.425
β_3	X_3	3.708	0.059	62.848

C. Analysis of variance results

Regression		
Degrees of freedom	3	
Sum of squares	3.913	
Mean squares	1.304	

Error		
Degrees of freedom	5	
Sum of squares	0.207	
Mean square	0.041	
Standard error of estimate	0.457	
F-value *	31.804	

* Since F=31.804 > 5.41 [$F_{0.05,3,5}$=5.41] we can reject the hypothesis that three parameters β_1,β_2 and β_3 are zero.

D. Determination of residuals

Number	Observed	Estimated	Residual
1	10.03	10.18	-0.15
2	10.36	10.59	-0.23
3	8.69	8.75	-0.06
4	10.13	10.12	0.01
5	10.22	10.14	0.08
6	10.07	9.84	0.23
7	10.49	10.25	0.24
8	11.06	11.07	-0.01
9	11.18	11.28	-0.10

C. Modification of function $d_2=\gamma_0(\delta, d_1/t)$

Analysis of the data and statistical characteristics (coefficient of determination R^2=0.795 ,coefficient of correlation r=0.892,standard deviation $S_{y/x1,x2}$=0.3750 and also the minimization of the mean square error(min MSE=0.094) and minimization of the mean absolute deviation (min MAD=0) show that this functional dependence $d_2=\gamma_0(\delta, d_1/t)$ better submits to the linear than non-linear ^{2} regression model with equation view of

$$Y_c=8.859 +4.192X_1-0.009(X_2/X_3) \text{ or } d_2=8.859+4.192\delta- 0.009(d_1/t)\ (8).$$

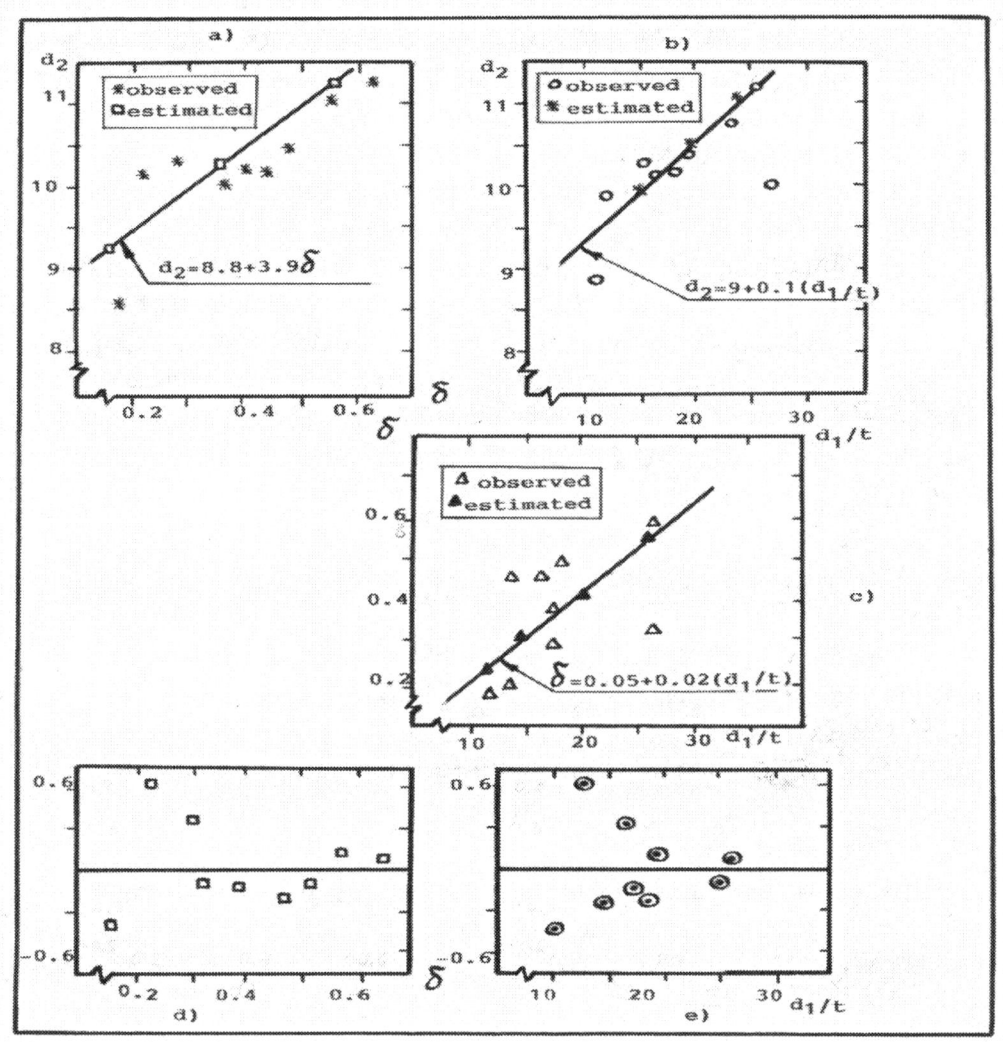

Figure 2

2– Non-linear regression model has view $d_2=8.19+8.02\delta-5.66\delta^2+1.18\cdot10^{-4}$ (d_1/t)

In Figure 2 (a) is shown the functional dependence of external diameter (d_2) stainless chip from its thickness (δ) ,i.e we have the function $d_2=\gamma(\delta)$. This functional dependence has the linear regression model with equation view of $Y_c=8.808+3.897X$ or $d_2=8.808+3.897\delta$. So ,we see from Figure 2(a) that with increasing of thickness (δ) stainless chip ,the value of its external diameter (d_2) increases considerably with regression equation $d_2=8.808+3.897\delta$.

In Figure 2(b) is shown the functional dependency of external diameter (d_2) stainless chip from the ratio of values (d_1/t),such as internal diameter (d_1) to clearance (t) between of chip wraps for this stainless chip ,i.e has a place the function view of $d_2=\phi(d_1/t)$. Analysis

151

of the data and also statistical characteristics (coefficient of determination $R^2=0.331$,coefficient of correlation r=0.575) ,minimization of the mean absolute deviation (min MAD=0) show that this functional dependence $d_2=\phi(d_1/t)$ better submits to the linear than non-linear[3] regression model with equation view of $\mathbf{d_2=8.99+0.07(d_1/t)}$ **(9)**.

As we see from Figure 2(b) that with increasing of ratio values (d_1/t) ,the value of external diameter (d_2) stainless chip increases considerably in accordance with regression equation view of $d_2=8.99+0.07 (d_1/t)$.

In Figure 2 (c) is shown the functional dependence of thickness (δ) stainless chip from the ratio of values (d_1/t) ,such as internal diameter (d_1) to clearance (t) between of chip wraps for this stainless chip ,i.e has a place the function view of $\delta=\phi_1(d_1/t)$.Analysis of the data and also the statistical characteristics (coefficient of determination $R^2=0.475$ and coefficient of correlation r=0.689) show that this functional dependence $\delta=\phi_1(d_1/t)$ better submits to the linear than non-linear[4] regression model with equation view of $\boldsymbol{\delta=0.046+0.019(d_1/t)}$ **(10)**.

As we see from Figure 2 (c) that with increasing of ratio (d_1/t) ,the value of thickness (δ) stainless chip increases considerably in accordance with the regression equation view of $\delta=0.046+0.019(d_1/t)$.

In Figure 2(d) and 2(e) are illustrated the residual plots (residual versus δ and d_1/t) of the above-named functional dependencies $d_2=\gamma(\delta)$ and $d_2=\phi(d_1/t)$ accordingly.

Analyzing the Figure 3(A) ,we see that functional surface $d_2=\gamma_0(\delta,d_1/t)$ is shown in view of three-dimensional drawing for the linear regression model with equation view of $Y_c=8.859+4.192X_1-0.009X_2{}'$ or $Y_c=8.859+4.192X_1-0.009(X_2/X_3)$ where $d_2=8.859+4.192\delta-0.009(d_1/t)$.

And Figure 3(B) shows that functional surface $d_2=\gamma_0(\delta,d_1/t)$ has view of trapezium on the face YOX'_2 with such coordinates of their points [sizes in millimeters]:

a($X_{1,1}=0, ,X_{2,1}=10.35, Y_1=9.19$); b($X_{1,2}=0, X_{2,2}=10.35, Y_2=0$); c($X_{1,3}=0 ,X_{2,3}=16.09, ,Y_3=0$); d($X_{1,4}=0, ,X_{2,4}=16.09, Y_4=10.64$).

3– Non-linear regression model has view of $d_2=6.397 (d_1/t)^{0.168}$

4 –Non-linear regression model has view of $\delta=-0.266+0.056(d_1/t) -0.001(d_1/t)^2$

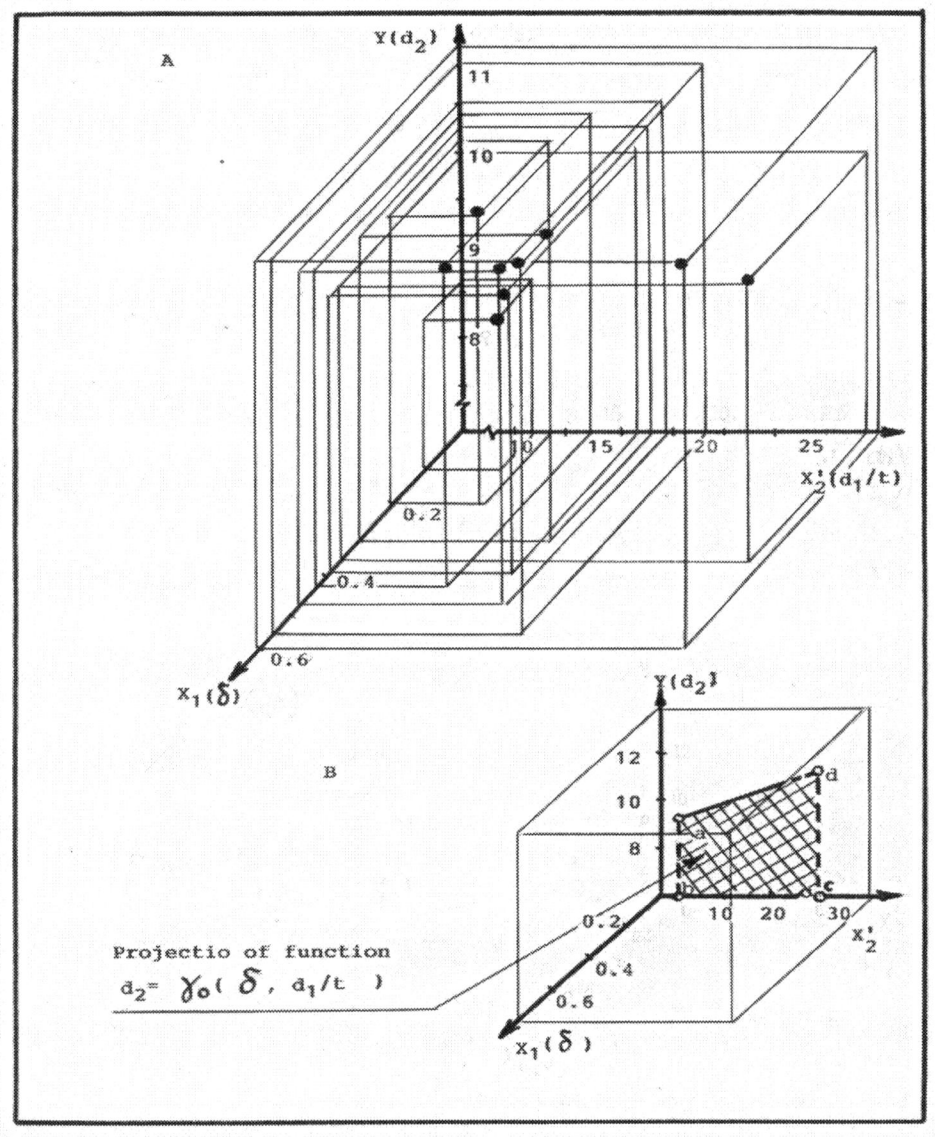

Figure 3

ANALYSIS OF FUNCTIONAL SURFACE $d_2 = \gamma_0 (\delta, d_1/t)$

From Figure 3(A) we see that functional surface $d_2 = \gamma_0(\delta, d_1/t)$,on which are situated the points of the linear regression equation view of $Y_c = 8.859 + 4.192X_1 - 0.009X'_2$ or $Y_c = 8.859 + 4.192X_1 - 0.009(X_2 / X_3)$ where $d_2 = 8.859 + 4.192\delta - 0.009 (d_1/t)$ has the following coordinates of their peaks in three- dimensional drawing (sizes in millimeters):
a' ($X_{1,1} = 0.10, X_{2,1} = 10.35, Y_1 = 9.19$);b'($X_{1,2} = 0.10, X_{2,2} = 10.35, Y_2 = 0$);   c'   ($X_{1,3} = 0.46$, ,$X_{2,3} = 16.09, Y_3 = 0$); d'($X_{1,4} = 0.46, X_{2,4} = 16.09, Y_4 = 10.64$).

- The module of vectors $|a'b'|$, $|b'c'|$, $|c'd'|$ and $|d'a'|$ could be calculated as:

$$|a'b'| = [(X_{1,2} - X_{1,1})^2 + (X_{2,2} - X_{2,1})^2 + (Y_2 - Y_1)^2]^{1/2} \quad (11)$$

$$|b'c'| = [(X_{1,3} - X_{1,2})^2 + (X_{2,3} - X_{2,2})^2 + (Y_3 - Y_2)^2]^{1/2} \quad (12)$$

$$|c'd'| = [(X_{1,4} - X_{1,3})^2 + (X_{2,4} - X_{2,3})^2 + (Y_4 - Y_3)^2]^{1/2} \quad (13)$$

$$|d'a'| = [(X_{1,4} - X_{1,1})^2 + (X_{2,4} - X_{2,1})^2 + (Y_4 - Y_1)^2]^{1/2} \quad (14)$$

At above-shown data ,we have the values: $|a'b'| = 9.19$, $|b'c'| = 5.75$, $|c'd'| = 10.64$ and $|d'a'| = 6.03$.

- Area of this functional surface is equal:

$$S_{a'b'c'd} = 1/2 \, [\, |b'c'| \cdot (|c'd'| - |a'b'|)] \cdot (|b'c'| \cdot |a'b'|) \quad (15)$$

At above-shown data , we have $S_{a'b'c'd'} = 55.15$ mm^2 .

All results of computations for the linear equation view of $Y_c = 8.859 + 4.192X_1 - 0.009X'_2$ or $Y_c = 8.859 + 4.192X_1 - 0.009(X_2/X_3)$ where $d_2 = 8.859 + 4.192\delta - 0.009(d_1/t)$ are given in Table 2.

Table 2 Evaluation of regression equation $Y_c = 8.859 + 4.192X_1 - 0.009X'_2$

A. Mean ,variance and standard deviation

Variable	Mean	Variance	Standard deviation
X_1	0.368	0.179	0.423
X'_2	16.969	248.827	15.774
Y	10.248	4.12	2.029

B. Results of multiple regression of Y on X_1 and X'_2

Parameter	Variable	Coefficients	Standard error	T-value
β_1	X_1	4.192	0.141	29.73
β_2	X'_2	-0.009	5.258	-0.002

C. Analysis of variance results

Regression
- Degrees of freedom 2
- Sum of squares 3.276
- Mean squares 1.638

Error
- Degrees of freedom 6
- Sum of squares 0.844
- Mean square 0.141
- Standard error of estimate 0.423
- F-value* 3.872

*Since $F = 3.872 < 5.14$ [$F_{0.05,2,6} = 5.14$] we cannot reject the hypothesis that both β_1 and β_2 are zero

D. Determination of residuals

Number	Observed	Estimated	Residual
1	10.03	10.14	-0.11
2	10.36	10.50	-0.14
3	8.69	9.19	-0.50
4	10.13	9.52	0.61
5	10.22	9.88	0.34
6	10.07	10.27	-0.20
7	10.49	10.64	-0.15
8	11.06	10.96	0.10
9	11.18	11.14	0.04

So, the parameters of this functional surface $d_2=\gamma_0(\delta,d_1/t)$ are the following:

1.Function $d_2=\gamma_0(\delta,d_1/t)$ better submits to the linear regression model with equation view of $Y_c=8.859+4.192X_1-0/009X'_2$ or $Y_c=8.859+4.192X_1-0.009(X_2 /X_3)$ where $d_2=8.859+4.192\delta-0.009(d_1/t)$ with such statistical characteristics :

- Coefficient of determination $R^2=0.795$
- Coefficient of correlation $r=0.892$
- Standard deviation $S_{y/x1,x'2}=0.375$
- Minimization of the mean square error (min MSE=0.094)
- Minimization of the mean absolute deviation (min MAD=0)

2.The total area of functional surface $d_2=\gamma_0(\delta,d_1/t)$ is equal $\sum S=55.15$ mm^2 with such coordinates of their peaks in three-dimensional drawing for this surface (sizes in millimeters): a'($X_{1,1}=0.10,X_{2,1}=10.35,Y_1=9.190$) ;b'($X_{1,2}=0.10,X_{2,2}=10.35,Y_2=0$); c' ($X_{1,3}=0.46,X_{2,3}=16.09,Y_3=0$) ,d' ($X_{1,4}=0.46 ,X_{2,4}=16.09, Y_4=10.64$).

D. Modification of function $d_2=\gamma_1(\delta, \delta/t)$

Analysis of the data and statistical characteristics (coefficient of determination $R^2=0.799$, coefficient of correlation $r=0.894$,standard deviation $S_{y/x1,x'2}=0.369$) and also the minimization of the mean square error (min MSE=0.091) and minimization of the mean absolute deviation (min MAD=0) show that this functional dependence $d_2=\gamma_1(\delta,\delta/t)$ better submits to the linear than non-linear [5] regression model with equation view of

$$\mathbf{Y_c=8.743+5.092X_1-0.297(X_1/X_3)} \text{ or } \mathbf{d_2=8.743+5.092\delta-0.297 (\delta/t)} \quad \mathbf{(16)}.$$

In Figure 4(a) is shown the functional dependence of external diameter stainless chip from its internal diameter (d_1) ,i.e we have the function view of $d_2=\varphi_3(d_1)$. This dependence has the non-linear regression model with equation view of $Y_c=5.728X^{0.359}$ or $d_2=5.728d_1^{0.359}$. So ,we see from Figure 4(a) that with increasing of internal diameter (d_1) stainless chip ,the value of its external diameter (d_2) increases considerably with regression equation $d_2=5.728d_1^{0.359}$.

In Figure 4(b) is shown the functional dependence of external diameter (d_2) stainless chip from the ratio of values (δ/t) ,such as thickness (δ) to clearance (t) between of chip wraps for this stainless chip, i.e we have the function view of $d_2=\phi_2(\delta/t)$. Analysis of the data and statistical characteristics (coefficient of determination $R^2=0.609$,coefficient of correlation $r=0.784$),minimization of the mean absolute deviation (min MAD=0) and scatter plots are given in Figure 4(b),show that this functional dependence $d_2=\phi_2(\delta/t)$ better submits to the linear than non-linear[6] regression model with equation view of $Y_c=9.193+0.848X'_2$ or $Y_c=9.193+0.848(X_1/X_3)$ where $d_2=9.133+0.848(\delta/t)$. As we see from Figure 4(b) that with increasing of ratio values (δ/t) ,the value of external diameter (d_2) for stainless chip increases considerably in accordance with regression equation view of $d_2=9.133+0.848(\delta/t)$.

---

5 – Non-linear regression model has equation view of $d_2=9.16+4.63\delta-3.85\delta^2-0.004(\delta/t)$
6 – Non-linear regression model has equation view of $d_2=10.16(\delta/t)^{0.085}$

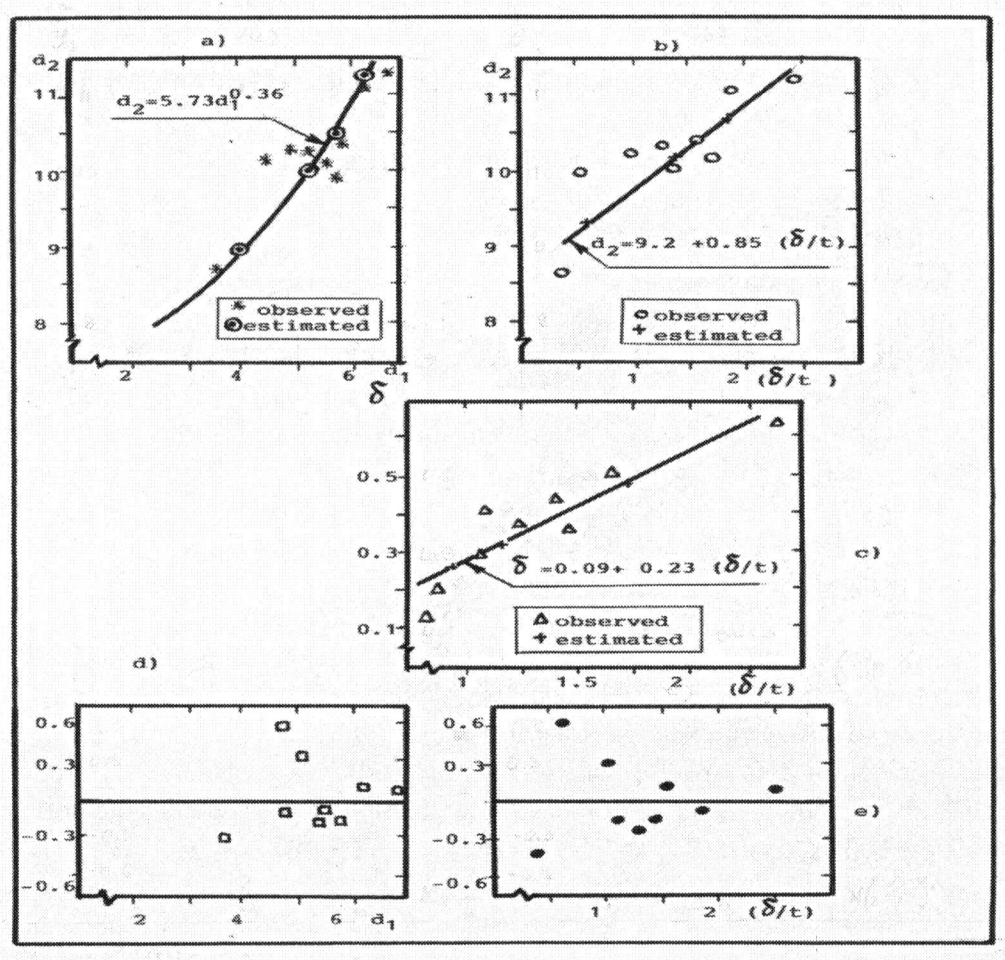

Figure 4

In Figure 4(c) is shown the functional dependence of thickness(δ) stainless chip from the ratio of values (δ/t) ,such as thickness (δ) of this chip to clearance (t) between of chip wraps ,i.e we have the function view of $\delta=\phi_3(\delta/t)$. Analysis of data and also the statistical characteristics(coefficient of determination R^2=1.0 and coefficient of correlation r=1.0 ,standard deviation $S_{y/x}$=0) and   also the minimization of the mean square error (min MSE=0) ,minimization of the mean absolute deviation (min MAD=0) show that this functional dependence $\delta=\phi_3(\delta/t)$ better submits to the linear than non-linear[7] regression model with equation view of   **Y_c=0.087+0.226X'_2** or **Y_c=0.087+0.226(X_1/X_3)**, where **δ=0.087+0.226 (δ/t)   (17) .**

In Figure 4(d) and 4(e) are illustrated residual plots (residual versus d_1 and residual versus (δ/t) the above-named functional dependencies $d_2=\phi_3(d_1)$ and $d_2=\phi_2(\delta/t)$.

7–Non-linear regression model has view of $\delta=0.314(\delta/t)^{0.809}$

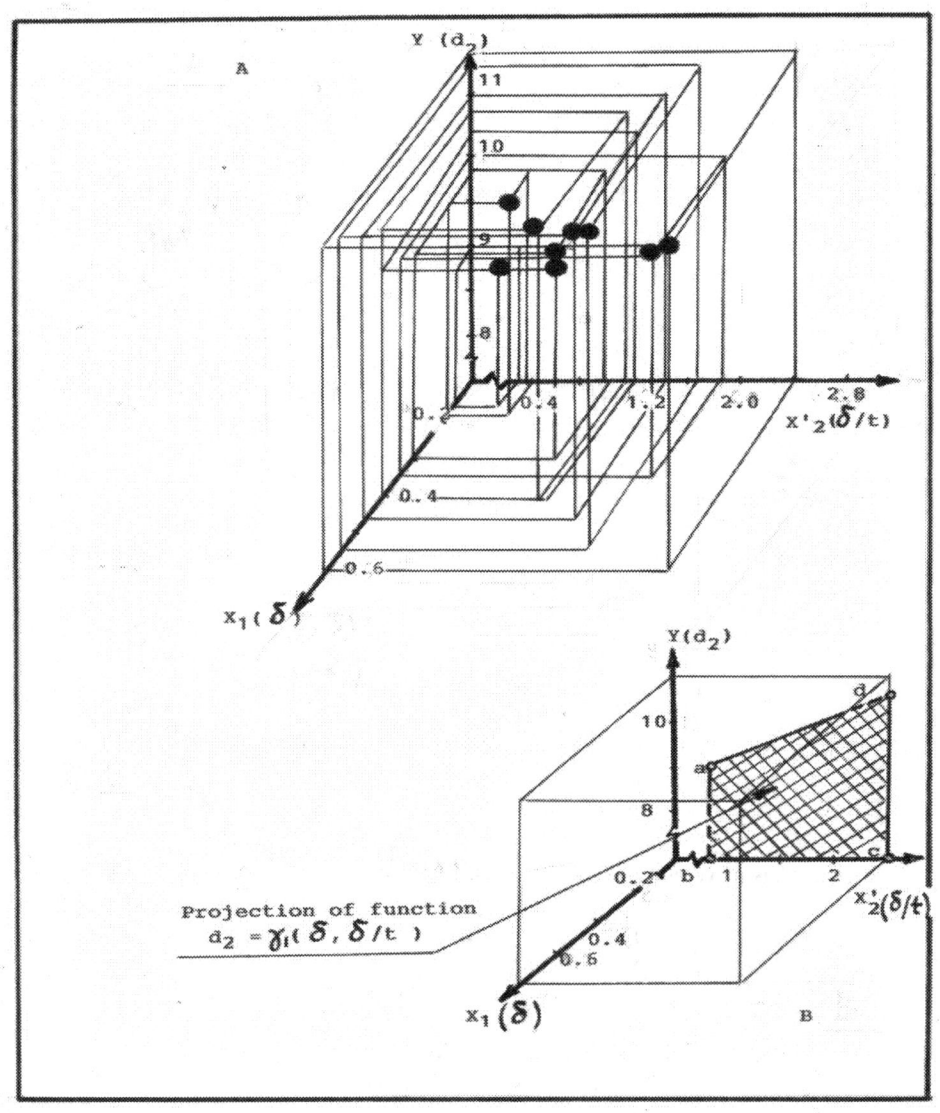

Figure 5

Analyzing the Figure 5(A) ,we see that functional surface $d_2=\gamma_1(\delta,\delta/t)$ is shown in view of three-dimensional drawing for the linear regression model with equation view of $Y_c=8.743+5.092X_1-0.297X'_2$ or $Y_c=8.743+5.092X_1-0.297(X_1/X_3)$ where $d_2=8.743+5.092\delta-0.297(\delta/t)$.

And Figure 5(B) shows that projection of functional surface $d_2=\gamma_1(\delta,\delta/t)$ has view of trapezium on the face YOX'_2 with such coordinates of their points(sizes in millimeters):

a ($X_{1,1}=0,X_{2,1}=0.29$,$Y_1=9.17$) ; b ($X_{1,2}=0,X_{2,2}=0.29$,$Y_2=0$); c ($X_{1,3}=0$,$X_{2,3}=2.4$, $Y_3=0$) ; d ($X_{1,4}=0$,$X_{2,4}=2.4$,$Y_4=11.09$).

157

Analysis of functional surface $d_2 = \gamma_1(\delta , \delta / t)$

From Figure 5(A) we see that functional surface $d_2 = \gamma_1(\delta , \delta / t)$, on which are situated the points of the linear regression equation   view of $Y_c = 8.743 + 5.092X_1 - 0.297X'_2$ or $Y_c = 8.743 + 5.092X_1 - 0.297(X_1/X_3)$ where $d_2 = 8.743 + 5.092\delta - 0.297 (\delta / t)$ has the following coordinates of their peaks in three-dimensional drawing (sizes in millimeters): a' ($X_{1,1} = 0.10$, $X_{2,1} = 0.29$, $Y_1 = 9.17$); b'($X_{1,2} = 0.10$, $X_{2,2} = 0.29$, $Y_2 = 0$); c'($X_{1,3} = 0.60, X_{2,3} = 2.40, Y_3 = 0$); d'($X_{1,4} = 0.60$, $X_{2,4} = 2.40$, $Y_4 = 11.09$). Referencing to formulas (11),(12),(13) ,(14) and (15) we have $|a'b'| = 9.17$, $|b'c'| = 2.17$, $|c'd'| = 11.09$, $|d'a'| = 2.89$ and $S_{a'b'c'd'} = 21.98$ mm^2 .

The result of computations for the linear equation view of $Y_c = 8,743 + 5.092X_1 - 0.297X'_2$ or $Y_c = 8.743 + 5.092X_1 - 0.297(X_1/X_3)$,where $d_2 = 8.743 + 5.092\delta - 0.297(\delta/t)$ are given in Table 3.

Table 3 Evaluation of regression equation $Y_c = 8.743 + 5.092X_1 - 0.297X'_2$

A. Mean ,variance and standard deviation

Variable	Mean	Variance	Standard deviation
X_1	0.368	0.209	0.457
X'_2	1.244	3.478	1.865
Y	10.248	4.120	2.029

B. Results of multiple regression of Y on X_1 and X'_2

Parameter	Variable	Coefficients	Standard error	T-value
β_1	X_1	5.092	0.152	33.50
β_2	X'_2	−0.297	0.622	−0.480

C.   Analysis of variance results

Regression
- Degrees of freedom 2
- Sum of squares   3.301
- Mean squares   1.651

Error
- Degrees of freedom   6
- Sum of squares   0.819
- Mean square   0.137
- Standard error of estimate   0.457
- F-value*   12.05

*Since F=12.05>5.14[$F_{0.05,2,6}$ =5.14] we can reject the hypothesis that both β_1 and β_2 are zero.

D.   Determination of residuals

Number	Observed	Estimated	Residual
1	10.03	10.07	-0.04
2	10.36	10.55	-0.19
3	8.69	9.17	-0.48
4	10.13	9.53	0.60
5	10.22	9.89	0.33
6	10.07	10.27	-0.20
7	10.49	10.66	-0.17
8	11.06	10.99	0.07
9	11.18	11.09	0.09

So ,the parameters of this functional surface $d_2 = \gamma_1(\delta, \delta/t)$ are the following :

1.Function $d_2 = \gamma_1(\delta, \delta/t)$ better submits to the linear regression model with equation view of $Y_c = 8.743 + 5.092X_1 - 0.297X'_2$ or $Y_c = 8.743 + 5.092X_1 - 0.297(X_1/X_3)$,where $d_2 = 8.743 + 5.092\delta - 0.297(\delta/t)$ with such statistical characteristics :

- Coefficient of determination $R^2 = 0.799$
- Coefficient of correlation $r = 0.894$
- Standard deviation $S_{y/x1,x'2} = 0.369$
- Minimization of the mean square error (min MSE=0.091)
- Minimization of the mean absolute deviation (min MAD=0).

2.The total area of functional surface $d_2 = \gamma_1(\delta, \delta/t)$ is equal $\sum S = 21.98$ mm^2 with such coordinates in three-dimensional drawing for this surface (sizes in millimeters):
a'($X_{1,1} = 0.10$, $X_{2,1} = 0.29$, $Y_1 = 9.17$) ;b' ($X_{1,2} = 0.10, X_{2,2} = 0.29, Y_2 = 0$) ; c' ($X_{1,3} = 0.60$, ,$X_{1,3} = 0.60, X_{2,3} = 2.40, Y_3 = 0$); d' ($X_{1,4} = 0.60, X_{2,4} = 2.40, Y_4 = 11.09$).

E. Modification of function $d_2 = \gamma_2(\delta, \delta/d_1)$

Analysis of data and the statistical characteristics (coefficient of determination $R^2 = 0.782$,coefficient of correlation $r = 0.884$,standard deviation $S_{y/x1,x'2} = 0.385$) and also minimization of the mean square error (min MSE=0.099) ,minimization of the mean absolute deviation (min MAD =0) show that this functional dependence $d_2 = \gamma_2(\delta, \delta/t)$ better submits to the non-linear than linear[8] regression model with equation view of
$$Y_c = 8.729 + 3.381X_1 - 0.01X_1^2 + 4X'_2 \quad \text{or} \quad Y_c = 8.729 + 3.381X_1 - 0.01X_1^2 + 4(X_1/X_2)$$
where $d_2 = 8.729 + 3.381\delta - 0.01\delta^2 + 4(\delta/d_1)$ (18) .

In Figure 6(a) is shown the functional dependence of external diameter (d_2) stainless chip from clearance (t) between of chip wraps ,i.e we have the function view of $d_2 = \gamma_3(t)$. This dependence has the non-linear regression model with equation view of $Y_c = 1.14X^{-0.56}$ or $t = 1.14d_2^{-0.56}$,where $d_2 = (0.877t)^{-1.786}$ (19) .

S ,we see from Figure 6(a) that with increasing of clearance (t) between of chip wraps for stainless chip ,the value of its external diameter (d_2) decreases in accordance with the regression equation view of $d_2 = (0.877t)^{-1.786}$.

In Figure 6(b) is shown the functional dependence of external diameter ($d_2$0 stainless chip from the ratio of values (δ/d_1) ,such as thickness (δ) to the internal diameter ($d_1$0 for this stainless chip ,i.e we have the function view of $d_2 = \phi_4(\delta/d_1)$.

Analysis of data and also the statistical characteristics(coefficient of determination $R^2 = 0.50$,coefficient of correlation $r = 0.707$) and scatter plots are given in Figure 6(b) show that this functional dependence $d_2 = \phi_4(\delta/d_1)$ better submits to the linear than non-linear[9] regression model with equation view of $Y_c = 8.97 + 18.5X'_2$ or
$$d_2 = 8.97 + 18.5(\delta/d_1) \quad (20).$$

8- Linear regression model has equation view of $d_2 = 8.73 + 3.38\delta + 4(\delta/d_1)$
9- Non-linear regression model has equation view of $d_2 = 14.76(\delta/d_1)^{0.134}$

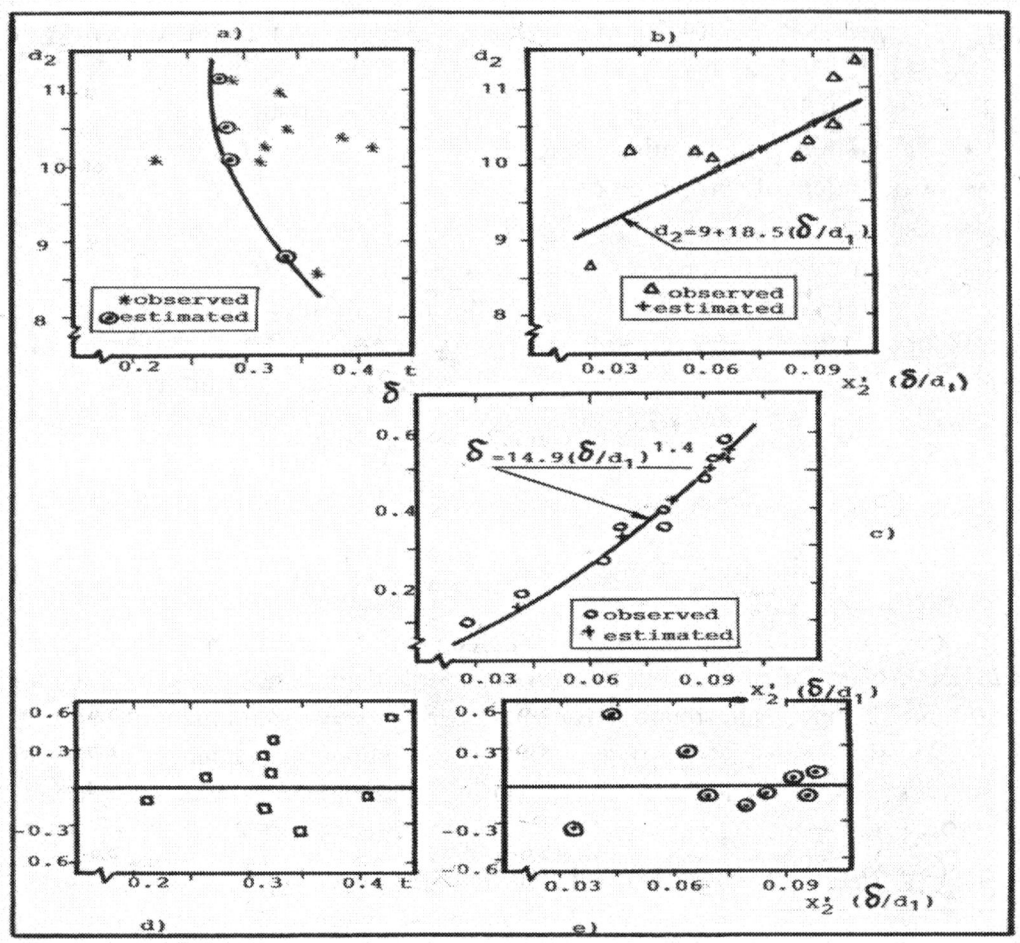

Figure 6

As we see from Figure 6(b) that with increasing of ratio values (δ /d_1) ,the value of external diameter (d_2) for stainless chip increases considerably in accordance with regression equation view of $d_2=8.97+18.5(\delta/d_1)$.

In Figure 6(c) is shown the functional dependence of thickness (δ) stainless chip from the ratio of values (δ/d_1) ,such as thickness (δ) of this chip to internal diameter (d_1) ,i.e we have the function view of $\delta=\phi_5(\delta/d_1)$.Analysis of data and the statistical characteristics (coefficient of determination R^2=1.0,coefficient of correlation r=1.0,standard deviation $S_{y/x'2}$=0.075 ,minimization of the mean square error (min MSE=0) ,and minimization of the mean absolute deviation (min MAD=0),show that this functional dependence $\delta=\phi_5(\delta/d_1)$ better submits to the non-linear than linear[10] regression model with equation view of $\mathbf{Y_c=14.93X'_2{}^{1.394}}$ or $\mathbf{Y_c=14.93(X_1 /X_2)^{1.394}}$, where $\boldsymbol{\delta=14.93(\delta/d_1)^{1.394}}$ **(21)**

.

10- Linear regression model has equation view of $\delta= -0.35+10.33(\delta/d_1)$

In Figure 6(d) and 6(e) are illustrated residual plots (residual versus t and residual versus δ/d_1) of the above-named functional dependencies $d_2=\phi_2(t)$ and $d_2=\phi_4$ (δ/d_1).

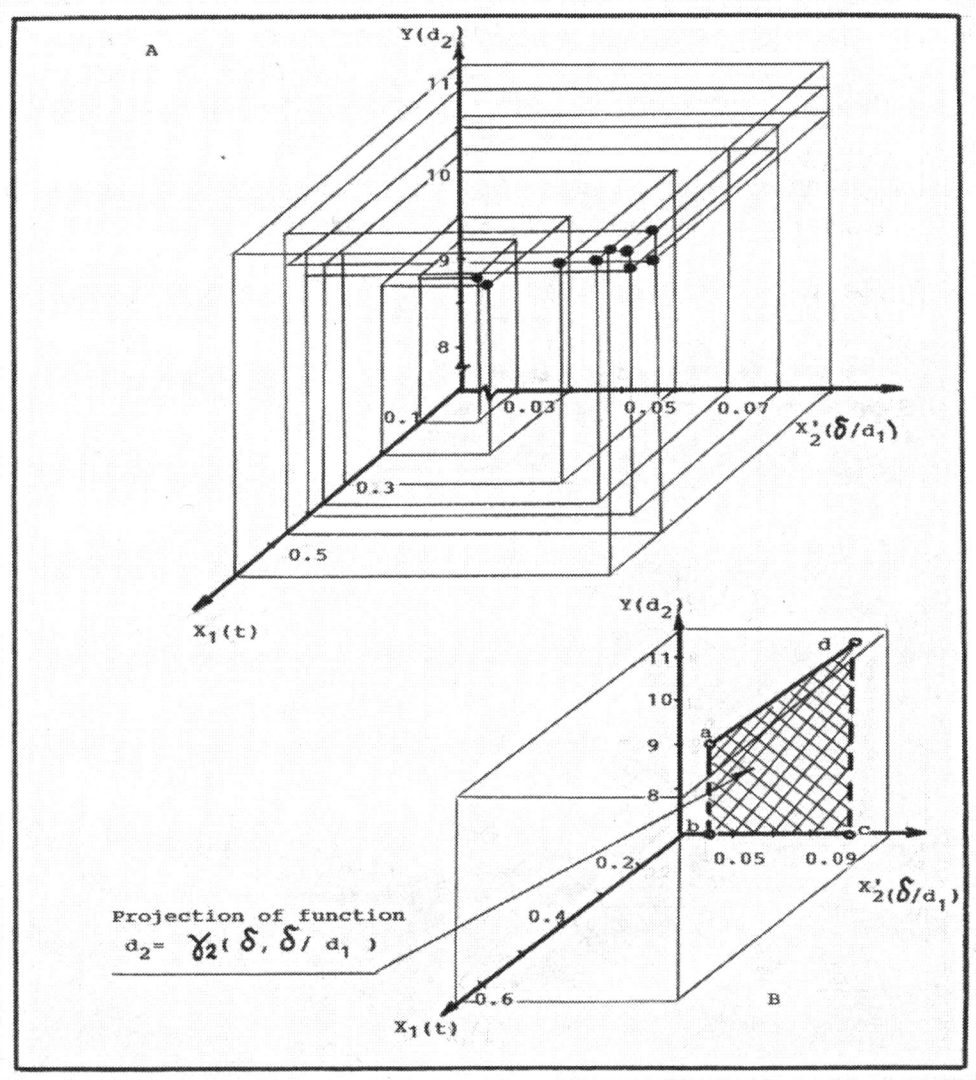

Figure 7

Analyzing the Figure 7(A) ,we see that functional surface $d_2=\gamma_2(\delta,\delta/d_1)$ is shown in view of three- dimensional drawing for the non-linear regression model with equation view of $Y_c=8.729+3.381X_1-0.01X_1^2+4X_2'$ or $Y_c=8.729+3.381X_1-0.01X_1^2+4(X_1/X_2)$,where $d_2=8.729+3.381\delta-0.01\delta_1^2+4(\delta/d_1)$.

Figure 7(B) shows that projection of this functional surface $d_2=\gamma_2(\delta,\delta/d_1)$on the face YOX$_2'$ has view of trapezium with such coordinates of their points (sizes in millimeters):
a ($X_{1,1}=0$,$X_{2,1}=0.03$,$Y_1=9.180$); b ($X_{1,2}=0$,$X_{2,2}=0.03$,$Y_2=0$); c($X_{1,3}=0$,$X_{2,3}=0.09$,$Y_3=0$); d ($X_{1,4}=0$,$X_{2,4}=0.09$,$Y_4=11.13$).

161

Analysis of functional surface $d_2=\gamma_2(\delta,\delta/d_1)$

From Figure 7(A) we see that functional surface $d_2=\gamma_2(\delta,\delta/d_1)$, on which the points are located of non-linear regression equation view of $Y_c=8.729+3.381X_1-0.01X_1^2+4X'_2$ or $Y_c=8.729+3.381X_1-0.01X_1^2+4(X_1/X_2)$, where $d_2=8.729+3.381\delta-0.01\delta^2+4(\delta/d_1)$ has a curvilinear character with the following coordinate of their peaks in three-dimensional drawing (sizes in millimeters): a'($X_{1,1}=0.10, X_{2,1}=0.03, Y_1=9.18$); b'($X_{1,2}=0.10, X_{2,2}=0.03, Y_2=0$); c'($X_{1,3}=0.60, X_{2,3}=0.09, Y_3=0$); d'($X_{1,4}=0.60, X_{2,4}=0.09, Y_4=11.13$).

In Figure 8 schematically is shown the functional; surface $d_2=\gamma_2(\delta,\delta/d_1)$ and graph for calculation of module for vectors $|a'b'|$, $|b'c'|$, $|c'd'|$ and $|d'a'|$.

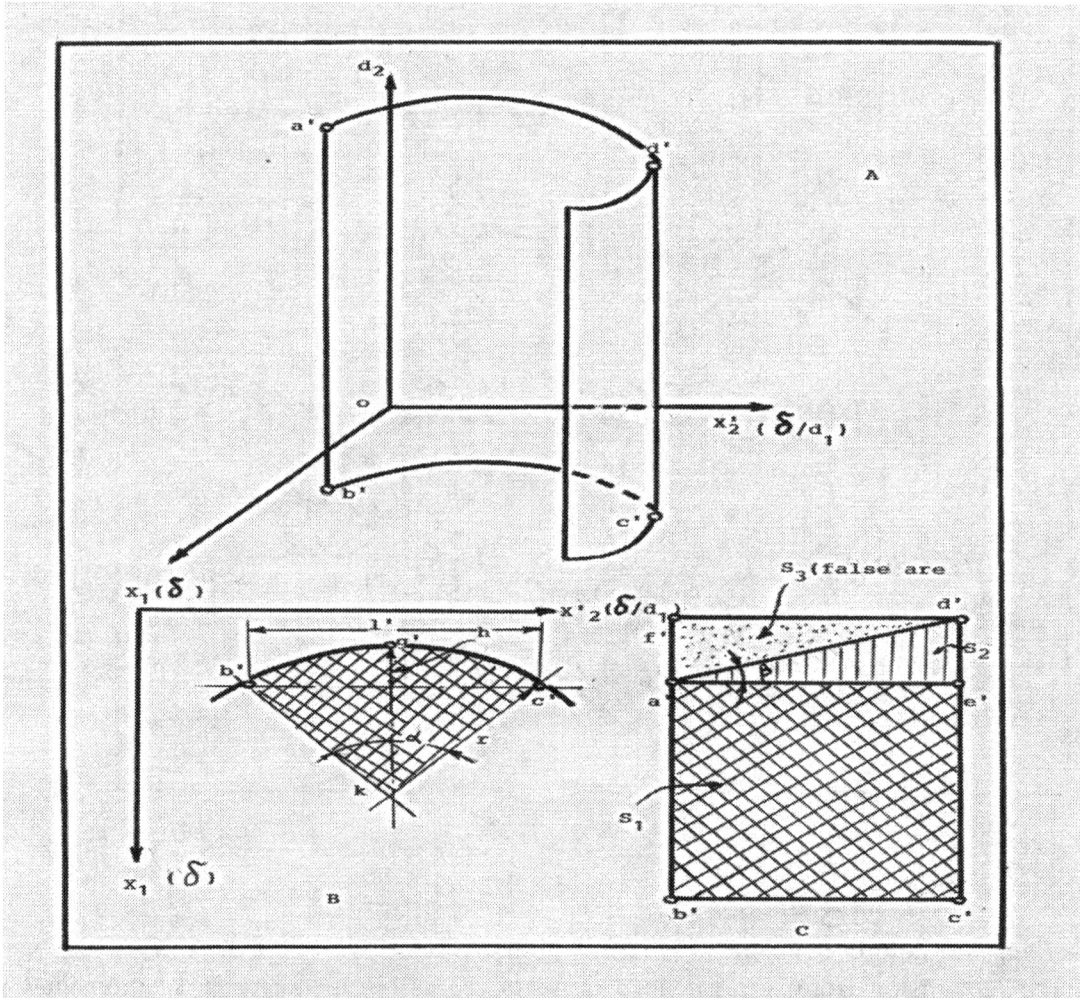

Figure 8 Graph for calculation of module vectors $|a'b'|$, $|b'c'|$, $|c'd'|$, $|a'd'|$ and area of functional surface $d_2=\gamma_2(\delta,\delta/d_1)$:
(A-total view of functional surface ; B-graph for calculation of module vector $|b'c'|$; C- graph for calculation of module vector $|a'd'|$ and area of this functional surface $d_2=\gamma_2(\delta,\delta/d_1)$.

162

1.The module of vector $|a'b'|$ and $|c'd'|$ can be calculated from the projection of this functional surface $d_2=\gamma_2(\delta,\delta/d_1)$ on the plane YOX'_2 (Figure 7B and 8A). So, the module of vector $|a'b'|=9.18$ and $|c'd'|=11.13$.

2.The module of vector $|b'c'|$ or it length can be calculated by formula $L=|b'c'|=0.01745\cdot r\cdot\alpha$,where L=length of arc (b'c'),r=radius of sector ,α=angle of sector. So , the module of vector $|kg'|$ is equal: $r=|kg'|=[(X_{1,6}-X_{1,5})^2+(X_{2,6}-X_{2,5})^2]^{1/2}$,

where coordinates of points g' and k: $g'(X_{1,6}=0.28 ,X_{2,6}=0.06)$; $k(X_{1,5}=0,X_{2,5}=0)$.At data $\alpha=82°$,$r=0.29$ we have the module of vector$|b'c'|=0.41$ and other parameters of circular segment which are equal: height $h=r[1-\cos(\alpha/2)]=0.07$,length of chord $l_1=2[h(2r-h)]^{1/2}=0.38$.

3.The module of vector $|a'd'|$ and area of this functional surface $d_2=\gamma_2(\delta,\delta/d_1)$,as shown in Figure 8 (C) can be calculated by the following way: the module of length for curve $|a'e'|=|b'c'|=0.41$ and the module of length $|d'e'|=|c'd'|-|a'b'|=1.95$. For calculation of area S_1,S_2 , S_3we find the angle β which is equal $\tan\beta=(|d'e'|)/|a'e'|$,

where $\beta=78°10'$and then the module of length for curve $|a'd'|=|d'e'|/\sin78°10'=1.99$. Coordinate of point f' $(X_{1,1}=0.1 ,X_{2,1}=0.03,Y_4=11.13$ and module of length $|f'd'|=|b'c'|=0.41$ and area of $S_2=S_3=S/2=0.5[(|f'd'|\cdot|d'e'|]=0.4mm^2$. For these conditions we have the area $S_1=(|a'b'|\cdot|b'c'|)=3.76mm^2$ and total area ΣS of functional surface $d_2=\gamma_2(\delta,\delta/d_1)$ is equal $\Sigma S=4.16mm^2$.

The results of computations for the non-linear equation view of $Y_c=8.729+3.381X_1-0.01X_1^2+4X'_2$ are given in Table 4.

Table 4 Evaluation of regression equation $Y_c=8.729+3.381X_1-0.01X_1^2+4X'_2$

A. Mean, variance and standard deviation			
Variable	Mean	Variance	Standard deviation
X_1	0.368	0.209	0.457
X'_2	0.069	0.002	0.045
Y	10.248	4.12	2.029

B. Results of multiple regression of Y on X_1 and X'_2				
Parameter	Variable	Coefficients	Standard error	T-value
β_1	X_1	3.381	0.152	22.24
β_2	X'_2	4	0.015	0.004

C. Analysis of variance results

Regression : Degrees of freedom 2;Sum of squares 3.232;Mean square 1.616

Error : Degrees of freedom 6 ;Sum of squares 0.888;Mean square 0.148

Standard error of estimate 0.457

F-value* F=10.92

*Since F=10.92> 5.14[$F_{0.05,2,6}=5.14$] we can reject the hypothesis that both β_1 and β_2 are zero.

D. Determination of residuals			
Number	Observed	Estimated	Residual
1	10.03	10.21	-0.18
2	10.36	10.47	-0.11
3	8.69	9.18	-0.49
4	10.13	9.49	0.64
5	10.22	9.90	0.32
6	10.07	10.29	-0.22
7	10.49	10.64	-0.15
8	11.06	10.92	0.14
9	11.18	11.13	0.05

So , the parameters of this functional surface $d_2=\gamma_2(\delta,\delta/d_1)$ are the following :

1.Function $d_2=\gamma_2(\delta,\delta/d_1)$ better submits to the non-linear regression model in view of curvilinear surface ,as is shown in Figure 7(A) and 8(A), and describes by equation view of $d_2=8.729+3.381\delta-0.01\delta^2+4(\delta/d_1)$.

2.The total area of functional surface $d_2=\gamma_2(\delta,\delta/d_1)$ is equal $\sum S=4.16 mm^2$ with such coordinates in three-dimensional drawing for this surface: $a'(X_{1,1}=0.1 , X_{2,1}=0.03 , , Y_1=9.18); b' (X_{1,2}=0.1 , X_{2,2}=0.03 , Y_2=0); c'(X_{1,3}=0.6 , X_{2,3}=0.09, Y_3=0); d' (X_{1,4}=0.6 , , X_{2,4}=0.09, Y_4=11.13).$

6.2 Dependence of external diameter (d_2) from thickness (δ), number of chip wraps (ω) and width (B) of this stainless chip, i.e $d_2=\alpha_2(\delta,\omega,B)$ and also from their modifications : $d_2=\gamma_3(\delta,\omega/B)$, $d_2=\gamma_4(\delta,\delta/B)$ and $d_2=\gamma_5(\delta,\delta/\omega)$.

A. Function $d_2=\alpha_2(\delta,\omega,B)$

Analysis of data and the statistical characteristics (coefficient of determination $R^2=0.802$, coefficient of correlation r=0.896, standard deviation $S_{y/x1,x2}=0.404$) and also the minimization of the mean square error (min MSE=0.091) and minimization of the mean absolute deviation (min MAD=0) show that this functional dependence $d_2=\alpha_2(\delta,\omega,B)$ better submits to the linear than non-linear[11] regression model with equation view of

$$Y_c=7.833+4.273X_1+0.237X_2+1.476\cdot10^{-4} X_3 \quad (22) \quad or \quad d_2=7.833+4.273 \delta+0.237\omega+1.476\cdot10^{-4} B .$$

In Figure 9 (a) is shown the functional dependence of external diameter (d_2) stainless chip from its thickness (δ) ,i.e we have the function $d_2=\gamma(\delta)$. This dependence has the linear regression model with equation view of $Y_c=8.808+3.897X$ or $d_2=8.808+3.897\delta$. So ,we see from Figure 9(a) that with increasing of thickness (δ) stainless chip ,the value of external diameter (d_2) increases accordingly with regression equation $d_2=8.808+3.897\delta$.

In Figure 9(b) is shown the functional dependence of external diameter (d_2)stainless chip from the number of its wraps (ω) ,i.e we have the function view of $d_2=\phi_1(\omega)$. This dependence, as was above-shown, better submits to the linear regression model and has the regression equation view of $Y_c=13.137-0.811X$ or $d_2=13.137-0.811\omega$. So ,we see from Figure 9(b) that with increasing of number chip wraps (ω) stainless chip ,the value of external diameter (d_2) decreases in accordance   with regression equation view of $d_2=13.137-0.811\omega$.

In Figure 9 (c) is shown the functional dependence of external diameter (d_2) stainless chip from width (B) ,i.e has a place the function view of $d_2=\phi_2(B)$. This dependence better submits   to the linear regression model with equation view of $Y_c=3.026+1.237X$ or $d_2=3.026+1.237B$. So ,we see from Figure 9(c) that with increasing of chip width (B) ,the value of external diameter (d_2) stainless chip increases   considerably with the regression equation view of $d_2=3.026+1.237B$.

---

11 – Non-linear regression model has equation view of $d_2= -52.3+45.1\delta-54.6\delta^2 +3.1\omega+7.6B$

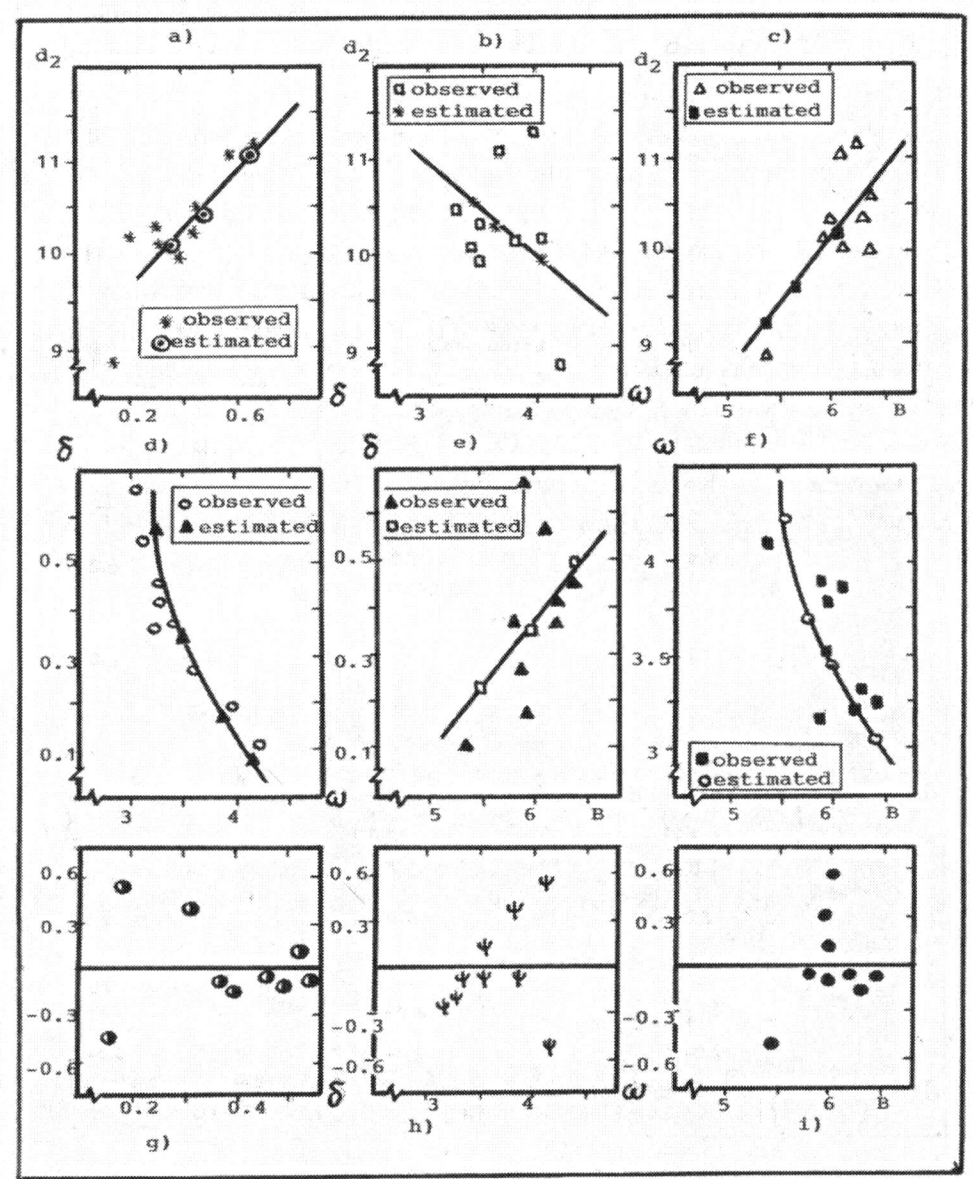

Figure 9

In Figure 9 (d) is shown the functional dependence of thickness (δ) stainless chip from its number chip wraps (ω) ,i.e we have the function view of $\delta=\alpha(\omega)$. This dependence better submits to the non-linear regression model with equation view of $Y_c=3.09X^{-0.123}$ or $\omega=3.09\delta^{-0.123}$, where $\delta =(0.323\omega)^{-8.13}$. So ,we see from Figure 9(d) that with increasing of number chip wraps (ω) ,the value of thickness (δ) stainless chip considerably decreases with regression equation $\delta= (0.323\omega)^{-8.13}$.

In Figure 9(e) is shown the functional dependence of thickness (δ) stainless chip from the width (B) of this chip ,i.e we have the function view of $\delta=f(B)$.

165

This dependence has the linear regression model with equation $Y_c = -1.236+0.275X$ or $\delta = -1.236+0.275B$. So, from Figure 9(e) we see that with increasing of chip width (B) ,the value of chip thickness (δ) increases considerably in accordance with regression equation $\delta = -1.236+0.275B$.

In Figure 9(f) is shown the functional dependence of number chip wraps (ω) stainless chip from width (B) of this chip, i.e we have the function view of $\omega=\gamma_2(B)$.This dependence has the non-linear regression model with equation view of $Y_c=703.1X^{-3.0}$ or $\omega=703.1B^{-3.0}$.So ,we see from Figure 9(f) that with increasing of chip width (B) ,the value of number chip wraps (ω) decreases considerably in accordance with regression equation view of $\omega=703.1B^{-3.0}$.In Figure 9(g),9(h) and 9(i) are illustrated the residual plots (residual versus δ,residual versus ω and residual versus B) of the above-named functional dependencies $d_2=\gamma(\delta)$,$d_2=\phi_1(\omega)$ and $d_2=\varphi_2(B)$ accordingly.

All results of computations for the linear equation view of $Y_c=7.833+4.273X_1+0.237X_2+1.476\cdot10^{-4}X_3$ are given in Table 5.

Table 5 Evaluation of regression equation $Y_c=7.83+4.27X_1+0.24X_2+1.48\cdot10^{-4}X_3$

A. Mean ,variance and standard deviatio			
Variable	Mean	Variance	Standard deviation
X_1	0.368	0.209	0.457
X_2	3.559	1.112	1.055
X_3	5.836	0.516	0.718
Y	10.248	4.120	2.029

B. Results of multiple regression of Y on X_1,X_2 and X_3				
Parameter	Variable	Coefficients	Standard error	T-value
β_1	X_1	4.273	0.152	28.112
β_2	X_2	0.237	0.352	0.673
β_3	X_3	$1.476\cdot10^{-4}$	0.239	0.001

C. Analysis of variance results

Regression

    Degrees of freedom   3

    Sum of squares   3.305

    Mean squares   1.102

Error

    Degrees of freedom   5

    Sum of squares   0.815

    Mean square   0.163

Standard error of estimate   0.457

F-value*   6.761

*Since $F=6.761>5.41$ $[F_{0.05,3,5}=5.41]$,we can reject the hypothesis that three parameters β_1,β_2 and β_3 are zero.

D. Determination of residuals

Number	Observed	Estimated	Residual
1	10.03	10.12	-0.09
2	10.36	10.40	-0.04
3	8.69	9.26	-0.57
4	10.13	9.55	0.58
5	10.22	9.89	0.33
6	10.07	10.18	-0.11
7	10.49	10.56	-0.07
8	11.06	10.98	0.08
9	11.18	11.29	-0.11

B. Modification function $d_2 = \gamma_3(\delta, \omega/B)$

Analysis of data and the statistical characteristics: coefficient of determination $R^2 = 0.824$, coefficient of correlation $r = 0.908$, standard deviation $S_{y/x1,x2} = 0.347$ and also the minimization of the mean square error (min MSE=0.08) and minimization of the mean absolute deviation (min MAD=0) show that this functional dependence $d_2 = \gamma_3(\delta, \omega/B)$ better submits to the linear than non-regression[12] model with equation view of $Y_c = 8.411 + 4.24X_1 + 0.383X'_2$ or $Y_c = 8.411 + 4.24X_1 + 0.383(X_2/X_3)$, where $d_2 = 8.411 + 4.24\delta + 0.383(\omega/B)$ (23).

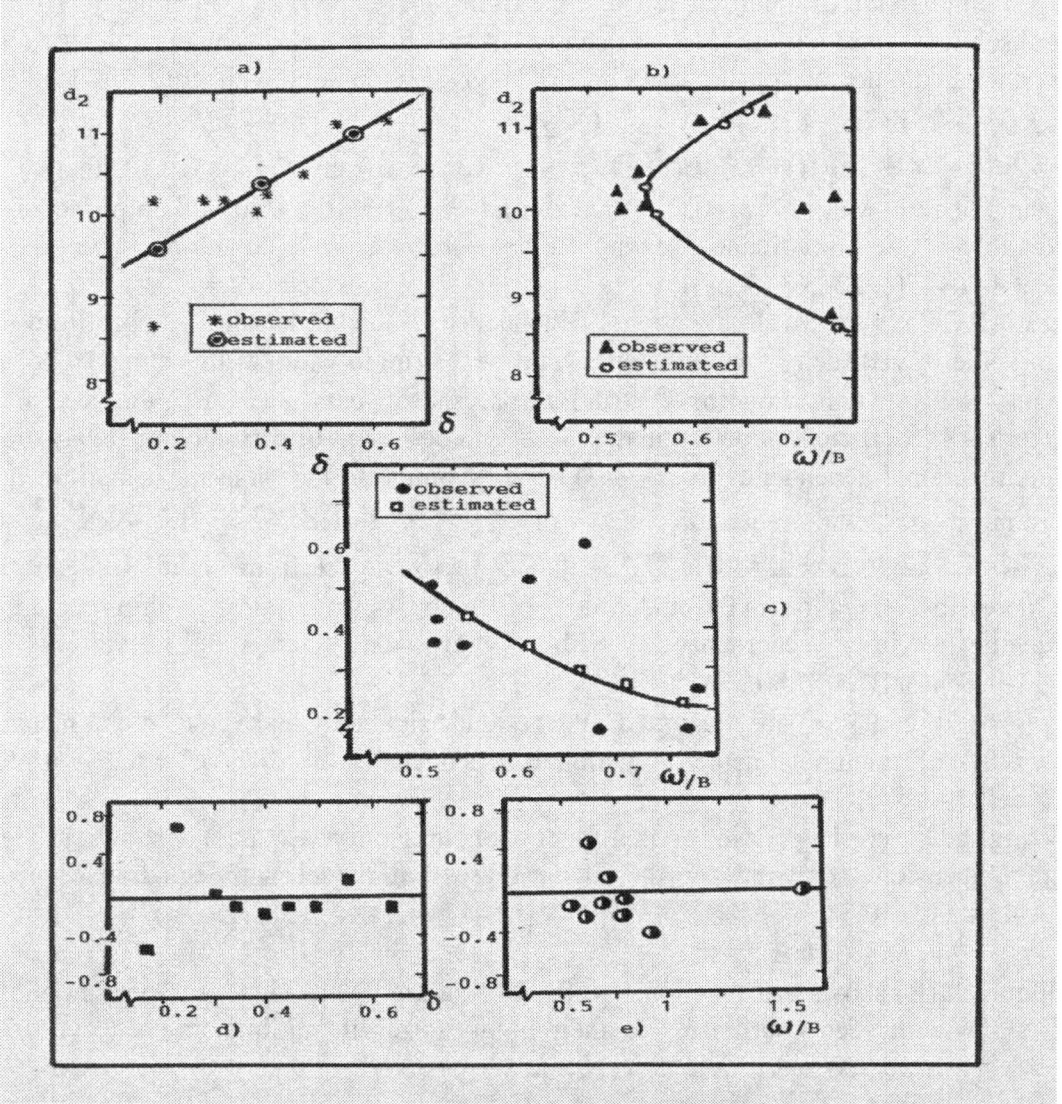

Figure 10

12- Non-regression model has equation view of $d_2 = 11.27 - 12.4\delta + 22.87\delta^2 - 0.1(\omega/B)$

In Figure 10(a) is shown the functional dependence of external diameter (d_2) stainless chip from its thickness (δ) ,i.e we have the function $d_2=\gamma(\delta)$.This dependence has the linear regression model with equation view of $Y_c=8.808+3.897X$ or $d_2=8.808+3.897\delta$. So, we see from Figure 10(a) that with increasing of thickness (δ) stainless chip ,the value of its external diameter (d_2) increases considerably with regression equation $d_2=8.808+3.897\delta$.

In Figure 10(b) is shown the functional dependency of external diameter (d_2) stainless chip from the ratio of values (ω/B) ,such as number of chip wraps to width (B) of this stainless chip ,i.e we have the function view of $d_2=\mu_1(\omega/B)$. Analysis of data, and also the statistical characteristics (coefficient of determination $R^2=0.20$,coefficient of correlation $r=0.45$), minimization of the mean square error (min MSE=0.384) and minimization of the mean absolute deviation (min MAD=0) show that this functional dependence $d_2=\mu_1(\omega/B)$ better submits to the non-linear than linear[13] regression model with equation view of

$$Y_c=13.564-7.487X'_2+3.346 \qquad (X'_2)^2 \qquad (\qquad 24 \qquad) \qquad \text{or}$$

$d_2=13.564-7.487(\omega/B)+3.346(\omega/B)^2$. So, we see from Figure 10(b) that with increasing of ratio values (ω/B), the of external diameter (d_2) stainless chip decreases considerably in accordance with the regression equation view of $d_2=13.564-7.487(\omega/B)+3.346(\omega/B)^2$.

In Figure 10(c) is shown the functional dependence of thickness stainless chip from the ratio of values (ω/B),such as the number of chip wraps (ω) to width of this chip(B) ,i.e has a place the function view of $\delta=\mu(\omega/B)$.Analysis of data and also the statistical characteristics (coefficient of determination $R^2=0.3$,coefficient of correlation $r=0.54$) show that this functional dependence $\delta=\mu_2(\omega/B)$ better submits to the non-linear than linear[14] regression model with equation view of $Y_c=1.214-1.854X'_2+0.788(X'_2)^2$ or

$$\delta=1.214-1.854(\omega/B)+0.788(\omega/B)^2 \qquad (\quad 25 \ ).$$ So ,we see from Figure 10 (c) that with increasing of ratio (ω/B) ,the value of thickness (δ) stainless chip decreases considerably in accordance with regression equation view of $\delta=1.214-1.854(\omega/B)+0.788(\omega/B)^2$.

In Figure 10(d) and 10(e) are illustrated the residual plots (residual versus δ and residual versus ω/B) of the above-named functional dependencies of $d_2=\gamma(\delta)$ and $d_2=\mu(\omega/B)$ accordingly.

Analyzing the Figure 11(A) ,we see that functional surface $d_2=\gamma_3(\delta,\omega/B)$ is shown in view of three-dimensional drawing for the linear regression model with equation view of $Y_c=8.411+4.24X_1+0.383X'_2$ or $Y_c=8.411+4.24X_1+0.383(X_2/X_3)$,where $d_2=8.411+4.24\delta+0.383(\omega/B)$.

And Figure 11(B) shows that functional surface $d_2=\gamma_3(\delta,\omega/B)$ has view of trapezium on the plane YOX_1 with such coordinates of their points (sizes in millimeters): a ($X_{1,1}=0.18$, ,$X_{2,1}=0$,$Y_1=9.44$);b ($X_{1,2}=0.18$,$X_{2,2}=0$,$Y_2=0$); c ($X_{1,3}=0.54$,$X_{2,3}=0$,$Y_3=0$); d ($X_{1,4}=0.54$, ,$X_{2,4}=0$, $Y_4=10.94$).

---

13 – Linear regression model has equation view of $d_2=10.45-0.28(\omega/B)$
14 -Linear regression model has equation view of $\delta=0.82-0.63(\omega/B)$

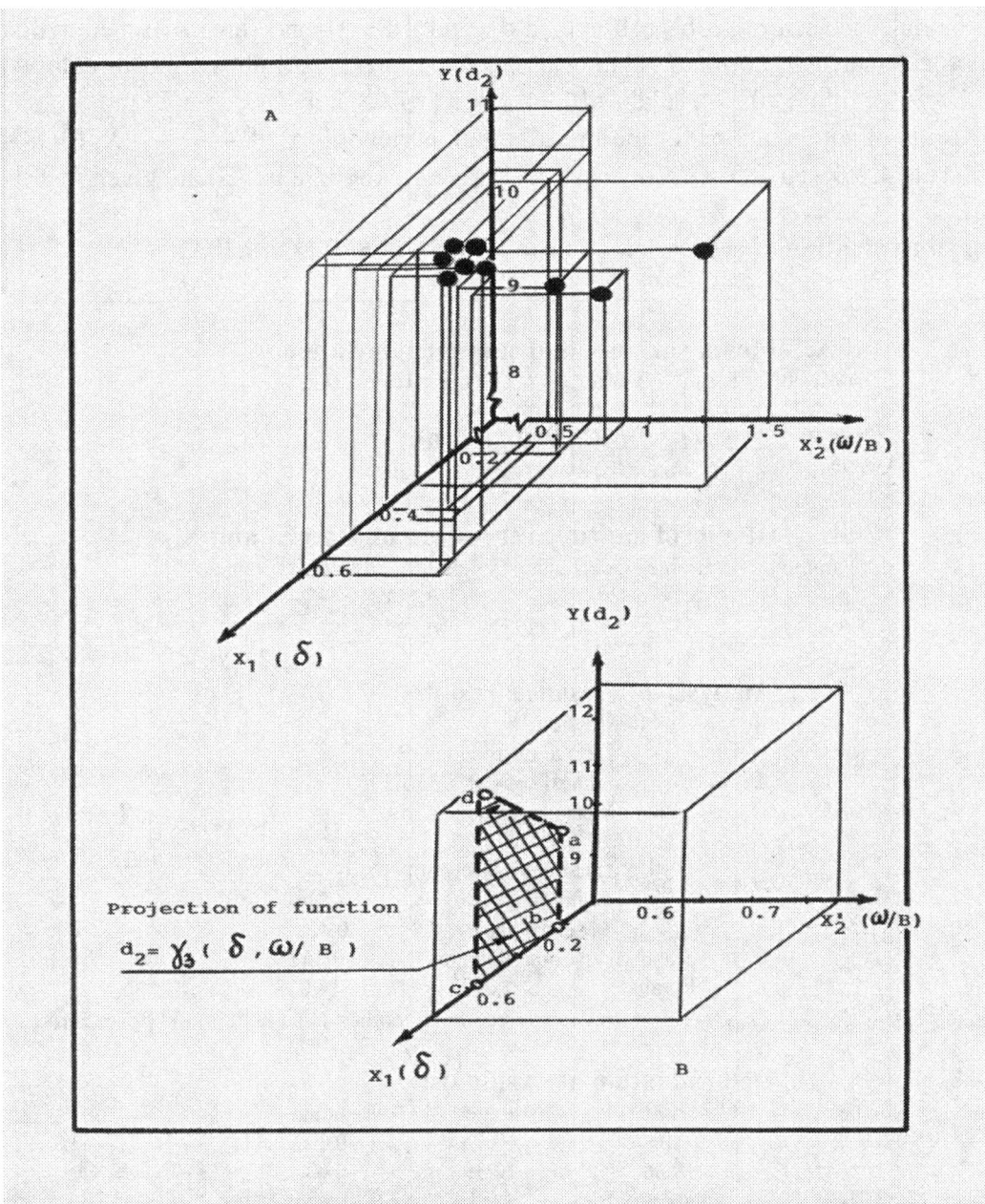

Figure 11

Analysis of functional surface $d_2 = \gamma_3(\delta, \omega/B)$

From Figure 11(A) we see that functional surface $d_2 = \gamma_3(\delta, \omega/B)$, on which are situated the points of the linear regression equation view of $Y_c = 8.411 + 4.24X_1 + 0.383X'_2$ or $Y_c = 8.411 + 4.24X_1 + 0.383(X_2/X_3)$, where $d_2 = 8.411 + 4.24\delta + 0.383(\omega/B)$ has the following coordinates of their peaks in three-dimensional drawing (sizes in millimeters): a'($X_{1,1}=0.18$, $X_{2,1}=0.69, Y_1=9.44$); b'($X_{1,2}=0.18, X_{2,2}=0.69, Y_2=0$); c'($X_{1,3}=0.54, X_{2,3}=0.62$, , $Y_3=0$); d' ($X_{1,4}=0.54, X_{2,4}=0.62, Y_4=10.94$).

The module of vectors $|a'b'|$, $|b'c'|$, $|c'd'|$ and $|d'a'|$ and area of functional surface can be calculated by formula (11),(12),(13),(14) and (15). At above-shown data we have $|a'b'|$=9.44, $|b'c'|$=0.37, , $|c'd'|$=10.94 , $|d'a'|$=1.28 and $S_{a'b'c'd'}$=3.77 mm^2.

All results of computations for the linear equation view of Y_c=8.411+4.24X_1+0.383X'_2 or Y_c=8.411+4.24X_1+0.383(X_2/X_3),where d_2=8.411+4.24δ+0.383(ω/B) are given in Table 6.

Table 6 Evaluation of regression equation Y_c=8.411+4.24X_1+0.383X'_2

A. Mean, variance and standard deviation

Variable	Mean	Variance	Standard deviation
X_1	0.368	0.209	0.457
X'_2	0.724	0.946	0.973
Y	10.248	4.121	2.030

B. Results of multiple regression of Y on X_1 and X'_2

Parameter	Variable	Coefficients	Standard error	T-value
β_1	X_1	4.24	0.151	28.08
β_2	X'_2	0.383	0.324	0.13

C. Analysis of variance results

Regression
Degree of freedom 2
Sum of squares   3.396
Mean squares   1.698

Error
Degrees of freedom   6
Sum of squares   0.724
Mean squares   0.121
Standard error of estimate   0.457
F-value*   14.033

*Since F=14.033>5.14 [$F_{0.05,2,6}$ =5.14] we can reject the hypothesis that both β_1 and β_2 are zero

D. Determination of residuals

Number	Observed	Estimated	Residual
1	10.03	10.15	-0.12
2	10.36	10.39	-0.03
3	8.69	9.14	-0.45
4	10.13	9.44	0.69
5	10.22	10.22	0
6	10.07	10.18	-0.11
7	10.49	10.56	-0.07
8	11.06	10.94	0.12
9	11.18	11.21	-0.03

1.Function $d_2=\gamma_3(\delta,\omega/B)$ better submits to the linear regression model with equation view of $Y_c=8.411+4.24X_1+0.383X'_2$ or $Y_c=8.411+4.24X_1+0.383(X_2/X_3)$, where $d_2=8.411+4.24\delta+0.383(\omega/B)$ with such statistical characteristics: coefficient of determination $R^2=0.824$,coefficient of correlation r=0.908,standard deviation $S_{y/x1,x'2}=0.347$,minimization of the mean square error (min MSE=0.08) and minimization of the mean absolute deviation (min MAD=0).

3.The total area of functional surface $d_2=\gamma_3(\delta,\omega/B)$ is equal $\sum S=3.77$ mm^2 with such coordinates of their points in three-dimensional drawing for this surface (sizes in millimeters): a'($X_{1,1}=0.18$,$X_{2,1}=0.69$,$Y_1=9.44$); b'($X_{1,2}=0.18$,$X_{2,2}=0.69$,$Y_2=0$); c'($X_{1,3}=0.54$,$X_{2,3}=0.62$,$Y_3=0$);d'($X_{1,4}=0.54$,$X_{2,4}=0.62$,$Y_4=10.94$).

C. Modification function $d_2=\gamma_4(\delta,\delta/B)$

Analysis of data and the statistical characteristics (coefficient of determination $R^2=0.797$,coefficient of correlation r=0.893,standard deviation $S_{y/x1,x'2}=0.373$) and also the minimization of the mean square error (min MSE=0.093),minimization of the mean absolute deviation (min MAD=0) show that this functional dependence $d_2=\gamma_4(\delta,\delta/B)$ better submits to the non-linear than linear[15] regression model with equation view of

$$\mathbf{Y_c=8.389+5.43X_1-3.669X_1^2+7X'_2} \quad or \quad \mathbf{Y_c=8.389+5.43X_1-3.669X_1^2+7(X_1/X_3)}$$

,where $\mathbf{d_2=8.389+5.43\delta-3.669\delta^2+7(\delta/B)}$ (26) .

In Figure 12(a) is shown the functional dependence of external diameter (d_2) stainless chip from its thickness (δ) , i.e we have the function view of $d_2=\gamma(\delta)$. This dependence has the linear regression model with equation view of $Y_c=8.808+3.897X$ or $d_2=8.808+3.897\delta$.So ,we see from Figure 12(a) that with increasing of thickness (δ) stainless chip ,the value of its external diameter (d_2) increases considerably with regression equation $d_2=8.808+3.897\delta$.

In Figure 12(b) is shown the functional dependency of external diameter (d_2) stainless chip from the ratio of values (δ/B) ,such as thickness (δ) of stainless chip to its width (B) ,i.e we have the function view of $d_2=\theta_1(\delta/B)$.Analysis of data and also the statistical characteristics (coefficient of determination $R^2=0.788$,coefficient of correlation r=0.888) show that this functional dependence $d_2=\theta_1(\delta/B)$ better submits to the linear than non-linear[16] regression model with equation view of $\mathbf{Y_c=8.82+23.67X'_2}$ or $\mathbf{d_2=8.82+23.67(\delta/B)}$ (27) .

As we see from Figure 12(b) that with increasing of ratio values (δ/B) ,the value of external diameter (d_2) stainless chip increases considerably in accordance with the regression equation view of $d_2=8.82+23.67(\delta/B)$.

In Figure 12(c) is shown the functional dependence of thickness (δ0 stainless chip from the ratio of values (δ/B) ,such as thickness (δ) of chip to width (B) of this chip, i.e has a place the function view of $\delta=\theta_2(\delta/B)$.

15- Linear regression model has equation view of $d_2=8.78+3.98\delta-0.01(\delta/B)$

16 – Non-linear regression model has equation view of $d_2=13.87(\delta/B)^{0.106}$

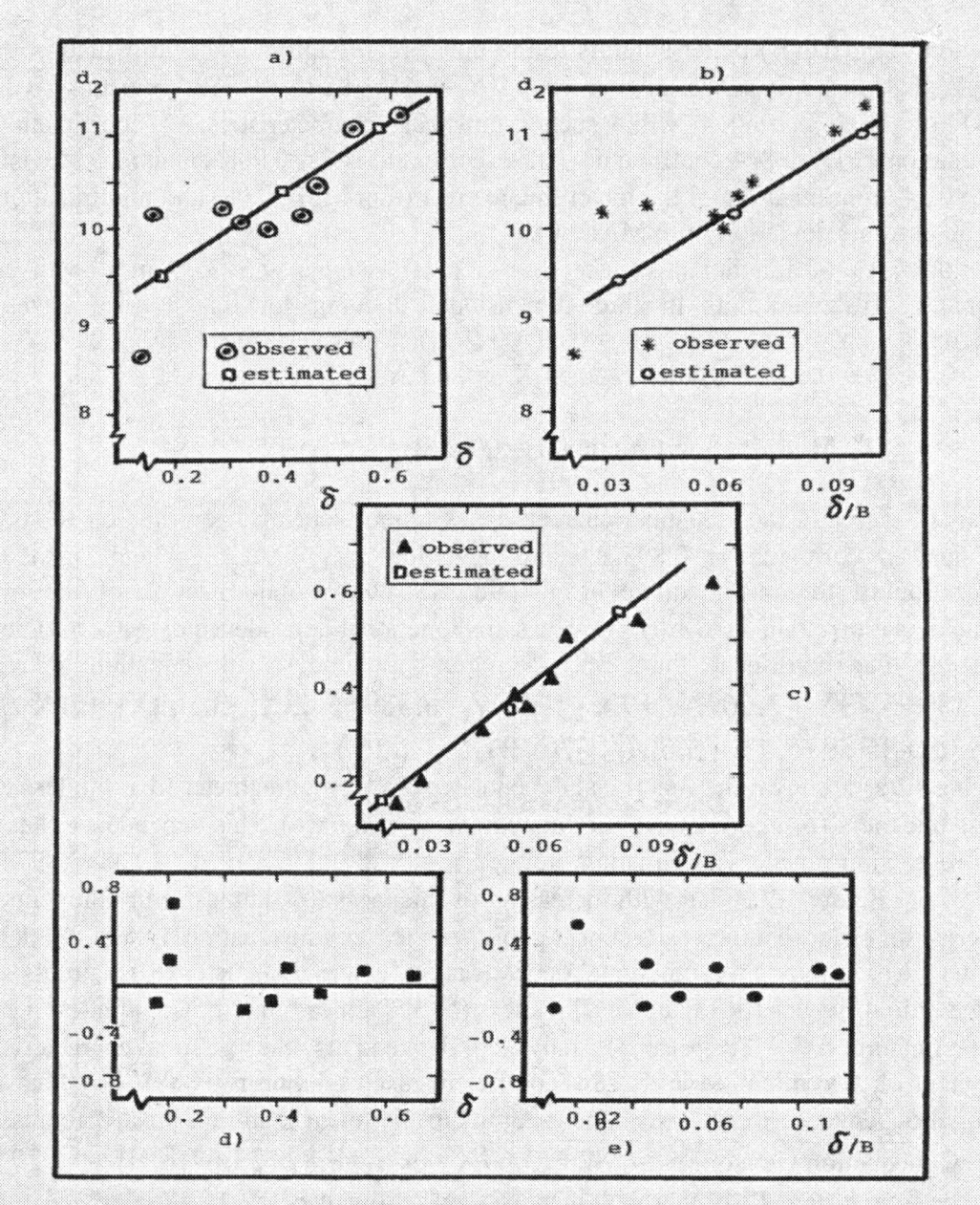

Figure 12

Analysis of data and also the statistical characteristics (coefficient of determination $R^2 = 1.0$ and coefficient of correlation $r = 1.0$) show that this functional dependence $\delta = \theta_2(\delta/B)$ better submits to the linear than non-linear[17] regression model with equation view of $\mathbf{Y_c} = \mathbf{-0.01 + 6X'_2}$, where $\delta = -0.01 + 6(\delta/B)$ (**28**) .

---

17 – Non-regression model has equation view of $\delta = 6.43(\delta/B)^{1.034}$

So ,we see from Figure 12 (c) that with increasing of ratio (δ/B) ,the value of thickness (δ) stainless chip increases considerably in accordance with regression equation view of $\delta = -0.01 + 6(\delta/B)$.

In Figure 12(d) and 12(e) are illustrated residual plots (residual versus δ and residual versus δ/B) of the above-named functional dependencies $d_2 = \gamma(\delta)$ and $d_2 = \theta_1(\delta/B)$ accordingly.

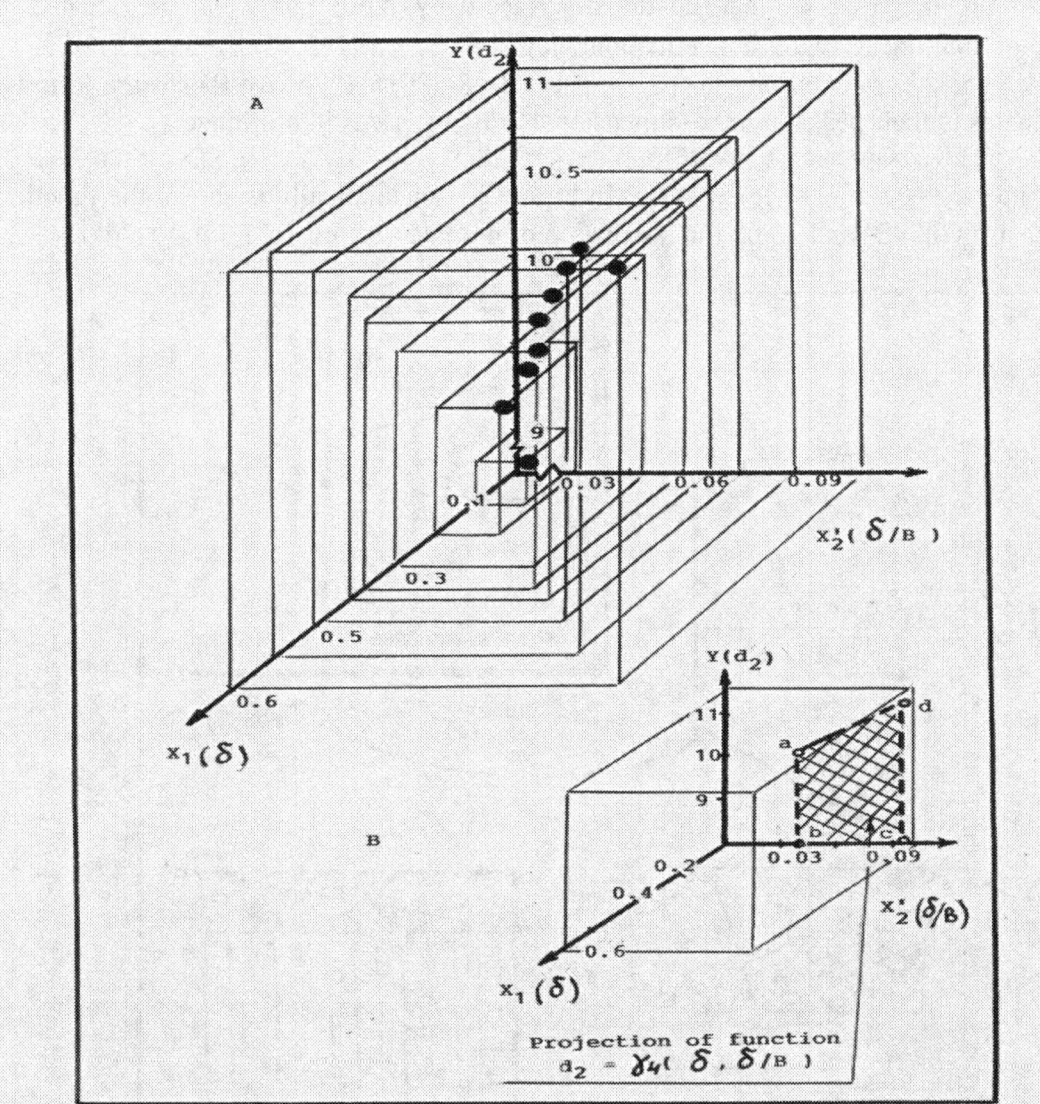

Figure 13

Analyzing the Figure 13(A) ,we see that functional surface $d_2 = \gamma_4(\delta, \delta/B)$ is shown in view of three-dimensional drawing for non-linear regression model with   equation view of $Y_c = 8.389 + 5.427X_1 - 3.669X12 + 7X'_2$ or $Y_c = 8.389 + 5.427X_1 - 3.669X_1^2 + 7(X_1/X_3)$,where $d_2 = 8.389 + 5.427\delta - 3.669\delta^2 + 7(\delta/B)$.

And Figure 13(B) shows that functional surface $d_2=\gamma_4(\delta,\delta/B)$ has view of trapezium on the plane YOX'_2 with such coordinate of their points (sizes in millimeters):
a($X_{1,1}=0$,$X_{2,1}=0.05$,$Y_1=9.96$);b($X_{1,2}=0$,$X_{2,2}=0.05$,$Y_2=0$);c($X_{1,3}=0$,$X_{2,3}=0.10$,$Y_3=0$); d($X_{1,4}=0$,$X_{2,4}=0.10$,$Y_4=11.05$).

Analysis of functional surface $d_2=\gamma_4(\delta,\delta/B)$

From Figure 13(A) we see that functional surface $d_2=\gamma_4(\delta,\delta/B)$,on which are situated the points of non-linear regression equation view of $Y_c=8.389+5.427X_1-3.669X_1^2+7X'_2$ or $Y_c=8.389+5.427X_1-3.669X_1^2+7(X_1/X_3)$,where $d_2=8.389+5.427\delta+7(\delta/B)$ has the following coordinates of their peaks in three-dimensional drawing (sizes in millimeters):
a' $(X_{1,1}=0.28$,$X_{2,1}=0.05$,$Y_1=9.96)$; b'$(X_{1,2}=0.28$,$X_{2,2}=0.05$,$Y_2=0)$;c'$(X_{1,3}=0.6$,$X_{2,3}=0.1$, ,$Y_3=0$);d'$(X_{1,4}=0.6$,$X_{2,4}=0.1$,$Y_4=11.05)$.In Figure 14 schematically is shown the functional surface $d_2=\gamma_4(\delta,\delta/B)$ and graph for calculation of module vectors $|b'c'|$ and $|a'd'|$.

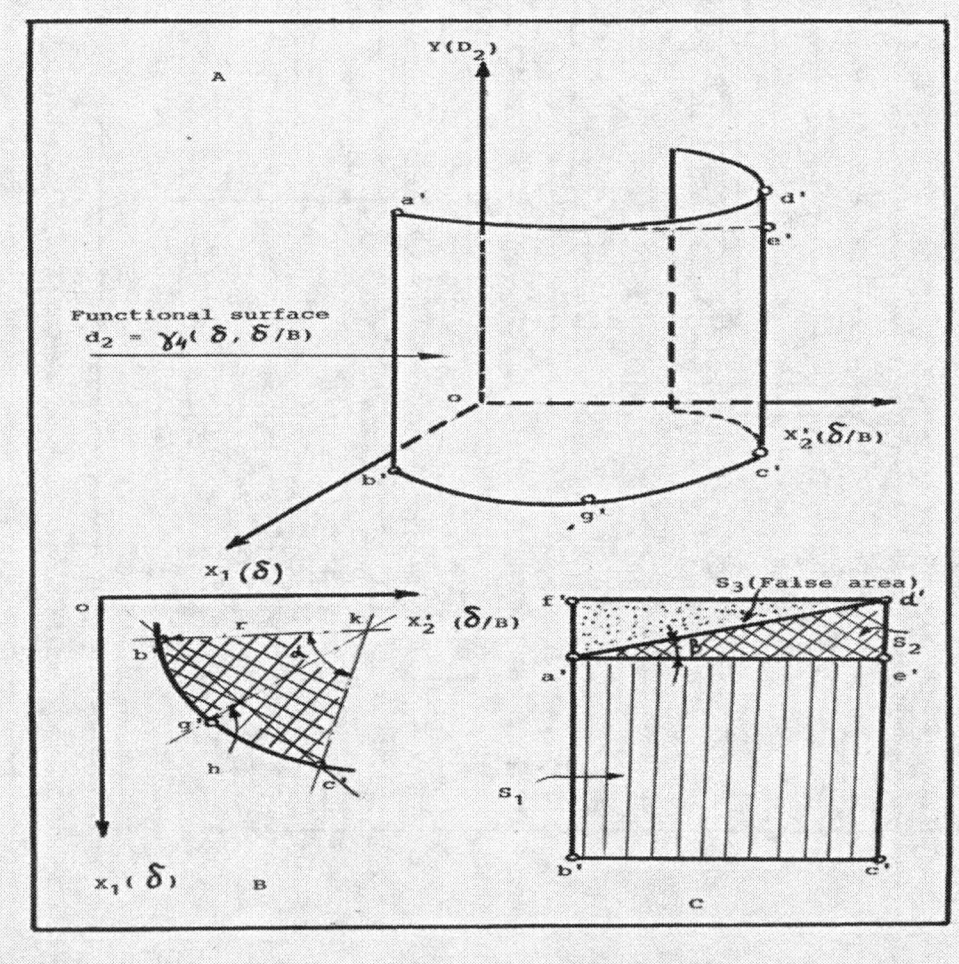

Figure 14 Graph for calculation of module vectors $|b'c'|$,$|a'd'|$ and area of functional surface $d_2=\gamma_4(\delta,\delta/B)$: (A-total view of functional surface, B-graph for calculation of module vector $|b'c'|$,C-graph for calculation of module vector $|a'd'|$ and area of this surface).

1.The module of vectors $|a'b'|$ and $|c'd'|$ can be calculated from the projection of this functional surface $d_2=\gamma_4(\delta,\delta/B)$ on the plane YOX'_2(Figure 13A and 14A) by formula (11) and (13).So, we have $|a'b'|=9.96$ and $|c'd'|=11.05$.

2.The module of vector $|b'c'|$ or it length can be calculated by formula $L=|b'c'|=0.01745r\alpha$,where L=length of arc(b'c'),r=radius of sector,α=angle of sector. So ,the module of vector $r=|kg'|=[(X_{1,6}-X_{1,5})^2+(X_{2,6}-X_{2,5})^2]^{1/2}$,where the coordinates of point g' and k: g'($X_{1,6}=0.46,X_{2,6}=0.08$) ,k ($X_{1,5}=0,X_{2,5}=0$). At data $\alpha=82°,r=0.46$ we have the module of vector $|b'c'|=0.66$ and other parameters of circular segment which are equal: height $h=r[1-\cos(\alpha/2)]=0.11$,the length of chord $l_1=2[h(2r-h)]^{1/2}=0.54$.
The graph for calculation of module vectors $|b'c'|$,$|a'd'|$ and area of functional surface $d_2=\gamma_4(\delta, \delta/B)$ is shown in Figure 14 (c) .

3.The module of vector $|a'd'|$ and area of this functional surface $d_2=\gamma_4(\delta,\delta/B)$,as shown in Figure 14 (c) ,can be calculated by the following way: the module of length for curve $|a'e'|=|b'c'|=0.66$ and module of length for $|d'e'|=|c'd'|-|a'b'|=1.09$.

4.For calculation of area $S_1,S_2$and S_3 we find the angle β which is equal: $\tan\beta=|d'e'|/|a'e'|$,where $\beta=59°$and the module of length for curve $|a'd'|=|d'e'|/\sin59°=1.27$.the coordinates of point f' is equal: f'($X_{1,1}=0.28,X_{2,1}=0.05$, ,$Y_4=11.050$ and module of length $|f'd'|=|b'c'|=0.66$ and area of $S_2=S_3=S/2=0.5(|f'd'|\cdot|d'e'|=0.36mm^2$. For these conditions , we have the area S_1 which is equal $S_1=|a'b'|\cdot|b'c'|=6.57mm^2$ and total area $\sum S$ for this functional surface $d_2=\gamma_4(\delta,\delta/B)$ is equal $\sum S=6.93$ mm^2.
All results of computation for the non-linear equation view of $Y_c=8.389+5.427X_1-3.669X_1^2+7X'_2$,where $d_2=8.389+5.427\delta-3.669\delta^2+7(\delta/B)$ are given in Table 7.
So, the parameters of this functional surface $d_2=\gamma_4$ ($\delta,\delta/B$) are the following:

1. Function $d_2=\gamma_4(\delta,\delta/B)$ better submits to the non-linear regression model in view of curvilinear surface ,as is shown in Figure 13 (A) and 14(A) describes by the equation view of $Y_c=8.389+5.427X_1-3.669X_1^2+7X'_2$ or $Y_c=8.389+5.427X_1-3.669X_1^2+7(X_1/X_3)$, where $d_2=8.389+5.427\delta-3.669\delta^2+7(\delta/B)$ with such statistical characteristics:
   - Coefficient of determination $R^2=0.797$
   - Coefficient of correlation r=0.893
   - Standard deviation $S_{y/x1,x'2}=0.373$
   - Minimization of the mean square error (min MSE=0.093)
   - Minimization of the mean absolute deviation (min MAD=0.038)
2.The total area of functional surface $d_2=\gamma_4(\delta,\delta/B)$ is equal $\sum S=6.93$ mm^2 with such coordinates of their points in three-dimensional drawing for this surface (sizes in millimeters):
a' ($X_{1,1}=0.28$,$X_{2,1}=0.05$,$Y_1=9.96$); b' ($X_{1,2}=0.28$,$X_{2,2}=0.05,Y_2=0$) ; c' ($X_{1,3}=0.60$, ,$X_{2,3}=0.10,Y_3=0$); d' ($X_{1,4}=0.60,X_{2,4}=0.10,Y_4=11.05$).

Table 7 Evaluation of regression equation $Y_c=8.389+5.427X_1-3.669X_1^2+7X'_2$

A. Mean , variance and standard deviation

Variable	Mean	Variance	Standard deviation
X_1	0.368	0.209	0.457
X'_2	0.063	0.006	0.077
Y	10.248	4.121	2.030

B. Results of multiple regression of Y on X_1 and X'_2

Parameter	Variable	Coefficients	Standard error	T-value
β_1	X_1	5.427	0.152	35.70
β_2	X'_2	7	0.026	269.23

C. Analysis of variance results

Regression :Degrees of freedom 2,sum of squares 3.286,mean squares 1.643
Error : Degrees of freedom 6, sum squares 0.834, mean square 0.139
Standard error of estimate 0.457
F-value * 11.82

*Since F=11.82> [$F_{0.05,2,6}$=5.14] we can reject the hypothesis that both β_1 and β_2 are zero.

D. Determination of residuals

Number	Observed	Estimated	Residual
1	10.03	10.31	-0.28
2	10.36	10.50	0.14
3	8.69	9.03	-0.34
4	10.13	9.47	0.66
5	10.22	9.96	0.26
6	10.07	10.31	- 0.24
7	10.49	10.63	-0.14
8	11.06	10.91	0.15
9	11.18	11.05	0.13

D. Modification of function $d_2=\gamma_5(\delta,\delta/\omega)$

Analysis of data and the statistical characteristics (coefficient of determination $R^2=0.789$,coefficient of correlation r=0.888,standard deviation $S_{y/x1,x'2}=0.381$) and also the minimization of the mean square error (min MSE=0.089) ,and minimization of the mean absolute deviation (min MAD=0) show that this functional dependence $d_2=\gamma_5(\delta,\delta/\omega)$ better submits to the linear than non-linear[18] regression model with equation view of
$$Y_c=8.773+2.855X_1+4X'_2 \qquad \text{or} \qquad Y_c=8.733+2.855X_1+4(X_1/X_2)$$ where $d_2=8.773+2.855\delta+4(\delta/\omega)$ (29) .

In Figure 15(a) is shown the functional dependence of external diameter (d_2) stainless chip from its thickness (δ) ,i.e we have the function $d_2=\gamma(\delta)$. This dependence has the linear regression model with equation view of $Y_c=8.808+3.897X$ or $d_2=8.808+3.897\delta$. So ,we see from Figure 15(a) that with increasing of thickness (δ) stainless chip ,the value of its external diameter (d_2) increases considerably with regression equation $d_2=8.808+3.897\delta$.

18– Non-linear regression model has equation view of $d_2=8.7-3\delta-0.01\delta^2+25(\delta/\omega)$

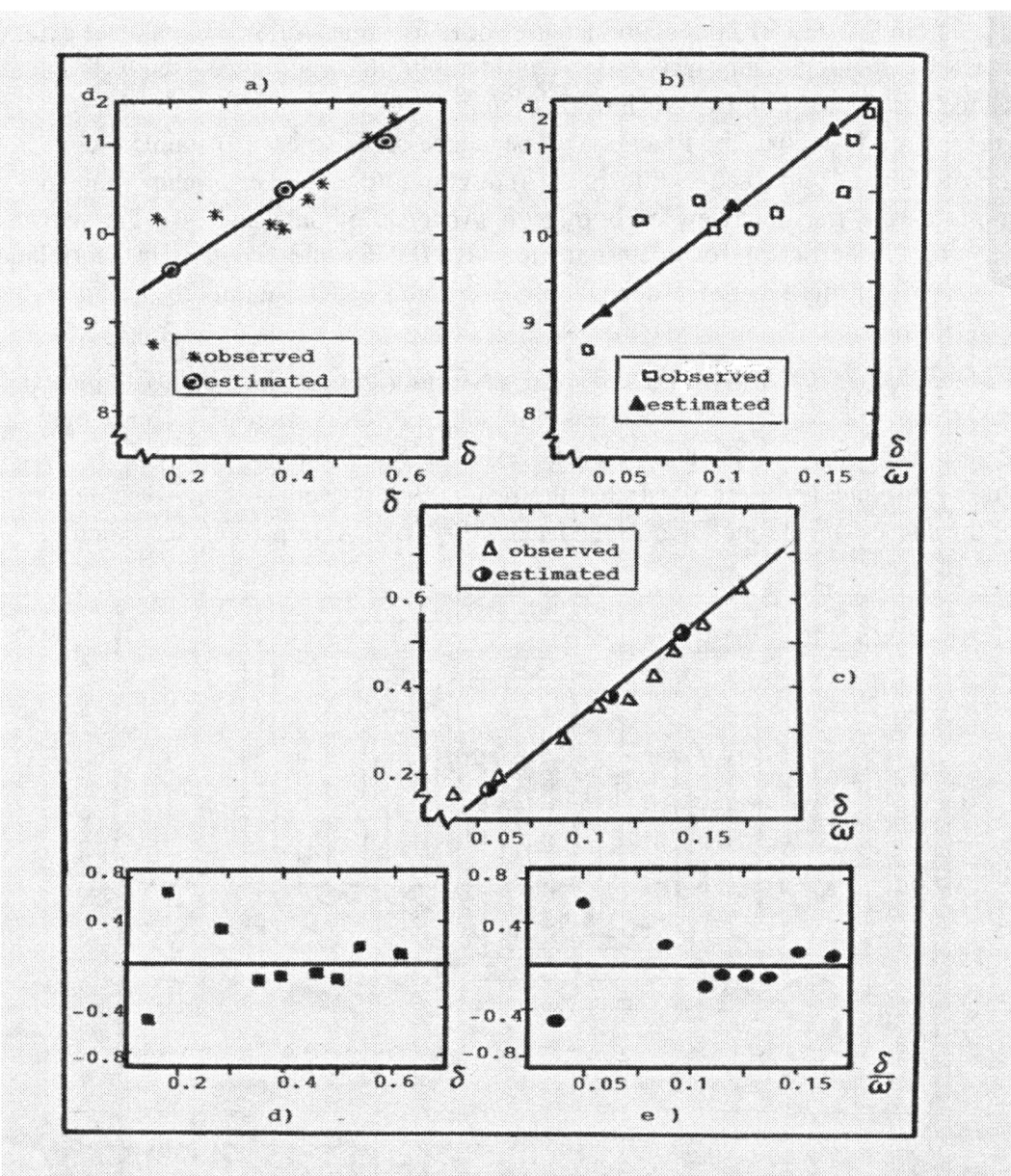

Figure 15

In Figure 15(b) is shown the functional dependency of external diameter (d_2) stainless chip from the ratio of values (δ/ω) ,such as thickness (δ) of stainless chip to its number of wraps (ω),i.e we have the function view of $d_2=\rho_1(\delta/\omega)$.Analysis of data and also the statistical characteristics(coefficient of determination R^2=0.805,coefficient of correlation r=0.897) shows that this functional dependence $d_2=\rho_1(\delta/\omega)$ better submits to the linear than non-linear[19] regression model with equation view of **Y_c=8.841+13.278 X'_2** or **d_2=8.841+13.278(δ/ω)** **(30).**

---

19 – Non-linear regression model has equation view of $d_2=12.71(\delta/\omega)^{0.091}$

As we see from Figure 15(b) that with increases of ratio values (δ/ω) ,the value of external diameter (d_2) stainless chip increases considerably in accordance with the linear regression equation view of $d_2=8.841+13.278(\delta/\omega)$.

In Figure 15(c) is shown the functional dependence of thickness (δ) stainless chip from the ratio of values (δ/ω) ,such as thickness (δ) of chip to the number of chip wraps(ω) ,i.e has a place the function view of $\delta=\rho_2(\delta/\omega)$. Analysis of data and also the statistical characteristics (coefficient of determination $R^2=0.976$ and coefficient correlation r=0.988)show that this functional dependence $\delta=\rho_2(\delta/\omega)$ better submits to the linear than non-linear[20] regression model with equation view of **$Y_c=0.015+3.333X'_2$**   or **$\delta=0.015+3.333(\delta/\omega)$**   **(31)** . So , we see from Figure 15 (c) that with increasing of ratio (δ/ω) ,the value of thickness (δ) stainless chip increases considerably in accordance with regression equation view of $\delta=0.015+3.333(\delta/\omega)$.

In Figure 15(d) and 15(e) are illustrated the residual plots (versus δ and residual versus δ/ω) of the above-named functional dependencies $d_2=\gamma(\delta)$ and $d_2=\rho_1(\delta/\omega)$ accordingly.

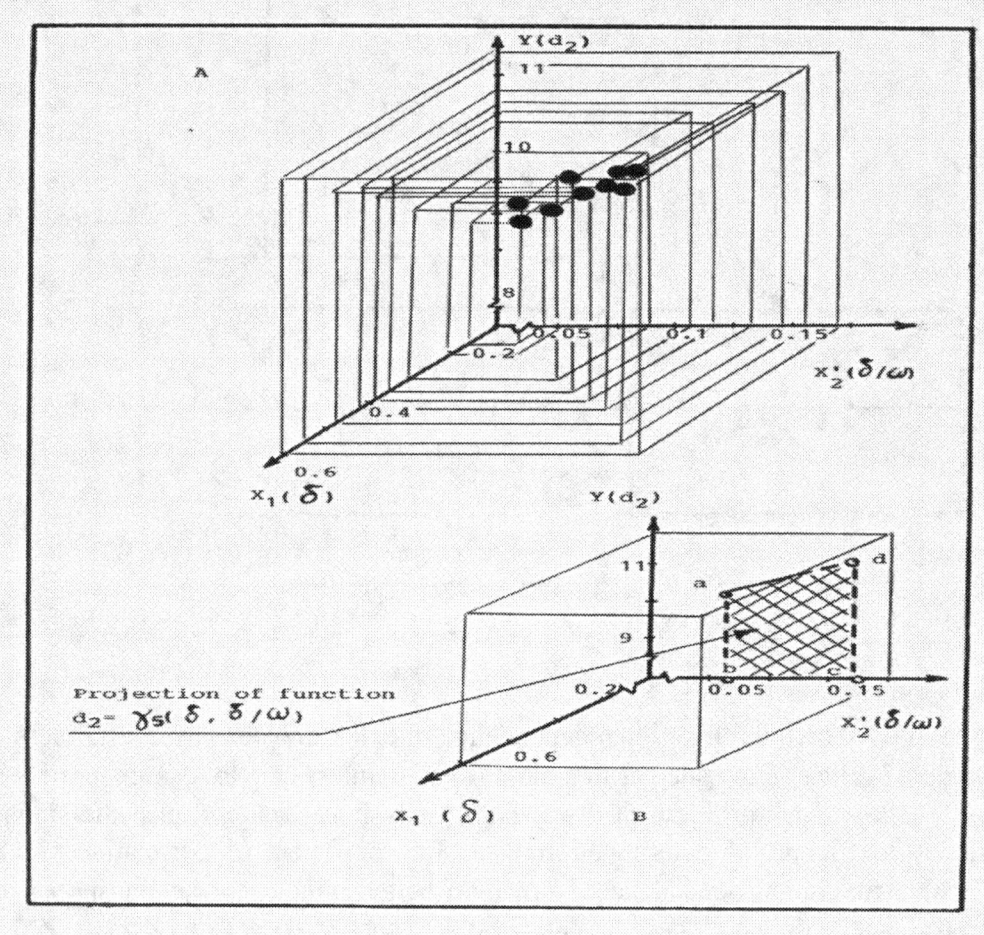

Figure 16

20– Non-linear regression model has equation view of $\delta=2.46(\delta/\omega)^{0.843}$

178

Analyzing the Figure 16(A) ,we see that functional surface $d_2=\gamma_5(\delta,\delta/\omega)$ is shown in view of three-dimensional drawing for the linear regression model with equation $Y_c=8.773+2.855X_1+4X'_2$ or $Y_c=8.773+2.855X_1+4(X_1/X_2)$, where $d_2=8.773+2.855\delta+$ $+4(\delta/\omega)$. And Figure 16(B) shows that functional surface $d_2=\gamma_5(\delta,\delta/\omega)$ has view of trapezium on the plane YOX_2 with such coordinates of their points(sizes in millimeters): a $(X_{1,1}=0 ,X_{2,1}=0.08 ,Y_1=9.89)$; b $(X_{1,2}=0 , X_{2,2}=0.08 , Y_2=0)$; c $(X_{1,3}=0,X_{2,3}=0.16,Y_3=0)$; d$(X_{1,4}=0,X_{2,4}=0.16,Y_4=11.13)$.

Analysis of functional surface $d_2=\gamma_5(\delta,\delta/\omega)$

From Figure 16(A) we see that functional surface $d_2=\gamma_5(\delta,\delta/\omega)$,on which are situated the points of the linear regression equation view of $Y_c=8.773+2.855X_1+4X'_2$ or $Y_c=8.773+2.855X_1+4(X_1/X_2)$,where $d_2=8.773+2.855\delta+4(\delta/\omega)$ has the following coordinates of their peaks in three-dimensional drawing (sizes in millimeters): a' $(X_{1,1}=0.28 ,X_{2,1}=0.08 ,Y_1=9.89)$; b' $(X_{1,2}=0.28 ,X_{2,2}=0.08 ,Y_2=0)$; c' $(X_{1,3}=0.60 ,X_{2,3}=0.16 ,Y_3=0)$; d' $(X_{1,4}=0.60 ,X_{2,4}=0.16 ,Y_4=11.13)$.The module of vectors can be calculated by formula (11),(12),(13), (14) and equal $|a'b'|=9.89$, $|b'c'|=0.33, |c'd'|=11.13$ and $|d'a'|=1.28$. Area of functional surface is equal by formula (15): $S_{a'b'c'd'}=3.46$ mm^2.All results of computations for the linear equation view of $Y_c=8.773+2.855X_1+4X'_2$ or $d_2= 8.773+2.855\delta+4(\delta/\omega)$ are given in Table 8.

Table 8 Evaluation of regression equation $Y_c=8.773+2.855X_1+4X'_2$

A. Mean ,variance and standard deviation			
Variable	Mean	Variance	Standard deviation
X_1	0.368	0.209	0.457
X'_2	0.106	0.016	0.126
Y	10.248	4.121	2.030

B.results of multiple regression of Y on X_1 and X'_2			
Parameter	Variable	Coefficients	Standard error	T-value
β_1	X_1	2.855	0.152	18.78
β_2	X'_2	4	0.042	95.24

C. Analysis of variance results

Regression : degree of freedom 2, sum of squares 3.25 ,mean squares 1.625

Error : degree of freedom 6, sum of squares 0.870,mean square 0.145

Standard error of estimate 0.457

F-value* 11.207

*Since F=11.207> 5.14 [$F_{0.05,2,6}=5.14$] we can reject the hypothesis that both $\beta_1$and β_2 are zero.

D. Determination of residuals

Number	Observed	Estimated	Residual
1	10.03	10.24	-0.21
2	10.36	10.49	-0.13
3	8.69	9.14	-0.45
4	10.13	9.49	0.64
5	10.22	9.89	0.33
6	10.07	10.27	-0.20
7	10.49	10.65	-0.16
8	11.06	10.92	0.14
9	11.18	11.13	0.05

So ,the parameters of this functional surface $d_2=\gamma_5(\delta,\delta/\omega)$ are the following:

1.Function $d_2=\gamma_5(\delta,\delta/\omega)$ better submits to the linear regression model with equation view of $Y_c=8.773+2.855X_1+4X'_2$ or $Y_c=8.773+2.855X_1+4(X_1/X_2)$,where $d_2=8.773+2.855\delta+4(\delta/\omega)$ with such statistical characteristics: coefficient of determination $R^2=0.789$,coefficient of correlation r=0.888,standard deviation $S_{y/x1,x'2}=0.381$,minimization of the mean square error (min MSE=0.089),minimization of the mean absolute deviation (min MAD=0).

2.The total area of functional surface $d_2=\gamma_5(\delta,\delta/\omega)$ is equal $\sum S=3.46mm^2$ with such coordinates of their points in three-dimensional drawing for this surface:
a'($X_{1,1}=0.28,X_{2,1}=0.08,Y_1=9.89$),b'($X_{1,2}=0.28,X_{2,2}=0.08,Y_2=0$),   c'($X_{1,3}=0.60,X_{2,3}=0.16,Y_3=0$), d'($X_{1,4}=0.60,X_{2,4}=0.16,Y_4=11.13$).

6.3 Dependence of external diameter (d_2) from thickness (δ) , internal diameter(d_1) and width(B) of this stainless chip, i.e $d_2=\alpha_3(\delta,d_1,B)$ and also from its modifications $d_2=\gamma_6(\delta,\delta/d_1)$,$d_2=\gamma_7(\delta,d_1/B)$.

A. Function $d_2=\alpha_3(\delta,d_1,B)$

Analysis of data and the statistical characteristics (coefficient of determination $R^2=0.917$,coefficient of correlation r=0.958, standard deviation $S_{y/x1,x2,x3}=0.261$),minimization of the mean square error (min MSE=0.038) and minimization of the mean absolute deviation (min MAD=0) show that this functional dependence $d_2=\alpha_3(\delta,d_1,B)$ better submits to the linear than non-linear[21] regression model with equation view of
$$Y_c=2.32-0.76X_1+0.89X_2+0.63X_3 \text{ or } d_2=2.32-0.76\delta+0.89d_1+0.63B \quad (32).$$
The functional dependencies $d_2=\gamma(\delta)$,$d_2=\varphi_2(B)$ and $\delta=f(B)$ are shown in Figure 9(a), 9(c), and 9(e) accordingly. And also in Figure 9(g),9(h) are illustrated   the residual plots (residual versus δ and residual versus B) of the above-named functional dependencies $d_2=\gamma(\delta)$ and $d_2=\varphi_2(B)$ accordingly.
In Figure 17(a) is shown the functional dependence of external diameter (d_2) stainless chip from its internal diameter (d_1) ,i.e   we have the function view of $d_2=\varphi_3(d_1)$. This dependence has the non-linear regression model with equation view of $Y_c=5.728X^{0.359}$ or $d_2=5.728d_1^{0.359}$. So ,we see from Figure 17(a) that with increasing of internal diameter (d_1) ,the value of external diameter (d_2) for this stainless chip considerably increases in accordance with the regression equation $d_2=5.728d_1^{0.359}$.
In Figure 17(b) is shown the functional dependence of thickness (δ)stainless chip from its internal diameter (d_1) ,i.e we have the function view of $\delta=\varphi(d_1)$. This dependence has the linear regression model with equation view of $Y_c=3.338+4.762X$ or $d_1=3.338+4.762\delta$,where $\delta=0.21d_1-0.7$ (33) . So ,we see from Figure 17(b) that with increasing of internal diameter (d_1) stainless chip ,the value of thickness (δ) for this chip increases considerably in accordance with regression equation view of $\delta=0.21d_1-0.70$.

21- Non-linear regression model has equation view of $d_2= -0.3+10.66\delta+3\delta^2-2.61d_1+3.33B$

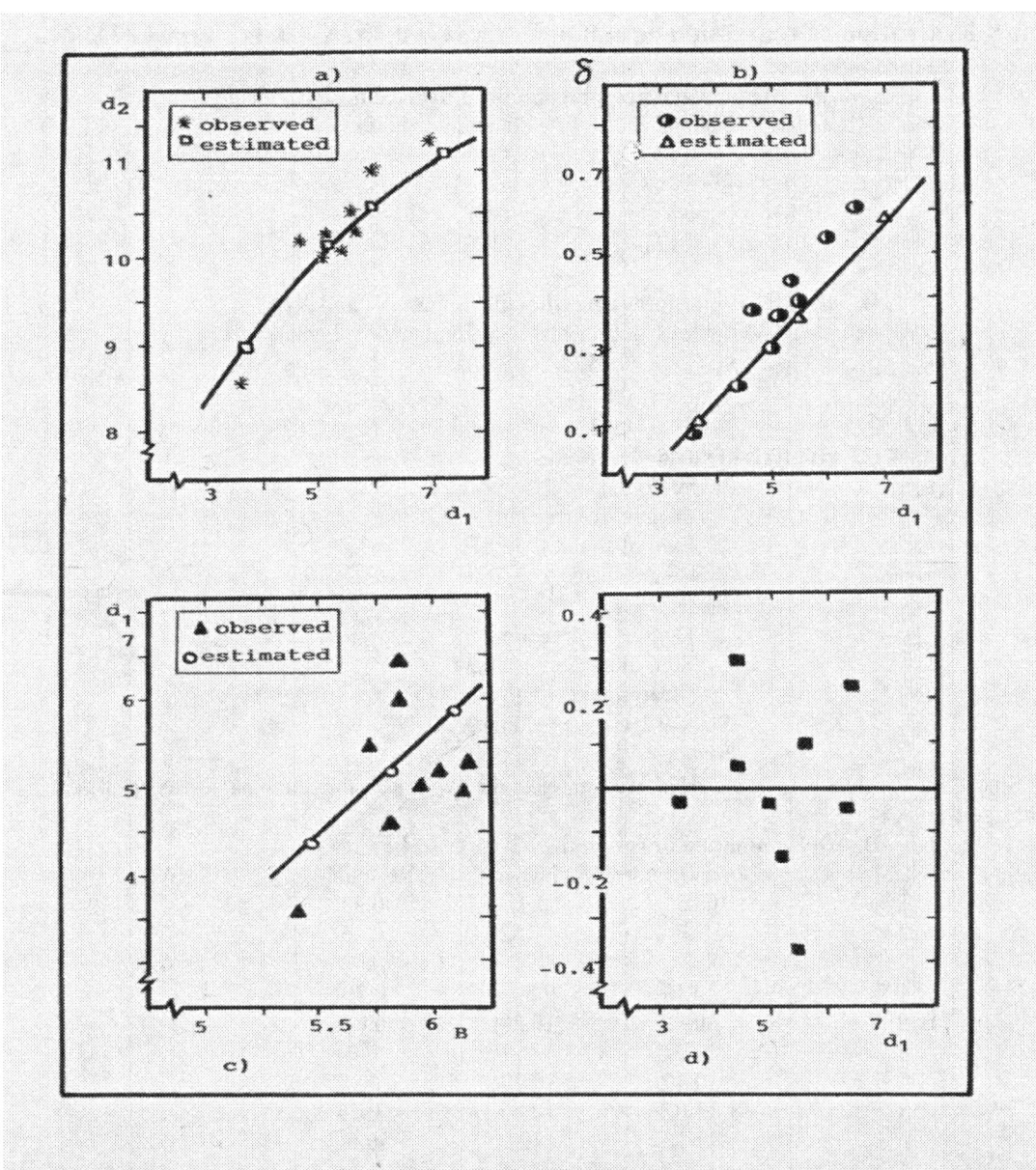

Figure 17

In Figure 17(c) is shown the functional dependence of internal diameter (d_1) stainless chip from its width (B) ,i.e we have the function view of $d_1=\varphi_1(B)$ this dependence has the linear regression model with equation view of $Y_c=0.06+0.863X$ or $d_1=0.06+0.863B$. So ,we see from Figure 17(c) that with increasing of width (B) stainless chip ,the value of internal diameter (d_1) increases considerably in accordance with regression equation view of $d_1=0.06+0.863B$.

In Figure 17(d) is illustrated the residual versus d_1 of the above-named dependency $d_2=\varphi_3(d_1)$. All results of computations for the linear equation view of $Y_c=2.321-0.757X_1+0.892X_2+0.627X_3$ or $d_2=2.321-0.757\delta+0.892d_1+0.627B$ are given in Table 9.

181

Table 9 Evaluation of regression equation $Y_c=2.321-0.757X_1+0.892X_2+0.627X_3$

A. Mean, variance and standard deviation

Variable	Mean	Variance	Standard deviation
X_1	0.368	0.209	0.457
X_2	5.098	5.608	2.368
X_3	5.836	0.514	0.717
Y	10.248	4.121	2.029

B. Results of multiple regression of Y on X_1, X_2 and X_3

Parameter	Variable	Coefficients	Standard error	T-value
β_1	X_1	-0.757	-0.152	4.98
β_2	X_2	0.892	0.789	1.13
β_3	X_3	0.627	0.239	2.620

C. Analysis of variance results

Regression
    Degrees of freedom 3
    Sum of squares 3.781
    Mean squares 1.260
Error
    Degrees of freedom 5
    Sum of squares 0.34
    Mean square 0.068
Standard error of estimate 0.457
F-value * 18.529

* Since F= 18.529 > [$F_{0.05,3,5}$ =5.41] , we can reject the hypothesis that three parameters β_1,β_2 and β_3 are zero.

D. Determination of residuals

Number	Observed	Estimated	Residual
1	10.03	10.41	-0.38
2	10.36	10.52	-0.16
3	8.69	8.75	-0.06
4	10.13	9.84	0.29
5	10.22	10.20	0.02
6	10.07	10.11	-0.04
7	10.49	10.39	0.10
8	11.06	10.80	0.26
9	11.18	11.22	-0.04

B. Modification function $d_2=\gamma_6(\delta,\delta/d_1)$

Analysis of data and the statistical characteristics (coefficient of determination $R^2=0.790$,coefficient of correlation r=0.89 ,standard deviation $S_{y/x1,x'2}=0.30$) and also the minimization of the mean square error (min MSE=0.096) and minimization of the mean absolute deviation (min MAD=0) show that this functional dependence $d_2=\gamma_6(\delta,\delta/d_1)$ better submits to the linear than non-linear[22] regression model with equation view of
$Y_c=8.742+3.713X_1+2X'_2$ or $Y_c=8.742+3.713X_1+2(X_1/X_2)$,where
$d_2=8.742+3.713\delta+2(\delta/d_1)$ (34) .

22-Non-linear regression model has equation view of $d_2=8.95+3.58\delta+\delta^2-2.5(\delta/d_1)$

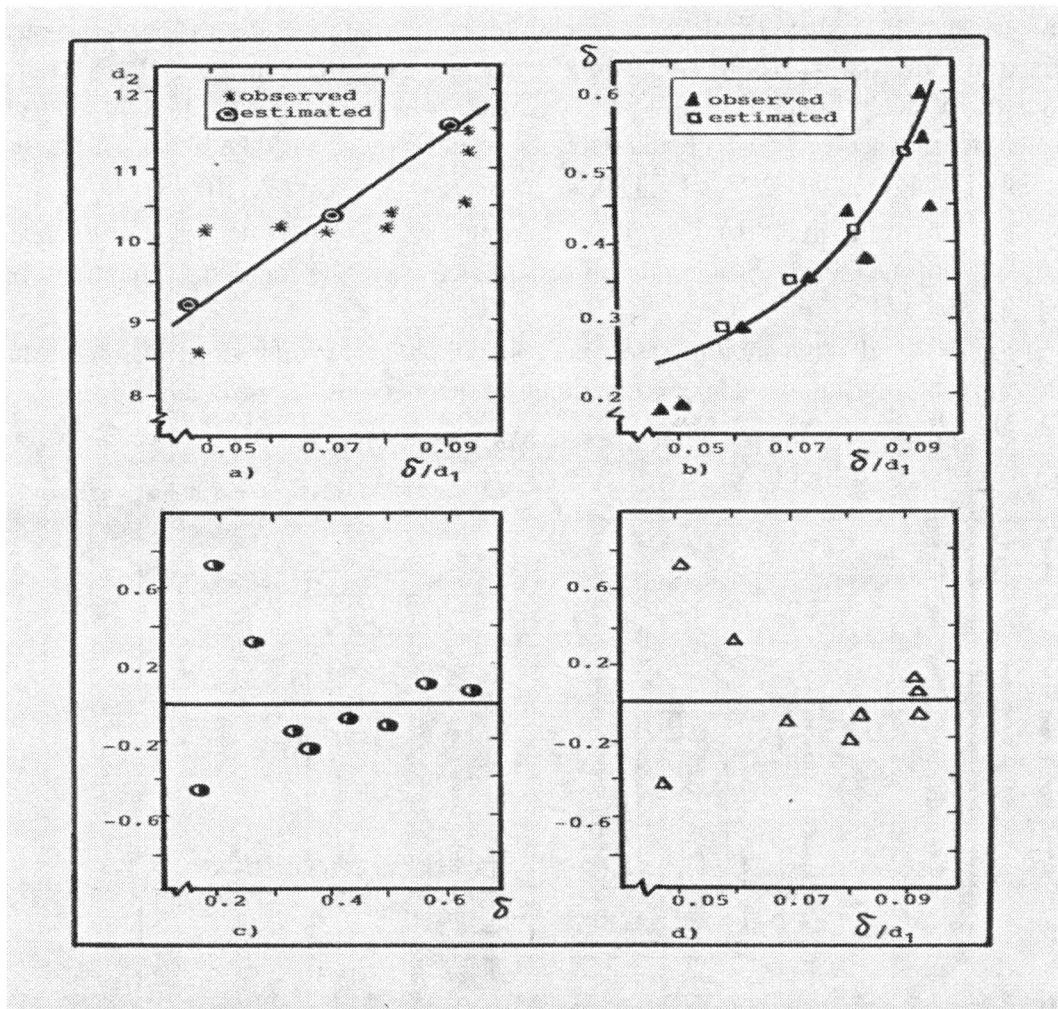

Figure 18

In Figure 18(a) is shown the functional dependency of external diameter (d_2) stainless chip from the ratio of values (δ/d_1) ,such as thickness (δ) of this chip to the internal diameter (d_1) of this stainless chip ,i.e we have the function view of $d_2 = \phi_1(\delta/d_1)$.Analysis of data and also the statistical characteristics(coefficient of determination $R^2=0.917$,coefficient of correlation r=0.946 ,standard deviation $S_{e/x'2}=0.254$) and minimization of the mean absolute deviation (min MAD=0) show that this functional dependence $d_2=\phi_1(\delta/d_1)$ better submits to the linear than non-linear [23] regression model with equation view of $\mathbf{Y_c=7.8+35X'_2}$ or $\mathbf{d_2=7.8+35(\delta/d_1)}$ **(34)**

As we see from Figure 18(a) that with increasing of ratio values (δ/d_1) ,the value of external diameter (d_2) stainless chip increases considerably in accordance with the regression equation view of $d_2=7.8+35(\delta/d_1)$.

23- Non –linear regression model has equation view of $d_2=15.17(\delta/d_1)^{0.145}$

In Figure 18(b) is shown the functional dependence of thickness (δ) stainless chip from the ratio of values (δ/d_1) , such as thickness (δ)chip to its internal diameter(d_1),i.e has a place the function view of $\delta=\phi_2(\delta/d_1)$.Analysis of data and also the statistical characteristics (coefficient of determination $R^2=1.0$ and coefficient of correlation r=1.0) show that this functional dependence $\delta=\phi_2(\delta/d_1)$ better submits to the non-linear than linear[24] regression model with equation view of $\mathbf{Y_c=17.18X'_2{}^{1.457}}$ or $\qquad \boldsymbol{\delta=17.18(\delta/d_1)^{1.457}}$ (35) . So ,we see from Figure 18(b) that with increasing of ratio (δ/d_1) ,the value of thickness (δ) stainless chip increases considerably in accordance with the regression equation view of $\delta=17.18\ (\delta/d_1)^{1.457}$.

In Figure 18 (c) and 18(d) are illustrated the residual plots (residual versus δ and residual versus δ/d_1) of the above-named functional dependencies $d_2=\gamma(\delta)$ and $d_2=\phi_1(\delta/d_1)$ accordingly.

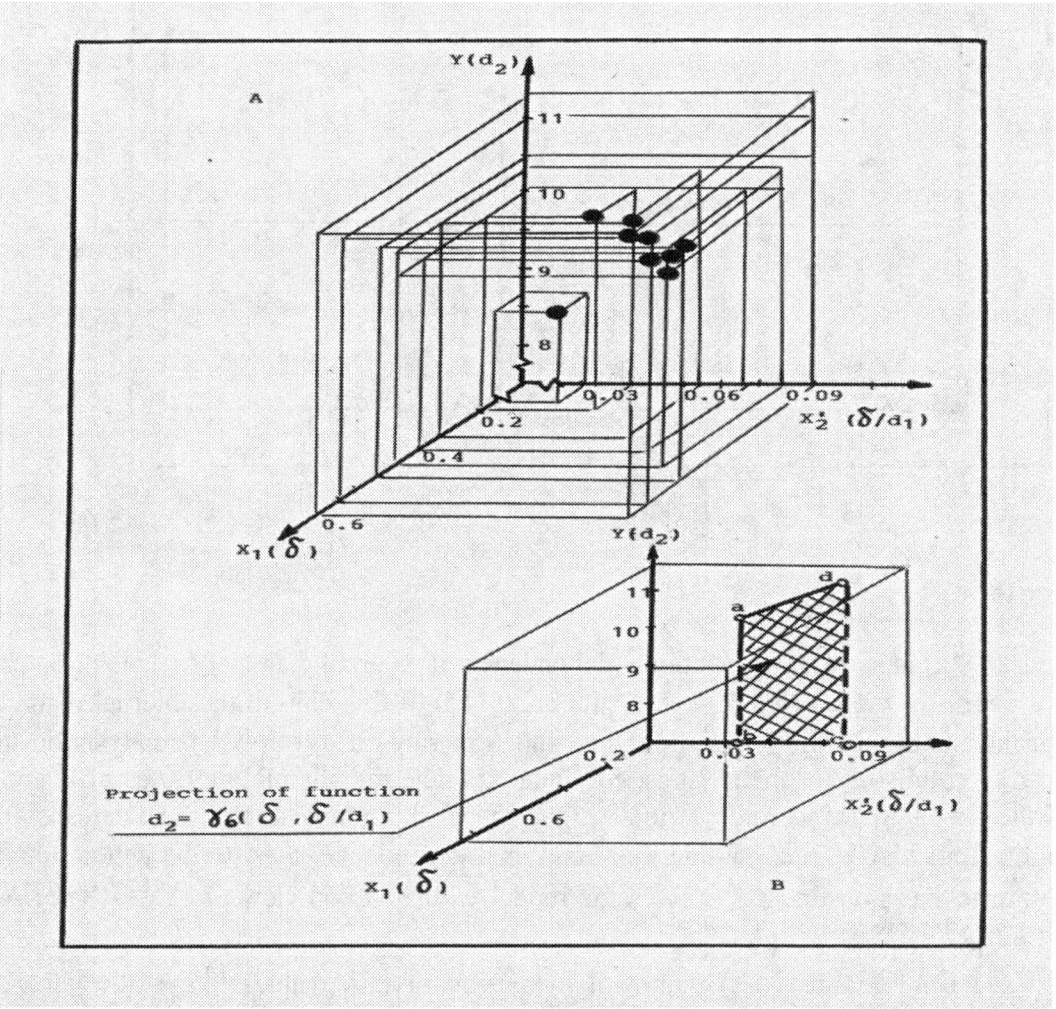

Figure 19

24-Linear regression model has equation view of $\delta=-0.26+9(\delta/d_1)$

184

Analyzing the Figure 19(A), we see that functional surface $d_2=\gamma_6(\delta,\delta/d_1)$ is shown in view of three-dimensional drawing for the linear regression model with equation view of $Y_c=8.74+3.71X_1+2X'_2$ or $Y_c=8.74+3.71X_1+2(X_1/X_2)$, where $d_2=8.74+3.71\delta+2(\delta/d_1)$.

And Figure 19(B) shows that this functional surface $d_2=\gamma_6(\delta,\delta/d_1)$ has view of trapezium on the plane YOX'_2 with such coordinates of their points: a ($X_{1,1}=0$, $X_{2,1}=0.07$, $Y_1=10.22$); b ($X_{1,2}=0$, $X_{2,2}=0.07$, $Y_2=0$), c ($X_{1,3}=0$, $X_{2,3}=0.09$, $Y_3=0$); d ($X_{1,4}=0$, $X_{2,4}=0.09$, $Y_4=11.15$).

Analysis of functional surface $d_2=\gamma_6(\delta,\delta/d_1)$

From Figure 19(A) we see that functional surface $d_2=\gamma_6(\delta,\delta/d_1)$ on which are situated the points of the linear regression equation view of $Y_c=8.742+3.713X_1+2X'_2$ or $Y_c=8.742+3.713X_1+2(X_1/X_2)$, where $d_2=8.742+3.713\delta+2(\delta/d_1)$ has the following coordinates of their peaks in three-dimensional drawing: a'($X_{1,1}=0.36$, $X_{2,1}=0.07$, $Y_1=10.22$) ; b' ($X_{1,2}=0.36$, $X_{2,2}=0.07$, $Y_2=0$) ; c'($X_{1,3}=0.60$, $X_{2,3}=0.09$, $Y_3=0$); d'($X_{1,4}=0.60$, $X_{2,4}=0.09$, $Y_4=11.15$).

The module of vectors $|a'b'|$, $|b'c'|$, $|c'd'|$, $|d'a'|$ and area of this functional surface can be calculated by formulas (11),(12),(13),(14) and (15). So, we have $|a'b'|=10.22$, $|b'c'|=0.24$, $|c'd'|=11.15$, $|d'a'|=0.96$ and $S_{a'b'c'd'}=2.56$ mm^2.

All results of computations for the linear equation view of $Y_c=8.742+3.713X_1+2X'_2$ are given in Table 10.

Table 10 Evaluation of regression equation $Y_c=8.742+3.713X_1+2X'_2$

A. Mean, variance and standard deviation

Variable	Mean	Variance	Standard deviation
X_1	0.368	0.209	0.457
X'_2	0.07	0.002	0.045
Y	10.248	4.121	2.030

B. Results of multiple regression of Y on X_1 and X'_2

Parameter	Variable	Coefficients	Standard error	T-value
β_1	X_1	3.713	0.152	24.59
β_2	X'_2	2.0	0.015	133.33

C. Analysis of variance results

Regression : Degree of freedom 2 ,sum of squares 3.254, mean squares 1.627
Error : Degree of freedom 6 , sum of squares 0.867 , mean squares 0.145
Standard error of estimate   0.457
F-value*   11.221

* Since F=11.221> [$F_{0.05,2,6}$=5.14] we can reject the hypothesis that both β_1 and β_2 are zero.

D. Determination of residual

Number	Observed	Estimated	Residual
1	10.03	10.22	-0.19
2	10.36	10.46	-0.10
3	8.69	9.17	-0.48
4	10.13	9.49	0.64
5	10.22	9.90	0.32
6	10.07	10.28	-0.21
7	10.49	10.63	-0.14
8	11.06	10.93	0.13
9	11.18	11.15	0.03

So ,the parameters of functional surface $d_2=\gamma_6(\delta,\delta/d_1)$ are the following:

1.Function $d_2=\gamma_6(\delta,\delta/d_1)$ better submits to the linear regression model with equation view of $Y_c=8.742+3.713X_1=2X'_2$ or $Y_c=8.742+3.713X_1+2(X_1/X_2)$,where $d_2=8.742+3.713\delta+ +2(\delta/d_1)$ with such statistical characteristics: coefficient of determination $R^2=0.79$, coefficient of correlation r=0.89,standard deviation $S_{y/x1,x'2}=0.310$,minimization of the mean square error)min MSE=0.096),minimization of the mean absolute deviation (min MAD=0).

2.The total area of functional surface $d_2=\gamma_6(\delta,\delta/d_1)$ is equal $\sum S=2.56$ mm^2 with such coordinates of their points in three-dimensional drawing for this surface : a' ($X_{1,1}=0.36$, $X_{2,1}=0.07$,$Y_1=10.22$) ; b' ($X_{1,2}=0.36$,$X_{2,2}=0.07$,$Y_2=0$); c' ($X_{1,3}=0.60$,$X_{2,3}=0.09$,$Y_3=0$); d' ($X_{1,4}=0.60$,$X_{2,4}=0.09$,$Y_4=11.15$).

C. Modification function $d_2=\gamma_7(\delta,d_1/B)$

Analysis of data and statistical characteristics (coefficient of determination $R^2=0.821$, coefficient of correlation r=0.906, standard deviation $S_{y/x1,x'2}=0.286$)and also the minimization of the mean square error (min MSE=0.082) and minimization of the mean absolute deviation (min MAD=0) show that this functional dependence $d_2=\gamma_7(\delta,d_1/B)$ better submits to the linear than non-linear [25] regression model with equation view of

$$Y_c=7.26+2.11X_1+2.52X'_2 \quad \text{or} \quad Y_c=7.26+2.11X_1+2.52(X_2/X_3) \quad ,\text{where}$$
$$d_2=7.26+2.11\delta+2.52(d_1/B) \quad (36) .$$

In Figure 20(a) is shown the functional dependency of external diameter (d_2) stainless chip from the ratio of values (d_1) ,such as internal diameter (d_1) of this chip to its width(B) ,i.e we have the function view of $d_2=\phi_3(d_1/B)$. Analysis of data and also the statistical characteristics (coefficient of determination $R^2=0.756$,coefficient of correlation r=0.869,standard deviation $S_{y/x'2}=0.334$)and minimization of the mean square error (min MSE=0.111) and minimization of the mean absolute deviation(min MAD=0.042) show that this functional dependence $d_2=\phi_3(d_1/B)$ better submits to the non-linear than linear[26] regression model with equation view of $Y_c=10.76X'_2{}^{0.368}$ or $d_2=10.76(d_1/B)^{0.368}$.

As we see from Figure 20(a) that with increasing of ratio values (d_1/B) ,the value of external diameter (d_2) stainless chip increases considerably in accordance with the regression equation view of $d_2=10.76(d_1/B)^{0.368}$ (37) .

In Figure 20(b) is shown the functional dependence of thickness (δ) stainless chip from the ratio of values (d_1/B) ,such as the internal diameter(d_1) of this chip to its width (B) ,i.e has a place the function view of $\delta=\phi_4(d_1/B)$.Analysis of data and also the statistical characteristics (coefficient of determination $R^2=0.735$ and coefficient of correlation r=0.850 and minimization of the mean square error (min MSE=0.006) and minimization of the mean absolute deviation (min MAD=0) show that this functional dependence $\delta=\phi_4(d_1/B)$ better submits to the linear than non-linear [27] regression model with equation view of $Y_c= -0.474+0.963X'_2$ or $\delta= -0.474+0.963(d_1/B)$ (38) .

25- Non-linear regression model has equation view of $d_2=8.16+8.28\delta-5.97\delta^2-0.02(d_1/B)$
26-Linear regression model has equation view of $d_2=5.44+5.5(d_1/B)$
27-Non-linear regression model has equation view of $\delta=0.5(d_1/B)^{2.875}$

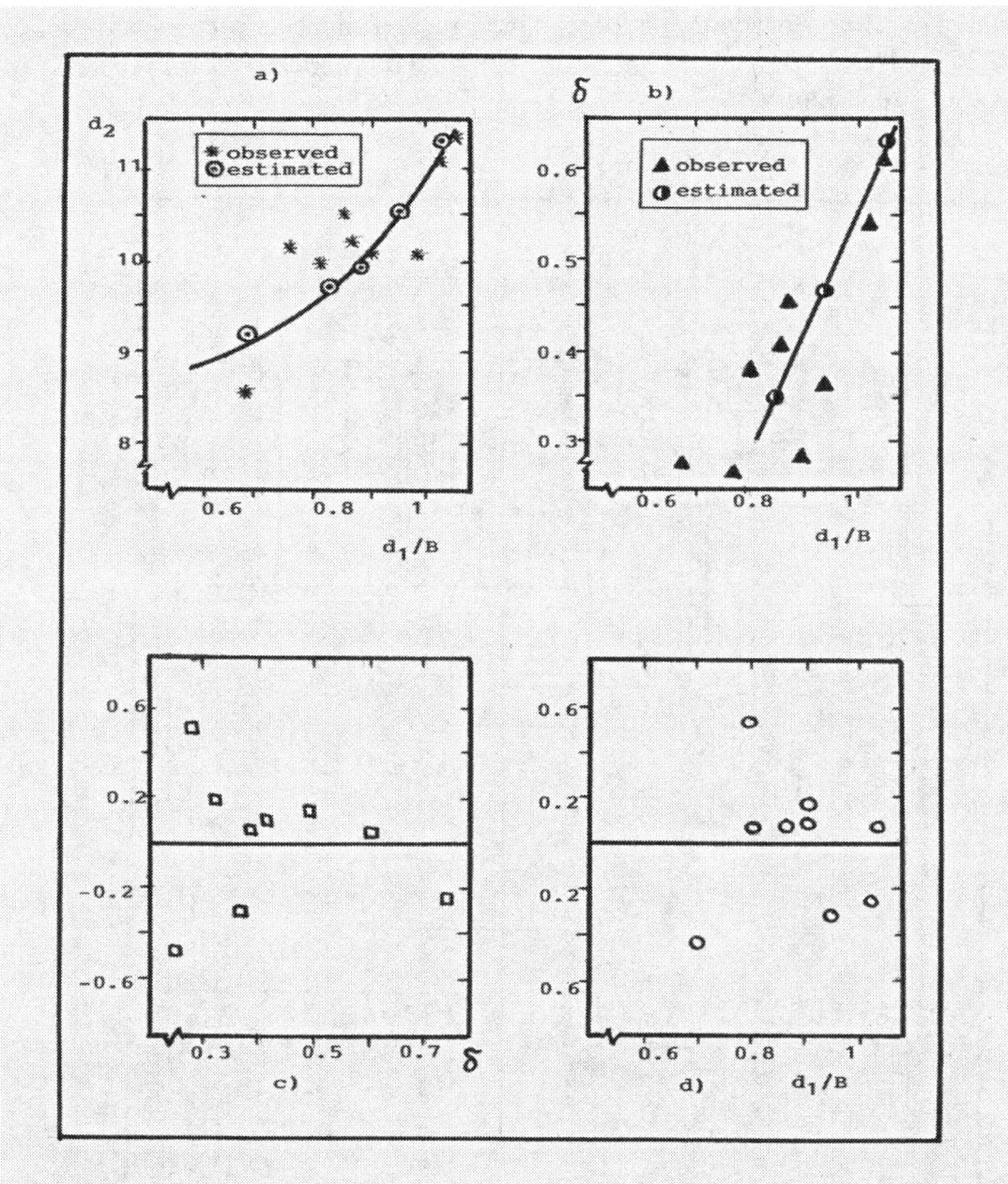

Figure 20

So ,we see from Figure 20(b) that with increasing of ratio (d_1/B) ,the value of thickness (δ) stainless chip increases considerably in accordance with the linear regression equation view of $\delta = -0.474 + 0.963(d_1/B)$.

In Figure 20 (c) and 20(d) are illustrated the residual plots (residual versus δ and residual versus d_1/B) of the above-named functional dependencies of $d_2 = \gamma(\delta)$ and $d_2 = \phi_3(d_1/B)$ accordingly.

Analyzing the Figure 21(A) , we see that functional surface $d_2=\gamma_7(\delta,d_1/B)$ is shown in view of three-dimensional drawing for the linear regression model with equation view of $Y_c=7.26+2.11X_1+2.52X'_2$ or $Y_c=7.26+2.11X_1+2.52(X_2/X_3)$,where $d_2=7.26+2.11\delta+2.52(d_1/B)$.

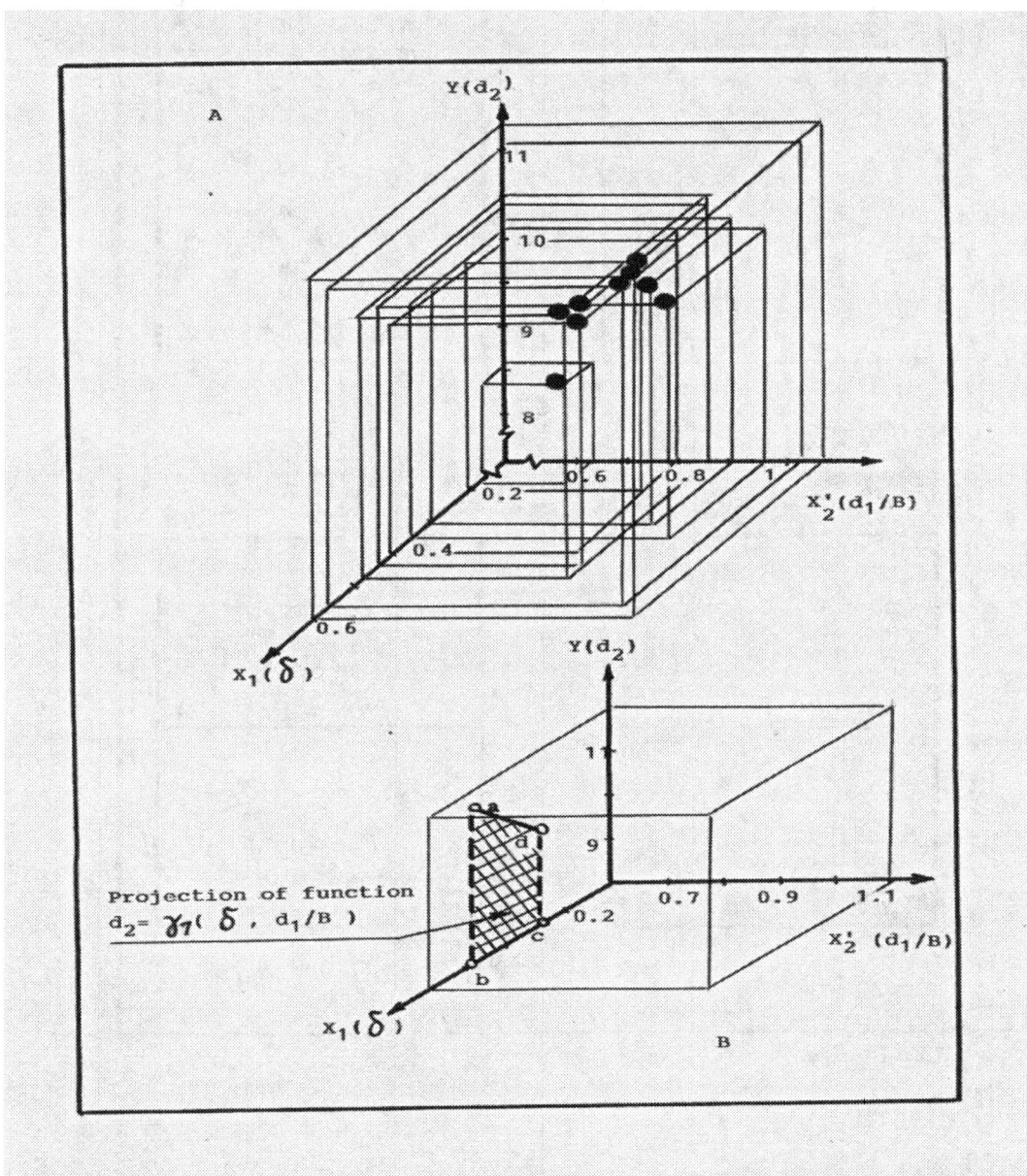

Figure 21

And besides the Figure 21(B) shows that functional surface $d_2=\gamma_7(\delta,d_1/B)$ has view of trapezium on the plane YOX_1 with such coordinates of their points (sizes in millimeters): a($X_{1,1}=0.60$,$X_{2,1}=0$, $Y_1=11.18$) ;b ($X_{1,2}=0.60$,$X_{2,2}=0$,$Y_2=0$) ; c ($X_{1,3}=0.36$,$X_{2,3}=0$,$Y_3=0$) ; d ($X_{1,4}=0.36$,$X_{2,4}=0$,$Y_4=10.39$).

Analysis of functional surface $d_2 = \gamma_7(\delta, d_1/B)$

From Figure 21(A) we see that functional surface $d_2 = \gamma_7(\delta, d_1/B)$ on which are situated the points of the linear regression equation view of $Y_c = 7.26 + 2.11X_1 + 2.52X'_2$ or $Y_c = 7.26 + 2.11X_1 + 2.52(X_2/X_3)$, where $d_2 = 7.26 + 2.11\delta + 2.52(d_1/B)$ has the following coordinates of their peaks in three- dimensional drawing (sizes in millimeters): a' $(X_{1,1} = 0.60, X_{2,1} = 1.12, Y_1 = 11.35)$;b' $(X_{1,2} = 0.60, X_{2,2} = 1.12, Y_2 = 00$; c' $(X_{1,3} = 0.36, X_{2,3} = 0.94, Y_3 = 0)$;
The module of vectors $|a'b'|, |b'c'|, |c'd'|, |d'a'|$ can be evaluated by formulas (11),(12),(13) and (14). At data we have the following values: $|a'b'| = 11.35$, $|b'c'| = 0.30, |c'd'| = 10.39, |d'a'| = 1.01$. Area of functional surface is equal : $S_{a'b'c'd'} = 0.5[|b'c'| \cdot (|a'b'| - |c'd'|)] + |b'c'| \cdot |c'd'| = 3.26mm^2$.
All results of computation for the linear equation view of $Y_c = 7.26 + 2.11X_1 + 2.52X'_2$ or $Y_c = 7.26 + 2.11X_1 + 2.52(X_2/X_3)$, where $d_2 = 7.26 + 2.11\delta + 2.52(d_1/B)$ are given in Table 11.

Table 11 Equation of regression equation $Y_c = 7.26 + 2.11X_1 + 2.52X'_2$

A. Mean, variance and standard deviation

Variable	Mean	Variance	Standard deviation
X_1	0.368	0.209	0.457
X'_2	0.874	0.163	0.404
Y	10.248	4.121	2.030

B.Results of multiple regression of Y on X_1 and X'_2

Parameter	Variable	Coefficients	Standard error	T-value
β_1	X_1	2.11	0.152	13.88
β_2	X'_2	2.52	0.135	18.67

C. Analysis of variance results

Regression : Degree of freedom 2, sum of squares 3.427,mean squares 1.714
Error: Degrees of freedom 6, sum of squares 0.694,mean squares 0.116
Standard error of estimate   0.457
F-value*   14.772
*Since $F = 14.772 > [F_{0.05,2,6} = 5.14]$,we can reject the hypothesis that both β_1 and β_2 are zero.

D. Determination of residual

Number	Observed	Estimated	Residual
1	10.03	10.39	-0.36
2	10.36	10.29	0.07
3	8.69	9.13	-0.44
4	10.13	9.61	0.52
5	10.22	10.02	0.20
6	10.07	9.98	0.09
7	10.49	10.35	0.14
8	11.06	11.05	0.01
9	11.18	11.35	-0.17

So ,the parameters of this functional surface $d_2=\gamma_7(\delta,d_1/B)$ are the following:

1.Function $d_2=\gamma_7(\delta,d_1/B)$ better submits to the linear regression model with equation view of $Y_c=7.26+2.11X_1+2.52X'_2$,where $d_2=7.26+2.11\delta+2.52$ (d_1/B) with such statistical characteristics: coefficient of determination $R^2=0.821$,coefficient of correlation $r=0.906$,standard deviation $S_{y/x1,x'2}=0.286$,minimization of the mean square error (min MSE=0.082),minimization of the mean absolute deviation (min MAD=0).

2.The total area of functional surface $d_2=\gamma_7(\delta,d_1/B)$ is equal $\sum S=3.26$ mm^2 with such coordinates of their points in three-dimensional drawing for this surface (sizes in millimeters): a' ($X_{1,1}=0.60,X_{2,1}=1.12,Y_1=11.35$) ,b'($X_{1,2}=0.60,X_{2,2}=1.12$,$Y_2=0$) ; c' ($X_{1,3}=0.36$,$X_{2,3}=0.94,Y_3=0$); d'($X_{1,4}=0.36,X_{2,4}=0.94$,$Y_4=10.39$).

6.4 Dependence of external diameter (d_2) from thickness (δ) of chip ,the value of clearance (t) between of chip wraps and its width(B) ,i.e $d_2=\alpha_4(\delta,t,B)$ and also from its of modifications $d_2=\gamma_8(\delta,\delta/t)$, $d_2=\gamma_9(\delta,t/B)$.

A. Function $d_2=\alpha_4(\delta,t,B)$

Analysis of data and the statistical characteristics (coefficient of determination $R^2=0.827$,coefficient of correlation $r=0.909$,standard deviation $S_{y/x1,x2,x3}=0.378$), minimization of the mean square error(min MSE=0.079) and minimization of the mean absolute deviation (min MAD=0) ,show that this functional dependence $d_2=\alpha_4(\delta,t,B)$ better submits to the linear than non-linear[28] regression model with equation view of

$$Y_c=7.653+4.405X_1+2.668X_2+0.022X_3 \text{ or } d_2=7.653+4.405\delta+2.668t+0.022B \text{ (39).}$$

The functional dependence $d_2=\gamma(\delta)$ is shown in Figure 22(a), and as was shown in chapter 4 of Figure 2 , has the linear regression model with equation view of $Y_c=8.808+3.897X$ or $d_2=8.808+3.897\delta$. So ,we see from Figure 22(a) that with increasing of thickness (δ) stainless chip ,the value of external diameter (d_2) for this chip considerably increases in accordance with the regression equation $d_2=8.808+3.897\delta$.

In Figure 22(b) is shown the functional dependence of thickness (δ) stainless chip from its clearance (t) between of chip wraps ,i.e we have the function view of $\delta=\phi(t)$. This dependence has the linear regression model with equation view of $Y_c=0.379-0.16X$ or $t=0.379-0.16\delta$,where $\delta=2.37-6.25t$. So, we see from Figure 22(b) that with increasing of clearance (t) between of chip wraps ,the value of thickness (δ) for this stainless chip considerably decreases in accordance with the regression equation $\delta=2.37-6.25t$.

In Figure 22 (c) is shown the functional dependence of external diameter (d_2) stainless chip from its clearance (t) between of chip wraps ,i.e we have the function view of $d_2=\gamma_3(t)$. This dependence has the non-linear regression model with equation view of $Y_c=1.14X^{-0.56}$ or $t=1.14d_2^{-0.56}$, where $d_2=(0.88t)^{-1.785}$.

---

28 –Non-linear regression model has equation view of $d_2=6.91+7.79\delta-4.45\delta^2+3.76\,t-0.003B$

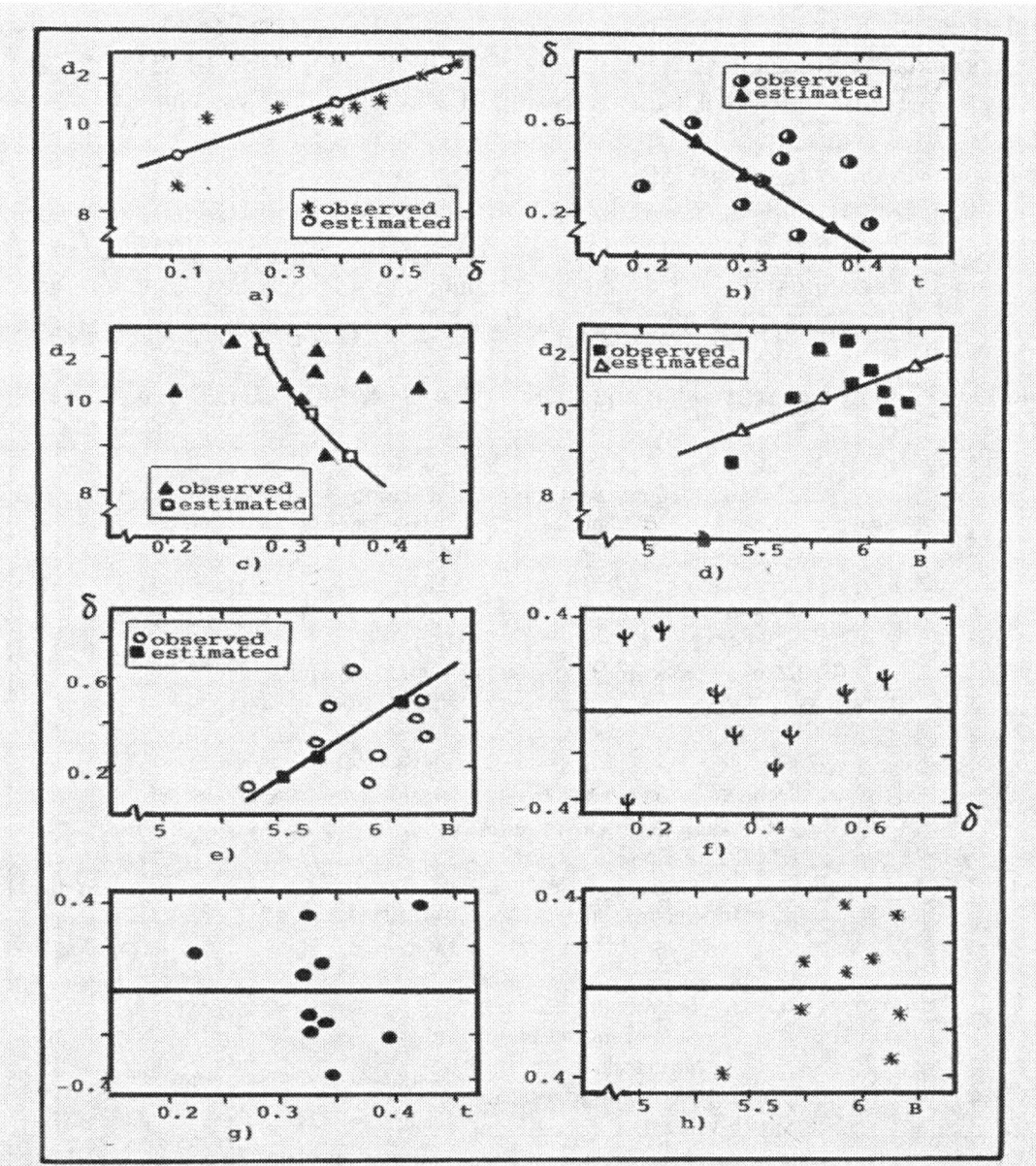

Figure 22

So ,we see from Figure 22(c) that with increasing of clearance (t) between of chip wraps stainless chip, the value of external diameter(d_2) for this chip decreases considerably in accordance with regression equation view of $d_2=(0.88t)^{-1.785}$.

In Figure 22(d) is shown the functional dependence of external diameter (d_2) stainless chip from its width (B),i.e we have   the function $d_2=\varphi_2(B)$. This dependence has the linear regression model with equation view of $Y_c=3.026+1.237X$ or $d_2=3.026+1.237B$. So ,we see from Figure 22(d) that with increasing of width (B) stainless chip ,the value of external diameter for this chip increases considerably in accordance with the regression equation view of $d_2=3.026+1.237B$.

191

In Figure 22(e) is shown the functional dependence of thickness (δ) stainless chip from its width(B),i.e we have the function view of $\delta=f(B)$. This dependence has the linear regression model with equation view of $Y_c = -1.236+0.275X$ or $\delta = -1.236+0.275B$. So ,we see from Figure 22(e) that with increasing of width (B) stainless chip ,the value of thickness (δ) for this chip increases considerably in accordance with the regression equation view of $\delta = -1.236+0.275B$.

The residual plots (residual versus of thickness δ, residual versus of clearance t between of chip wraps and residual versus of width B for this chip) are illustrated in Figures 22(f) ,22(g) and 22(h) accordingly. All results of computations for the linear equation view of $Y_c=7.653+4.405X_1+2.668X_2+0.022X_3$ or $d_2=7.653+4.405\delta+2.668t+0.022B$ are given in Table 12.

Table 12 Evaluation of regression equation $Y_c=7.653+4.405X_1+2.668X_2+0.022X_3$

A. Mean , variance and standard deviation

Variable	Mean	Variance	Standard deviation
X_1	0.368	0.209	0.457
X_2	0.317	0.030	0.173
X_3	5.836	0.515	0.718
Y	10.248	4.121	2.030

B. Results of multiple regression of Y on X_1,X_2 and X_3

Parameter	Variable	Coefficients	Standard error	T-value
β_1	X_1	4.405	0.152	28.98
β_2	X_2	2.668	0.058	46
β_3	X_3	0.022	0.239	0.092

C . Analysis of variance results

Regression
- Degrees of freedom 3
- Sum of squares 3.408
- Mean squares 1.136

Error
- Degrees of freedom 5
- Sum of squares 0.713
- Mean square 0.143

Standard error of estimate 0.457

F-value* 7.944

*Since F=7.944 > [$F_{0.05\ ,3,5}$=5.41] we can reject the hypothesis that three parameters β_1,β_2 and β_3 are zero.

D. Determination of residuals

Number	Observed	Estimated	Residual
1	10.03	9.93	0.10
2	10.36	10.65	-0.29
3	8.69	9.12	-0.43
4	10.13	9.69	0.44
5	10.22	9.81	0.41
6	10.07	10.24	-0.17
7	10.49	10.67	-0.18
8	11.06	11.01	0.05
9	11.18	11.09	0.09

B. Modification function $d_2=\gamma_8(\delta,\delta/t)$

Analysis of data and the statistical characteristics (coefficient of determination $R^2=0.801$, coefficient of correlation $r=0.895$, standard deviation $S_{y/x1,x'2}=0.369$) and also the minimization of the mean square error (min MSE=0.091) and minimization of the mean absolute deviation (min MAD=0) show that this functional dependence $d_2=\gamma_8(\delta,\delta/t)$ better submits to the linear than non-linear [29] regression model with equation view of $Y_c=8.743+5.077X_1-0.292X'_2$ or $Y_c=8.743+5.077X_1-0.292(X_1/X_2)$, where $d_2=8.743+5.077\delta-0.292(\delta/t)$ (40).

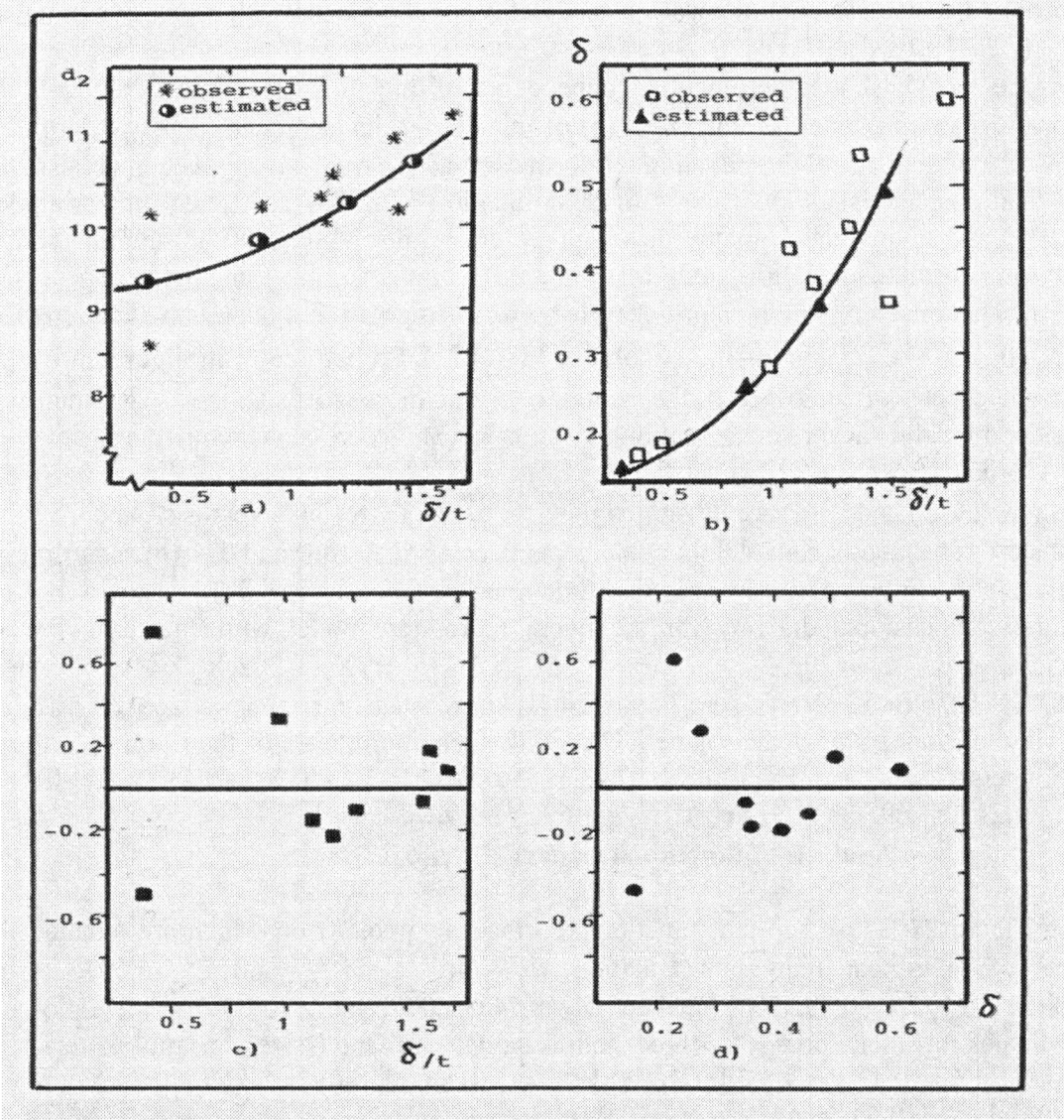

Figure 23

29- *Non-linear regression model has equation view of $d_2=9.38+0.91\delta+6.25\delta^2-0.36(\delta/t)$*

In Figure 23(a) is shown the functional dependency of external diameter (d_2 stainless chip from the ratio of values (δ/t) ,such as thickness (δ) of this chip to its clearance (t) between of chip wraps ,i.e we have the function view of $d_2=\phi_5(\delta / t)$ .Analysis of data and also the statistical characteristics (coefficient of determination $R^2=0.669$,coefficient of correlation $r=0.818$,standard deviation $S_{y/x'2}=0.441$)and minimization of the mean square error (min MSE=0.152) and minimization of the mean absolute deviation (min MAD=0.033),show that this functional dependence $d_2=\phi_5(\delta/t)$ better submits to the non-linear than linear [30] regression model with equation view of $\mathbf{Y_c=10.16X'_2{}^{0.087}}$ or $\mathbf{d_2=10.16(\delta/t)^{0.087}}$ **(41).**
As we see from Figure 23(a) that with increasing of ratio values (δ/t),the value of external diameter (d_2) stainless chip increases considerably in accordance with regression equation view of $d_2=10.16X'_2{}^{0.087}$.
In Figure 23(b) is shown the functional dependence of thickness (δ) stainless chip from the ratio of values (δ/t) ,such as thickness (δ) of this chip to its clearance (t) between of chip wraps ,i.e has a place the function view of $\delta=\phi_6(\delta/t)$. Analysis of data and also the statistical characteristics (coefficient of determination $R^2=0.882$,coefficient of correlation $r=0.939$) and minimization of the mean square error (min MSE=0.003) ,minimization of the mean absolute deviation (min MAD=0.003) show that this functional dependence $\delta=\phi_6(\delta/t)$ better submits to the non-linear than linear[31] regression model with equation view of $\mathbf{Y_c=0.316X'_2{}^{0.804}}$ or $\mathbf{\delta=0.316(\delta/t)^{0.804}}$ **(42).** So ,we see from Figure 23(b) that with increasing of ratio (δ/t) ,the value of thickness (δ) stainless chip increases considerably in accordance with the non-linear regression equation view of $\delta=0.316(\delta/t)^{0.804}$.
In Figure 23(c) and 23(d) are illustrated the residual plots (residual versus δ/t and residual versus δ) of the above-named functional dependencies $d_2=\phi_5(\delta/t)$ and $d_2=\gamma(\delta)$ accordingly. Analyzing the Figure 24(A) ,we see that functional surface $d_2=\gamma_8(\delta,\delta/t)$ is shown in view of three- dimensional drawing for the linear regression model with equation view of $Y_c=8.743+5.077X_1-0.292X'_2$ or $Y_c=8.743+5.077X_1-0.292(X_1/X_2)$,where $d_2=8,743+5.077\delta-0.292(\delta/t)$. And Figure 24(B) shows that functional surface $d_2=\gamma_8(\delta,\delta/t)$ has view of trapezium on the plane YOX_2 with such coordinates of their points (sizes in millimeters): a $(X_{1,1}=0,X_{2,1}=0.43,Y_1=9.53)$; $b(X_{1,2}=0,X_{2,2}=0.43$,$Y_2=0)$; $c(X_{1,3}=0,X_{2,3}=1.69,Y_3=0)$,d $(X_{1,4}=0 ,X_{2,4}=1.69, Y_4=10.99)$.

Analysis of functional surface $d_2=\gamma_8(\delta,\delta/t)$

From Figure 24(B) we see that functional surface $d_2=\gamma_8(\delta,\delta/t)$ on which are situated the points of the linear regression equation view of $Y_c=8.743+5.077X_1-0.292X'_2$ or $Y_c=8.743+5.077X_1-0.292(X_1/X_2)$,where $d_2=8.743+5.077\delta-0.292(\delta/t)$,has the following coordinates of their peaks in three- dimensional drawing (sizes in millimeters): a' $(X_{1,1}=0.18,X_{2,1}=0.43,Y_1=9.53)$;b'$(X_{1,2}=0.18$,$X_{2,2}=0.43$,$Y_2=0)$;c' $(X_{1,3}=0.54,X_{2,3}=1.69,$,$Y_3=0)$; d'$(X_{1,4}=0.54 ,X_{2,4}=1.69 ,Y_4=10.99)$.

---

30 – Linear regression model has the equation view of $d_2=9.19+0.85(\delta/t)$
31 – Linear regression model has the equation view of $\delta=0.09+0.23(\delta/t)$

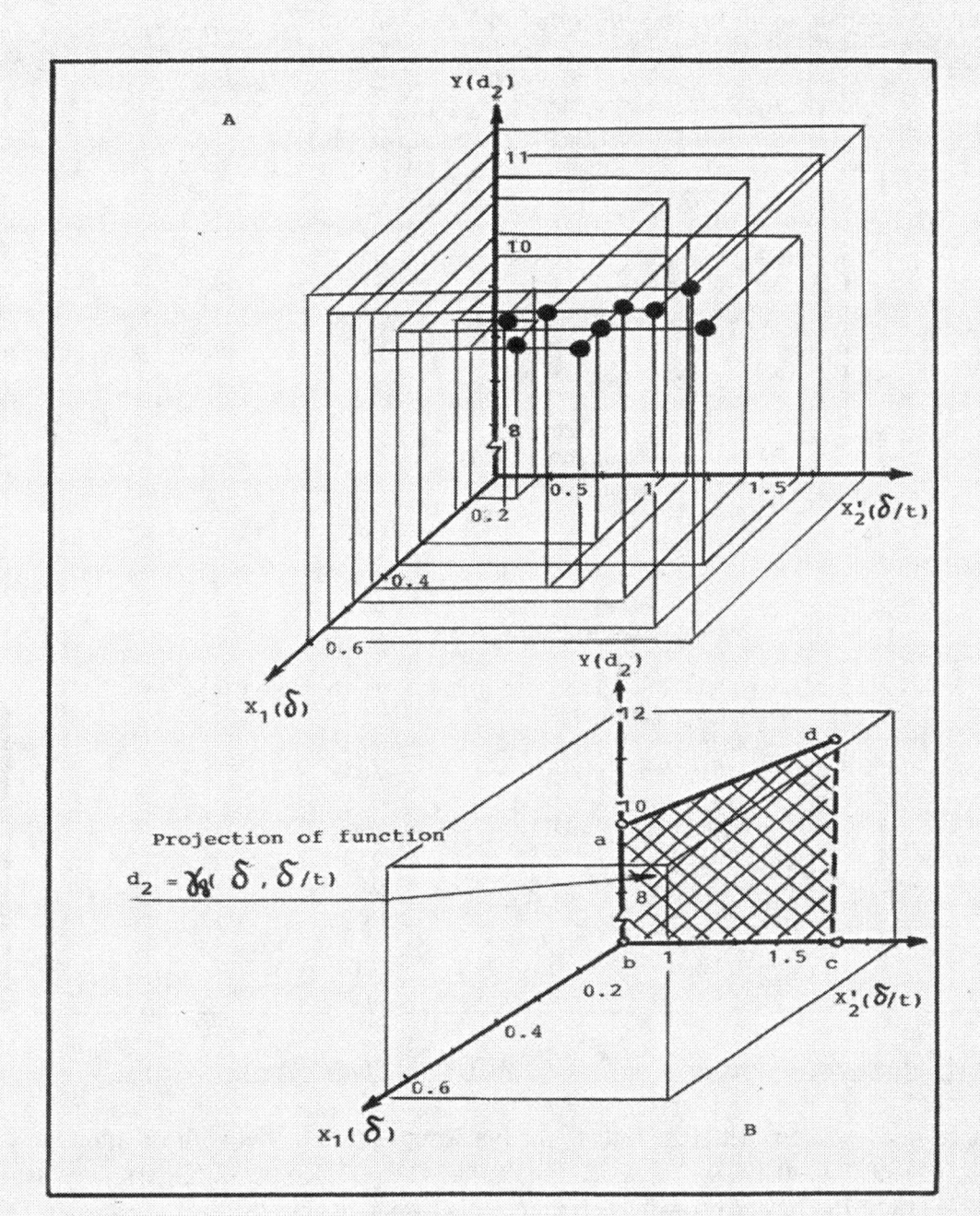

Figure 24

The module of vectors $|a'b'|, |b'c'|, |c'd'|, |d'a'|$ and area of functional surface can be calculated by formula (11),(12),(13),(14) and (15).At data we have $|a'b'|=9.53$, $|b'c'|=1.31$, $|c'd'|= 10.99, |d'a'|=1.96$ and $S_{a'b'c'd'}=13.44mm^2$.

All results of computation for the linear equation view of $Y_c=8.743+5.077X_1-0.292X'_2$ or $Y_c=8.743+5.077X_1-0.292(X_1/X_2)$,where $d_2=8.743+5.077\delta-0.292(\delta/t)$ are given in Table 13.

195

Table 13 Evaluation of regression equation $Y_c = 8.743 + 5.077X_1 - 0.292X'_2$

A. Mean ,variance and standard deviation

Variable	Mean	Variance	Standard deviation
X_1	0.368	0.209	0.457
X'_2	1.242	3.485	1.867
Y	10.248	4.121	2.030

B. Results of multiple regression of Y on X_1 and X'_2

Parameter	Variable	Coefficients	Standard error	T-value
β_1	X_1	5.077	0.152	33.401
β_2	X'_2	-0.292	0.622	-0.469

C. Analysis of variance results

Regression
- Degree of freedom 2
- Sum of squares 3.302
- Mean squares 1.651

Error
- Degree of freedom 6
- Sum of squares 0.819
- Mean squares 0.137

Standard error of estimate 0.457

F-value* 12.05

*Since $F=12.05 > [F_{0.05,2,6}=5.14]$,we can reject the hypothesis that both β_1 and β_2 are zero.

D. Determination of residual

Number	Observed	Estimated	Residual
1	10.03	10.07	-0.04
2	10.36	10.55	-0.19
3	8.69	9.17	-0.48
4	10.13	9.53	0.60
5	10.22	9.89	0.33
6	10.07	10.27	-0.20
7	10.49	10.66	-0.17
8	11.06	10.99	0.07
9	11.18	11.09	0.09

So ,the parameters of functional surface $d_2 = \gamma_8(\delta, \delta/t)$ are the following:

1.Function $d_2 = \gamma_8(\delta, \delta/t)$ better submits to the linear regression model with equation view of $Y_c = 8.743 + 5.077X_1 - 0.292X'_2$ or $Y_c = 8.743 + 5.077X_1 - 0.292(X_1/X_2)$,where $d_2 = 8.743 + 5.077\delta - 0.292 (\delta/t)$ with such statistical characteristics:

- Coefficient of determination $R^2 = 0.801$
- Coefficient of correlation $r = 0.895$
- Standard deviation $S_{y/x1,x'2} = 0.369$
- Minimization of the mean square error (min MSE=0.091)
- Minimization of the mean absolute deviation (min MAD=0).

2.The total area of functional surface $d_2 = \gamma_8(\delta, \delta/t)$ is equal $\sum S = 13.44 mm^2$ with such coordinates of their points in three-dimensional drawing (sizes in millimeters) for this surface: a'($X_{1,1}=0.18, X_{2,1}=0.43, Y_1=9.53$);b'($X_{1,2}=0.18, X_{2,2}=0.43, Y_2=0$);c'($X_{1,3}=0.54, ,X_{2,3}=1.69, Y_3=0$); d'($X_{1,4}=0.54, X_{2,4}=1.69, Y_4=10.99$).

C. Modification function $d_2 = \gamma_9(\delta, t/B)$

Analysis of data and the statistical characteristics (coefficient of determination $R^2=0.803$, coefficient of correlation $r=0.896$, standard deviation $S_{y/x1,x'2}=0.368$) and also the minimization of the mean square error (min MSE=0.090) and minimization of the mean absolute (min MAD =0.004) show that this functional dependence $d_2=\gamma_9(\delta,t/B)$ better submits to the linear than non-linear regression[32] regression model with equation view of

$$Y_c=8.68+4.03X_1+1.6X'_2 \quad \text{or} \quad Y_c=8.68+4.03X_1+1.6(X_2/X_3) \quad , \text{where}$$

$$d_2=8.68+4.03\delta+1.6(t/B) \quad (43).$$

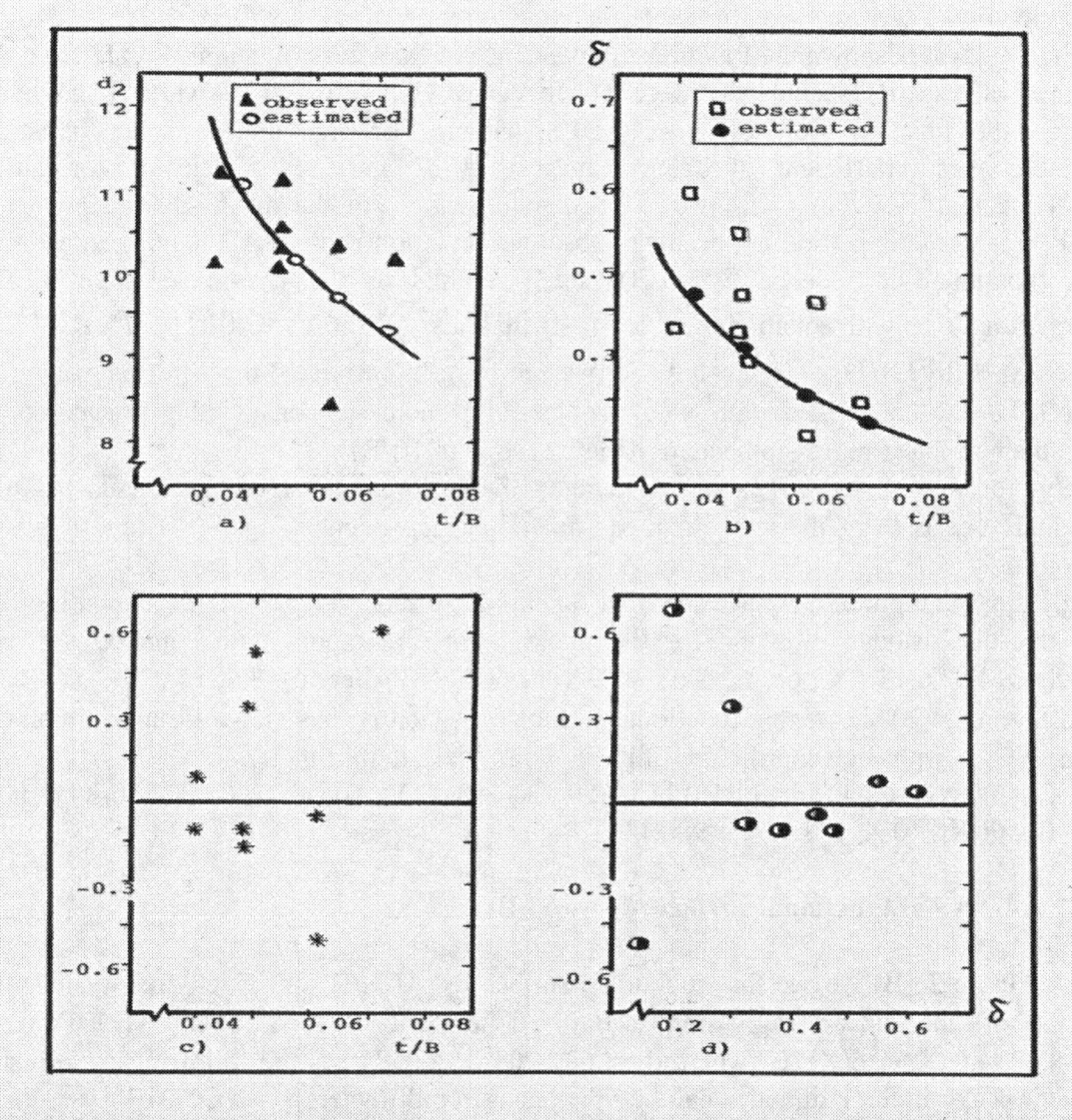

Figure 25

32- Non-linear regression model has equation view of $d_2=9.29+1.47\delta+2.64\delta^2-0.05(t/B)$

In Figure 25(a) is shown the functional dependency of external diameter (d_2) stainless chip from the ratio of values (t/B) ,such as clearance (t) between of chip wraps to its width (B) ,i.e we have the function view of $d_2=\phi_7$(t/B).Analysis of data and also the statistical characteristics (coefficient of determination R^2=0.14,coefficient of correlation r=0.37,standard deviation $S_{y/x'2}$=0.711) and minimization of the mean square error (min MSE=0.394) ,minimization of the mean absolute deviation (min MAD=0.032) show that this functional dependence $d_2=\phi_7$(t/B) better submits to the non-linear than linear[33] regression model with equation view of $\mathbf{Y_c=4.79X'_2{}^{-0.255}}$ or $\mathbf{Y_c=4.79(X_2/X_3)^{-0.255}}$,where $\mathbf{d_2=4.79(t/B)^{-0.255}}$ (44) . As we see from Figure 25(a) that with increasing of ratio values (t/B) , the value of external diameter (d_2) stainless chip decreases considerably in accordance with regression equation view of $d_2=4.79(t/B)^{-0.255}$.

In Figure 25(b) is shown the functional dependence of thickness (δ) stainless chip from the ratio of values (t/B) ,such as clearance (t) between of chip wraps to its width (B),i.e has a place the function view of $\delta=\phi_8$(t/B) .Analysis of data and also the statistical characteristics (coefficient of determination R^2=0.17 and coefficient of correlation r=0.41,standard deviation $S_{y/x'2}$=0.158)and minimization of the mean square error (min MSE=0.020),minimization of the mean absolute deviation (min MAD=0.068) show that this functional dependence $\delta=\phi_8$(t/B) better submits to the non-linear than linear[34] regression model with equation view of $\mathbf{Y_c=0.001X'_2{}^{-1.902}}$ or $\mathbf{Y_c=0.001(X_2/X_3)^{-1.902}}$,where $\boldsymbol{\delta=0.001(t/B)^{-1.902}}$ (45). So ,we see from Figure 25(b) that with increasing of ratio (t/B) ,the value of thickness (δ) stainless chip increases considerably in accordance with the non-linear regression equation view of $\delta=0.001(t/B)^{-1.902}$.

In Figure 25(c) and 25(d) are illustrated the residual plots(residual versus t/B and residuals versus δ) of the above-named functional dependence $d_2=\phi_7$(t/B) and $d_2=\gamma(\delta)$ accordingly.

Analyzing the Figure 26(A) ,we see that functional surface $d_2=\gamma_9(\delta,$t/B) is shown in view of three-dimensional drawing for the linear regression model with equation view of $Y_c=8.68+4.03X_1+1.6X'_2$ or $Y_c=8.68+4.03X_1+1.6(X_2/X_3)$,where $d_2=8.68+4.03\delta+1.6$(t/B).

And Figure 26(B) shows that functional surface $d_2=\gamma_9(\delta,$t/B) has view of trapezium on the plane YOX'_2 with such coordinates of their points (sizes in millimeters):
a($X_{1,1}$ =0,$X_{2,1}$=0.04 ,Y_1=11.16);b($X_{1,2}$=0 ,$X_{2,2}$=0.04,Y_2=0);c($X_{1,3}$=0 ,$X_{2,3}$=0,$X_{2,3}$=0.07, ,Y_3=0); d($X_{1,4}$=0,$X_{2,4}$=0.07,Y_4=9.52).

Analysis of functional surface $d_2=\gamma_9(\delta,$t/B)

From Figure 26(B) we see that functional surface $d_2=\gamma_9(\delta,$t/B) on which are situated the points of the linear regression equation view of $Y_c=8.68+4.03X_1+1.6X'_2$ or $Y_c=8.68+4.03X_1+1.6(X_2/X_3)$,where $d_2=8.68+4.03\delta+1.6$(t/B) has the following coordinates of their peaks in three-dimensional drawing (sizes in millimeters):
a'($X_{1,1}$=0.60,$X_{2,1}$=0.04,Y_1=11.16); b'($X_{1,2}$=0.04,Y_2=0);c'($X_{1,3}$=0.18,$X_{2,3}$=0.07,Y_3=0); d'($X_{1,4}$=0.18,$X_{2,4}$=0.07,Y_4=9.52).

33-Linear regression model has the equation view of $d_2=10.26-0.18$(t/B)
34-Linear regression model has the equation view of $\delta=0.47-2$(t/B)

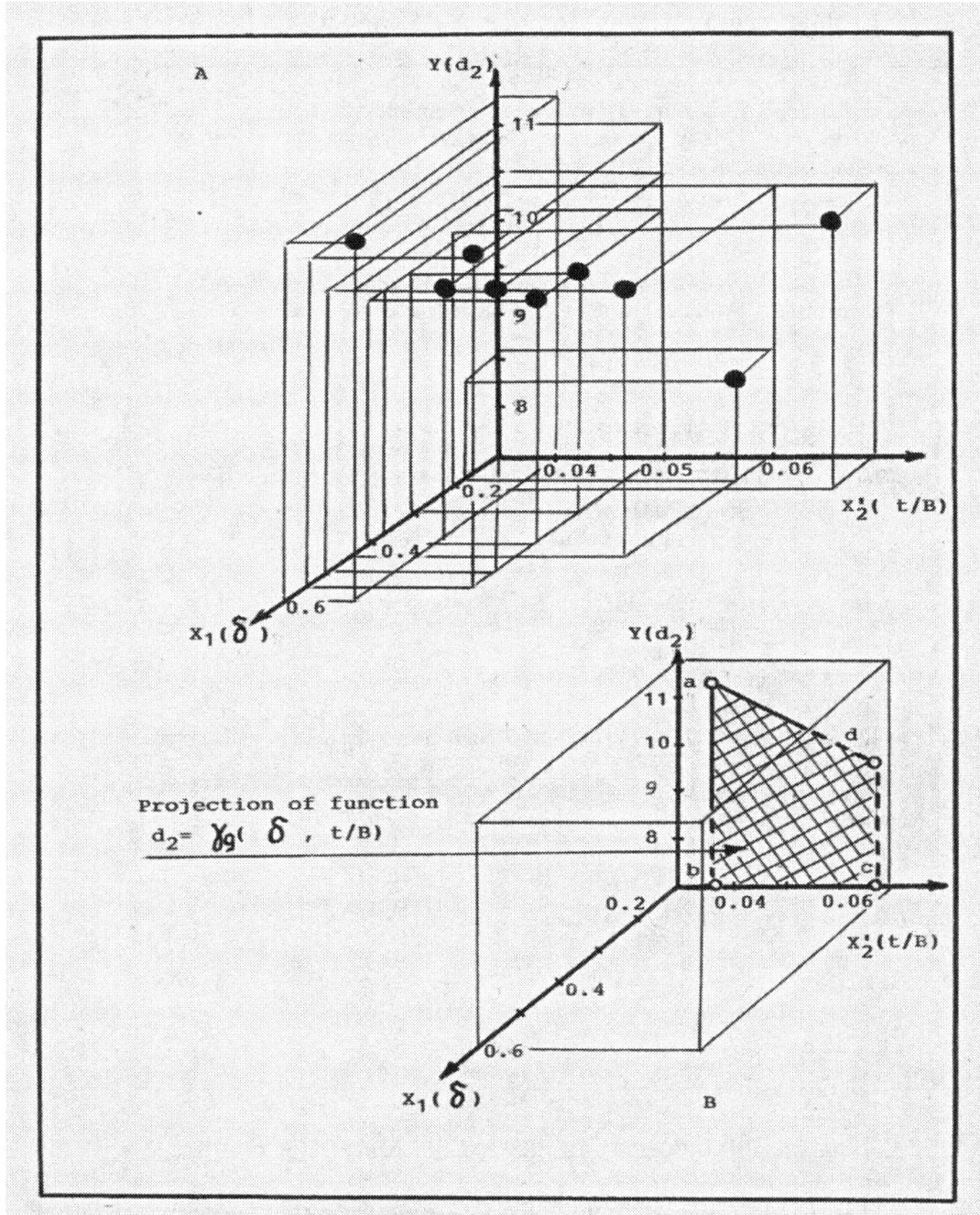

Figure 26

The module of vectors $|a'b'|$, $|b'c'|$, $|c'd'|$ and $|d'a'|$ can be calculated by formulas (11),(12),(13) and (14). Area of functional surface is equal: $\mathbf{S_{a'b'c'd'}=0.5[\,|a'b'|\,-}$ $\mathbf{-\,|c'd'|\cdot|b'c'|\,]+|b'c'|\cdot|c'd'|}$ **(46)** . At data we have the values: $|a'b'|$=11.16 , $|b'c'|$=0.42, $|c'd'|$=9.52 , $|d'a'|$=1.69 and $S_{a'b'c'd'}$=4.34 mm^2.

All results of computations for the linear equation view of Y_c=8.68+4.03X_1+1.6X'_2 or Y_c=8.68+4.03X_1+1.6(X_2/X_3) ,where d_2=8.68+4.03δ+1.6 (t /B) are given in Table 14.

199

Table 14 Evaluation of regression equation $Y_c=8.68+4.03X_1+1.6X'_2$

A. Mean , variance and standard deviation

Variable	Mean	Variance	Standard deviation
X_1	0.368	0.209	0.457
X'_2	0.052	0	0
Y	10.248	4.121	2.030

B. Results of multiple regression of Y on X_1 and X'_2

Parameter	Variable	Coefficients	Standard error	T-value
β_1	X_1	4.03	0.152	26.513
β_2	X'_2	1.60	0	0

C. Analysis of variance results

Regression
    Degrees of freedom 2
    Sum of squares 3.309
    Mean squares   1.655
Error
    Degrees of freedom 6
    Sum of squares   0.812
    Mean squares   0.135
Standard error of estimate 0.457
F-value *   12.26

* Since F=12.26> [$F_{0.05,2,6}$=5.14] ,we can reject the hypothesis that both β_1 and β_2 are zero.

D. Determination of residuals

Number	Observed	Estimated	Residual
1	10.03	10.17	-0.14
2	10.36	10.47	-0.11
3	8.69	9.18	-0.49
4	10.13	9.52	0.61
5	10.22	9.89	0.33
6	10.07	10.25	-0.18
7	10.49	10.61	-0.12
8	11.06	10.94	0.12
9	11.18	11.16	0.02

So, the parameters of this functional surface $d_2=\gamma_9(\delta,t/B)$ are the following:

1.Function $d_2=\gamma_9(\delta,t/B)$ better submits to the linear regression model with equation view of $Y_c=8.68+4.03X_1+1.6X'_2$ or $Y_c=8.68+4.03X_1+1.6(X_2/X_3)$, where $d_2=8.68+4.03\delta+1.6(t/B)$ with such statistical characteristics: coefficient of determination R^2=0.803,coefficient   of correlation r=o.896,standard deviation $S_{y/x1,x'2}$=0.368, minimization of the mean square error (min MSE=0.090) , minimization of the mean absolute deviation (min MAD=0).
2.The total area of functional surface $d_2=\gamma_9(\delta,t/B)$ is equal $\sum S$=4.34 mm^2 with such coordinates of their points in three-dimensional drawing for this surface(sizes in millimeters): a'($X_{1,1}$=0.60, ,$X_{2,1}$=0.04, Y_1=11.16); b'($X_{1,2}$=0.60 ,$X_{2,2}$=0.04 ,Y_2=0); c'($X_{1,3}$=0.18,$X_{2,3}$=0.07,Y_3=0); d' ($X_{1,4}$=0.18,$X_{2,4}$=0.07 ,Y_4=9.52).

6.5 Dependence of external diameter (d_2) from clearance (t) between of chip wraps, number (ω) of chip wraps and width(B) of this stainless chip, i.e $d_2=\alpha_5(t,\omega,B)$ and also from their modification: $d_2=\gamma_{10}(t, t/\omega)$,$d_2=\gamma_{11}(t ,t/B)$, $d_2=\gamma_{12}(t ,\omega/B)$.

A. Function $d_2=\alpha_5(t ,\omega,B)$

Analysis of data and the statistical characteristics (coefficient of determination $R^2=0.26$,coefficient of correlation $r=0.510$,standard deviation $S_{y/x1,x2,x3}=0.781$), minimization of the mean square error (min MSE=0.339) and minimization of the mean absolute deviation (min MAD=0) show that this functional dependence $d_2=\alpha_5(t,\omega,B)$ better submits tot the non-linear than linear[35] regression model with equation view of $Y_c=4.094-3.069X_1-0.462X_1{}^2-0.021X_2+1.242X_3$, where $d_2=4.094 -3.069t --0.462 t^2-0.021\omega+1.242B$ (47).

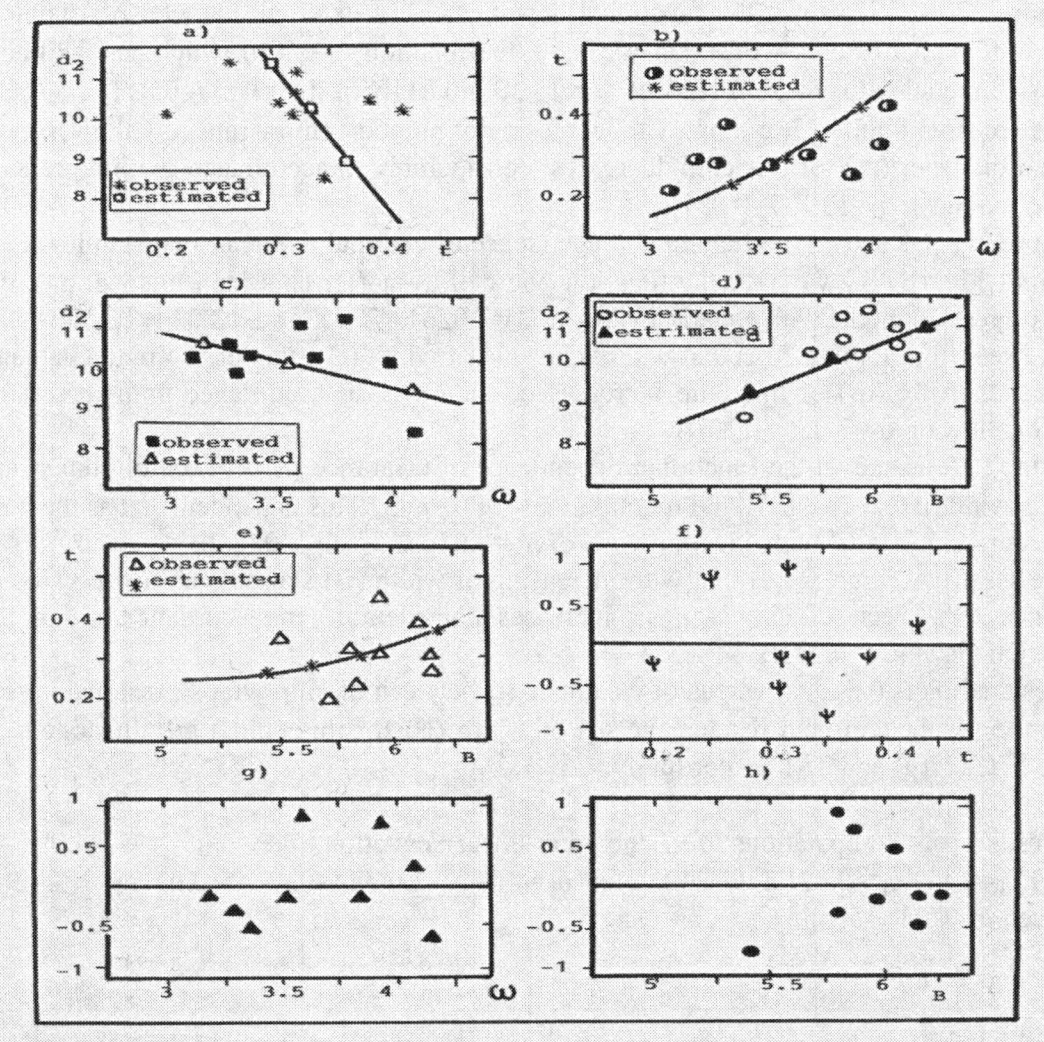

Figure 27

35-Linear regression model has equation view of $d_2=2.31-3.97t+0.17\omega+1.47B$

The functional dependence $d_2=\gamma_3(t)$ is shown in Figure 27(a) and has the linear regression model with equation view of $Y_c=0.515-0.019X$ or $t=0.515-0.019d_2$, where $d_2=27.11-52.63t$. So, we see from Figure 27(a) that with increasing of clearance (t) between of chip wraps, the value of external diameter (d_20 for this chip considerably decreases in accordance with regression equation view of $d_2=27.11-52.63t$.

In Figure 27(b) is shown the functional dependence of clearance (t) between of chip wraps from the number of wraps (ω),i.e we have the function view of $t=\alpha_1(\omega)$. This dependence has the non-linear regression model with equation view of $Y_c=4.227X^{0.152}$ or $\omega=4.227t^{0.152}$,where $\mathbf{t=(0.237\omega)^{6.579}}$ **(48)** .

So, we see from Figure 27(b) that with increasing of number wraps (ω) for this stainless chip , the value of clearance (t) between of chip wraps considerably increases in accordance with regression equation view of $t=(0.237\omega)^{6.579}$.

In Figure 27 (c) is shown the functional dependence of external diameter (d_2)stainless chip from number of chip wraps ,i.e we have the function $d_2=\phi_1(\omega)$ which has the linear regression model with equation view of $Y_c=13.137-0.811X$ or $d_2=13.137-0.811\omega$.

So, we see from Figure 27(c) that with increasing of number chip wraps (ω) ,the value of external diameter (d_2) for this chip decreases considerably in accordance with regression equation view of $d_2=13.137-0.811\omega$.

In Figure 27(d) is shown the functional dependence of external diameter (d_2) stainless chip from its width (B) ,i.e we have the function view of $d_2=\varphi_2(B)$.This dependence has the linear regression model with equation view of $Y_c=3.026+1.237X$ or $d_2=3.026+1.237B$.

So , we see from Figure 27(d) that with increasing of width (B) stainless chip, the value of external diameter (d_2) of this chip increases considerably in accordance with regression equation view of $d_2=3.026+1.237B$.

In Figure 27(e)is shown the functional dependence of clearance (t) between of chip wraps from its width(B) ,i.e we have the function view of $t=\gamma_1(B)$.This dependence has the non-linear regression model with equation view of $Y_c=0.032X^{1.296}$ or $t=0.032B^{1.296}$.

So ,we see from Figure 27(e) that with increasing of width (B) stainless chip ,the value of clearance (t)between of chip wraps increases considerably in accordance with the regression view of $t=0.032B^{1.296}$.

The residual plots (residual versus of clearance (t) between of chip wraps , residual versus of number wraps (ω) and residual versus of width (B) for this chip) are illustrated in Figure 27(f),27(g) and 27(h) accordingly.

All results of computations for the non-linear equation view of $Y_c=4.094$ $-3.069X_1-0.462X_1^2-0.021X_2+1.242X_3$ or $d_2=4.094-3.069t-0.462t^2-0.021\omega+1.242$ B are given in Table 15.

Table 15 Evaluation of regression equation $Y_c=4.09-3.07X_1-0.46X_1^2-0.02X_2+1.24X_3$

A. Mean, variance and standard deviation

Variable	Mean	Variance	Standard deviation
X_1	0.317	0.031	0.176
X_2	3.559	1.113	1.055
X_3	5.836	0.515	0.718
Y	10.248	4.121	2.030

B. Results of multiple regression of Y on X_1,X_2 and X_3

Parameter	Variable	Coefficients	Standard error	T-value
β_1	X_1	-3.069	0.059	-52.02
β_2	X_2	-0.021	0.352	-0.06
β_3	X_3	1.242	0.239	5.19

C. Analysis of variance results

Regression

Degrees of freedom 3

Sum of squares 1.072

Mean squares 0.357

Error

Degrees of freedom 5

Som of squares 3.049

Mean square 0.609

Standard error of estimate 0.176

F-value* 0.59

*Since F=0.58< [$F_{0.05,3,5}$=5.41] we can not reject the hypothesis that three parameters β_1,β_2 and β_3 are zero.

D. Determination of residuals

Number	Observed	Estimated	Residuals
1	10.03	10.46	-0.43
2	10.36	10.44	-0.08
3	8.69	9.58	-0.89
4	10.13	9.83	0.30
5	10.22	10.28	-0.06
6	10.07	10.63	-0.56
7	10.49	10.57	-0.08
8	11.06	10.07	0.99
9	11.18	10.36	0.82

B. Modification function $d_2=\gamma_{10}(t, t/\omega)$

Analysis of data and the statistical characteristics (coefficient of determination R^2=0.17, coefficient of correlation r=0.41, standard deviation $S_{y/x1,x'2}$=0.701) and also the minimization of the mean square error (min MSE=0.380), minimization of the mean absolute (min MAD=0) show that this functional dependence $d_2=\gamma_{10}(t ,t/\omega)$ does not have the good correlation and we can admit that this function better submits to the linear than non-linear[36] regression model and has the equation view of $Y_c=10.97-5.84X_1+12.5X'_2$ or $Y_c=10.97-5.84X_1+12.5(X_1/X_2)$, where $d_2=10.97-5.84t+12.5(t/\omega)$ (49) .

36-Non-linear regression equation has view of $d_2=10.91-2.61t+2.43t^2-0.93(t/\omega)$

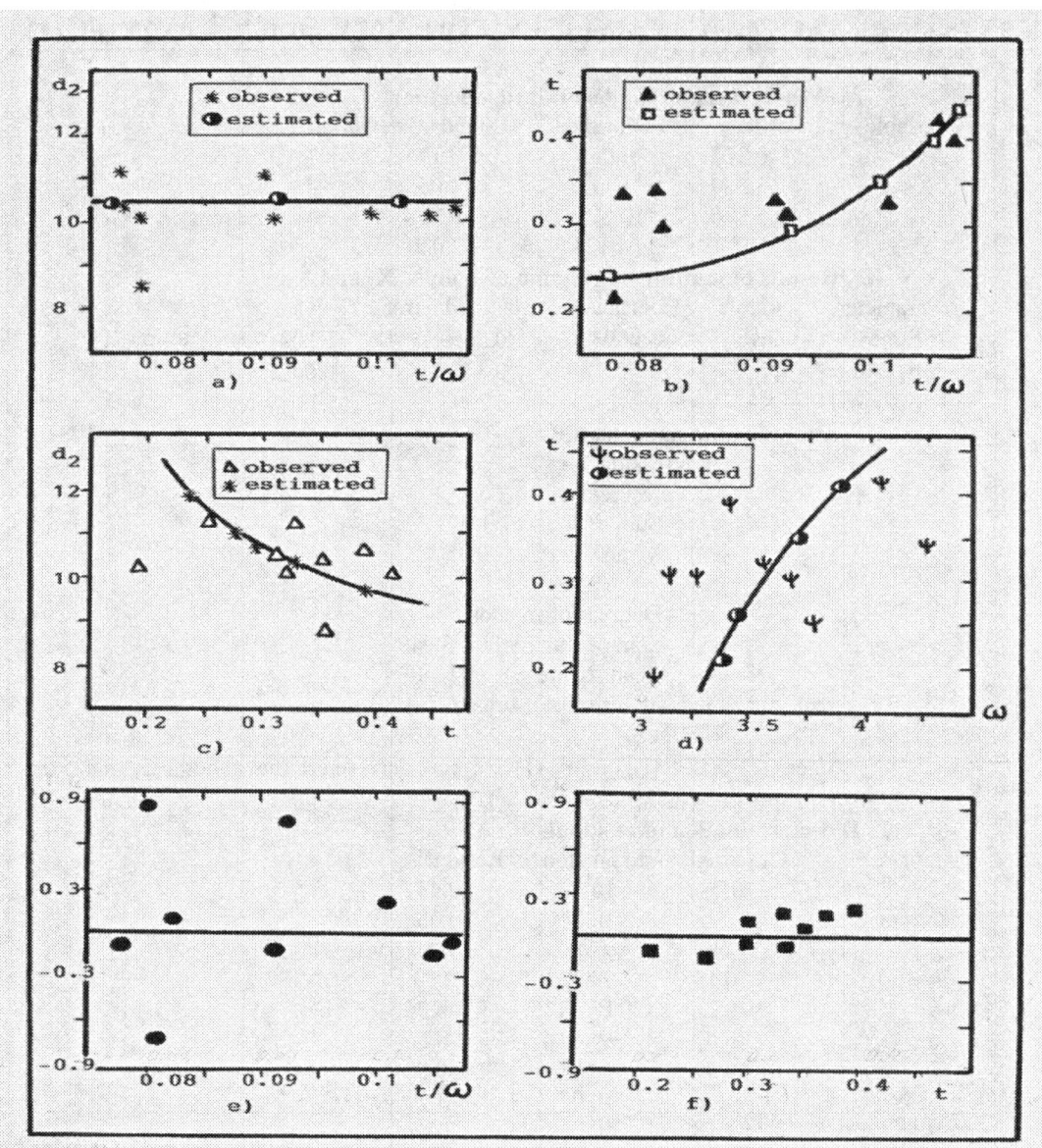

Figure 28

In Figure 28(a) is shown the functional dependency of external diameter (d_2) stainless chip from the ratio of values (t/ω) ,such as clearance (t) between of chip wraps to its number of wraps (ω) ,i.e we have the function view of $d_2=\phi_9(t/\omega)$. Analyzing of data and the scatter plots shown in Figure 28(a),and also the statistical characteristics (coefficient of determination $R^2=0$,coefficient of correlation r=0)show that this functional dependence $d_2=\phi_9(t/\omega)$ better submits to the linear than non-linear[37] regression model with equation view of $Y_c=9.8+5X'_2$ or $Y_c=9.8+5(X_1/X_2)$, where $d_2=9.8+5(t/\omega)$ **(50)** .

37- Non-linear regression model has equation view of $d_2=8.45(t/\omega)^{-0.078}$

So, we see from Figure 28(a) that practically between of two parameters, such as external diameter (d_2) and ratio (t/ω), the correlation absents.

In Figure 28(b) is shown the functional dependence of clearance (t) between of chip wraps from the ratio of its values (t/ω),i.e has a place the function view of $t=\phi_{10}(t/\omega)$. Analysis of data and also the statistical characteristics (coefficient of determination $R^2=0.709$,coefficient of correlation r=0.842,standard deviation $S_{y/x'2}=0.036$) and minimization of the mean square error (min MSE=0) and minimization of the mean absolute deviation (min MAD =0.019) show that this functional dependence $t=\phi_{10}(t/\omega)$ better submits to the non-linear than linear[38] regression model with equation view of

$$\mathbf{Y_c=4.256X'_2{}^{1.078}} \quad \text{or} \quad \mathbf{Y_c=4.256(X_1/X_2)^{1.078}}, \text{ where } \mathbf{t=4.256(t/\omega)^{1.078}} \quad (\mathbf{51}).$$

So ,we see from Figure 28(b) that with increasing of ratio (t/ω),the value of clearance (t) between of chip wraps increases considerably in accordance with the non-linear regression equation view of $t=4.256(t/\omega)^{1.078}$.

In Figure 28(c) is shown the functional dependence of external diameter (d_2) from the clearance (t) between of chip wraps . This functional dependence $t=\gamma_3(d_2)$ has the non-linear regression model with equation view of $Y_c=1.14X^{-0.56}$ or $t=1.14d_2^{-0.56}$,where

$$\mathbf{d_2=(0.88t)^{-1.785}} \quad (\mathbf{52}).$$

Analysis of data shown in Figure 28 (c) shows that with increasing of clearance (t) between of chip wraps ,the value of external diameter (d_2) decreases considerably in accordance with non-linear regression equation view of $d_2=(0.888t)^{-1.785}$.

In Figure 28(d) is shown the functional dependence of clearance (t) between of chip wraps from the number of wraps (ω) for this chip ,i.e has a place the function view of $t=\alpha_1(\omega)$. This functional dependence has non-linear regression model with equation view of $Y_c=4.227X^{0.152}$ or $\omega=4.227t^{0.152}$, where $\mathbf{t=(0.24\omega)^{6.578}}$ ($\mathbf{53}$) .

So , we see from Figure 28(d) that with increasing of number chip wraps (ω) ,the value of clearance (t) between of chip wraps increases considerably according to non-linear regression equation view of $t=(0.24\omega)^{6.578}$.

In Figure 28(e) and 28(f) are illustrated the residual plots (residual versus t/ω and residual versus t) of functional dependencies $d_2=\phi_9(t/\omega)$ and $d_2=\gamma_3(t)$ accordingly.

Analyzing the Figure 29(A) , we see that functional surface $d_2=\gamma_{10}(t,t/\omega)$ is shown in view of three-dimensional drawing for the linear regression model with equation view of $Y_c=10.97-5.84X_1+12.5X'_2$ or $Y_c=10.97-5.84X_1+12.5(X_1/X_2)$,where $d_2=10.97-5.84t+12.5(t/\omega)$.

And Figure 29(B) shows that functional surface $d_2=\gamma_{10}(t,t/\omega)$ has view of trapezium on the plane YOX'_2 with such coordinates of their points (sizes in millimeters):
a ($X_{1,1}=0$, $X_{2,1}=0.07$,$Y_1=10.6$); b ($X_{1,2}=0$,$X_{2,2}=0.07$,$Y_2=0$); c ($X_{1,3}=0$, ,$X_{2,3}=0.11$,$Y_3=0$) ; d ($X_{1,4}=0$,$X_{2,4}=0.11$,$Y_4=9.90$).

---

38 – Linear regression model has equation view of $t = -0.4 +8 (t/\omega)$

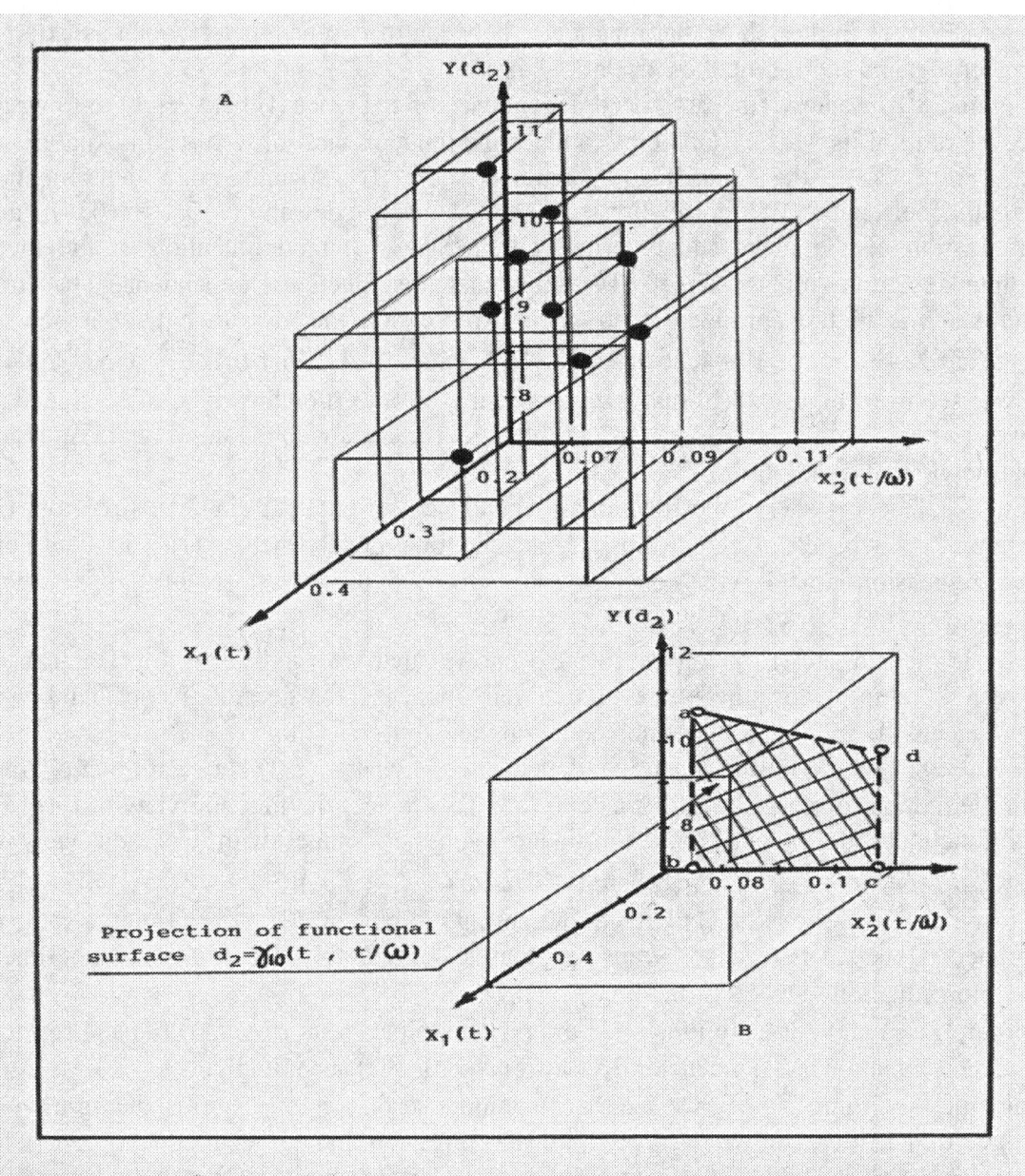

Figure 29

Analysis of functional surface $d_2=\gamma_{10}(t,t/\omega)$

From Figure 29 (B) we see that functional surface $d_2=\gamma_{10}(t,t/\omega)$,on which are situated the points of the linear regression equation view of $Y_c=10.97-5.84X_1+12.5X'_2$ or $Y_c=10.97-5.84X_1+12.5(X_1/X_2)$,where $d_2=10.97-5.84t+12.5(t/\omega)$ has the following coordinates of their peaks in three-dimensional drawing: a'($X_{1,1}=0.21$,$X_{2,1}=0.07,Y_1=10.6$); b' ($X_{1,1}=0.21,X_{2,2}=0.07,Y_2=0$); c'($X_{1,3}=0.42,X_{2,3}=0.11,Y_3=0$); d' ($X_{1,4}=0.42,X_{2,4}=0.11, Y_4=9.90$).

The module of vectors $|a'b'|$, $|b'c'|$, $|c'd'|$, $|d'a'|$ and area of functional surface $S_{a'b'c'd'}$ can be determined by formulas (11),(12),(13),(14) and (46). So, at data we have the values : $|a'b'|$=10.6 , $|b'c'|$=0.21, $|c'd'|$=9.90 , $|d'a'|$=0.73 and $S_{a'b'c'd'}$=2.15mm^2.All results of computation for the linear equation view of Y_c=10.97−5.84X_1+12.5X'_2 or Y_c=10.97−5.84X_1+12.5(X_1/X_2) ,where d_2=10.97−5.84t+12.5(t/ω) are given in Table 16.

Table 16 Evaluation of regression equation Y_c=10.97−5.84X_1+12.5X'_2

A. Mean , variance and standard deviation			
Variable	Mean	Variance	Standard deviation
X_1	0.317	0.031	0.176
X'_2	0.09	0	0
Y	10.248	4.121	2.030

B. Results of multiple regression of Y on X_1 and X'_2				
Parameter	Variable	Coefficients	Standard error	T-value
$β_1$	X_1	-5.84	0.059	-98.98
$β_2$	X'_2	12.50	0	0

C. Analysis of variance results

Regression
  Degrees of freedom 2
  Sum of squares 0.699
  Mean squares 0.349
Error
  Degrees of freedom 6
  Sum of squares 3.422
  Mean squares 0.570
  Standard error of estimate 0.176
  F-value* 0.612

* Since F=0.612< [$F_{0.05,2,6}$=5.14] , we can not reject the hypothesis that both $β_1$and $β_2$ are zero.

D. Determination of residual

Number	Observed	Estimated	Residual
1	10.03	10.60	-0.57
2	10.36	10.25	0.11
3	8.69	9.98	-1.29
4	10.13	9.90	0.23
5	10.22	10.22	0
6	10.07	10.29	-0.22
7	10.49	10.35	0.14
8	11.06	10.23	0.83
9	11.18	10.39	0.79

So, the parameters of this functional surface $d_2=\gamma_{10}(t,t/\omega)$ are the following:
1.This function better submits to the linear regression model with equation view of Y_c=10.97−5.84X_1+12.5X'_2,where d_2=10.97−5.84t+12.5(t/ω) with such statistical characteristics: coefficient of determination R^2=0.17,coefficient of correlation r=0.41,standard deviation $S_{y/x1,x'2}$=0.701,minimization of the mean square error (min MSE=0.380), minimization of the mean absolute deviation (min MAD=0).
2.The total area of this functional surface is equal $\sum S$=2.15mm^2 with such coordinates of their points in three-dimensional drawing: a'($X_{1,1}$=0.21,$X_{2,1}$=0.07,Y_1=10.60);b'($X_{1,2}$=0.21 , $X_{2,2}$=0.07,Y_2=0);c'($X_{1,3}$=0.42,$X_{2,3}$=0.11,Y_3=0);d'($X_{1,4}$=0.42,$X_{2,4}$=0.11,Y_4=9.90).

C. Modification function $d_2=\gamma_{11}(t, t/B)$

Analysis of data and the statistical characteristics (coefficient of determination $R^2=0.05$, coefficient of correlation $r=0.22$, standard deviation $S_{y/x1,x'2}=0.80$) and also the minimization of the mean square error (min MSE=0.436) and minimization of the mean absolute deviation (min MAD=0) show that this functional dependence $d_2=\gamma_{11}(t, t/B)$ does not the good correlation and the same time we can also admit that this function better submit to the linear than non-linear[39] regression model with equation view of

$$Y_c=11.05-2.3X_1-1.25X'_2 \quad \text{or} \quad Y_c=11.05-2.3X_1-1.25(X_1/X_2) \quad , \quad \text{where}$$

$$d_2=11.05-2.3t-1.25(t/B) \quad (54).$$

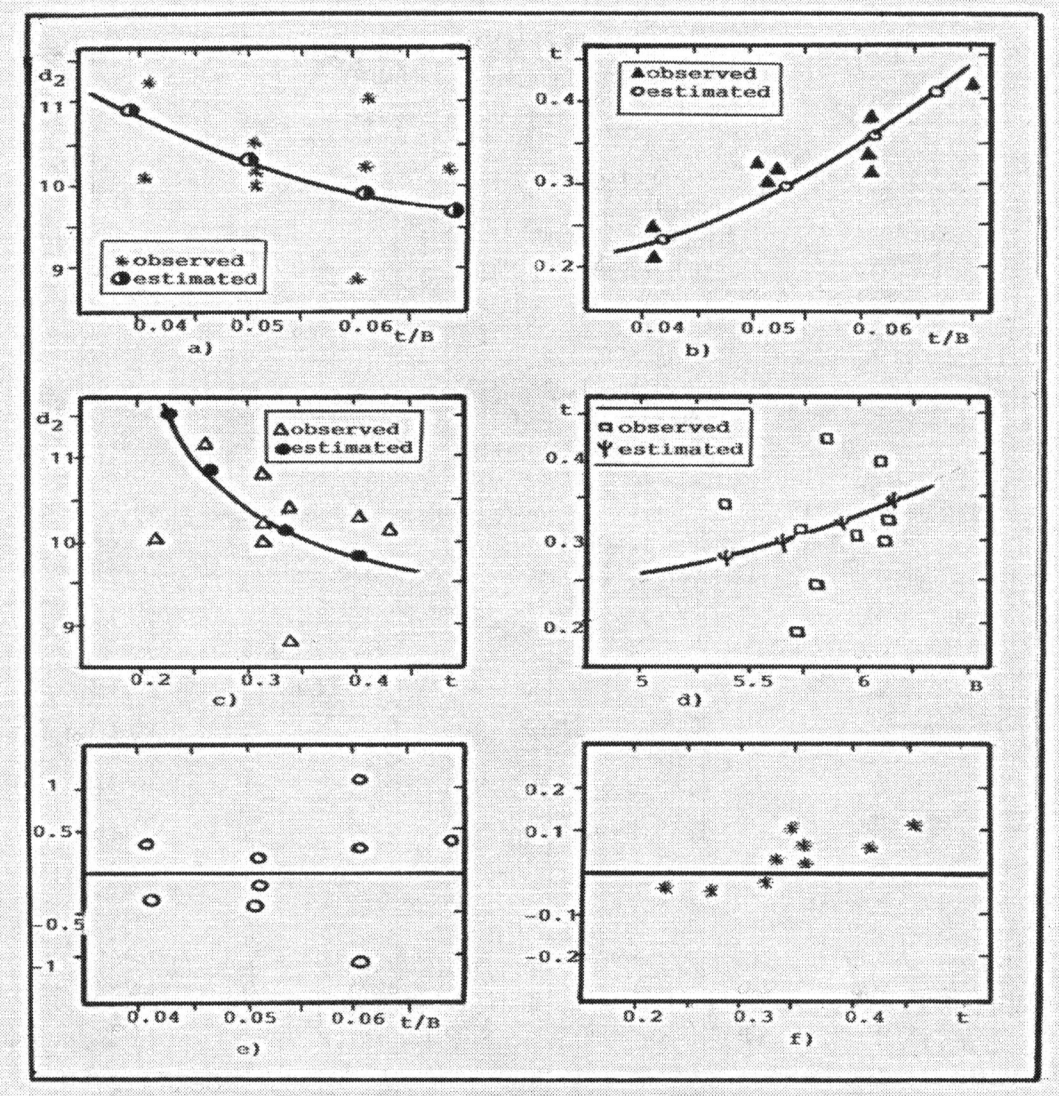

Figure 30

39- Non-linear regression model has equation view of $d_2=11.23-3.43t+2t^2-2(t/B)$

208

In Figure 30(a) is shown the functional dependency of external diameter (d_2) stainless chip from the ratio of values (t/B) , such as clearance (t) between of chip wraps to its width (B) ,i.e we have the function view of $d_2=\phi_{11}$(t/B).

Analyzing of data and also the scatter plots shown in Figure 30(a), and the statistical characteristics (coefficient of determination R^2=0.04,coefficient of correlation r=0.20) show that this functional dependence $d_2=\phi_{11}$(t/B) better submits to the non-linear than linear[40] regression model with equation view of $\mathbf{Y_c=5.96X'_2{}^{-0.183}}$ or $\mathbf{Y_c=5.96(X_1/X_2)^{-0.183}}$, where $\mathbf{d_2=5.96(t/B)^{-0.183}}$ (55) .

From Figure 30(a) we see that between two parameters, such as external diameter (d_2) and ratio (t/B) there is small correlation (r=0.20). And this function $d_2=\phi_{11}$(t/B) can be expressed in view of non-linear regression equation $d_2=5.96(t/B)^{-0.183}$ which shows that with increasing of ratio (t/B) , the value of external diameter (d_2) considerably decreases.

In Figure 30(b) is shown the functional dependence of clearance (t) between of chip wraps from its ratio (t/B).i.e has a place the function view of $t=\phi_{12}$(t/B).

Analysis of data and also the statistical characteristics (coefficient of determination R^2=1.0,coefficient of correlation r=1.0,standard deviation $S_{y/x'2}$=0.053) and minimization of the mean square error (min MSE=0) and minimization of the mean absolute deviation (min MAD=0) show that this functional dependence $t=\phi_{12}$(t/B) better submits to the non-linear than linear[41] regression model with equation view of $\mathbf{Y_c=4.753X'_2{}^{0.926}}$ or $\mathbf{Y_c=4.753(X_1/X_2)^{0.926}}$, where $\mathbf{t=4.753(t/B)^{0.926}}$ (56).

So, we see from Figure 30(b) that with increasing of (t/B) ,the value of clearance (t) between of chip wraps increases considerably in accordance with non-linear regression equation view of $t=4.753(t/B)^{0.926}$.

In Figure 30(c) is shown the functional dependence of external diameter (d_2) from the clearance (t) between of chip wraps .So ,we see from Figure 30(c) that this functional dependence $t=\gamma_3(d_2)$ has the non-linear regression model with equation view of $Y_c=1.14X^{-0.56}$ or $t=1.14d_2{}^{-0.56}$,where $d_2=(0.88t)^{-1.785}$.

Analysis of data shown in Figure 30(c) indicate on the fact that with increasing of clearance (t) between of chip wraps ,the value of external diameter (d_2) decreases considerably in accordance with the non-linear regression equation view of $d_2=(0.88t)^{-1.785}$.

In Figure 30(d) is shown the functional dependence of clearance (t) between of chip wraps from width (B) of this chip ,i.e has a place the function view of $t=\gamma_1(B)$.From Figure 30(d) we see that this functional dependence has the non-linear regression model with equation view of $Y_c=0.032X^{1.296}$ or $t=0.032B^{1.296}$. So ,we see from Figure 30(d) that with increasing of width(B) stainless chip ,the value of clearance (t) between of chip wraps increases considerably in accordance with non-linear regression equation view of $t=0.032B^{1.296}$.

In Figure 30(e) and 30(f) are illustrated the residual plots (residual versus t/B and residual versus t) of the functional dependencies $d_2=\phi_{11}$(t/B) and $d_2=\gamma_3$(t) accordingly.

---

40 – Linear regression equation has view of d_2=10.45–3.75(t/B)
41- Linear regression model has equation view of t=0.27+(t/B)

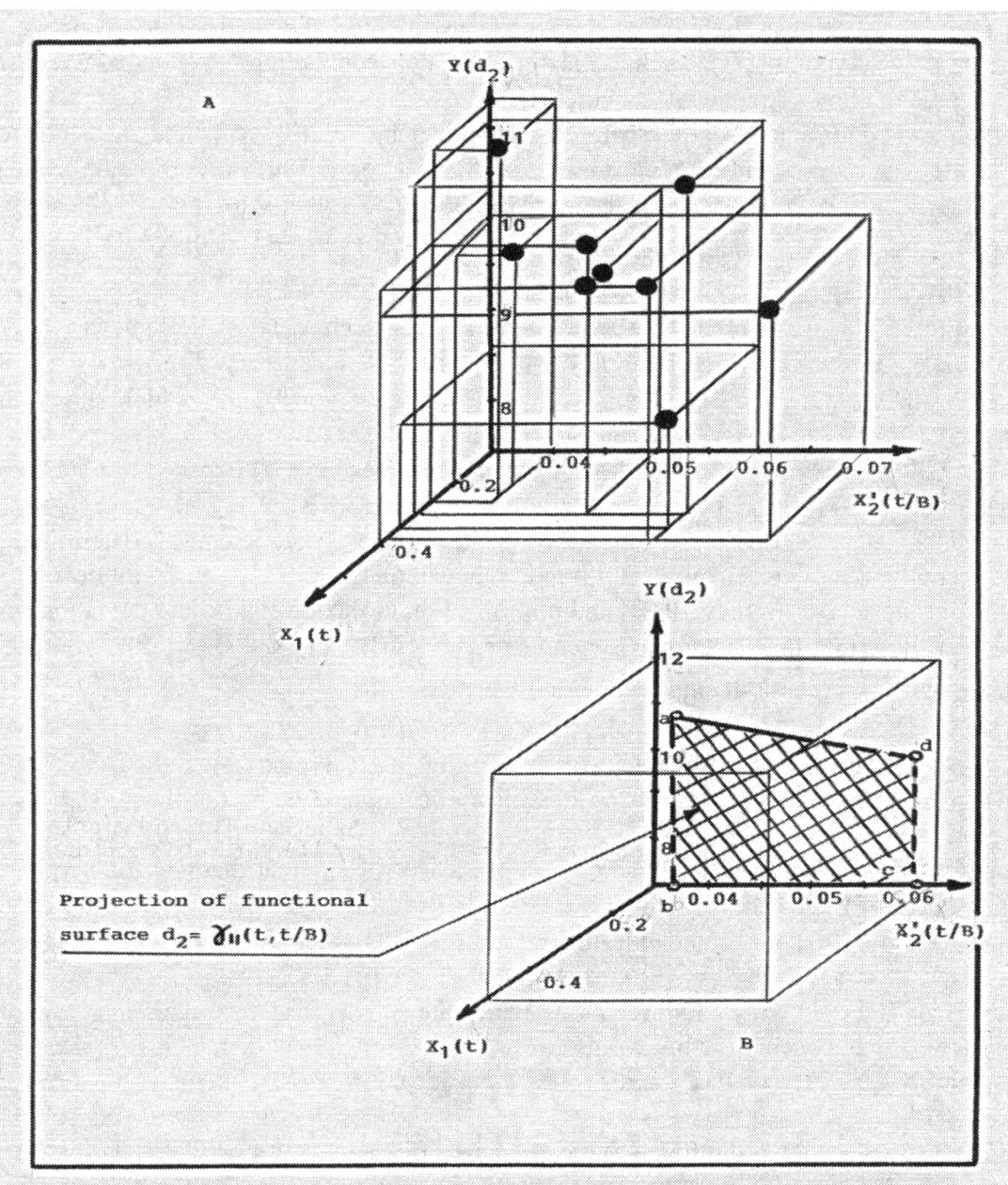

Figure 31

Analyzing the Figure 31(A) ,we see that functional surface $d_2=\gamma_{11}(t,t/B)$ is shown in view of three-dimensional drawing for the linear regression model with equation view of $Y_c=11.05-2.3X_1-1.25X'_2$ or $Y_c=11.05-2.3X_1-1.259(X_1/X_2)$,where $d_2=11.05-2.3t-1.25(t/B)$.

And Figure 31 (B) shows that this functional surface $d_2=\gamma_{11}$ (t, t/B) has view of trapezium on the place YOX'_2 with such coordinates of their points (sizes in millimeters):

a $(X_{1,1}=0 ,X_{2,1}=0.04,Y_1=10.51)$; b $(X_{1,2}=0,X_{2,2}=0.04,Y_2=0)$; c $(X_{1,3}=0 ,X_{2,3}=0.06,Y_3=0)$; d$(X_{1,4}=0,X_{2,4}=0.06,Y_4=10.09)$.

Analysis of functional surface $d_2=\gamma_{11}(t, t/B)$

Form Figure 31(B) we see that functional surface $d_2=\gamma_{11}(t, t/B)$ on which are situated the points of the linear regression equation view of $Y_c=11.05 -2.3X_2 -1.25X'_2$ or $Y_c=11.05-2.3X_1 -1.25(X_1/X_2)$, where $d_2=11.05-2.3t-1.25(t/B)$ has the following coordinates of their peaks in three-dimensional drawing(sizes in millimeters):
a'($X_{1,1}=0.21, X_{2,1}=0.04$, $Y_1=10.51$) ,b'($X_{1,2}=0.21, X_{2,2}=0.04, Y_2=0$);c'($X_{1,3}=0.38, X_{2,3}=0.06$, $Y_3=0$); d' ($X_{1,4}=0.38$, $X_{2,4}=0.06$, $Y_4=10.09$).
The module of vectors $\left|a'b'\right|, \left|b'c'\right|, \left|c'd'\right|, \left|d'a'\right|$ and area of functional surface can be expressed by formulas (11),(12),(13),(14) and (46). At data we have the following module of values: $\left|a'b'\right|=10.51, \left|b'c'\right|=0.17, \left|c'd'\right|=10.09, \left|d'a'\right|=0.45$ and area of this functional surface $S_{a'b'c'd'}=1.75$ mm^2.

All results of computation for the linear equation view of $Y_c=11.05-2.3X_1-1.25X'_2$ or $Y_c=11.05-2.3X_1-1.25(X_1/X_2)$,where $d_2=11.05-2.3t-1.25(t/B)$ are given in Table 17.

Table 17 Evaluation of regression equation $Y_c=11.05-2.3X_1-1.25X'_2$

A. Mean , variance and standard deviation			
Variable	Mean	Variance	Standard deviation
X_1	0.317	0.031	0.176
X'_2	0.050	0	0
Y	10.248	4.121	2.030

B. Results of multiple regression of Y on X_1 and X'_2				
Parameter	Variable	Coefficients	Standard error	T-value
β_1	X_1	-2.3	0.059	-38.98
β_2	X'_2	-1.25	0	0

C. Analysis of variance results

Regression
Degrees of freedom   2
Sum of squares   0.196
Mean squares   0.098
Error
Degrees of freedom 6
Sum of squares   3.925
Mean squares   0.654
Standard error of estimate   0.176
F-value*   0.15

*Since F=0.15<[$F_{0.05,2,6}$=5.14] we can not reject the hypothesis that both β_1 and β_2 are zero

D. Determination of residual			
Number	Observed	Estimated	Residual
1	10.03	10.51	-0.48
2	10.36	10.09	0.27
3	8.69	10.19	-1.50
4	10.13	9.99	0.14
5	10.22	10.29	-0.07
6	10.07	10.27	-0.20
7	10.49	10.25	0.24
8	11.06	10.24	0.82
9	11.18	10.42	0.76

1.Function $d_2=\gamma_{11}(t, t/B)$ better submits to the linear regression model with equation view of $Y_c=11.05-2.3X_1-1.25X'_2$ or $Y_c=11.05-2.3X_1-1.25(X_1/X_2)$,where $d_2=11.05-2.3t-1.25(t/B)$ with such statistical characteristics : the coefficient of determination is not so big and equal $R^2=0.05$ and also this concerns to the coefficient of correlation r=0.22,standard deviation $S_{y,x1,x;2}=0.80$),minimization of the mean square error (min MSE=0.436) and also minimization of the mean absolute deviation (min MAD=0).

2.The total area of functional surface $d_2=\gamma_{11}(t, t/B)$ is equal $\sum S=1.75mm^2$ with such coordinates of their points in three- dimensional drawing for this surface (sizes in millimeters): a '($X_{1,1}=0.21,X_{2,1}=0.04$,$Y_1=10.51$);b'($X_{1,2}=0.21,X_{2,2}=0.04$, $Y_2=0$); c'($X_{1,3}=0.38,X_{2,3}=0.06,Y_3=0$);d'($X_{1,4}=0.38,X_{2,4}=0.06,Y_4=10.09$).

D. Modification function $d_2=\gamma_{12}(t ,\omega/B)$

Analysis of data and statistical characteristics (coefficient of determination $R^2=0.236$,coefficient of correlation r=0.486 ,standard deviation $S_{y/x1,x'2}=0.724$)and also the minimization of the mean square error (min MSE=0.349) and minimization of the mean absolute deviation(min MAD=0) show that this functional dependence $d_2=\gamma_{12}(t,\omega/B)$ does not have the good correlation and we can admit that this function better submits to the non-linear than linear[42] regression model with equation view of **$Y_c=12.688-0.355X_1-1.531X_1^2-3.531X'_2$ or $Y_c=12.688-0.355X_1-1.531X_1^2-$ $-3.531(X_2/X_3)$,where $d_2=12.688-0.355t-1.531t^2-3.531(\omega/B)$ (57).**

In Figure 32(a) is shown the functional dependency of external diameter (d_2) stainless chip from the ratio of values (ω/B) ,such as number of wraps (ω) to width (B) of chip ,i.e we have the function view of $d_2=\phi_{13}(\omega/B)$.

Analyzing of data and also the scatter plots shown in Figure 32(a) ,and the statistical characteristics (coefficient of determination $R^2=0.213$,coefficient of correlation r=0.461) show that this functional dependence $d_2=\phi_{13}(\omega/B)$ better submits to the linear than non-linear[43] regression model with equation view of **$Y_c=12.588-3.809X'_2$** or **$Y_c=12.588-3.809(X_2/X_3)$**,where **$d_2=12.588-3.809(\omega/B)$ (58).**

And as we see from Figure 32(a) that between two parameters, such as external diameter (d_2) and ratio (ω/B) there is small correlation (r=0.461). And this function $d_2=\phi_{13}(\omega/B)$ can be expressed in view of linear regression equation $d_2 =12.588-3.809(\omega/B)$ which shows that with increasing of ratio (ω/B), the value of external diameter (d_2) considerably decreases.

42- Linear regression model has equation view of $d_2=12.81-1.34t-3.48(\omega/B)$
43-Non-linear regression model has equation view of $d_2=8.91(\omega/B)^{-0.276}$

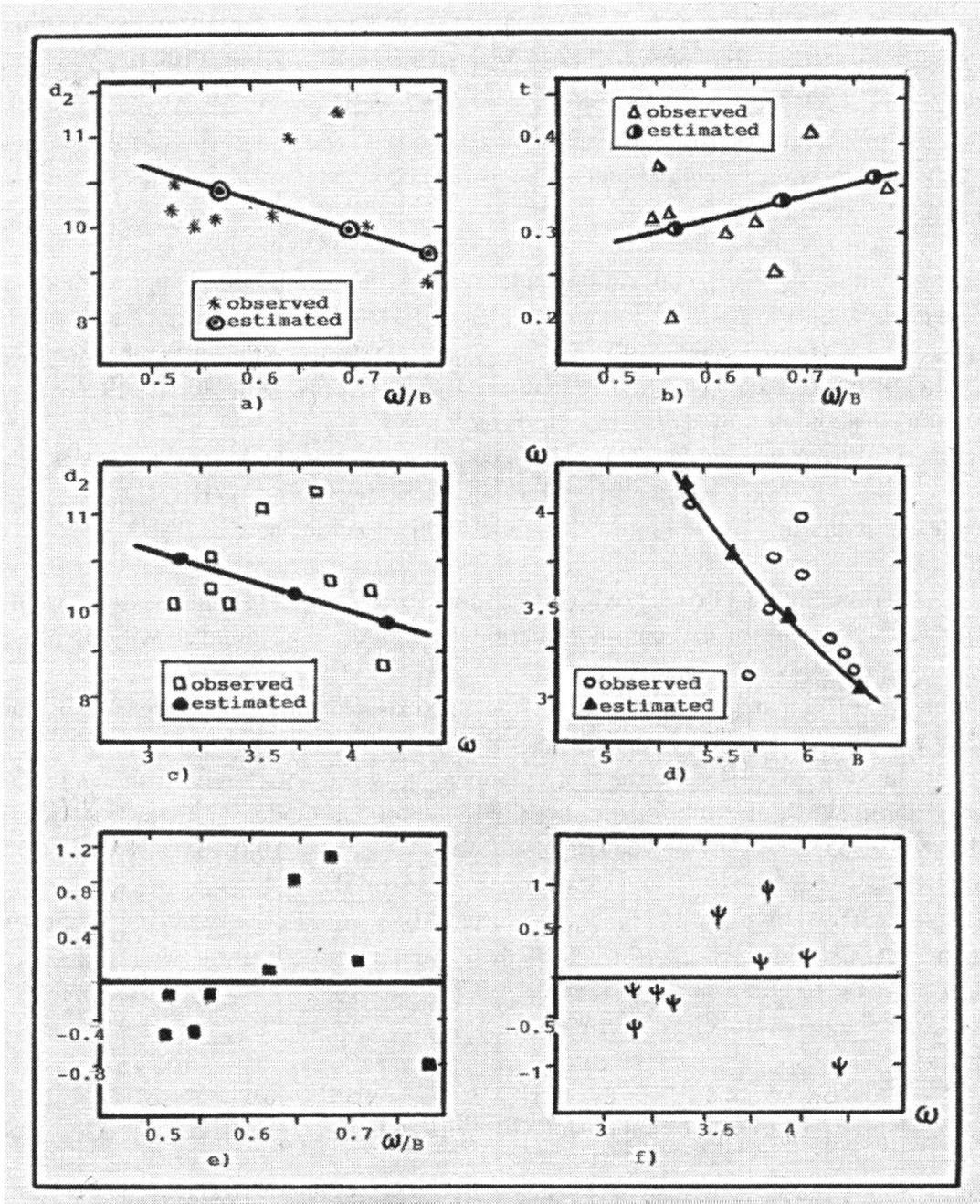

Figure 32

In Figure 32(b) is shown the functional dependence of clearance (t) between of chip wraps from its ratio (ω/B) such as the number of wraps (ω) to width chip (B) ,i.e has a place the function view of $t=\phi_{14}(\omega/B)$. Analysis of data and also the statistical characteristics show that this functional dependence does not have the good correlation between of two parameters, such as clearance (t) and ratio (ω/B).

But at the same time this functional dependence $t=\phi_{14}(\omega/B)$ better submits to the linear than non-linear regression model[44] with equation view of **$Y_c=0.239+0.129X'_2$** or **$Y_c=0.239+0.129(X_2/X_3)$**, where **$t=0.239+0.129(\omega/B)$** (**59**).

So ,we see from Figure 32(b) that with increasing of ratio (ω/B),the value of clearance (t) between of chip wraps increases considerably in accordance with the linear regression equation $t=0.239+0.129(\omega/B)$.

In Figure 32(c) is shown the functional dependence of external diameters (d_2) from the number of wraps(ω).This functional dependence $d_2=\phi_1(\omega)$ has the linear regression model with equation view of $Y_c=13.137-0.811X$ or $d_2=13.137-0.811\omega$.

Analysis of data given in Figure 32 (c) shows that with increasing of number wraps(ω) ,the value of external diameter (d_2) decreases considerably in accordance with the linear regression equation view of $d_2=13.137-0.811\omega$.

In Figure 32(d) is shown the functional dependence of number wraps (ω) for stainless chip from its width (B) ,i.e has a place the function view of $\omega=\gamma_2(B)$. This functional dependence as the non-linear regression model with equation view of $Y_c=703.1X^{-3.0}$ or $\omega=703.1B^{-3.0}$.

So , we see from Figure 32(d) that with increasing of width (B) for stainless chip ,the value of number wraps (ω) for this chip considerably decreases in accordance with non-linear regression equation view of $\omega=703.1B^{-3.0}$.

In Figure 32(e) and 32(f) are illustrated the residual plots (residual versus ω/B) and residual versus ω) of functional dependencies $d_2=\phi_{13}(\omega/B)$ and $d_2=\phi_1(\omega)$ accordingly.

Analyzing the Figure 33(A) , we see that functional surface $d_2=\gamma_{12}(t,\omega/B)$ is shown in view of three-dimensional drawing for the non-linear regression model with equation view of $Y_c=12.688-0.355X_1-1.531X_1^2-3.531X'_2$ or $Y_c=12.688-0.355X_1-1.531X_1^2-3.531(X_2/X_3)$,where $d_2=12.688-0.355t-1.531t^2-3.531(\omega/B)$.

And Figure 33(B) shows that this functional surface $d_2=\gamma_{12}(t,\omega/B)$ has view of trapezium on the plane YOX'_2 with such coordinates of their points (sizes in millimeters): a $(X_{1,1}=0$,$X_{2,1}=0.53$,$Y_1=10.56)$; b $(X_{1,2}=0,X_{2,2}=0.53$,$Y_2=0)$;c$(X_{1,3}=0$,$X_{2,3}=0.66,Y_3=0)$;d $(X_{1,4}=0,X_{2,4}=0.66$,$Y_4=10.18)$.

Analysis of functional surface $d_2=\gamma_{12}(t,\omega/B)$

From Figure 33(A) we see that functional surface $d_2=\gamma_{12}(t,\omega/B)$ on which are situated of the non-linear regression equation view of $Y_c=12.688-0.355X_1-1.531X_1^2-3.531X'_2$ or $Y_c=12.688-0.355X_1-1.531X_1^2-3.531(X_2/X_3)$,where $d_2=12.688-0.355t-1.531t^2-3.531(\omega/B)$ has the following coordinates of their peaks in three-dimensional drawing(sizes in millimeters):

a' ($X_{1,1}=0.31$,$X_{2,1}=0.53,Y_1=10.56)$; b' ($X_{1,2}=0.31$,$X_{2,2}=0.53$,$Y_2=0)$; c' ($X_{1,3}=0.25$, ,$X_{2,3}=0.66,Y_3=0)$; d' $(X_{1,4}=0.25,X_{2,4}=0.66,Y_4=10.18)$.

44-Non-linear regression model has equation view of $t=0.37(\omega/B)^{0.321}$

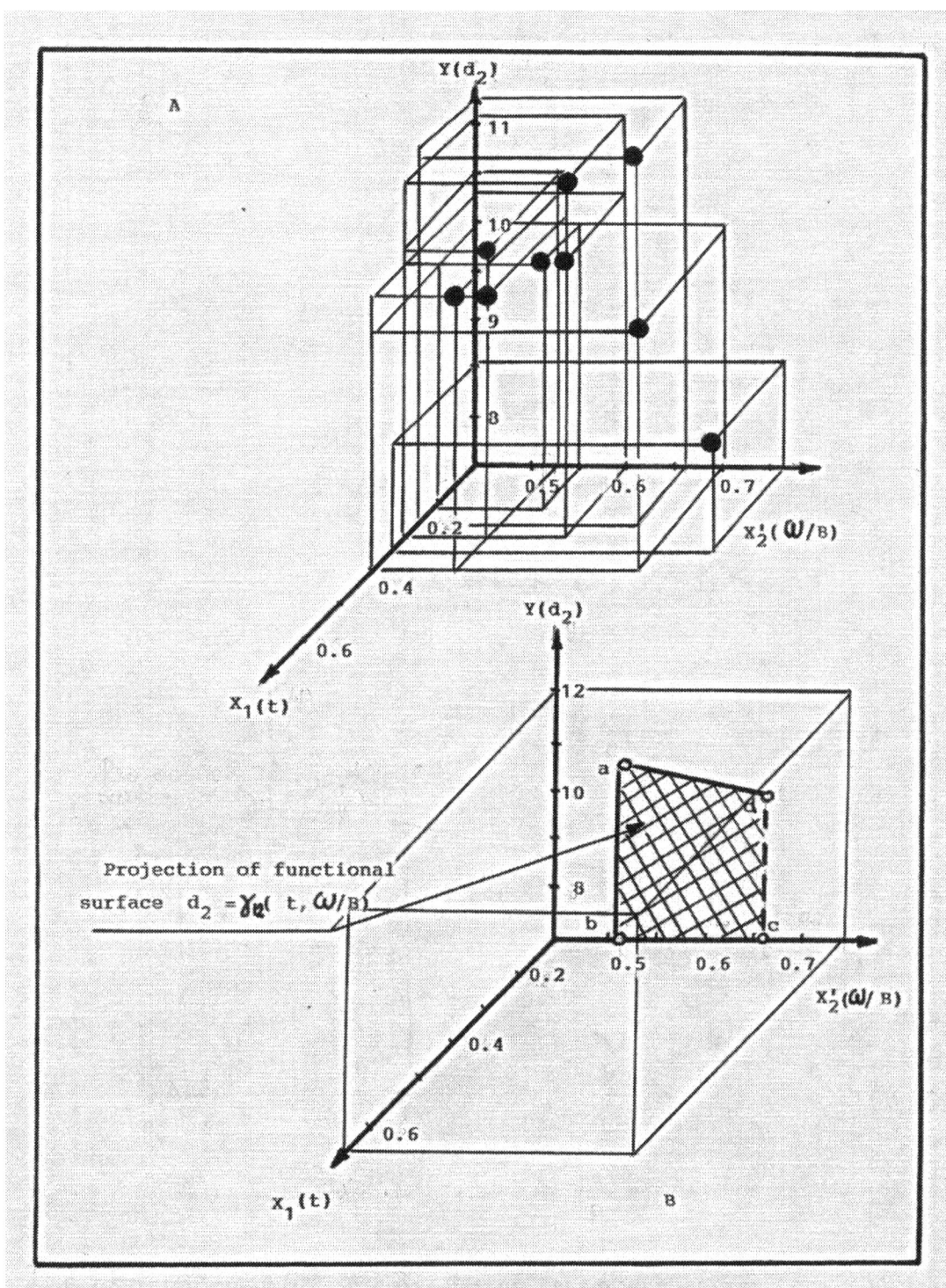

Figure 33

In Figure 34 schematically is shown the functional surface $d_2 = \gamma_{12}(t, \omega/B)$ and graph for calculation of module for vectors $|b'c'|$ and $|a'd'|$.

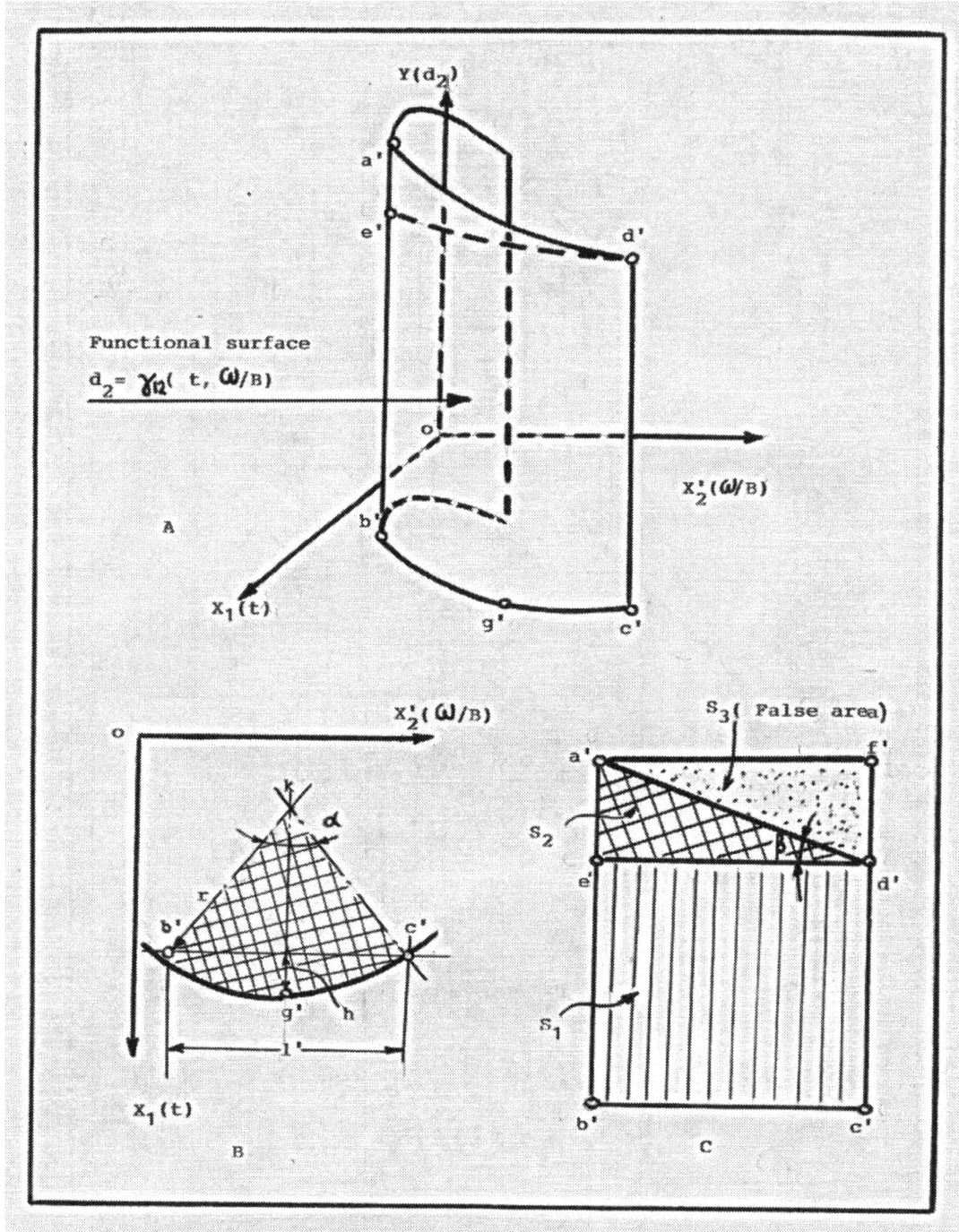

Figure 34 Graph for calculation of module vectors $|b'c'|$, $|a'd'|$,and area of functional surface $d_2 = \gamma_{12}(t, \omega/B)$:

    A-total view of functional surface;

    B- graph for calculation of module vector $|b'c'|$;

    C- graph for calculation of module vector $|a'd'|$ and area of this functional surface.

1. The module of vector $|a'b'|$ and $|c'd'|$ can be calculated from the projection of this functional surface $d_2=\gamma_{12}(t,\omega/B)$ on the plane YOX'_2(Figures 33B and 34A). So, the module of vectors $|a'b'|$ and $|c'd'|$ are equal: $|a'b'|=10.56, |c'd'|=10.18$.

2. The module of vectors $|b'c'|$ or it length can be calculated by formula $L=|b'c'|=0.01745\cdot r\cdot\alpha$, where L=length of arc (b'c'),r=radius of sector,α=angle of sector. So, the module of vector $|kg'|$ is equal: $r=|kg'|=[(X_{1,6}-X_{1,5})^2+(X_{2,6}-X_{2,5})^2]^{1/2}$, where the coordinates of points g' and k are equal: $g'(X_{1,6}=0.32, X_{2,6}=0.62); k(X_{1,5}=0, X_{2,5}=0)$. At data $\alpha=82°$, $r=0.70$, we have the module of vector $|b'c'|=1.0$ and other parameters of circular segment which are equal: *height $h=r[1-\cos(\alpha/2)=0.17$ and the length of chord $l_1= 2[h(2r-h)]^{1/2}=0.91$. The graph for calculation of module vectors $|b'c'|, |a'd'|$ and area of functional surface $d_2=\gamma_{12}(t,\omega/B)$ is shown in Figure 34(B) and 34 (C).

3. The module of vector $|a'd'|$ and area of this functional surface can be calculated by the following way: module of length for curve $|d'e'|=|b'c'|=1.0$ and the module of length for $|a'e'|=|a'b'|-|c'd'|=0.38$. For calculation of area S_1,S_2 and S_3 we find the angle β which is equal: $\tan\beta= =(|a'e'|/(|d'e'|=0.38$, where $\beta=20°50'$ and then the module of length for curve $|a'd'|$ is equal: $|a'd'|=(|a'e'|)/\sin20°50'=1.07$. The coordinates of point f' is equal : $f'(X_{1,1}=0.25, X_{2,1}=0.66,Y_1=10.56$ and area of $S_2=S_3=0.5S=0.5(|a'e'|\cdot|d'e'|=0.19mm^2$. For these conditions, we have the area of $S_1=|c'd'|\cdot|b'c'|=10.18$ mm^2 and total area $\sum S$ of this functional surface $d_2=\gamma_{12}(t,\omega/B)$ is equal $\sum S=10.37mm^2$. All results of computation for the non-linear equation view of $Y_c=12.688-0.355X_1-1.531X_1^2-3.531X'_2$ are given in Table 18.

Table 18 Evaluation of regression equation $Y_c=12.688-0.355X_1-1.531X_1^2-3.531X'_2$

A. Mean ,variance and standard deviation			
Variable	Mean	Variance	Standard deviation
X_1	0.317	0.031	0.176
X'_2	0.614	0.063	0.251
Y	10.248	4.121	2.030

B. Results of multiple regression of Y on X_1 and X'_2				
Parameter	Variable	Coefficients	Standard error	T-value
β_1	X_1	-0.355	0.059	-6.017
β_2	X'_2	-3.531	0.084	-42.036

C. Analysis of variance results

Regression: Degrees of freedom 2,sum of squares 0.972,mean squares 0.486

Error : Degrees of freedom 6 ,sum of squares 3.149,mean squares 0.525

F-value *   0.930

*Since F=0.93 < [F$_{0.05,2,6}$=5.14] we can not reject the hypothesis that both β_1 and β_2 are zero.

D. Determination of residual			
Number	Observed	Estimated	Residual
1	10.03	10.57	-0.54
2	10.36	10.46	-0.10
3	8.69	9.60	-0.91
4	10.13	9.83	0.30
5	10.22	10.25	-0.03
6	10.07	10.56	-0.49
7	10.49	10.55	-0.02
8	11.06	10.23	0.83
9	11.18	10.18	1.0

1.Function $d_2=\gamma_{12}(t,\omega/B)$ better submits to the non-linear regression model in view of curvilinear surface, as shown in Figures 33(A) and 34(A), and describes by equation view of $Y_c=12.688-0.355X_1-1.531X_1^2-3.531X'_2$ or $Y_c=12.69-0.36X_1-1.53X_1^2-3.53(X_2/X_3)$,where $d_2=12.688-0.355t-1.531t^2-3.531(\omega/B)$ with such statistical characteristics: coefficient of determination $R^2=0.236$,coefficient of correlation $r=0.486$,standard deviation $S_{y/x1,x'2}=0.724$,minimization of the mean square error (min MSE=0.349)and minimization of the mean square error (min MAD=0).

2.The total area of functional surface $d_2=\gamma_{12}(t,\omega/B)$ is equal $\sum S=10.37$ mm^2 with such coordinates of their points in three-dimensional drawing for this surface (sizes in millimeters): a' $(X_{1,1}=0.31$,$X_{2,1}=0.53,Y_1=10.56)$;b'$(X_{1,2}=0.31$,$X_{2,2}=0.53,Y_2=0)$; ,c'$(X_{1,3}=0.25,X_{2,3}=0.66,Y_3=0)$; d'$(X_{1,4}=0.25,X_{2,4}=0.66$,$Y_4=10.18)$.

6.6 Dependence of external diameter (d_2) from internal diameter (d_1) ,clearance (t) between of stainless chip and number (ω) of wraps ,i.e $d_2=\alpha_6(d_1,t,\omega)$ and also from their modifications: $d_2=\rho_2(d_1,d_1/t)$,$d_2=\rho_3(d_1,d_1/\omega)$ and $d_2=\rho_4(d_1,t/\omega)$.

A. Function $d_2=\alpha_6(d_1,t,\omega)$

Analysis of data and the statistical characteristics (coefficient of determination $R^2=0.956$,coefficient of correlation $r=0.979$, standard deviation $S_{y/x1,x2,x3}=0.184$),minimization of the mean square error(min MSE= 0.019) and minimization of the mean absolute deviation (min MAD=0) show that this functional dependence $d_2=\alpha_6(d_1,t,\omega)$ better submits to the linear than non-linear[45] regression model with equation view of

$$Y_c=4.621+0.93X_1+3.769X_2-0.087X_3 \quad \text{or} \quad d_2=4.621+0.93d_1+ +3.769t-0.087\omega$$
$$(\ 60 \ ) \ .$$

The functional dependence $d_2=\varphi_3(d_1)$ is shown in Figure 35(a) and this dependence has the non-linear regression model with equation view of $Y_c=5.728X^{0.359}$ or $d_2=5.728d_1^{0.359}$. So ,we see from Figure 35(a) that with increasing of internal diameter (d_1) of stainless chip ,the value of external diameter (d_2) for this chip considerably increases in accordance with regression equation $d_2=5.728d_1^{0.359}$.

In Figure 35(b) is shown the functional dependence of internal diameter (d_1) from the clearance (t) between of chip wraps ,i.e we have the functional dependence view of $d_1=\gamma_1(t)$. This functional dependence has the non-linear regression model with equation view of $Y_c=0.828X^{-0.605}$ or $t=0.828d_1^{-0.605}$,where $d_1=(1.208t)^{-1.653}$. So ,we see from Figure 35(b) that with increasing of the clearance (t) between of chip wraps ,the value of internal diameter (d_1) for this chip considerably decreases in accordance with regression equation $d_1=(1.208t)^{-1.653}$.

In Figure 35(c) is shown the functional dependence of internal diameter (d_1) stainless chip from the number wraps (ω) of this chip ,i.e we have the function view of $d_1=\varphi_3(\omega)$. This function has the non-linear regression model with equation view of $Y_c=20.32X^{-1.102}$ or $d_1=20.32\omega^{-1.102}$.

45- Non-linear regression model has equation view of $d_2= -3.73+3.68d_1-0.27d_1^2+1.52t+0.54\omega$

So , we see from Figure 35 (c) that with increasing of the number wraps (ω) for stainless chip ,the value of internal diameter (d_1) of this chip decreases considerably in accordance with the regression equation view of $d_1 = 20.32\omega^{-1.102}$. The residual plot of function $d_2 = \varphi_3(d_1)$ (residual versus of internal diameter d_1) is shown in Figure 35(d)).

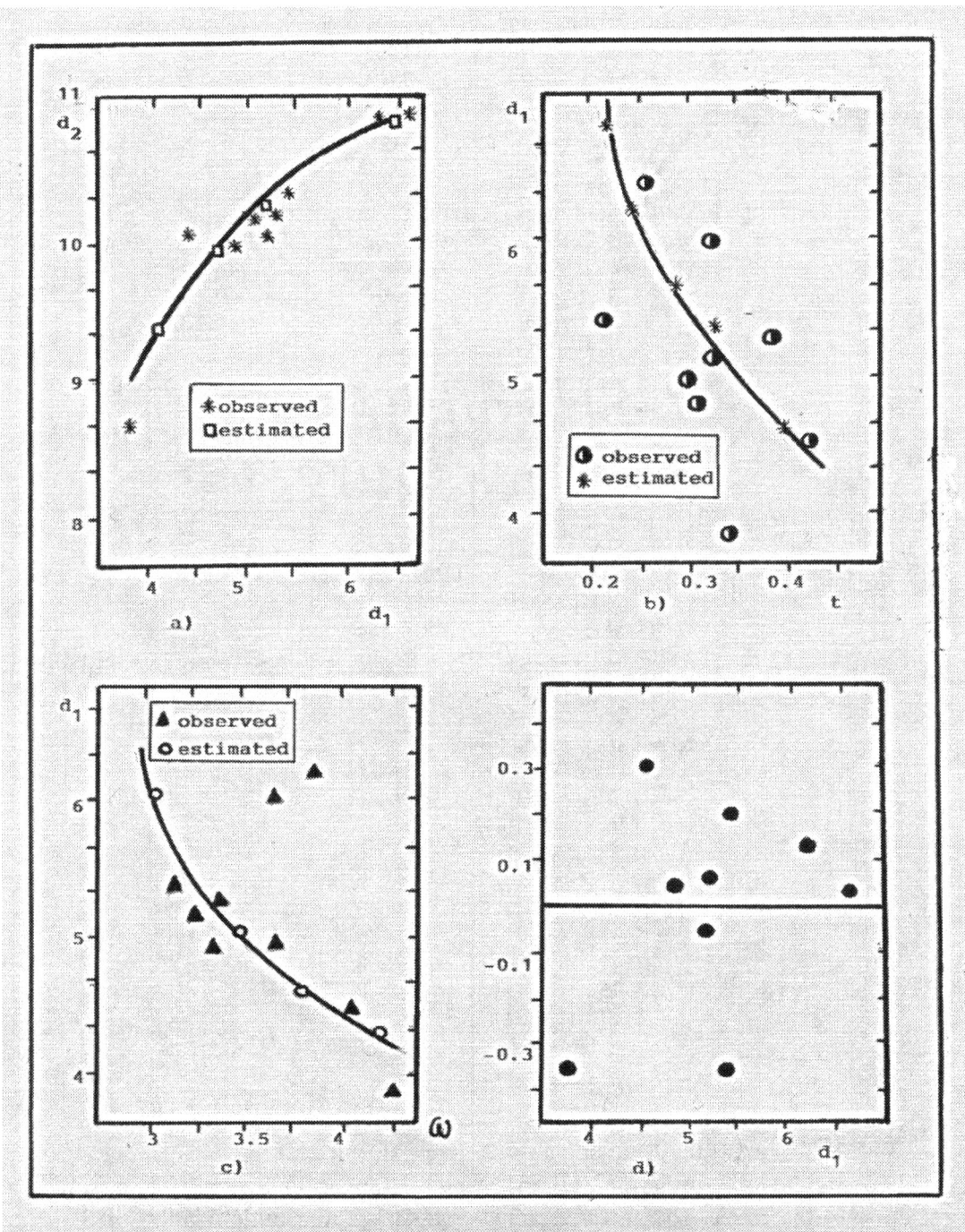

Figure 35

219

All results of computations for the linear equation view of $Y_c=4.621+0.93X_1+3.769X_2-0.087X_3$ or $d_2=4.621+0.93d_1+3.769t-0.087\omega$ are given in Table 19.

Table 19 Evaluation of regression equation $Y_c=4.621+0.93X_1+3.769X_2-0.087X_3$

A. Mean ,variance and standard deviation

Variable	Mean	Variance	Standard deviation
X_1	5.098	5.612	2.369
X_2	0.317	0.031	0.176
X_3	3.559	1.113	1.055
Y	10.248	4.121	2.030

B. Results of multiple regression of Y on X_1,X_2 and X_3

Parameter	Variable	Coefficients	Standard error	T-value
β_1	X_1	0.93	0.789	1.179
β_2	X_2	3.769	0.059	63.881
β_3	X_3	−0.087	0.352	−0.247

C. Analysis of variance results

Regression
- Degrees of freedom 3
- Sum of squares   3.95
- Mean squares   1.317

Error
- Degrees of freedom 5
- Sum of squares   0.171
- Mean squares   0.034

F-value *   38.735

*Since F=38.735> [$F_{0.05,3,5}$=5.41] we are able to reject the   hypothesis that parameters β_1,β_3 and β_3 are zero.

D. Determination of residuals

Number	Observed	Estimated	Residual
1	10.03	10.12	-0.09
2	10.36	10.62	-0.26
3	8.69	8.81	-0.12
4	10.13	10.05	0.08
5	10.22	10.07	0.15
6	10.07	9.91	0.16
7	10.49	10.34	0.15
8	11.06	11.06	0
9	11.18	11.23	-0.05

B. Modification function $d_2=\rho_2(d_1,d_1/t)$

Analysis of data and the statistical characteristics (coefficient of determination R^2=0.314,coefficient of correlation r=0.561,standard deviation $S_{y/x1,x'2}$=0.686) and also minimization of the mean square error (min MSE=0.313)and minimization of the mean absolute deviation (min MAD=0) show that this functional dependence $d_2=\rho_2(d_1,d_1/t)$

better submits to the linear than non-linear[46] regression model with equation view of

$$Y_c = 9.026 - 0.01X_1 + 0.075X'_2 \quad \text{or} \quad Y_c = 9.026 - 0.01X_1 + 0.075(X_1/X_2), \quad \text{where}$$

$$d_2 = 9.026 - 0.01d_1 + 0.075(d_1/t) \qquad (61).$$

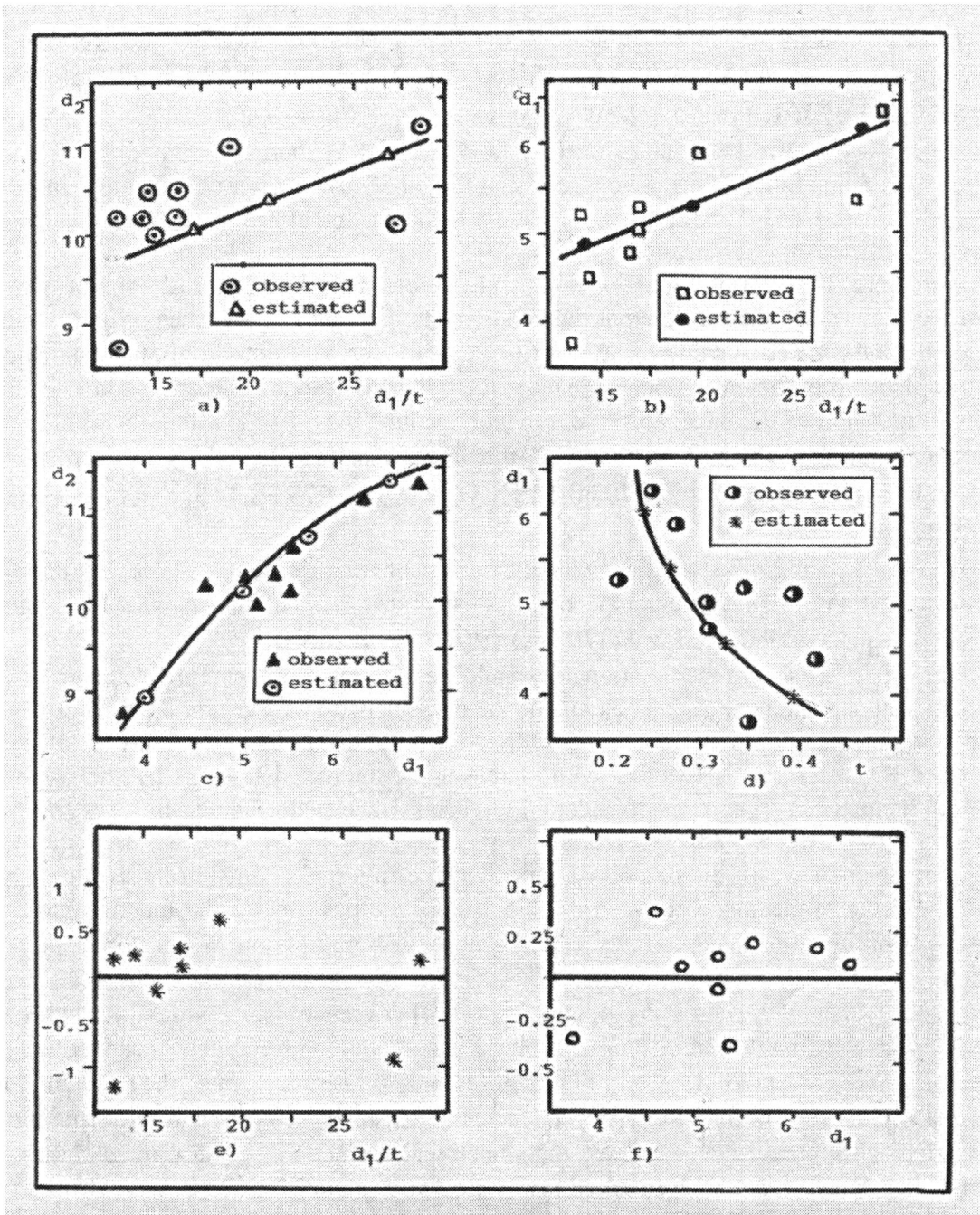

Figure 36

46- *Non-linear regression model has equation view of* $d_2 = -0.10 - 0.002d_1 + 43.26d_1^2 + 0.35(d_1/t)$

In Figure 36(a) is shown the functional dependence of external diameter (d_2) stainless chip from the ratio of values (d_1/t) ,such as internal diameter (d_1) to clearance (t) between of chip wraps ,i.e we have the function view of $d_2=\phi_{15}(d_1/t)$. Analyzing of data and also the scatter plots shown in Figure 36(a) and the statistical characteristics (coefficient of determination $R^2=0.33$,coefficient of correlation r=0.575) show that this functional dependence $d_2=\phi_{15}(d_1/t)$ better submits to the linear than non-linear[47] regression model with equation view of $\mathbf{Y_c=8.992+0.074X'_2}$ or $\mathbf{Y_c=8.992+0.074(X_1/X_2)}$, where $\mathbf{d_2=8.992+0.074(d_1/t)}$ (62) . As we see from Figure 36(a) that between of two parameters ,such as external diameter (d_2) and ratio (d_1/t) there is average correlation (r=0.575) and this function can be expressed in view of linear regression equation in view of $d_2=8.992+0.074(d_1/t)$ which shows that with increasing of ratio (d_1/t) ,the value of external diameter (d_2) increases considerably.

In Figure 36(b) is shown the functional dependence of internal diameter (d_1) stainless chip from the ratio (d_1/t), such as internal diameter (d_1) to clearance (t) between of chip wraps, i.e we have the function view of $d_1=\phi_{16}(d_1/t)$ .Analyzing of data and also the statistical characteristics (coefficient of determination $R^2=0.695$,coefficient of correlation r=0.834) and minimization of the mean absolute deviation (min MAD=0) show that this functional dependence $d_1=\phi_{16}(d_1/t)$ better submits to the linear than non-linear[48] regression model with equation view of $\mathbf{Y_c=3.079+0.119X'_2}$ or $\mathbf{Y_c=3.079+0.119(X_1/X_2)}$, where $\mathbf{d_1=3.079+0.119(d_1/t).}$

So ,we see from Figure 36(b) that with increasing of ratio (d_1/t) ,the value of internal diameter (d_1) for stainless chip increases considerably in accordance with the linear regression equation view of $\mathbf{d_1=3.079+0.119(d_1/t)}$ (63) .

In Figure 36(c) is shown the functional dependence of external diameter (d_2) stainless chip from its internal diameter(d_1). This functional dependence has the non-linear regression model with equation view of $Y_c=5.728X^{0.359}$,where $d_2=5.728d_1^{0.359}$.Analysis of data shown in Figure 36 (c) shows that with increasing of internal diameter (d_1) ,the value of external diameter (d_2) increases considerably in accordance with the non-linear regression equation view of $d_2=5.728d_1^{0.359}$.

In Figure 36(d) is shown the functional dependence of internal diameter (d_1) stainless chip from the value of clearance between of chip wraps ,i.e has a place the function view of $d_1=\gamma_4(t)$. This functional dependence has the non-linear regression model with equation view of $Y_c=0.828X^{-0.605}$ or $t=0.828d_1^{-0.605}$, where $d_1=(1.208t)^{-1.653}$. So ,we see from Figure 36(d) that with increasing of clearance (t) between of chip wraps ,the value of internal diameter (d_1) decreases considerably in accordance with the non-linear regression equation view of $d_1=(1.208t)^{-1.653}$.In Figure 36(e) and 36(f) are illustrated the residual plots (residual versus d_1 of functional dependencies $d_2=\phi_{15}(d_1/t)$ and $d_2=\varphi_3(d_1)$ accordingly. Analyzing the Figure 37(A) , we see that functional surface $d_2=\rho_2(d_1,d_1/t)$ is shown in three-dimensional drawing for the linear regression model with equation view of $Y_c=9.026-0.01X_1+0.075X'_2$ or $Y_c=9.026-0.01X_1+0.075(X_1/X_2)$,where $d_2=9.026-0.01d_1+$ $+0.075(d_1/t)$.

47-Non-linear regression equation has view of $d_2=6.3(d_1/t)^{0.174}$
48-Non-linear regression model has equation view of $d_1=1.5(d_1/t)^{0.435}$

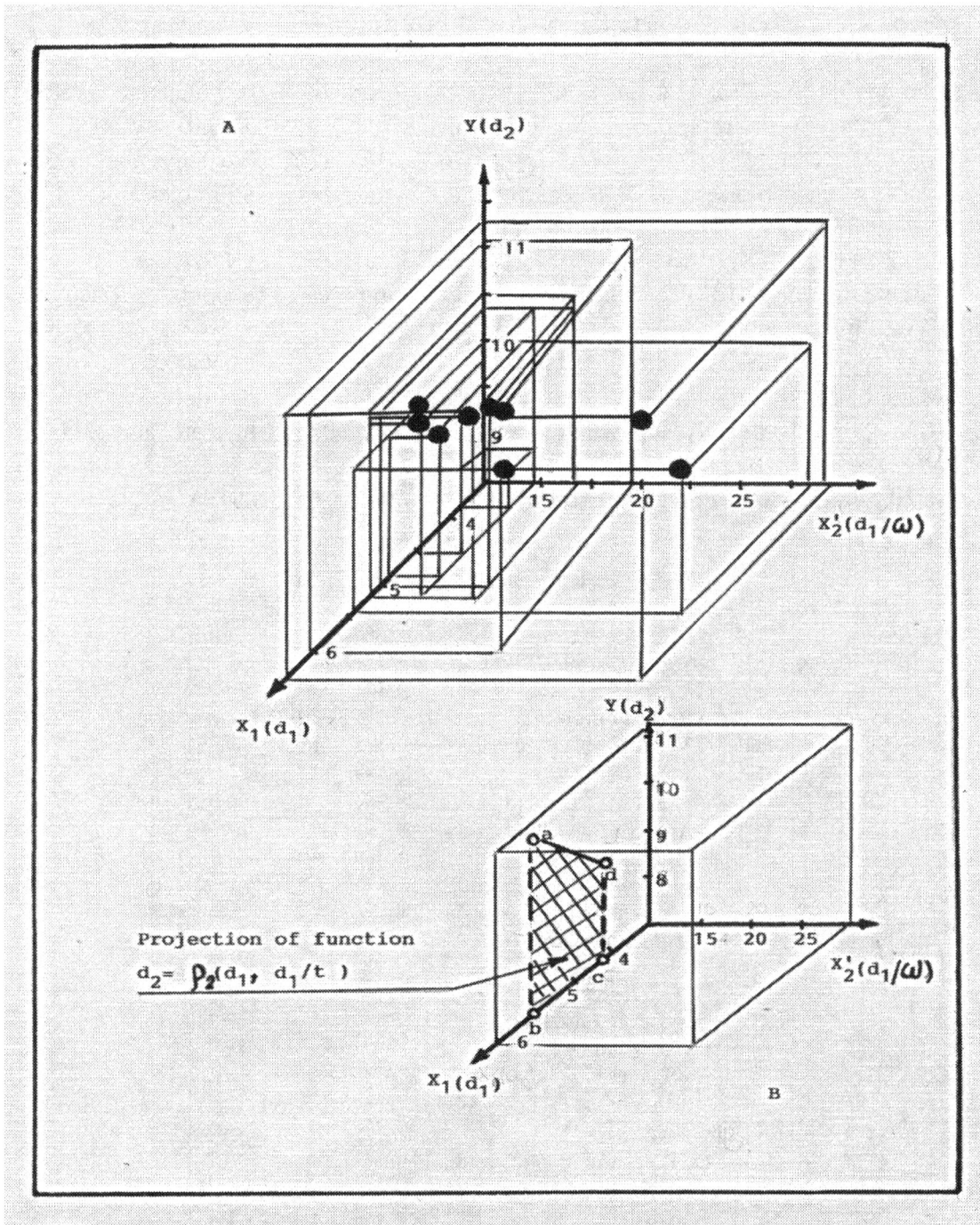

Figure 37

And Figure 37(B) shows that this functional surface $d_2=\rho_2(d_1,d_1/t)$ has view of trapezium on the place YOX_1 with such coordinates of their points (sizes in millimeters):
a $(X_{1,1}=5.96$, $X_{2,1}=0$,$Y_1=10.36)$; b$(X_{1,2}=5.96$,$X_{2,2}=0$,$Y_2=0)$,c$(X_{1,3}=4.51,X_{2,3}=0$,$Y_3=0)$; d$(X_{1,4}=4.51$;$X_{2,4}=0$,$Y_4=9.79)$.

223

Analysis of functional surface $d_2=\rho_2(d_1,d_1/t)$

From Figure 37(A) we see that functional surface $d_2=\rho_2(d_1,d_1/t)$,on which are situated the points of the linear regression equation view of $Y_c=9.026-0.01X_1+0.075X'_2$ or $Y_c=9.026-0.01X_1+0.075(X_1/X_2)$,where $d_2=9.026-0.01X_1+0.075(d_1/t)$, has the following coordinates of their peaks in three- dimensional drawing (sizes in millimeters):

a'($X_{1,1}$ =5.96 ,$X_{2,1}$=18.63,Y_1=10.36);b'($X_{1,2}$=5.96 ,$X_{2,2}$=18.63,Y_2=0),c'($X_{1,3}$=4.51 , ,$X_{2,3}$=10.74,Y_3=0);d'($X_{1,4}$=4.51 ,$X_{2,4}$=10.74 ,Y_4=9.79).

The module of vectors $|a'b'|$, $|b'c'|$, $|c'd'|$, $|d'a'|$ and area of functional surface can be calculated by formulas (11),(12),(13),(14) and (46). At data we have the values $|a'b'|$=10.36 , $|b'c'|$=7.89, $|c'd'|$=9.74, $|d'a'|$=8.05 and $S_{a'b'c'd'}$ =79.29 mm^2.

All results of computation for the linear equation view of $Y_c=9.026-0.01X_1+0.075X'_2$ or $Y_c=9.026-0.01X_1+0.075(X_1/X_2)$, where $d_2=9.026-0.01d_1+0.075(d_1/t)$ are given in Table 20.

Table 20 Evaluation of regression equation $Y_c=9.026-0.01X_1+0.075X'_2$

A. Mean ,variance and standard deviation

Variable	Mean	Variance	Standard deviation
X_1	5.098	5.608	2.368
X'_2	16.969	249.925	15.809
Y	10.248	4.121	2.030

B. Results of multiple regression of Y on X_1 and X'_2

Parameter	Variable	Coefficients	Standard error	T-value
β_1	X_1	-0.01	0.789	-0.013
β_2	X'_2	0.075	5.269	0.014

C. Analysis of variance results

Regression
    Degrees of freedom 2
    Sum of squares 1.297
    Mean squares 0.649

Error
    Degrees of freedom 6
    Sum of squares 2.824
    Mean squares 0.471

F-value *   1.379

* Since F=1.379 < [$F_{0.05,2,6}$=5.14] ,we cannot reject the hypothesis that both β_1 and β_2 are zero.

D. Determination of residual

Number	Observed	Estimated	Residual
1	10.03	10.89	-0.86
2	10.36	10.01	0.35
3	8.69	9.77	-1.08
4	10.13	9.79	0.34
5	10.22	10.22	0
6	10.07	10.13	-0.06
7	10.49	10.18	0.31
8	11.06	10.36	0.70
9	11.18	10.89	0.29

So ,the parameters of this functional surface $d_2=\rho_2(d_1,d_1/t)$ are the following:

1.Function $d_2=\rho_2(d_1,d_1/t)$ better submits to the linear regression model with equation view of $Y_c=9.026-0.01X_1+0.075X'_2$ or $Y_c=9.026-0.01X_1+0.075(X_1/X_2)$,where $d_2=9.026-0.01X_1+0.075(d_1/t)$ with such statistical characteristics : coefficient   of determination $R^2=0.314$,coefficient correlation r=0.561,standard deviation $S_{y/x1,x'2}=0.686$, minimization of the mean square error (min MSE=0.313),minimization of the mean absolute deviation (min MAD=0).

2.The total area of functional surface $d_2=\rho_2(d_1,d_1/t$) is equal $\sum S=79.29$ mm^2 with such coordinates of their points in three- dimensional drawing for this surface (sizes in millimeters): a' ($X_{1,1}=5.96$,$X_{2,1}=18.63$,$Y_1=10.36$) , b'($X_{1,2}=5.96$,$X_{2,2}=18.63$,$Y_2=0$) , c'($X_{1,3}=4.51$,$X_{2,3}=10.74$,$Y_3=0$); d'($X_{1,4}=4.51$,$X_{2,4}=10.74$,$Y_4=9.79$).

C. Modification function $d_2=\rho_3(d_1,d_1/\omega)$

Analysis of data and statistical characteristics (coefficient of determination $R^2=0.889$,coefficient of correlation r=0.943,standard deviation $S_{y/x1,x'2}=0.276$) and   also minimization of the mean square error (min MSE=0.051) ,minimization of the mean absolute deviation (min MAD=0) show that this functional dependence $d_2=\rho_3(d_1,d_1/\omega)$ has the good correlation and we can admit that this function better submit to the non-linear than linear[49]   regression   model   with   equation   view   of
$$\mathbf{Y_c=6.296+0.747X_1+0.003X_1{}^2+0.045X'_2}\quad \text{or}\quad \mathbf{Y_c=6.296+0.747X_1+0.003X_1{}^2+}$$
$$\mathbf{+0.045(X_1/X_2)}\text{ ,where }\mathbf{d_2=6.296+0.747d_1+0.003d_1{}^2+0.045(d_1/\omega)}\quad (\ 64\ ).$$
In Figure 38(a) is shown the functional dependence of external diameter (d_2) stainless chip from the ratio of values (d_1/ω) ,such as internal diameter (d_1) to the number of chip wraps (ω),i.e we have the function view of $d_2=\phi_{17}(d_1/\omega)$.

Analyzing of data and also the scatter plots shown in Figure 38(a), and the statistical characteristics (coefficient of determination $R^2=0.722$,coefficient of correlation r=0.849) show that this functional dependence $d_2=\phi_{17}(d_1/\omega)$ better submits to the non-linear than linear[50] regression   model   with   equation   view   of $\mathbf{Y_c=9.354X'_2{}^{0.25}}$ or
$$\mathbf{Y_c=9.354(X_1/X_2)^{0.25}}\text{ , where }\mathbf{d_2=9.354(\ d_1/\omega)^{0.25}}\quad (\ 65\ ).$$
And as we see from Figure 38(a) that between two parameters, such as external diameter (d_2) and ratio (d_1/ω) there is the good correlation (r=0.849). And this function $d_2=\phi_{17}(d_1/\omega)$ can be expressed in view of the non-linear regression equation $d_2=9.354(d_1/\omega)^{0.25}$ which shows that with increasing of ratio (d_1/ω) ,the value of external diameter (d_2) considerably increases .

In Figure 38(b) is shown the functional dependence of internal diameter (d_1) stainless chip from the ratio (d_1/ω) ,such as   internal diameter (d_1) to the number of chip wraps ,i.e we have the function view of $d_1=\phi_{18}(d_1/\omega)$.

49- The linear regression model has equation view of $d_2=6.12+0.8d_1-0.14(d_1/\omega)$
50-The linear regression model has equation view of $d_2=7.33+2.01(d_1/\omega)$

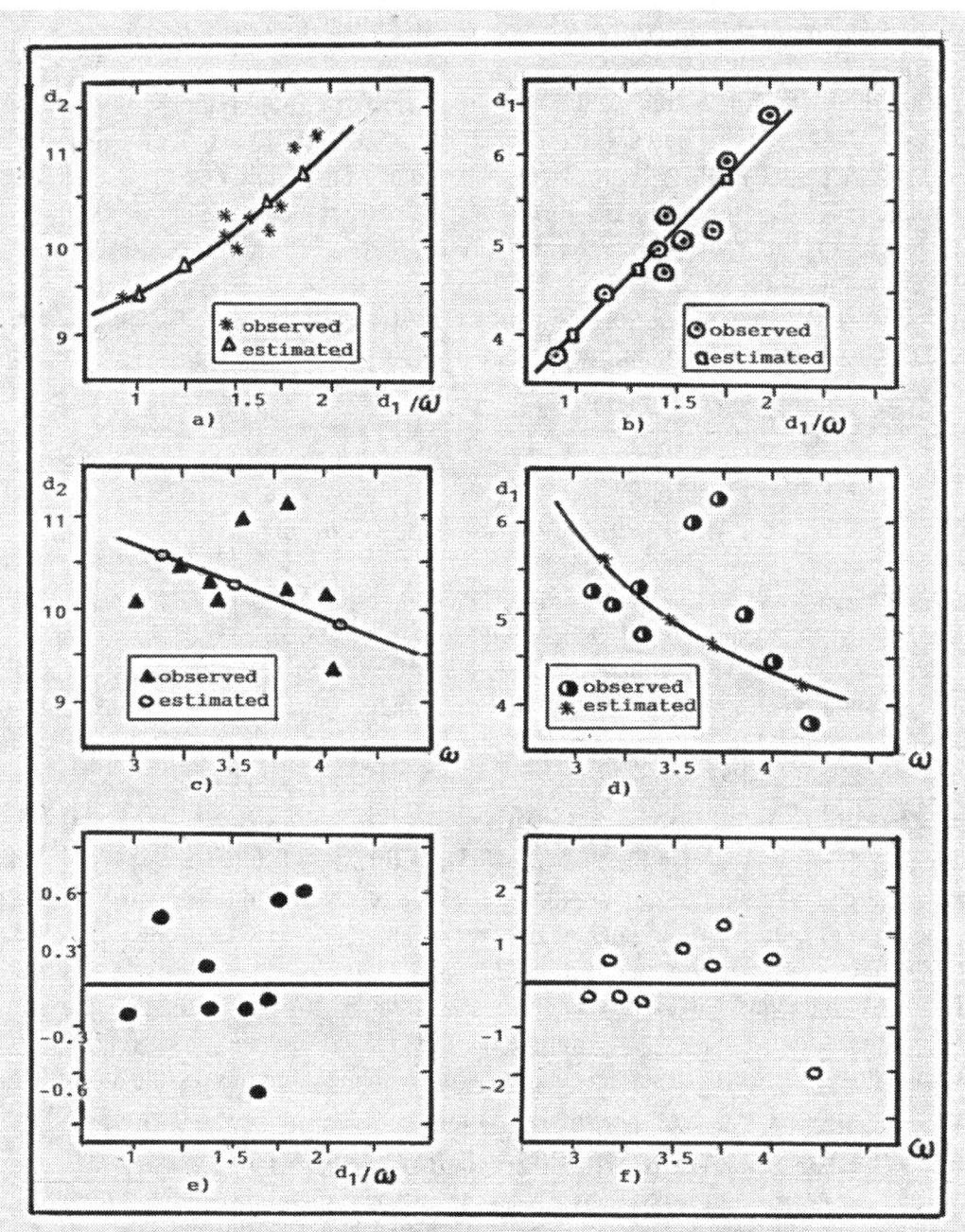

Figure 38

Analyzing of data and also the statistical characteristics (coefficient of determination $R^2=0.82$, coefficient of correlation $r=0.906$, standard deviation $S_{y/x'2}=0.379$) and minimization of the mean square error (min MSE=0.112) ,and minimization of the mean absolute deviation (min MAD= 0) show that this function $d_1= \phi_{18}(d_1/\omega)$ better submits to the linear than non-linear[51] regression model with equation view of $Y_c=1.43+2.52X'_2$ or $Y_c=1.43+2.52(X_1/X_2)$, where $d_1=1.43+2.52(d_1/\omega)$ **(66)**

51- Non-linear regression model has equation view of $d_1=49(d_1/\omega)^{0.659}$

226

So, we see from Figure 38(b) that with increasing of ratio (d_1/ω) ,the value of internal diameter (d_1) for stainless chip increases considerably in accordance with the linear regression equation view of $d_1 = 1.434 + 2.522(d_1/\omega)$.

In Figure 38(c) is shown the functional dependence of external diameter (d_2) stainless chip from the number of chip wraps (ω). This functional dependence has the linear regression model with equation view of $Y_c = 13.137 - 0.811X$ or $d_2 = 13.137 - 0.811\omega$. Analysis of data shown in Figure 38 (c) shows that with increasing of number wraps (ω) ,the value of external diameter (d_2) decreases considerably in accordance with the linear regression equation view of $d_2 = 13.137 - 0.811\omega$.

In Figure 39(d) is shown the functional dependence of internal diameter (d_1) stainless chip from the number of chip wraps (ω) ,i.e has a place the function view of $d_1 = \phi_3(\omega)$. This functional dependence has the non-linear regression model with equation view of $Y_c = 20.32X^{-1.102}$ or $d_1 = 20.32\omega^{-1.102}$.So ,we see from Figure 32(d0 that with increasing of number wraps(ω) for stainless chip , the value of internal diameter (d_1) decreases considerably in accordance with the non-linear regression equation view of $d_1 = 20.32\omega^{-1.102}$.In Figure 38(e) and 38(f) are illustrated the residual plots (residual versus d_1/ω) and residual versus ω) of functional dependencies $d_2 = \phi_{17}(d_1/\omega)$ and $d_2 = \phi_1(\omega)$ accordingly.

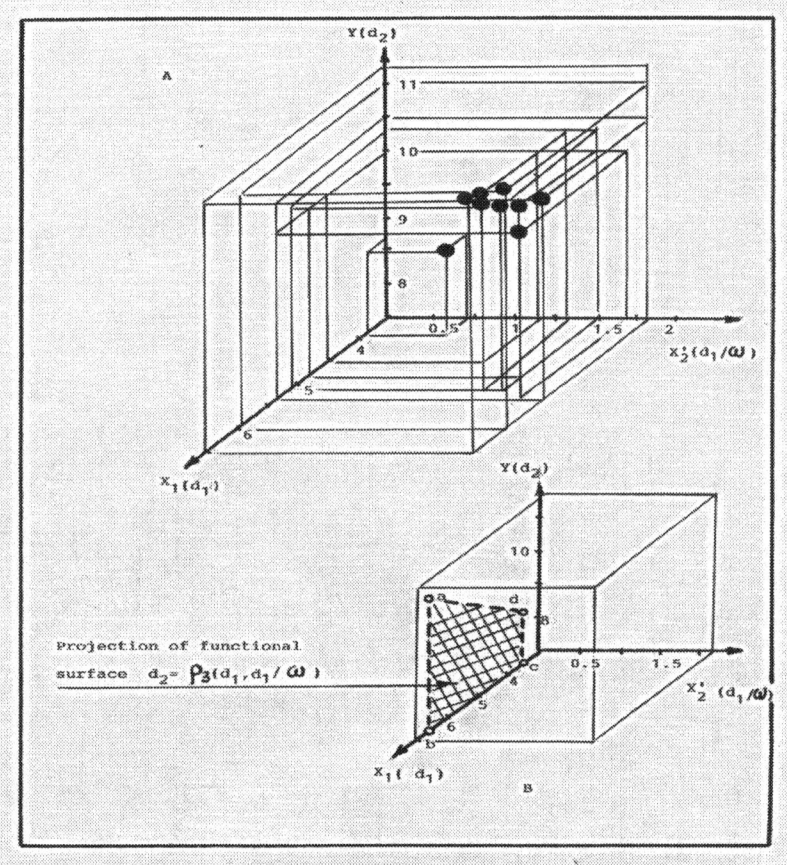

Figure 39

Analyzing the Figure 39(A) we see that functional surface $d_2=\rho_3(d_1,\ d_1\ /\omega)$ is shown in three-dimensional drawing for the non-linear regression model with equation view of $Y_c=6.296+0.747X_1+0.003X_1^2+0.045X'_2$ or $Y_c=6.296+0.747X_1+0.003X_1^2+0.045(X_1/X_2)$,where $d_2=6.296+0.747d_1+0.003d_1^2+0.045(d_1/\omega)$.

And Figure 39(B) shows that this functional surface $d_2=\rho_3(d_1,d_1/\omega)$ has view of trapezium on the place YOX_1 with such coordinates of their points (sizes in millimeters):
$a(X_{1,1}=6.44,X_{2,1}=0,Y_1=11.31)$; $b(X_{1,2}=6.44,X_{2,2}=0,Y_2=0)$; $c(X_{1,3}=3.52,X_{2,3}=0,Y_3=0)$;
$d(X_{1,4}=3.52,X_{2,4}=0,Y_4=8.99)$.

Analysis of functional surface $d_2=\rho_3(d_1,d_1/\omega)$

From Figure 39(A) we see that functional surface $d_2=\rho_3(d_1,d_1/\omega)$,on which the points are situated of the non-linear regression equation view of $Y_c=6.296+0.747X_1+0.003X_1^2+$ $+0.045X'_2$ or $Y_c=6.296+0.747X_1+0.003X_1^2+0.045(X_1/X_2)$,where $d_2=6.296+0.747d_1+$ $+0.003d_1^2+0.045(d_1/\omega)$ has the following coordinates of their peaks in three-dimensional drawing (sizes in millimeters): a' $(X_{1,1}=6.44,X_{2,1}=1.71,Y_1=11.31)$; b'$(X_{1,2}=6.44$, $X_{2,2}=1.71,Y_2=0)$;c'$(X_{1,3}=3.52,X_{2,3}=0.83,Y_3=0)$; d'$(X_{1,4}=3.52,X_{2,4}=0.83,Y_4=8.99)$.

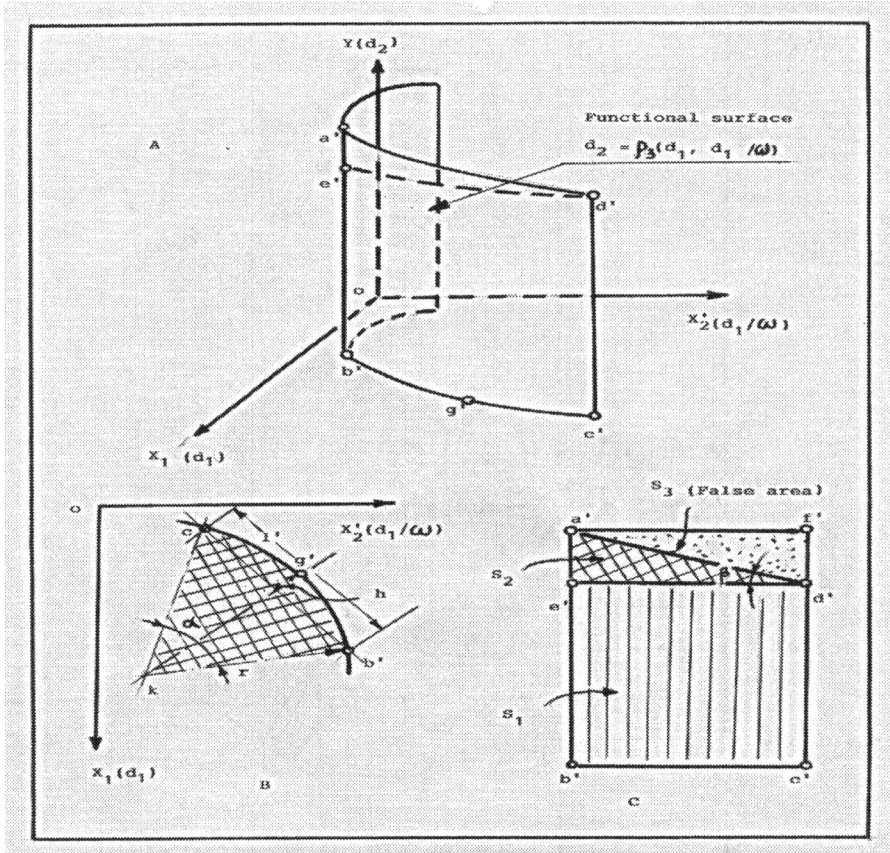

Figure 40 Graph for calculation of module vectors $|b'c'|$, $|a'd'|$ and area of functional surface $d_2=\rho_3(d_1,d_1/\omega)$: A-total view of this functional surface ; B-graph for calculation of module vector $|b'c'|$; C-graph for calculation of module vector $|a'd'|$ and area of this surface.

228

In Figure 40 schematically is shown the functional surface $d_2=\rho_3(d_1,d_1/\omega)$ and graph for calculation of module for vectors $|b'c'|$ and $|a'd'|$.

1. The module of vectors $|a'b'|$ and $|c'd'|$ can be calculated from the projection of this functional surface $d_2=\rho_3(d_1,d_1/\omega)$ on the plane YOX_2, as shown in Figure 39B and Figure 40A with using of formulas (11),(13). So, the module of vectors are equal: $|a'b'|=11.31$ and $|c'd'|=8.99$.

2. The module of vector $|b'c'|$ or it length can be calculated by formula $L=|b'c'|=0.01745 \cdot r \cdot \alpha$, where L=length of arc (b'c'), r=radius of sector, α=angle of sector. So, the module of vector $r=|kg'|=[(X_{1,6}-X_{1,5})^2+(X_{2,6}-X_{2,5})^2]^{1/2}=4.96$, where the coordinates of points g' and k are equal: $g'(X_{1,6}=4.74, X_{2,6}=1.46)$, $k(X_{1,5}=0, X_{2,5}=0)$. At data $\alpha=82°, r=4.96$ we have the module of vector $|b'c'|=7.10$ and other parameters of circular segment which are equal: height $h=r[1-\cos(\alpha/2)]=1.24$ and length of chord $l_1=2[h(2r-h)]^{1/2}=5.89$.

3. The graph for calculation of module vectors $|b'c'|, |a'd'|$ and area of functional surface $d_2=\rho_3(d_1,d_1/\omega)$ is shown in Figure 40(B) and 40(C). The module of vector $|a'd'|$ and area of this functional surface can be calculated by the following way: the module of length for curve $|d'e'|=|b'c'|$, i.e $|d'e'|=7.10$ and module of length for $|a'e'|=|a'b'|-|c'd'|=2.32$.

3. For calculation of area S_1, S_2 and S_3 we find the angle β which is equal: $\tan\beta=|a'e'|/|d'e'|=0.327$ and $\beta=18°10'$ and then the module of length for curve $|a'd'|$ is equal : $|a'd'|=|a'e'|/\sin18°10'=7.44$. The coordinates of point f' is equal: f' $(X_{1,1}=3.52, X_{2,1}=0.83, Y_1=11.31)$ and area of $S_2=S_3=0.5S=0.5[|a'e'| \cdot |d'e'|]=8.24$. For these conditions, we have the area S_1 which is equal $S_1=|c'd'| \cdot |b'c'|=63.83mm^2$ and total area $\sum S$ of this functional surface $d_2=\rho_3(d_1,d_1/\omega)$ is equal $\sum S=72.12mm^2$.

All results of computation for the non-linear equation view of $Y_c=6.296+0.747X_1+0.003X_1^2+0.045X_2'$ or $Y_c=6.296+0.747X_1+0.003X_1^2+0.045(X_1/X_2)$, where $d_2=6.296+0.747d_1+0.003d_1^2+0.045(d_1/\omega)$ are given in Table 21.

So, the parameters of this functional surface $d_2=\rho_3(d_1,d_1/\omega)$ are the following:

1. Function $d_2=\rho_3(d_1,d_1/\omega)$ better submits to the non-linear regression model in view of curvilinear surface, as shown in Figure 39(A) and Figure 40(A) and describes by equation view of $Y_c=6.296+0.747X_1+0.003X_1^2+0.045X'_2$ or $Y_c=6.296+0.747X_1+0.003X_1^2+0.045(X_1/X_2)$, where $d_2=6.296+0.747d_1+0.003d_1^2+0.045(d_1/\omega)$ with such statistical characteristics:

- Coefficient of determination $R^2=0.889$
- Coefficient of correlation $r=0.943$
- Standard deviation $S_{y/x1,x'2}=0.276$
- Minimization of the mean square error (min MSE=0.051)
- Minimization of the mean absolute deviation (min MAD=0).

2. The total area of functional surface $d_2=\rho_3(d_1,d_1/\omega)$ is equal $\sum S=72.12mm^2$ with such coordinates of their points in three-dimensional drawing for this surface (sizes in millimeters): a'($X_{1,1}=6.44, X_{2,1}=1.71, Y_1=11.31$); b'($X_{1,2}=6.44, X_{2,2}=1.71, Y_2=0$); c'($X_{1,3}=3.53, X_{2,3}=0.83, Y_3=0$); d'($X_{1,4}=3.52, X_{2,4}=0.83, Y_4=8.99$).

Table 21 Evaluation of regression equation $d_2=6.296+0.747d_1+0.003d_1{}^2+0.045(d_1/\omega)$

A . Mean ,variance and standard deviation

Variable	Mean	Variance	Standard deviation
X_1	5.098	5.608	2.368
X'_2	14.533	0.716	0.846
Y	10.248	4.121	2.030

B.   Results of multiple regression of Y on X_1 and X'_2

Parameter	Variable	Coefficients	Standard error	T-value
β_1	X_1	0.747	0.789	0.946
β_2	X'_2	0.045	0.282	0.159

C.   Analysis of variance results

Regeression

    Degrees of freedom 2

    Sum of squares   3.664

    Mean   squares   1.832

Error

    Degrees of freedom   6

    Sum of squares   0.457

    Mean squares   0.076

F-value*   24.105

* Since F= 24.105>[$F_{0.05,2,6}$=5.14] ,we can reject the hypothesis that both $\beta_1$and β_2 are zero.

D.   Determination of residual

Number	Observed	Estimated	Residual
1	10.03	10.46	-0.43
2	10.36	10.35	0.01
3	8.69	8.99	-0.30
4	10.13	9.78	0.35
5	10.22	10.15	0.07
6	10.07	9.97	0.10
7	10.49	10.29	0.20
8	11.06	10.93	0.13
9	11.18	11.31	-0.13

D. Modification function $d_2=\rho_4(d_1,t/\omega)$

Analysis of data and the statistical characteristics (coefficient of determination R^2=0.917,coefficient of correlation r=0.958,standard deviation $S_{y/x',x'2}$=0.238) and minimization of the mean square error (min MSE=0.026) ,minimization of the mean absolute deviation (min MAD=0.004) show that this functional dependence $d_2=\rho_4(d_1,t/\omega)$ has the good correlation and we can admit that this function better submits to the linear than non-linear[52] regression model with equation view of **$Y_c=4.47+0.85X_1+16X'_2$** or **$Y_c=4.47+0.85X_1+16(X_2/X_3)$, where $d_2=4.47+0.85d_1+16(t /\omega)$ (67)   .**

52- Non-linear regression model has equation view of $d_2=3.55+1.83d_1-0.1d_1{}^2+ t /\omega$

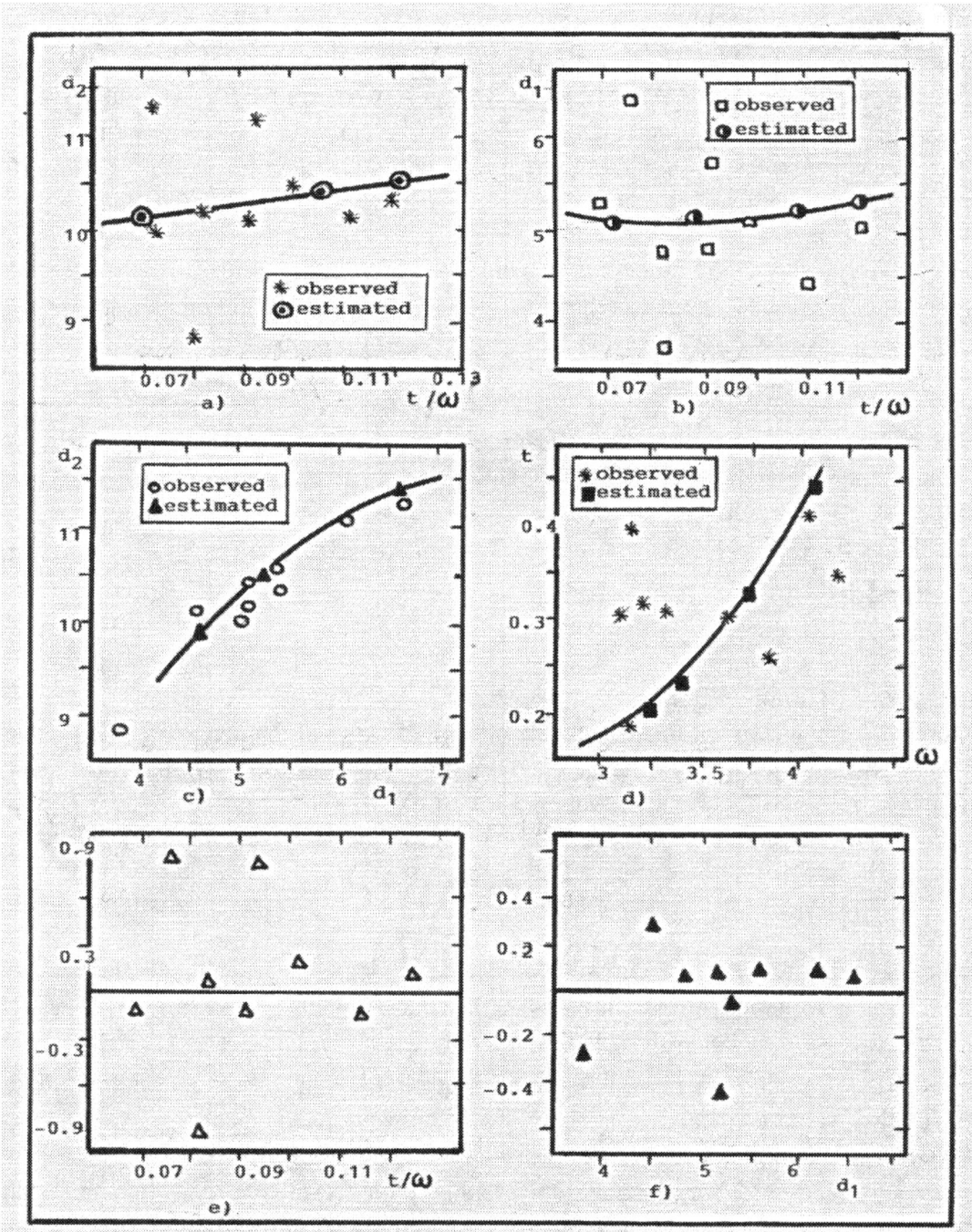

Figure 41

In Figure 41(a) is shown the functional dependence of external diameter (d_2) stainless chip from the ratio of values (t/ω), such as clearance (t) between of chip wraps to the number of wraps (ω) for this stainless chip, i.e we have the function view of $d_2 = \phi_{19}(t/\omega)$.

Analyzing of data and also the scatter plots shown in Figure 41(a), and the statistical characteristics (small coefficient of determination $R^2 = 0.05$ and coefficient of correlation $r = 0.22$) show that this functional dependence $d_2 = \phi_{19}(t/\omega)$ does not have the good correlation but better submits to the linear than non-linear[53] regression model with equation of view $Y_c=9.98+3X'_2$ or $Y_c=9.98+3(X_2/X_3)$, where $d_2=9.98+3(t/\omega)$.

And as we see from Figure 41(a) that between of two parameters ,such as external diameter (d_2) and ratio (t/ω) there is not the good correlation ($r=0.22$) and this function $d_2=\phi_{19}(t/\omega)$ can be expressed in view of the linear regression equation $d_2=9.98+3(t/\omega)$ (68) which shows that with increasing of ratio (t/ω) ,the value of external diameter (d_20 increases considerably.

In Figure 41(b) is shown the functional dependence of internal diameter (d_1) stainless chip from the ratio (t/ω) ,such values as the clearance (t) between of chip wraps to the number of wraps (ω) ,i.e we have the function view of $d_1=\phi_{20}(t/\omega)$.Analyzing of data and also the statistical characteristics (coefficient of determination $R^2=0$,coefficient of correlation $r=0$) show that this functional dependence does not have the good correlation, but comparing analysis of minimization of the mean square error (min MSE=0.618) and minimization of the mean absolute deviation (min MAD=0.084) shows that this function $d_1=\phi_{20}(t/\omega)$ better submits to the non-linear than linear[54]regression model with equation view of $Y_c=5.476-5.01X'_2+19(X_2')^2$ or $Y_c=5.476-5.01X_2'+19(X_2/X_3)^2$,where $d_1=5.476-5.01(t/\omega)+19(t/\omega)^2$ (69).

So ,we see from Figure 41(b) that with increasing of ratio (t/ω),the value of internal diameter (d_1) for stainless chip increases in accordance with the non-linear equation view of $d_1=5.476-5.01(t/\omega)+19(t/\omega)^2$.

In Figure 41 (c) is shown the functional dependence of external diameter (d_2) stainless chip from its internal diameter (d_1),i.e has a place the function view of $d_2=\varphi_3(d_1)$ which has the non-linear regression model with equation view of $Y_c=5.73X^{0.359}$ or $d_2=5.73d_1^{0.359}$.So,we see from Figure 41(c) that with increasing of internal diameter (d_1) ,the value of external diameter (d_2) increases considerably and this regression model better submits to the non-linear regression model in accordance with the non-linear equation view of $d_2=5.73d_1^{0.359}$.

In Figure 41(d) is shown the functional dependence of clearance (t) between of chip wraps from the number of wraps (ω) ,i.e has a place the function view of $t=\alpha_1(\omega)$ which has the non-linear regression model with equation view of $Y_c=4.227X^{0.152}$ or $\omega=4.227t^{0.152}$,where $t=(0.24\omega)^{6.579}$ (70). So , we see from Figure 41(d) that with increasing of number wraps (ω),the value of clearance (t) between of chip wraps considerably increases in accordance with the non-linear regression equation view of $t==(0.24\omega)^{6.579}$.

In Figure 41(e) and 41(f) are illustrated the residual plots (residual versus t/ω and residual versus d_1) of functional dependencies $d_2=\phi_{19}(t/\omega)$ and $d_2=\varphi_3(d_1)$ accordingly.

Analyzing the Figure 42(A) we see that functional surface $d_2=\rho_4(d_1,t/\omega)$ is shown in three-dimensional drawing for the linear regression model with equation view of $Y_c=4.47+0.85X_1+16X'_2$ or $Y_c=4.47+0.85X_1+16(X_2/X_3)$,where $d_2=4.47+0.85d_1+16 (t/\omega)$.

53- *Non-linear regression model has equation view of $d_2=10.26(t/\omega)^{0.002}$*
54- *Linear regression model has equation view of $d_1=7.17-23(t / \omega)$*

And Figure 42(B) shows that this functional surface $d_2=\rho_4(d_1,t/\omega)$ has view of trapezium on the face YOX_1 with such coordinates of their points (sizes in millimeters): $a(X_{1,1}=5.96\ ,X_{2,1}=0,Y_1=10.98)$;$b(X_{1,2}=0,\ X_{2,2}=0,Y_2=0)$;$c(X_{1,3}=3.52\ ,X_{2,3}=0,Y_3=0)$; $d(X_{1,4}=3.52,X_{2,4}=0,Y_4=8.74)$.

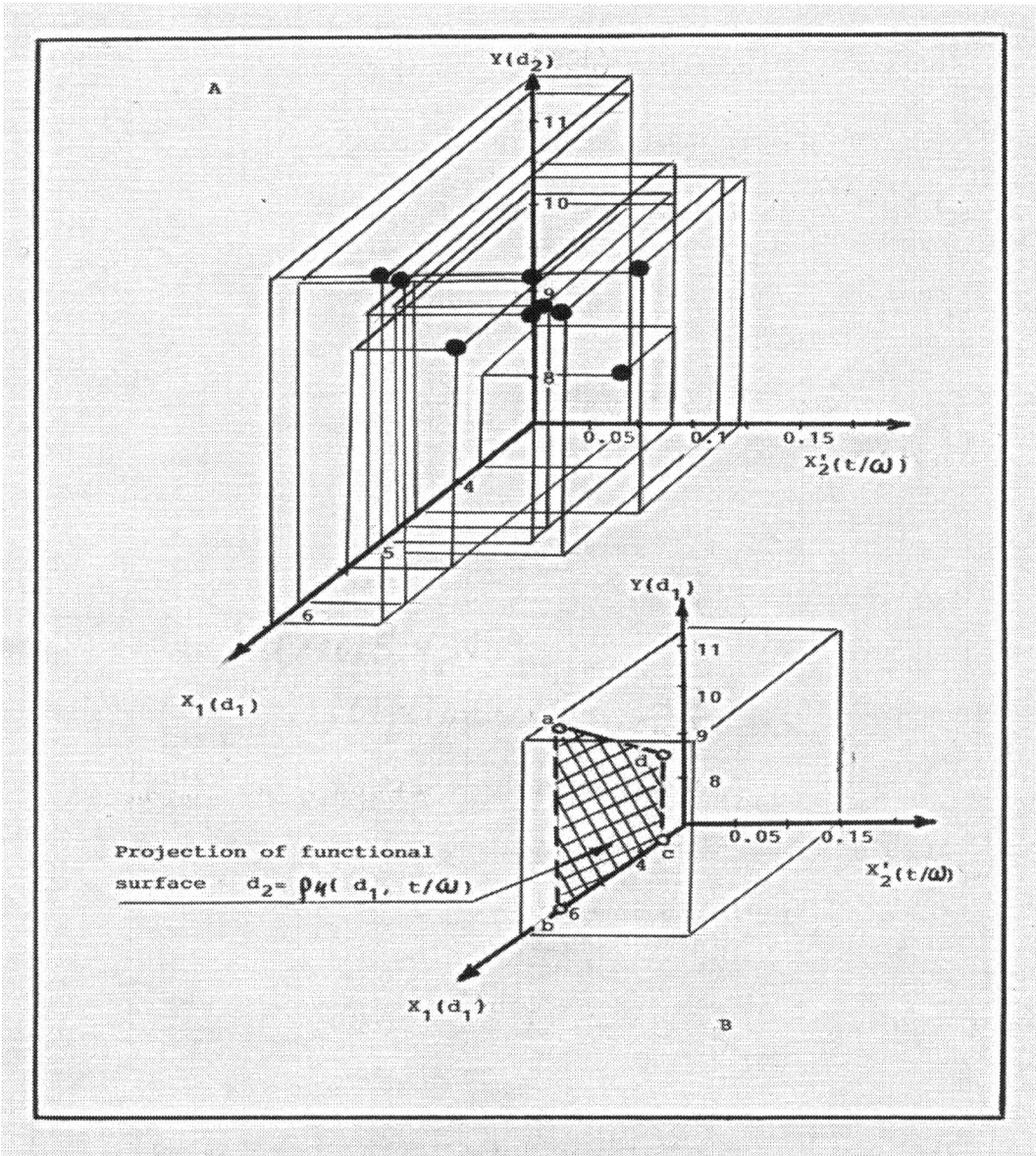

Figure 42

Analysis of functional surface $d_2=\rho_4(d_1,t/\omega)$

From Figure 42(B) we see that functional surface $d_2=\rho_4(d_1,t/\omega)$,on which are situated the points of the linear regression equation view of $Y_c=4.47+0.85X_1 +16X'_2$ or $Y_c=4.47+0.85X_1+16(X_2/X_3)$, where $d_2=4.47+0.85d_1+16(\ t\ /\omega\ )$ has the following coordinates of their peaks in three-dimensional drawing (sizes in millimeters):

a'($X_{1,1}$=5.96, $X_{2,1}$=0.09, Y_1=10.98); b'($X_{1,2}$=5.96, $X_{2,2}$=0.09, Y_2=0); c'($X_{1,3}$=3.52, $X_{2,3}$=0.08, Y_3=0); d' ($X_{1,4}$=3.52, $X_{2,4}$=0.08, Y_4=8.74). The module of vectors $|a'b'|$, $|b'c'|$, $|c'd'|$, $|d'a'|$ and area of this functional surface can be calculated by formulas (11),(12),(13),(14) and (46). At data we have the values: $|a'b'|$=10.98, $|b'c'|$=2.44, $|c'd'|$=8.74, $|d'a'|$=3.31 and $S_{a'b'c'd'}$=24.06 mm^2.

All results of computation for the equation view of Y_c=4.47+0.85X_1+16X'_2 or Y_c=4.47 + +0.85X_1+16(X_2/X_3), where d_2=4.47+0.85d_1+16(t/ω) are given in Table 22.

Table 22 Evaluation of regression equation Y_c=4.47+0.85X_1+16X'_2

A. Mean , variance and standard deviation

Variable	Mean	Variance	Standard deviation
X_1	5.098	5.608	2.368
X'_2	0.09	0.002	0.049
Y	10.248	4.121	2.030

B. Results of multiple regression of Y on X_1 and X'_2

Parameter	Variable	Coefficients	Standard error	T-value
β_1	X_1	0.85	0.789	1.077
β_2	X'_2	16	0.016	1000

C. Analysis of variance results

Regression
- Degrees of freedom 2
- Sum of squares 3.781
- Mean squares 1.891

Error
- Degrees of freedom 6
- Sum of squares 0.340
- Mean squares 0.057

F-value* 33.175

* Since F=33.175>[$F_{0.05,2,6}$=5.14] ,we can reject the hypothesis that β_1 and β_2 are zero.

D. Determination of residual

Number	Obsrerved	Estimated	Residual
1	10.03	10.15	-0.12
2	10.36	10.83	-0.47
3	8.69	8.74	-0.05
4	10.13	10.06	0.07
5	10.22	9.98	0.24
6	10.07	9.94	0.13
7	10.49	10.45	0.04
8	11.06	10.98	0.08
9	11.18	11.06	0.12

So, the parameters of this functional surface d_2=ρ_4(d_1,t /ω) are the following:

1.Function d_2=ρ_4(d_1, t/ω) better submits to the linear regression model with equation view of Y_c=4.47+0.85X_1+16X'_2 or Y_c=4.47+0.85X_1+16(X_2/X_3) ,where d_2=4.47+0.85d_1+16(t/ω) with such statistical characteristics:

coefficient of determination R^2=0.917 ,coefficient of correlation r=0.958,standard deviation $S_{y/x1,x'2}$=0.238,minimization of the mean square error (min MSE=0.026),minimization of the mean absolute deviation (min MAD=0.004).

2.The total area of functional surface $d_2=\rho_4(d_1,t/\omega)$ is equal $\sum S$=24.06 mm^2 with such coordinates of their points in three-dimensional drawing for this surface (sizes in millimeters): a' ($X_{1,1}$=5.96,$X_{2,1}$=0.09,Y_1=10.98); b'($X_{1,2}$=5.96,$X_{2,2}$=0.09 ,Y_2=0); c'($X_{1,3}$=3.52,$X_{2,3}$=0.08,Y_3=0) ;d'($X_{1,4}$=3.52,$X_{2,4}$=0.08,Y_4=8.74).

6.7 Dependence of external diameter (d_2) from width (B) of stainless chip ,internal diameter (d_1) and clearance (t) between of chip wraps , i.e $d_2=\alpha_7(B,d_1,t)$ and also from its modifications : $d_2=\rho_5(B,B/d_1),d_2=\rho_6(B,B/t)$, $d_2=\rho_7(B,d_1/t)$.

A. Function $d_2=\alpha_7(B,d_1,t)$

Analysis of data and statistical characteristics (coefficient of determination R^2=0.905,coefficient of correlation r=0.951,standard deviation $S_{y/x1,x2,x3}$=0.280), minimization of the mean square error (min MSE=0.044) and minimization of the mean absolute deviation (min MAD=0.002) show that this functional dependence $d_2=\alpha_7(B,d_1,t)$ better submits to the linear than non-linear[55] regression model with equation view of

Y_c=1.962+0.149X_1+1.144X_2+5X_3 or **d_2=1.962+0.149B+ +1.144d_1+5t** **(71)** .

The functional dependencies $d_2=\varphi_2(B)$,$B=\varphi_1(d_1)$ and $d_1=\varphi_3(d_2)$ are shown in Figures 43(a),43(b) and 43(c). In Figure 43 (d) is shown the residual plot of function $d_2=\varphi_2(B)$. The functional dependence $d_2=\varphi_2(B)$, as shown in Figure 43(a), has the linear regression model with equation view of Y_c=3.026+1.237X ,where d_2=3.026+1.237B. So, we see from Figure 43(a) that with increasing of width (B) stainless chip, the value of external diameter (d_2) considerably increases in accordance with the regression equation d_2=3.026+1.237B.

In Figure 43(b) is shown the functional dependence of width (B) stainless chip from its internal diameter (d_1) ,i.e we have the function view of $B=\varphi_1(d_1)$. This functional dependence has the linear regression model with equation view of Y_c=0.06+0.863X or d_1=0.06+0.863B ,where **B=1.16d_1−0.07** **(72)** . So ,we see from Figure 43 (b) that with increasing of internal diameter (d_1) ,the value of width (B) stainless chip considerably increases in accordance with the linear regression equation view of B=1.16d_1−0.07.

In Figure 43 (c) is shown the functional dependence of internal diameter (d_1) stainless chip from its external diameter (d_2) ,i.e we have the function view of $d_1=\varphi_3(d_2)$ .This function has the non-linear regression model with equation view of Y_c=5.728$X^{0.359}$ or d_2=5.728$d_1^{0.359}$,where **d_1=(0.175d_2)$^{2.786}$** **(73)** . So ,we see from Figure 43(c) that with increasing of external diameter (d_2) stainless chip ,the value of internal diameter (d_1) for this chip increases considerably in accordance with the regression equation view of d_1=(0.175d_2)$^{2.786}$.

---

55 – Non-linear regression model has the equation view of d_2=5.9+0.62B+0.06B^2+0.004d_1–3.64 t

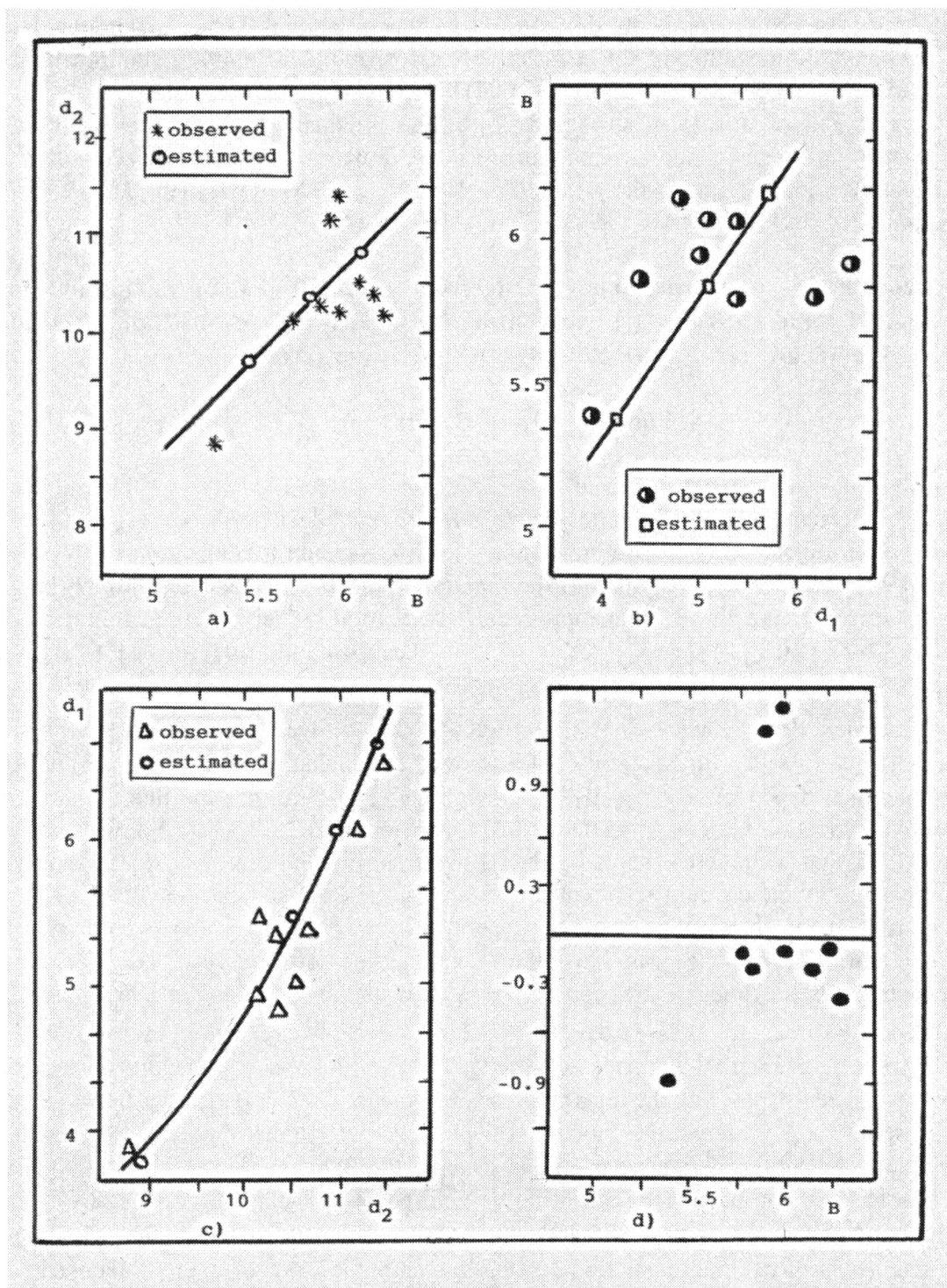

Figure 43

All results of computations for the linear equation view of $Y_c = 1.962 + 0.149X_1 + 1.144X_2 + 5.0 X_3$ or $d_2 = 1.962 + 0.149B + 1.144d_1 + 5.0 t$ are given in Table 23.

Table 23Evaluation of regression equation $Y_c=1.962+0.149X_1+1.144X_2+5.0\ X_3$

A. Mean , variance and standard deviation

Variable	Mean	Variance	Standard deviation
X_1	5.836	0.515	0.718
X_2	5.098	5.612	2.369
X_3	0.317	0.031	0.176
Y	10.248	4.121	2.030

B. Results of multiple regression of Y on X_1,X_2 and X_3

Parameter	Variable	Coefficients	Standard error	T-value
β_1	X_1	0.149	0.239	0.623
β_2	X_2	1.144	0.789	1.449
β_3	X_3	5.0	0.059	84.745

C. Analysis of variance results

Regression

    Degrees of freedom 3

    Sum of squares   3.728

    Mean squares   1.243

Error

    Degrees of freedom 5

    Sum of squares   0.393

    Mean squares   0.079

    F-value *   15.734

* Since F=15.734 >[$F_{0.05,3,5}$=5.41] we are able to reject the hypothesis that parameters β_1,β_2 and β_3 are zero.

D. Determination of residuals

Number	Observed	Estimated	Residual
1	10.03	9.99	0.01
2	10.36	10.75	-0.39
3	8.69	8.49	0.20
4	10.13	10.08	0.05
5	10.22	10.03	0.19
6	10.07	9.85	0.22
7	10.49	10.36	0.13
8	11.06	11.23	-0.17
9	11.18	11.44	-0.26

B. Modification function $d_2=\rho_5(B,B/d_1)$

Analysis of data and statistical characteristics (coefficient of determination R^2=0.908,coefficient of correlation r=0.953 , standard deviation $S_{y/x1,x'2}$=0.252) and also minimization of the mean square error (min MSE=0.042) ,minimization of the mean absolute deviation (min MAD=0.004) show that this functional dependence $d_2=\rho_5(B,B/d_1)$ has the good correlation and admit that this function better submits to the linear than non-linear[56] regression model with equation view of $Y_c=9.331+0.812X_1-3.261X'_2$ or $Y_c=9.331+0.812X_1-3.261(X_1/X_2)$, where $d_2=9.331+0.812B-3.261(B/d_1)$ (74).

56- Non-linear regression model has equation view of $d_2=33.3-6.91B+0.63B^2-3.61(B/d_1)$

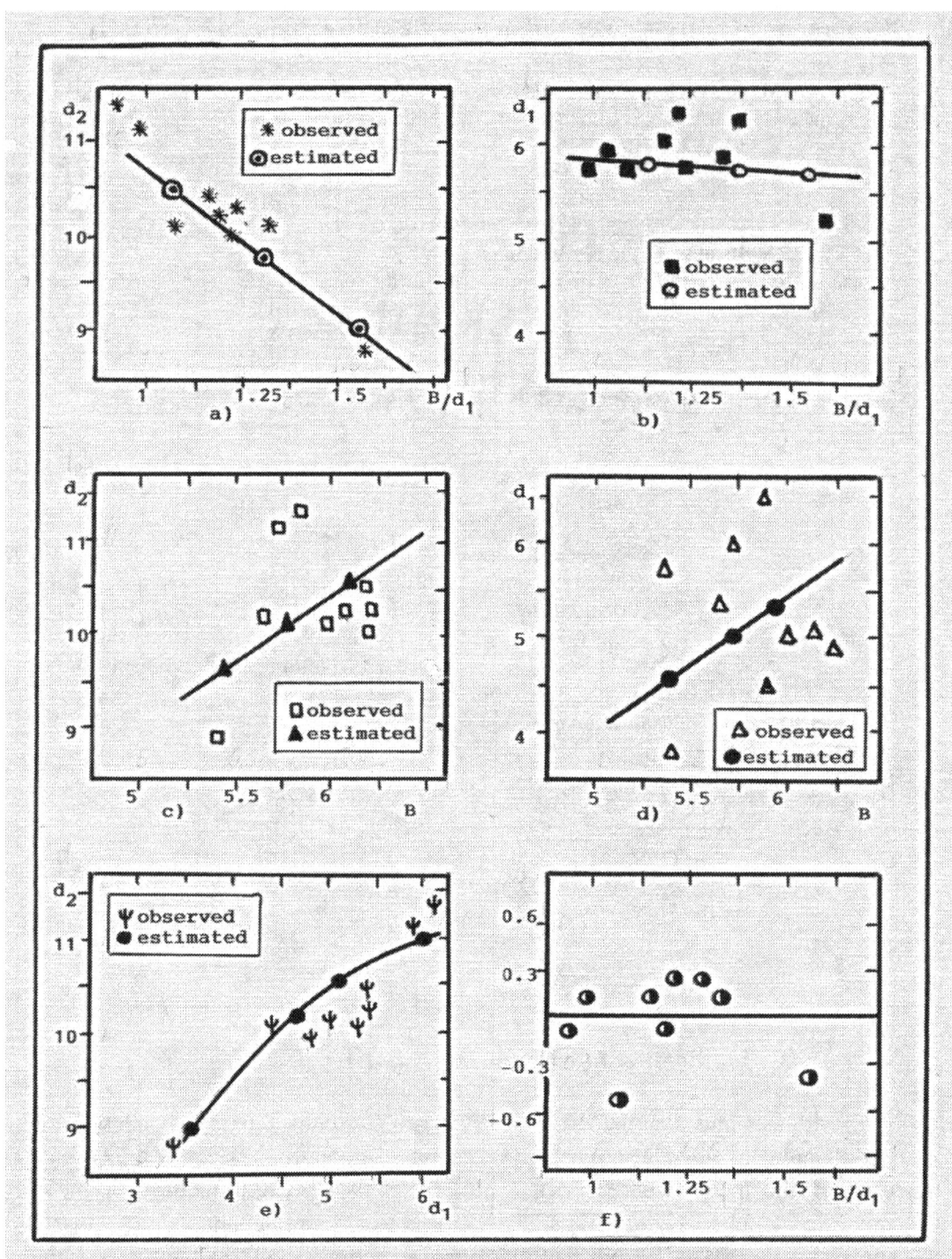

Figure 44

In Figure 44(a) is shown the functional dependence of external diameter (d_2) stainless chip from the ratio of values (B/d_1) ,such as width (B) of this chip to its internal diameter (d_1) ,i.e we have the function view of $d_2=\phi_{21}(B/d_1)$. Analyzing of data and also the scatter plots shown in Figure 44(a) ,and the statistical characteristics(coefficient of determination R^2=0.832 , coefficient of correlation r=0.912), minimization of the mean square error (min MSE=0.071),minimization of the mean absolute deviation (min MAD=0) show that this functional dependence $d_2=\phi_{21}(B/d_1)$ better submits to the linear than non-linear[57] regression model with equation view of **$Y_c=14.283-3.443X'_2$** or **$Y_c=14.283-3.443(X_1/X_2)$** ,where **$d_2=14.283-3.443(B/d_1)$** **(75)**. And as we see from Figure 44(a) that between two parameters, such as external diameter (d_2) and ratio (B/d_1) there is the good correlation (r=0.912). And this function $d_2=\phi_{21}(B/d_1)$ can be expressed in view of the linear regression equation $d_2=14.283-3.443(B/d_1($ which shows that with increasing of ratio (B/d_1) ,the value of external diameter(d_2) decreases considerably.

In Figure 44(b) is shown the functional dependence of internal diameter (d_1) stainless chip from the ration (B/d_1) ,such values as width of this chip (B) to its internal diameter (d_1) ,i.e we have the function view of $d_1=\phi_{22}(B/d_1)$. Analyzing of data and the statistical characteristics (coefficient of determination R^2=0.03,coefficient of correlation r=0.19) show that this functional dependence $d_1=\phi_{22}(B/d_1)$ does not have the good correlation ,but compare the characteristics we can admit that it better submits to the linear than non-linear[58] regression model with equation view of **$Y_c=6.08-0.208X'_2$** or **$Y_c=6.08-0.208(X_2/X_1)$**, where **$d_1=6.08-0.208(B/d_1)$** **(76)**. So ,we see from Figure 44 (b) that with increasing of ratio (B/d_1) ,the value of internal diameter (d_1) for this stainless chip decreases non-considerably in accordance with the linear regression equation view of $d_1=6.08-0.208(B/d_1)$.

In Figure 44(c) is shown the functional dependence of external diameter (d_2) stainless chip from its width (B). This function $d_2=\varphi_2(B)$ has the linear regression model with equation view of $Y_c=3.026+1.237X$ or $d_2=3.026+1.237B$.Analysis of data shown in Figure 44(c) shows that with increasing of width(B) stainless chip ,the value of external diameter (d_2) increases considerably in accordance with the linear regression equation view of $d_2=3.026+1.237B$.

In Figure 44(d) is shown the functional dependence of internal diameter (d_1) stainless chip from the width (B) of this chip ,i.e has a place the function view of $d_1=\varphi_1(B)$.And as we see this functional dependence has the linear regression model with equation view of $Y_c=0.06+0.863X$ or $d_1=0.06+0.863B$. So, we see from Figure 44(d) that with increasing of width (B) stainless chip ,the value of internal diameter (d_1) increases considerably in accordance with the linear regression equation view of $d_1=0.06+0.863B$.

In Figure 44(e) is shown the functional dependence of external diameter (d_2) stainless chip from internal diameter (d_1) ,i.e has a place the function view of $d_2=\varphi_3(d_1)$. This functional dependence has the non-linear regression model with equation view of $Y_c=5.728X^{0.359}$ or $d_2=5.728d_1^{0.359}$. So, we see from Figure 44(e) that with increasing of internal diameter (d_1) stainless chip ,the value of external diameter (d_2) increases considerably in accordance with the non-linear regression equation view of $d_2=5.728d_1^{0.359}$.

In Figure 44(f) is illustrated the residual plot (residual versus B/d_1) of functional dependency $d_2=\phi_{21}(B/d_1)$.

57-Non-linear regression model has the equation view of $d_2=10.74(B/d_1)^{-0.35}$
58-Non-linear regression model has the equation view of $d_1=5.88(B/d_1)^{-0.053}$

Analyzing the Figure 45(A) we see that functional surface $d_2=\rho_5(B,B/d_1)$ is shown in three-dimensional drawing for the linear regression model with equation view of $Y_c=9.331+0.812X_1-3.261X'_2$ or $Y_c=9.331+0.812X_1-3.261(X_1/X_2)$,where $d_2=9.331+0.812B-3.261(B/d_1)$.

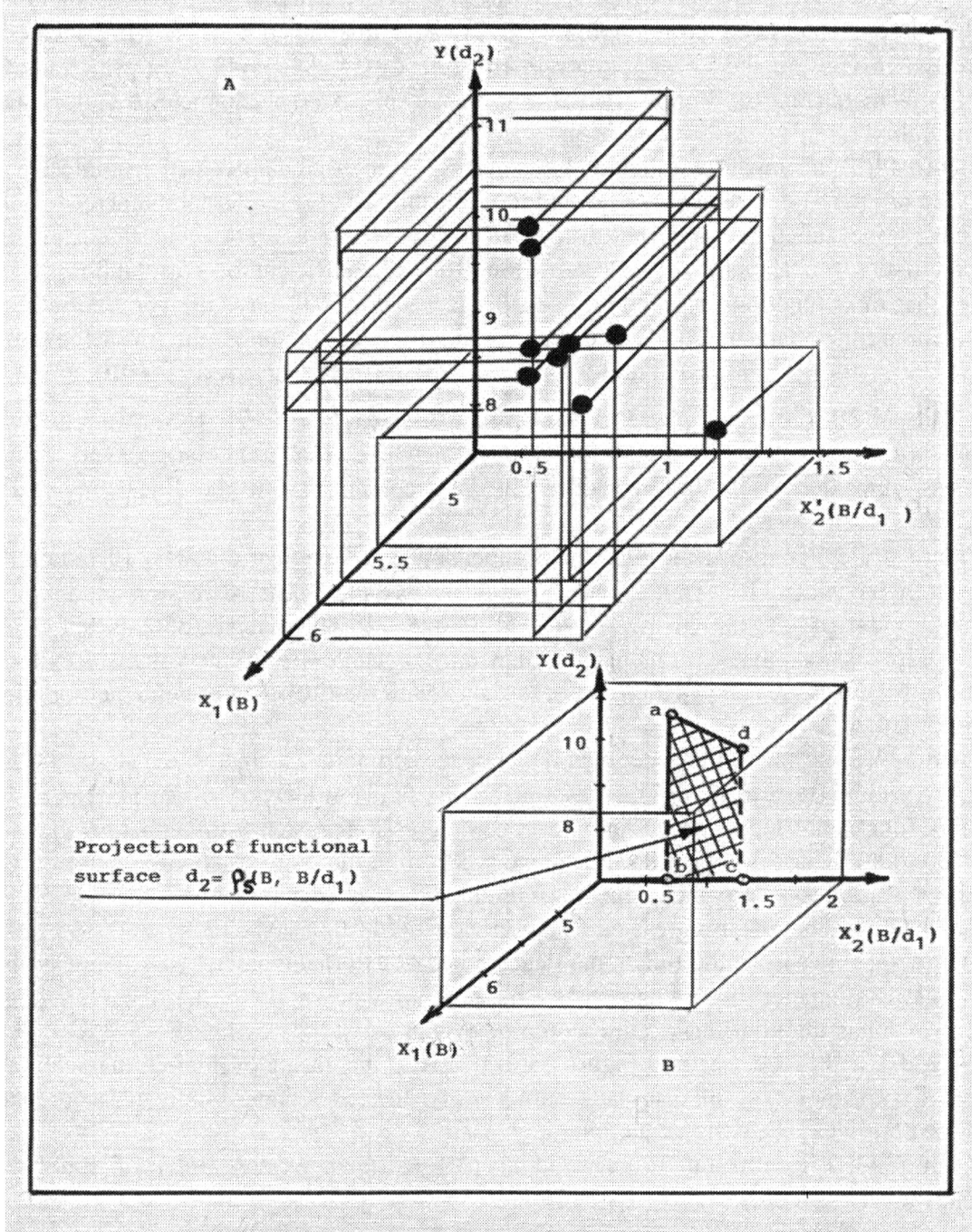

Figure 45

And Figure 45(B) shows that this functional surface $d_2=\rho_5(B,B/d_1)$ has view of trapezium on the plane YOX'_2 with such coordinates of their points (sizes in millimeters): a $(X_{1,1}=0, X_{2,1}=0.956, Y_1=10.84)$; b $(X_{1,2}=0, X_{2,2}=0.956, Y_2=0)$; c $(X_{1,3}=0, X_{2,3}=1.291, Y_3=0)$; d $(X_{1,4}=0, X_{2,4}=1.291, Y_4=10.09)$.

Analysis of functional surface $d_2=\rho_5(B,B/d_1)$

From Figure 45(A) we see that functional surface $d_2=\rho_5(B,B/d_1)$,on which are situated the points of the linear regression equation view of $Y_c=9.331+0.812X_1-3.261X'_2$ or $Y_c=9.331+0.812X_1-3.261(X_1/X_2)$,where $d_2=9.331+0.812B-3.261(B/d_1)$ has the following coordinates of their peaks in three-dimensional drawing (sizes in millimeters): a' $(X_{1,1}=5.70, X_{2,1}=0.956, Y_1=10.56)$; b' $(X_{1,2}=5.70, X_{2,2}=0.956, Y_2=0)$; c' $(X_{1,3}=6.12, X_{2,3}=1.291, Y_3=0)$; d' $(X_{1,4}=6.12, X_{2,4}=1.291, Y_4=10.09)$.
The module of vectors $|a'b'|, |b'c'|, |c'd'|, |d'a'|$ and area of functional surface can be determined by formulas (11),(12),(13),(14) and (46). At data we have the values: $|a'b'|=10.56$, $|b'c'|=0.54$, $|c'd'|=10.09$, $|d'a'|=0.71$ and $S_{a'b'c'd'}=2.85$ mm^2.
All results of computations for the linear equation view of $Y_c=9.331+0.812X_1-3.261X'_2$ or $d_2=9.331+0.812B-3.261(B/d_1)$ are given in Table 24.

Table 24 Evaluation of regression equation $Y_c=9.331+0.812X_1-3.261X'_2$

A. Mean ,variance and standard deviation			
Variable	Mean	Variance	Standard deviation
X_1	5.836	0.515	0.718
X'_2	1.172	0.288	0.537
Y	10.248	4.121	2.030

B. Results of multiple regression of Y on X_1 and X'_2				
Parameter	Variable	Coefficients	Standard error	T-value
β_1	X_1	0.812	0.239	3.398
β_2	X'_2	-3.261	0.179	-18.218

C. Analysis of variance results

Regression
Degrees of freedom 2
Sum of squares 3.741
Mean squares 1.871
Error
Degrees of freedom 6
Sum of squares 0.380
Mean squares 0.063
F-value * 29.698

* Since $F=29.698> [F_{0.05,2,6}=5.14]$ we can reject the hypothesis that both β_1 and β_2 are zero.

D. Determination of residuals

Number	Observed	Estimated	Residual
1	10.03	10.50	-0.47
2	10.36	10.48	-0.12
3	8.69	8.72	- 0.03
4	10.13	9.84	0.29
5	10.22	10.25	-0.03
6	10.07	10.09	0.02
7	10.49	10.42	0.07
8	11.06	10.84	0.22
9	11.18	11.09	0.09

So, the parameters of this functional surface $d_2=\rho_5(B,B/d_1)$ are the following:

1.Function $d_2=\rho_5(B,B/d_1)$ better submits to the linear regression model with equation view of $Y_c=9.331+0.812X_1-3.261X'_2$ or $Y_c=9.331+0.812X_1-3.261(X_1/X_2)$,where $d_2=9.331+0.812B-3.261(B/d_1)$ with such statistical characteristics:

coefficient of determination $R^2=0.908$,coefficient of correlation r=0.953, standard deviation $S_{y/x',x'2}=0.252$, minimization of the mean square error (min MSE=0.042) and minimization of the mean absolute deviation (min MAD=0.004).

2.The total area of functional surface $d_2=\rho_5(B,B/d_1)$ is equal $\sum S=2.85$ mm^2 with such coordinates   of their points in three-dimensional drawing for this surface(sizes in millimeters):   a'   ($X_{1,1}=5.70$,$X_{2,1}=0.956$,$Y_1=10.56$);b'($X_{1,2}=5.70$,$X_{2,2}=0.956$,$Y_2=0$); c'($X_{1,3}=6.12$,$X_{2,3}=1.291$,$Y_3=0$); d'($X_{1,4}=6.12$,$X_{2,4}=1.291$,$Y_4=10.09$).

C. Modification function $d_2=\rho_6(B,B/t)$

Analysis of data and the statistical characteristics (coefficient of determination $R^2=0.264$,coefficient of correlation r=0.514,standard deviation $S_{y/x1,x'2}=0.710$) and also minimization of the mean square error (min MSE=0.337) and minimization of the mean absolute deviation (min MAD=0.003) show that this functional dependence $d_2=\rho_6(B,B/t)$ better submits to the   linear than non-linear[59] regression model with equation view of **$Y_c=2.035+1.247X_1+0.049X'_2$**   or **$Y_c=2.035+1.247X_1+ +0.049(X_1/X_2)$** , where **$d_2=2.035+1.247B+0.49(B/t)$ (77) .**

In Figure 46(a) is shown the functional dependence of external diameter (d_2) stainless chip from the ratio of values (B/t),such as width (B) of this chip to its clearance (t) between of chip wraps ,i.e we have the function view of $d_2=\phi_{23}(B/t)$.Analyzing of data and also the scatter plots   and statistical characteristics (coefficient of determination $R^2=0.15$,coefficient of correlation r=0.387),minimization of the mean square   error   (min MSE=0.389),minimization of the mean absolute deviation (min MAD=0.060) show that this   functional dependence $d_2=\phi_{23}(B/t)$ better submits   to the non-linear than linear [60] regression model with equation of view **$Y_c=6.03X'_2{}^{0.18}$** or **$Y_c=6.03(X_1/X_2)^{0.18}$** ,where **$d_2=6.03(B/t)^{0.18}$ (78).**

And as we see from Figure 46(a) that between two parameters ,such as external diameter (d_2) and ratio (B/t) there is not the good correlation (r=0.387) and this function can be expressed in view of non-linear regression equation $d_2=6.03(B/t)^{0.18}$ which shows that with increasing of ratio (B/t) ,the value   of   external diameter (d_2) increases non-considerably.

And besides in Figure 46(a) is shown the functional dependence of internal diameter (d_1) stainless chip from the ratio (B/t) such values as width(B) of this chip to its clearance (t) between of chip wraps ,i.e we have the function view of $d_1=\phi_{24}(B/t)$.Analyzing of data we see also that there is not the good correlation between of two parameters (d_1 and B/t) because the coefficients of determinations (R^2) and correlation (r) in both cases are equal zero.

59- Non-linear regression model has the equation view of $d_2=9.11-1.08B+0.19B^2+0.05(B/t)$
60- Linear regression model has the equation view of $d_2=9.35+0.05(B/t)$

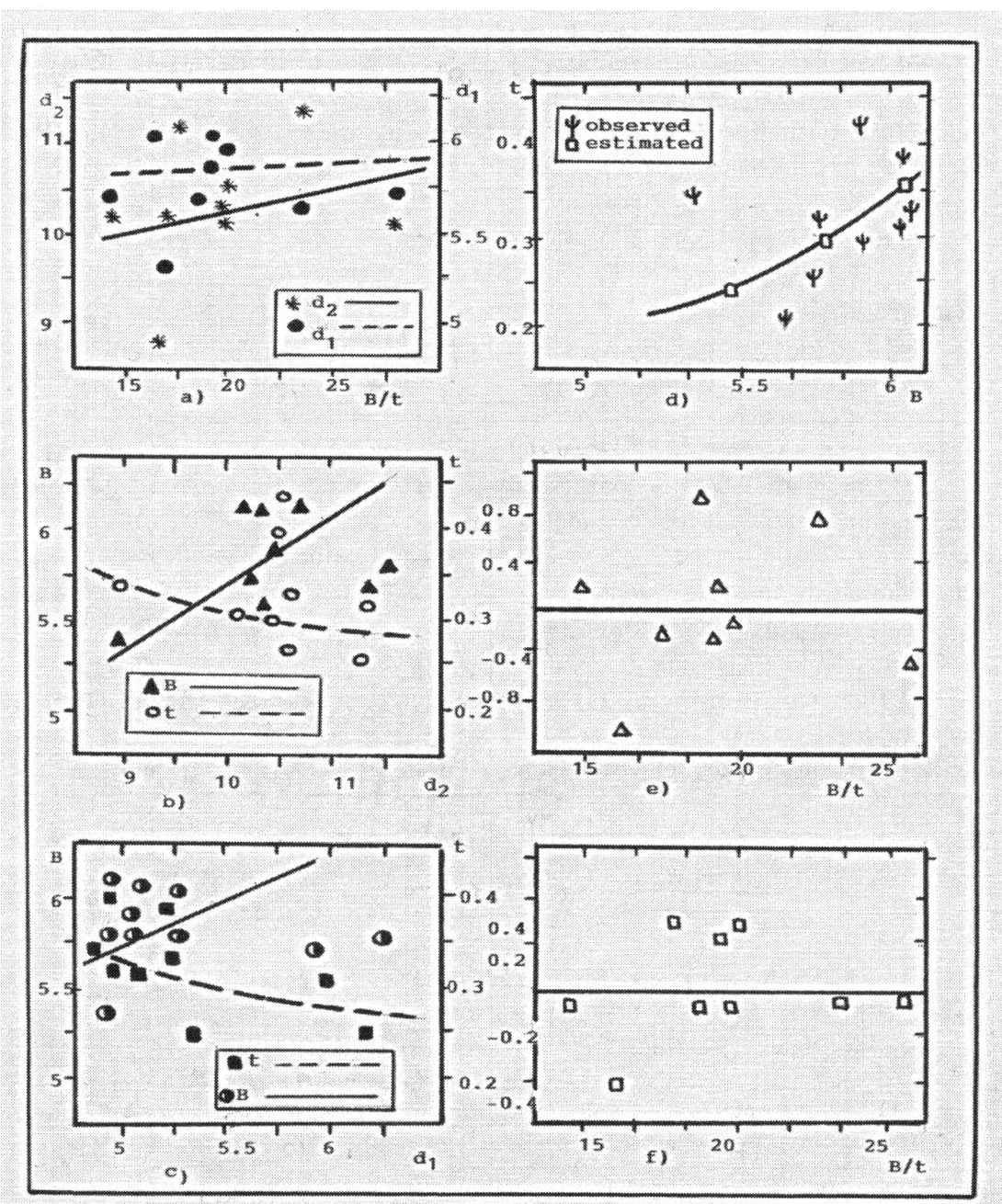

Figure 46

But the comparative analysis of minimization of the mean square error (min MSE) and minimization of the mean absolute deviation (min MAD) show that this functional dependence $d_1 = \phi_{24}(B/t)$ better submits to the linear than non-linear[61] regression model with equation $\mathbf{Y_c = 5.859 - 0.001X'_2}$ or $\mathbf{Y_c = 5.859 - 0.0019(X_1/X_2)}$,where $\mathbf{d_1 = 5.859 - 0.001(B/t)}$ (**79**).

61- Non-linear regression model has the equation view of $d_1 = 7.26(B/t)^{-0.075}$

So , we see from Figure 46(a) that with increasing of ratio (B/t) ,the value of internal diameter (d_1) increases non-considerably in accordance with the linear regression equation view of $d_1=5.859-0.001(B/t)$.

In Figure 46(b) is shown the functional dependence of width (B) stainless chip from its external diameter (d_2). This functional dependence has the linear regression model with equation view of $Y_c=3.026+1.237B$ or $d_2=3.026+1.237B$,where **B=0.81d_2−2.45 (80).**

Analysis of data shown in Figure 46(b) shows that with increasing of external diameter (d_2) ,the value of width(B) increases considerably in accordance with the linear regression equation view of $B= 0.81d_2-2.45$.

And besides in Figure 46(b) is shown the functional dependence of clearance (t) between of chip wraps from external diameter (d_2) of stainless chip. This functional dependence $t=\gamma_3(d_2)$ has the non-linear regression model with equation view of $Y_c=1.14X^{-0.56}$,where $t=1.14d_2^{-0.56}$. And besides we see also from Figure 46(b) that value of clearance (t) between of chip wraps decreases considerably with increasing of external diameter (d_2) in accordance with the non-linear regression equation view of $t=1.14d_2^{-0.56}$.

In Figure 46(c) is shown the functional dependence of width(B) stainless chip from its internal diameter(d_1) and this function $B= \varphi_1(d_1)$ has the linear regression model with equation view of $Y_c=0.06+0.863X$ or $d_1=0.06+0.863B$,where **B=1.16d_1−0.07 (81).**

And besides we see that value of width (B) stainless chip increases considerably with increasing of internal diameter (d_1) in accordance with the linear regression equation view of $B=1.16d_1-0.07$.

In Figure 46 (c) also is shown the functional dependence of clearance (t) between of chip wraps from its internal diameter(d_1) and this function $t=\gamma_4(d_1)$ has the non-linear regression model with equation view of $Y_c=0.828X^{-0.605}$,where $t=0.828d_1^{-0.605}$. And besides we see that value of clearance (t) between of chip wraps decreases considerably with increasing of internal diameter (d_1) in accordance with the non-linear regression equation view of $t=0.828d_1^{-0.605}$.

In Figure 46(d) is shown the functional dependence of clearance (t) between of chip wraps stainless chip from its width (B). This function $t=\gamma_1(B)$ has the non-linear regression model with the equation view of $Y_c=0.032X^{1.296}$,where $t=0.032B^{1.296}$. And besides we see that value of clearance (t) between of chip wraps increases considerably with increasing of width (B) this stainless chip in accordance with the non-linear regression equation view of $t=0.032B^{1.296}$.

In Figure 46(e) and 46(f) are illustrated the residual plots (residual versus B/t) of functional dependencies $d_2=\phi_{23}(B/t)$ and $d_1=\phi_{24}(B/t)$ accordingly.

Analyzing the Figure 47(A) we see that functional surface $d_2=\rho_6(B,B/t)$ is shown in three-dimensional drawing for the linear regression model with equation view of $Y_c=2.035+1.247X_1+0.049X'_2$ or $Y_c=2.035+1.247X_1+0.049(X_1/X_2)$,where $d_2=2.035+1.247B+0.049(B/t)$.

And Figure 47(B) shows that this functional surface $d_2=\rho_6(B,B/t)$ has view of trapezium on the place YOX'$_2$ with such coordinates of their points (sizes in millimeters):
a ($X_{1,1}=0$,$X_{2,1}=16.21$,$Y_1=10.51$); b ($X_{1,2}=0$,$X_{2,2} =16.21$,$Y_2=0$) ; c ($X_{1,3}=0$,$X_{2,3}=23$,$Y_3=0$) ; d ($X_{1,4}=0$, $X_{2,4}=23$, $Y_4=10.34$).

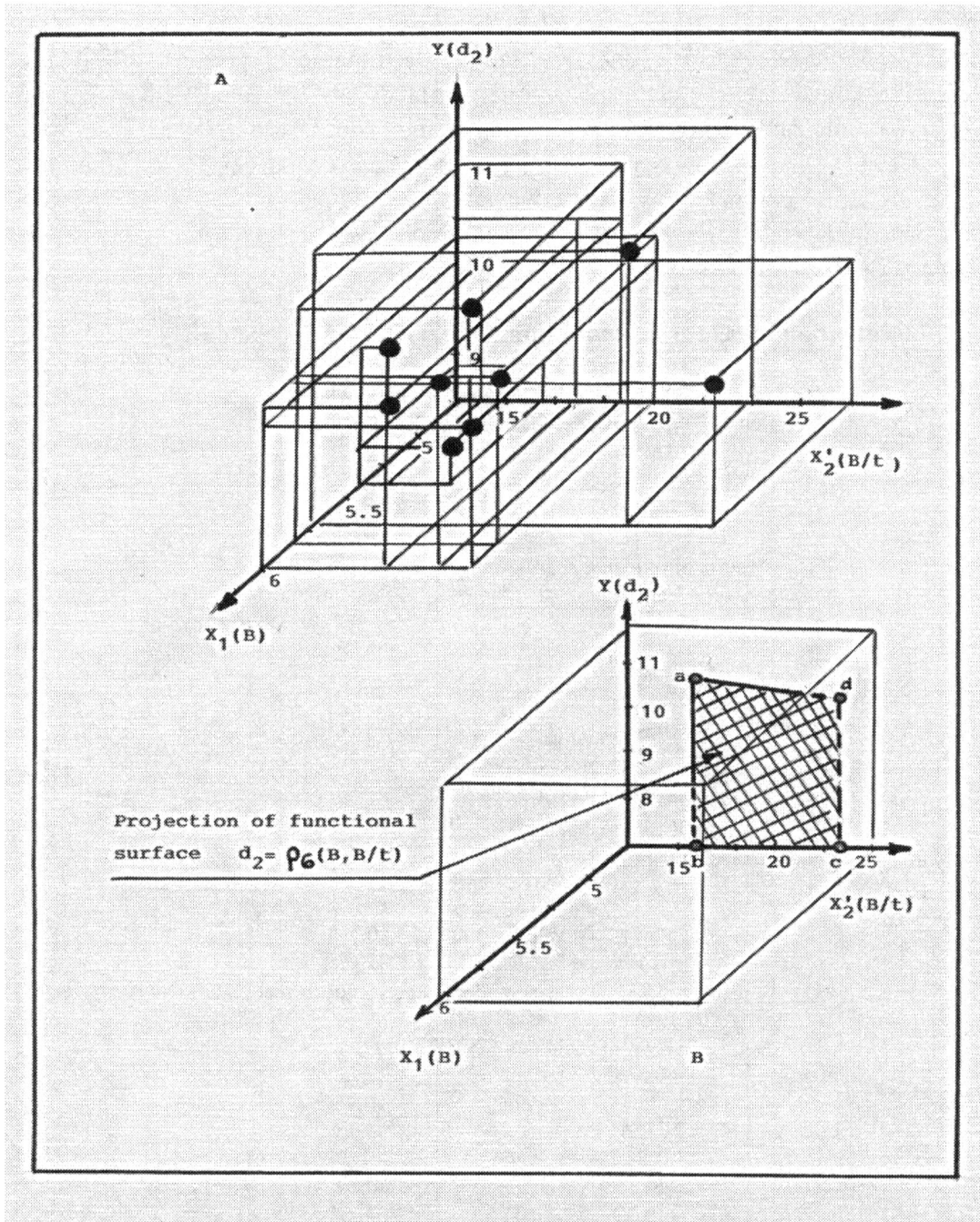

Figure 47

Analysis of functional surface $d_2=\rho_6(B,B/t)$

From Figure 47(A) we see that functional surface $d_2=\rho_6(B,B/t)$,on which are situated the points of the linear regression equation view of $Y_c=2.035+1.247X_1+0.049X'_2$ or $Y_c=2.035+1.247X_1+0.049(X_1/X_2)$,where $d_2=2.035+1.247B+0.049(B/t)$ has the following coordinates of their peaks in three-dimensional drawing (sizes in millimeters):

a'($X_{1,1}$=6.16 ,$X_{2,1}$=16.21 , Y_1=10.510 ; b' ($X_{1,2}$=6.16 ,$X_{2,2}$=16.21 , Y_2=0); c'($X_{1,3}$=5.75, $X_{2,3}$=23,Y_3=0) , d'($X_{1,4}$=5.75,$X_{2,4}$=23,Y_4=10.34). The module of vectors $|a'b'|$,$|b'c'|$, $|c'd'|$,$|d'a'|$ and area of this functional surface can be calculated by formulas (11),(12),(13),(14) and (46). At data we have the values: $|a'b'|$=10.51, $|b'c'|$=6.59, $|c'd'|$=10.34 , $|d'a'|$=6.82 and area of functional surface is equal $S_{a'b'c'd'}$=68.70 mm^2. All results of computations for the linear regression equation view of Y_c=2/035+1.247X_1+0.049X'_2 or Y_c=2.035+1.247X_1+0.049(X_1/X_2) ,where d_2=2.035+1.247B+0.049(B/t) are given in Table 25.

Table 25 Evaluation of regression equation Y_c=2.035+1.247X_1+0.049X'_2

A. Mean ,variance and standard deviation

Variable	Mean	Variance	Standard deviation
X_1	5.836	0.515	0.718
X'_2	19.11	130.326	11.416
Y	10.248	4.121	2.030

B. Results of multiple regression of Y on X_1 and X'_2

Parameter	Variable	Coefficients	Standard error	T-value
β_1	X_1	1.247	0.239	5.220
β_2	X'_2	0.049	3.805	0.010

C. Analysis of variance results

Regression
Degrees of freedom 2
Sum of squares   1.089
Mean squares   0.545
Error
Degrees of freedom 6
Sum of squares   3.032
Mean squares   0.505
F-value *   1.079

* Since F=1.079 <[$F_{0.05,2,6}$=5.14] ,we can not reject the hypothesis that both β_1 and β_2 are zero.

D. Determination of residuals

Number	Observed	Estimated	Residual
1	10.03	10.49	-0.46
2	10.36	10.51	-0.15
3	8.69	9.49	-0.80
4	10.13	9.94	0.19
5	10.22	10.25	-0.03
6	10.07	10.64	-0.57
7	10.49	10.58	-0.09
8	11.06	10.02	1.04
9	11.18	10.34	0.84

So ,the parameters of this functional surface d_2=ρ_6(B,B/t) are the following:
1.Function d_2=ρ_6(B,B/t) better submits to the linear regression model with equation view of Y_c=2.035+1.247X_1+0.049X'_2 or Y_c=2.035+1.247X_1+0.049(X_1/X_2) ,where d_2=2.035+1.247B+0.049(B/t) with such statistical characteristics:

coefficient of determination R^2=0.264,coefficient of correlation r=0.514, standard deviation $S_{y/x',x'2}$=0.710,minimization of the mean square error (min MSE=0.337) and minimization of the mean absolute deviation (min MAD=0.003).

2.The total area of functional surface $d_2=\rho_6(B,B/t)$ is equal $\sum S$=68.70 mm^2 with such coordinates of their peaks in three-dimensional drawing for this surface (sizes in millimeters): a'($X_{1,1}$=6.16 ,$X_{2,1}$=16.21 ,Y_1=10.51); b'($X_{1,2}$=6.16 ,$X_{2,2}$=16.21 ,Y_2=0); c' ($X_{1,3}$=5.75,$X_{2,3}$=23,Y_3=0); d'($X_{1,4}$=5.75,$X_{2,4}$=23 ,Y_4=10.34).

D. Modification function $d_2=\rho_7(B,d_1/t)$

Analysis of data and the statistical characteristics (coefficient of correlation r=0.809,standard deviation $S_{y/x1,x'2}$=0.488) and also minimization of the mean square error(min MSE=0.159) and minimization of the mean absolute deviation (min MAD=0.09) show that this functional dependence $d_2=\rho_7(B,d_1/t)$ better submits to the non-linear than linear[62] regression model with equation view of

$$Y_c=7.151-0.64X_1+0.16X_1^2+0.081X'_2 \quad \text{or} \quad Y_c=7.151-0.64X_1+0.16X_1^2+$$
$$+0.081(X_2/X_3),\text{ where } d_2=7.151-0.64B+0.16B^2+0.081(d_1/t) \quad (82).$$

In Figure 48(a) is shown the functional dependence of external diameter (d_2) stainless chip from the ratio of values (d_1/t),such as internal diameter (d_1) of this chip to its clearance (t) between of wraps ,i.e we have the function view of $d_2=\phi_{24}(d_1/t)$. Analyzing of data and also the scatter plots, and the statistical characteristics (coefficient of determination R^2=0.33 , coefficient of correlation r=0.575),minimization of the mean square error (min MSE=0.306) and minimization of the mean absolute deviation (min MAD=0) show that this functional dependence $d_2=\phi_{24}(d_1/t)$ better submits to the linear than non-linear[63] regression model with equation view of $Y_c=8.992+0.074X'_2$ or $Y_c=8.992+0.074(X_1/X_2)$,where $d_2=8.992+0.074(d_1/t)$ (83).

And as we see from Figure 48(a) that between of two parameters , such as external diameter (d_2) and ratio (d_1/t) there is the average correlation (r=0.575). And this function $d_2=\phi_{24}(d_1/t)$ can be expressed in view of the linear regression equation $d_2=8.992+0.074(d_1/t)$ which shows that with increasing of ratio (d_1/t) ,the value of external diameter (d_2) increases considerably.

In Figure 48(b) is shown the functional dependence of internal diameter (d_1) stainless chip from the ratio of values (d_1/t) , such as internal diameter (d_1) of this chip to its clearance (t) between of wraps ,i.e we have the function view of $d_1=\phi_{25}(d_1/t)$. Analyzing of data and also the scatter plots , statistical characteristics (coefficient of determination R^2=0.667,coefficient of correlation r=0.817,standard deviation $S_{y/x'2}$=0.517),minimization of the mean square error (min MSE=0.208), minimization of the mean absolute deviation (min MAD=0.022) show that this functional dependence $d_1=\phi_{25}(d_1/t)$ better submits to the non-linear than linear[64] regression model with equation view of $Y_c=1.542X'_2{}^{0.425}$ or $Y_c=1.542(X_1/X_2)^{0.425}$,where $d_1=1.542(d_1/t)^{0.425}$ (84).

62-Linear regression model has the equation view of $d_2=1.34+1.31B+0.08(d_1/t)$
63-Non-linear regression model has the equation view of $d_2=6.38(d_1/t)^{0.16}$
64-Linear regression model has the equation view of $d_1=3.08+0.12(d_1/t)$

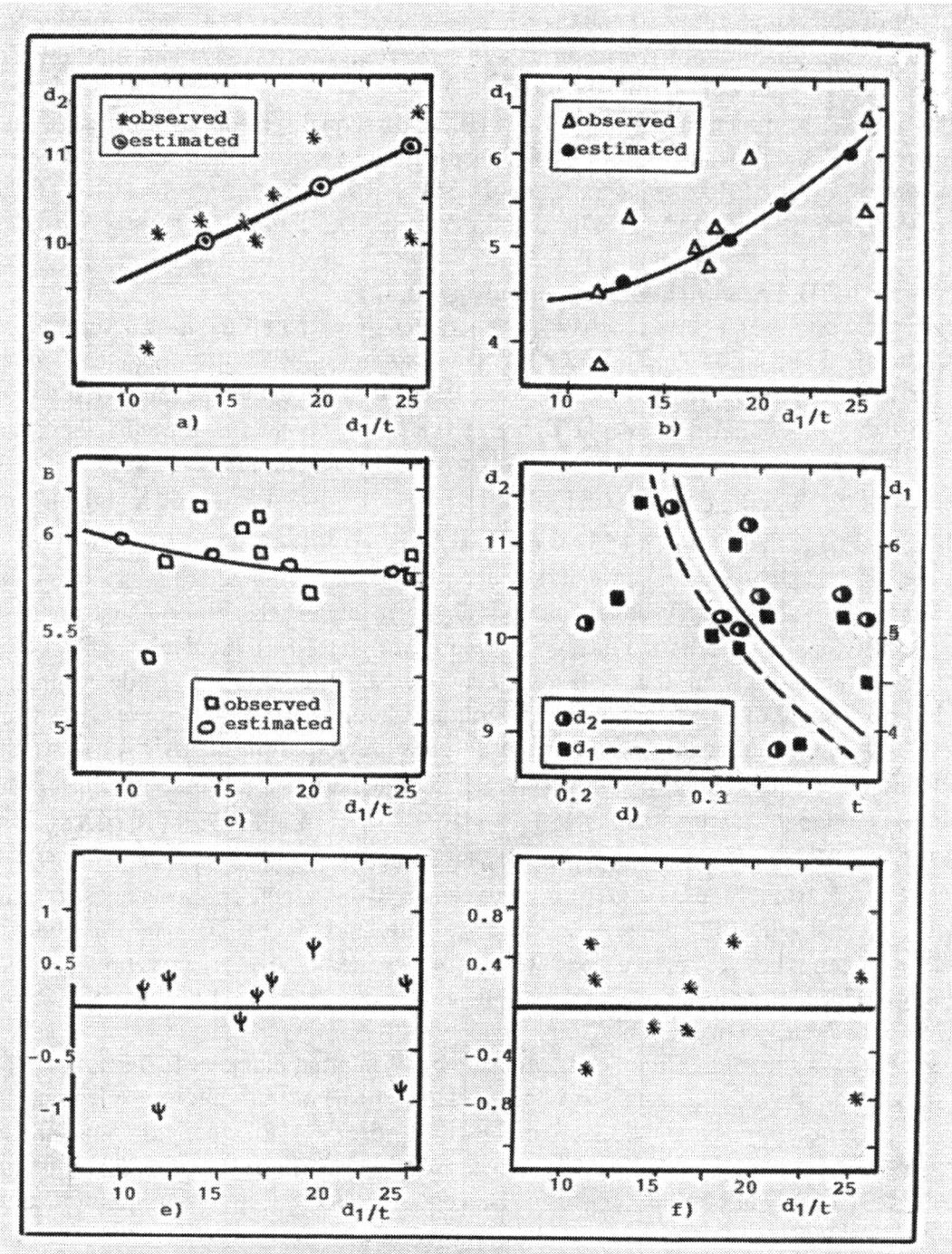

Figure 48

 And as we see from Figure 48(b) that between of two parameters, such as internal diameter
(d_1) and ratio (d_1/t) there is the good correlation (r=0.817).And this function $d_1=\phi_{25}(d_1/t)$
can be expressed in view of the non-linear regression equation $d_1=1.542(d_1/t)^{0.425}$ which
shows that with increasing of ratio(d_1/t) ,the value of internal diameter (d_1) increases
considerably.

In Figure 48(c) is shown the functional dependence of width(B) stainless chip from the ratio of values (d_1/t) ,such as internal diameter (d_1) of this chip to its clearance (t) between of wraps ,i.e we have the function view of $B=\phi_{26}(d_1/t)$.

Analyzing of data and comparing of the minimization of the mean square error (min MSE) and minimization of the mean absolute (min MAD)show that this functional dependence $B=\phi_{26}(d_1/t)$ better submits to the non-linear than linear[65] regression model with equation view of $Y_c=6.252X'^{-0.025}_2$ or $Y_c=6.252(X_1/X_2)^{-0.25}$,where $B=6.252(d_1/t)^{-0.025}$ (85).

And as we see from Figure 48 (c) that between of two parameters ,such as width (B) of stainless chip and ratio (d_1/t) absent the correlation (r=0).But this function $B=\phi_{26}(d_1/t)$ can be expressed in view of non-linear regression equation $B=6.252(d_1/t)^{-0.025}$ which shows that with increasing of ratio (d_1/t) , the value of width (B) stainless chip decreases non-considerably.

In Figure 48(d) also is shown the functional dependence of external diameter (d_2) stainless chip from its clearance (t) between of wraps. So ,we see from Figure 48(d) that this functional dependence $d_2=\gamma_3(t)$ has the non-linear regression model with equation view of $Y_c=1.14X^{-0.56}$ or $t=1.14d_2^{-0.56}$,where $d_2=(0.88t)^{-1.786}$.

Analysis of data shown in Figure 48(d) shows that with increasing of clearance (t) between of chip wraps ,the value of external diameter (d_2) decreases considerably in accordance with the non-linear regression equation view of $d_2=(0.88t)^{-1.786}$.

And besides in Figure 48(d) is shown the functional dependence of internal diameter (d_1) from the clearance (t) between of chip wraps. So, we see from Figure 48(d) that this functional dependence $d_1=\gamma_4(t)$ has the non-linear regression model with equation view of $Y_c=0.828X^{-0.605}$ or $t=0.828d_1^{-0.605}$,where $d_1=(1.208t)^{-1.653}$.

And as we see from Figure 48(d) , the value of internal diameter (d_1) decreases considerably with increasing of clearance (t) between of chip wraps in accordance with the non-linear regression equation view of $d_1=(1.208t)^{-1.653}$.

In Figure 48(e) and 48(f) are illustrated the residual plots (residual versus d_1/t) of functional dependencies $d_2=\phi_{24}(d_1/t)$ and $d_1=\phi_{25}(d_1/t)$ accordingly.

Analyzing the Figure 49(A) we see that functional surface $d_2=\rho_7(B, d_1/t)$ is shown in three-dimensional drawing for the non-linear regression model with equation view of $Y_c=7.151-0.64X_1+0.16X_1^2+0.081X'_2$ or $Y_c=7.151-0.64X_1+0.16X_1^2+0.081(X_2/X_3)$,where $d_2=7.151-0.64B+0.16B^2+0.081(d_1/t)$.

And Figure 49(B) shows that this functional surface $d_2=\rho_7(B,d_1/t)$ has view of trapezium on the place YOX'_2 with such coordinates of their points (sizes in millimeters): a ($X_{1,1}=0$,$X_{2,1}=10.73$,$Y_1=9.68$) ; b($X_{1,2}=0$,$X_{2,2}=10.73$,$Y_2=0$) ; c ($X_{1,3}=0$, $X_{2,3}=25.76$, ,$Y_3=0$) ; d ($X_{1,4}=0$,$X_{2,4}=25.76$,$Y_4=10.85$).

65- Linear regression model has the equation view of $B=6.12 -0.02 (d_1/t)$

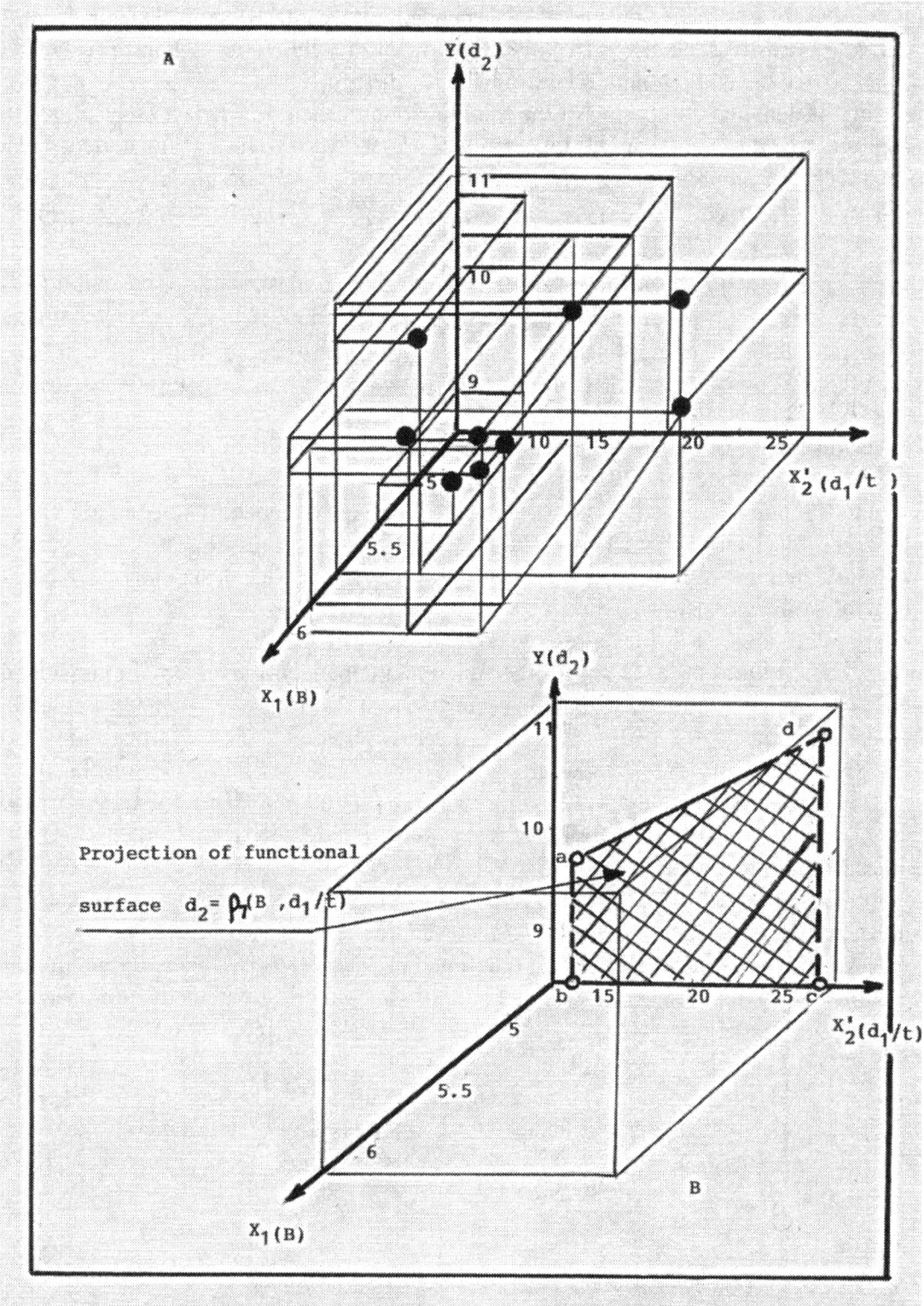

Figure 49

250

From Figure 49(A) we see that functional surface $d_2=\rho_7(B,d_1/t)$,on which the points are situated of the non-linear regression equation view of $Y_c=7.151-0.64X_1+0.16X_1^2+ +0.081X'_2$ or $Y_c=7.151-0.64X_1+0.16X_1^2+0.081(X_2/X_3)$,where $d_2=7.151-0.64B+0.16B^2+ +0.081(d_1/t)$ has the following coordinates of their peaks in three- dimensional drawing (sizes in millimeters): a' ($X_{1,1}=5.79$, $X_{2,1}=10.73$, $Y_1=9.68$); b'($X_{1,2}=5.79$, $X_{2,2}=10.73$, ,$Y_2=0$); c' ($X_{1,3}=5.75$, $X_{2,3}=25.76$,$Y_3=0$); d'($X_{1,4}=5.75$, $X_{2,4}=25.76$, $Y_4=10.85$).

In Figure 50 schematically is shown the functional surface $d_2=\rho_7(B,d_1/t)$ and graph for calculation of module for vectors $|b'c'|$ and $|a'd'|$.

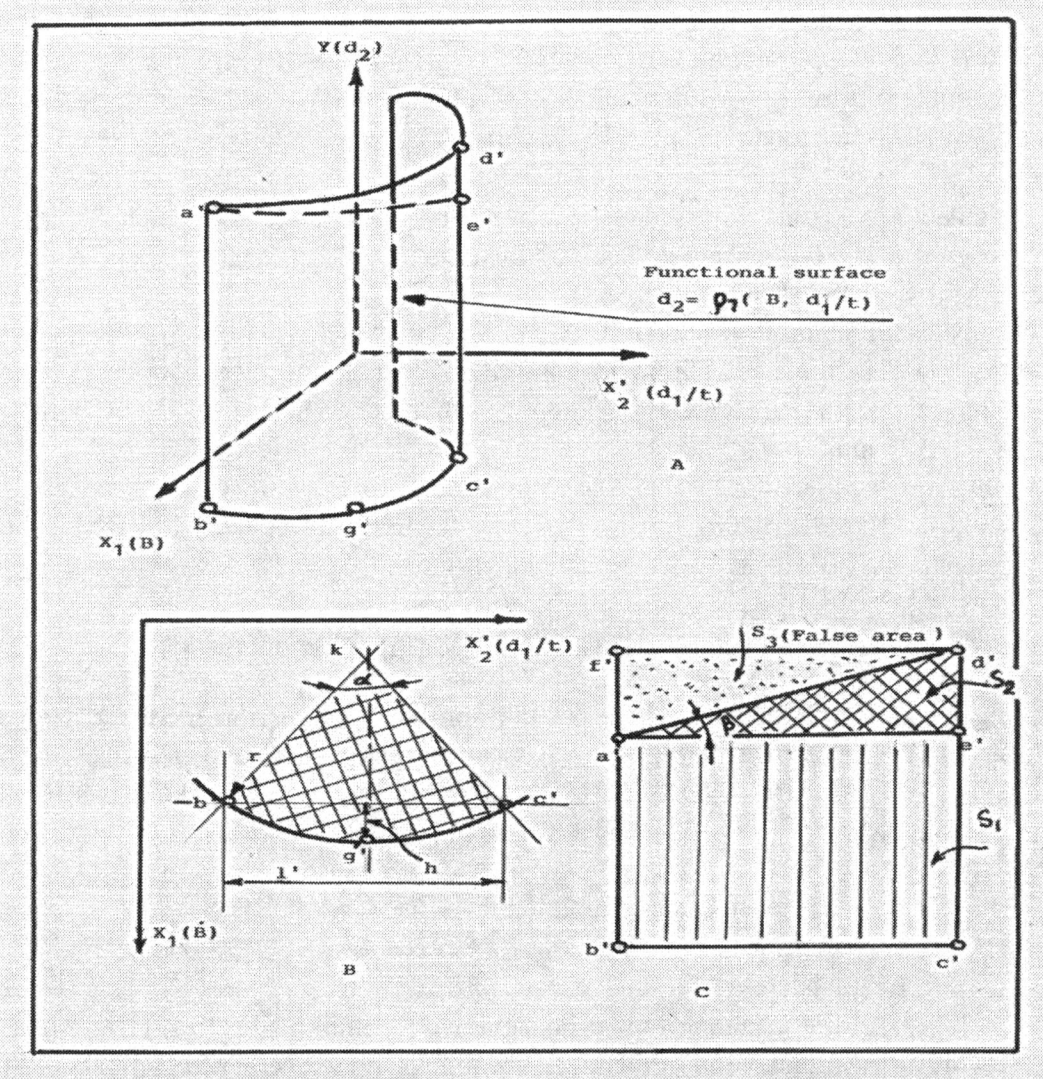

Figure 50   Graph for calculation of module vectors $|b'c'|$, $|a'd'|$ and area of functional surface $d_2=\rho_7(B,d_1/t)$:   A- total view of functional surface ; B-graph for calculation of module vector $|b'c'|$; C-graph for calculation of module vector $|a'd'|$ and area of this surface.

1.The module of vector $|a'b'|$ and $|c'd'|$ can be calculated from the projection of this functional surface $d_2=\rho_7(B,d_1/t)$ on the plane YOX'_2, as shown in Figure 49. So, the module of vectors $|a'b'|$ and $|c'd'|$ are determined by formula (11),(13): $|a'b'|=9.68$, $|c'd'|=10.85$.

2.The module of vector $|b'c'|$ or it length can be calculated by formula $L=|b'c'|=0.01745r\alpha$,where L=length of arc (b'c'),r=radius of sector ,α=angle of sector. So, the module of vector $r=|kg'|=[(X_{1,6}-X_{1,5})^2+(X_{2,6}-X_{2,5})^2]^{1/2}=17.21$,where the coordinates of points g' and k are equal: $g'(X_{1,6}=6.10, X_{2,6}=16.09)$; $k(X_{1,5}=0,X_{2,5}=0)$. At data $\alpha=95°$,r=17.21 we have the module of vector $|b'c'|=28.53$ and other parameters of circular segment which are equal: height $h=r[1-\cos(\alpha/2)]=5.50$, the length of chord $l_1=2[h(2r-h)]^{1/2}=25.22$.

3.The graph for calculation of module vectors $|b'c'|$,$|a'd'|$ and area of functional surface $d_2=\rho_7(B,d_1/t)$ is shown in Figure 50(B) and 50(C). The module of vector $|a'd'|$ and area of this functional surface $d_2=\rho_7(B,d_1/t)$,as shown in Figure 50(C) ,can be calculated by the following way: the module $|a'e'|=|b'c'|=28.53$ and module of length for $|d'e'|=|c'd'|-|a'b'|=1.17$.

4.For calculation of area S_1,S_2 and S_3 we find the angle β which is equal: $\tan\beta=|d'e'|/|a'e'|=0.041$,where $\beta=2°4'$ and then the module of length for curve $|a'd'|$ is equal : $|a'd'|=|d'e'|/\sin2°4'=28.54$.

The coordinates of point f' is equal: $f'(X_{1,1}=5.79, X_{2,1}=10.73,Y_4=10.85)$ and area of $S_2=S_3=0.5$ $(|a'e'|\cdot|d'e'|=16.69mm^2$. For these conditions we have the area $S_1=|a'b'|\cdot|b'c'|=276.17mm^2$ and total area $\sum S$ of functional surface $d_2=\rho_7(B,d_1/t)$ is equal: $\sum S= 292.86$ mm^2.

All results of computations for the non-linear regression equation view of $Y_c=7.151-0.64X_1+0.16X_1^2+0.081X'_2$ or $Y_c=7.151-0.64X_1+0.16X_1^2+0.081(X_2/X_3)$,where $d_2=7.151-0.64B+0.16B^2+0.081(d_1/t)$ are given in Table 26.

So, the parameters of this functional surface $d_2=\rho_7(B,d_1/t)$ are the following:

1.Function $d_2=\rho_7(B,d_1/t)$ better submits to the non-linear regression model in view of curvilinear surface ,as shown in Figure 49(A) and Figure 50(A) and describes by the equation view of $Y_c=7.151-0.64X_1+0.16X_1^2+0.081X'_2$ or $Y_c=7.151-0.64X_1=0.16X_1^2+ +0.081(X_1/X_3)$,where $d_2=7.151-0.64B+0.16B^2+0.081(d_1/t)$ with such statistical characteristics:
- Coefficient of determination $R^2=0.653$
- Coefficient of correlation r=0.809
- Standard deviation $S_{y/x',x'2}=0.488$
- Minimization of the mean square error (min MSE=0.159)
- Minimization of the mean absolute deviation (min MAD=0.09).

2.The total area of functional surface $d_2=\rho_7(B,d_1/t)$ is equal $\sum S=292.86$ mm^2 with such coordinates of their points in three-dimensional drawing for this surface (sizes in millimeters): $a'(X_{1,1}=5.79,X_{2,1}=10.73,Y_1=9.68)$;$b'(X_{1,2}=5.79,X_{2,2}=10.73,Y_2=0)$, $c'(X_{1,3}=5.75, X_{2,3}=25.76,Y_3=0)$; d' $(X_{1,4}=5.75,X_{2,4}=25.76, Y_4=10.85)$.

Table 26 Evaluation of regression equation $Y_c=7.151-0.64X_1+0.16X_1^2+0.081X'_2$

A. Mean, variance and standard deviation

Variable	Mean	Variance	Standard deviation
X_1	5.836	0.515	0.718
X'_2	16.968	250.05	15.812
Y	10.248	4.121	2.030

B. results of multiple regression of Y on X_1 and X'_2

Parameter	Variable	Coefficients	Standard error	T-value
β_1	X_1	-0.64	0.239	-2.68
β_2	X'_2	0.081	5.271	0.02

C. Analysis of variance results

Regression
- Degrees of freedom 2
- Sum of squares   2.694
- Mean squares   1.347

Error
- Degrees of freedom 6
- Sum of squares   1.427
- Mean squares   0.238

F-value *   5.659

* Since F=5.659 >[$F_{0.05,2,6}$=5.14] ,we can reject the hypothesis that both β_1 and β_2 are zero.

D. Determination of residuals

Number	Observed	Estimated	Residual
1	10.03	10.78	-0.75
2	10.36	10.39	-0.03
3	8.69	9.17	-0.48
4	10.13	9.68	0.45
5	10.22	10.19	0.03
6	10.07	10.46	-0.39
7	10.49	10.50	-0.01
8	11.06	10.21	0.85
9	11.18	10.85	0.33

In Table 27 is given the summary graphical function of "*Multiple Regression Analysis for External diameter of stainless chip* "in dependence from some general its parameters in function of $Y_c=\gamma(X_{i,1},X_{i,2},X_{i,3})$.

Table27Graphical function of " Multiple Regression Analysis for external diameter"

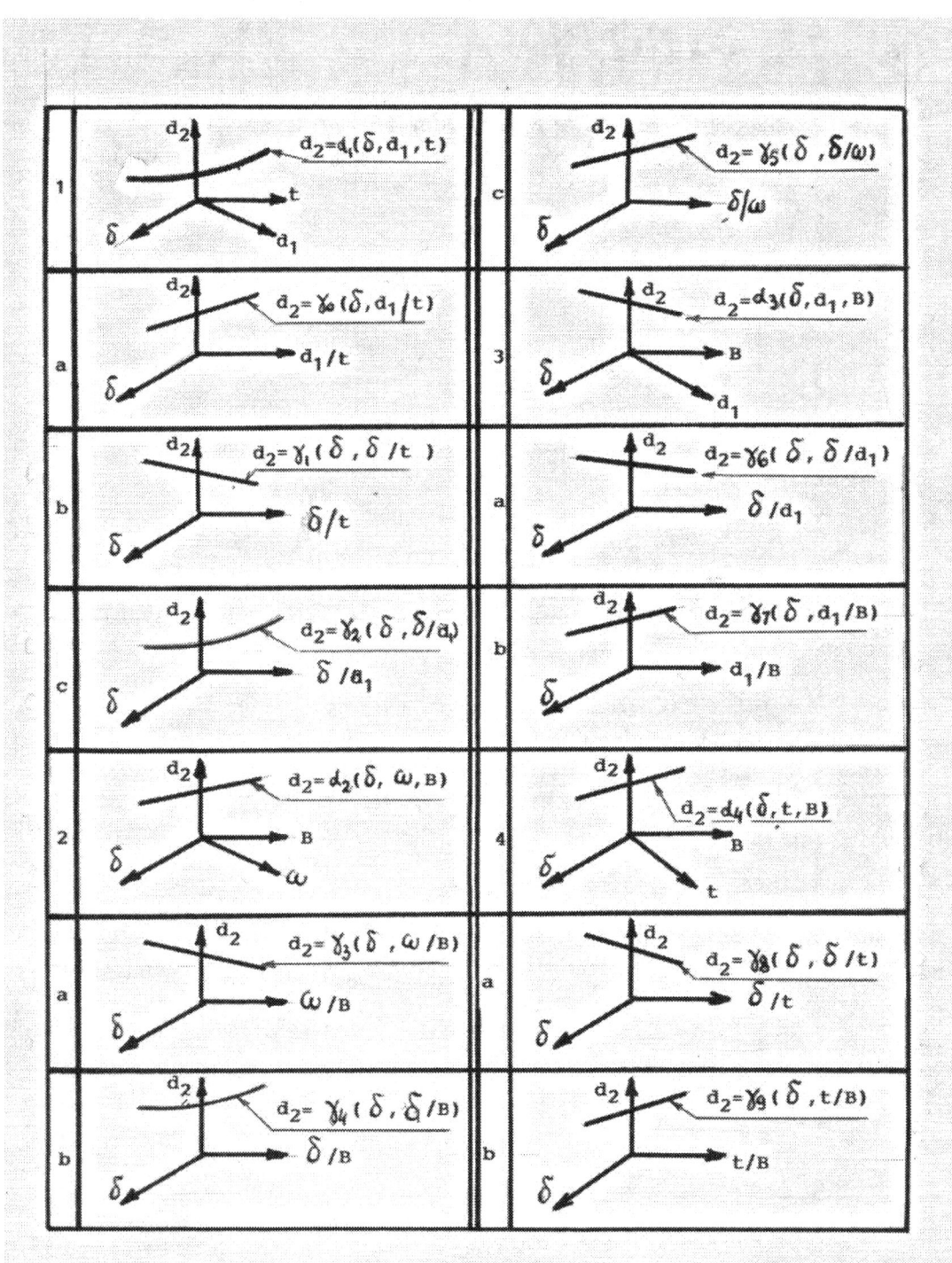

Table 27 (continue)

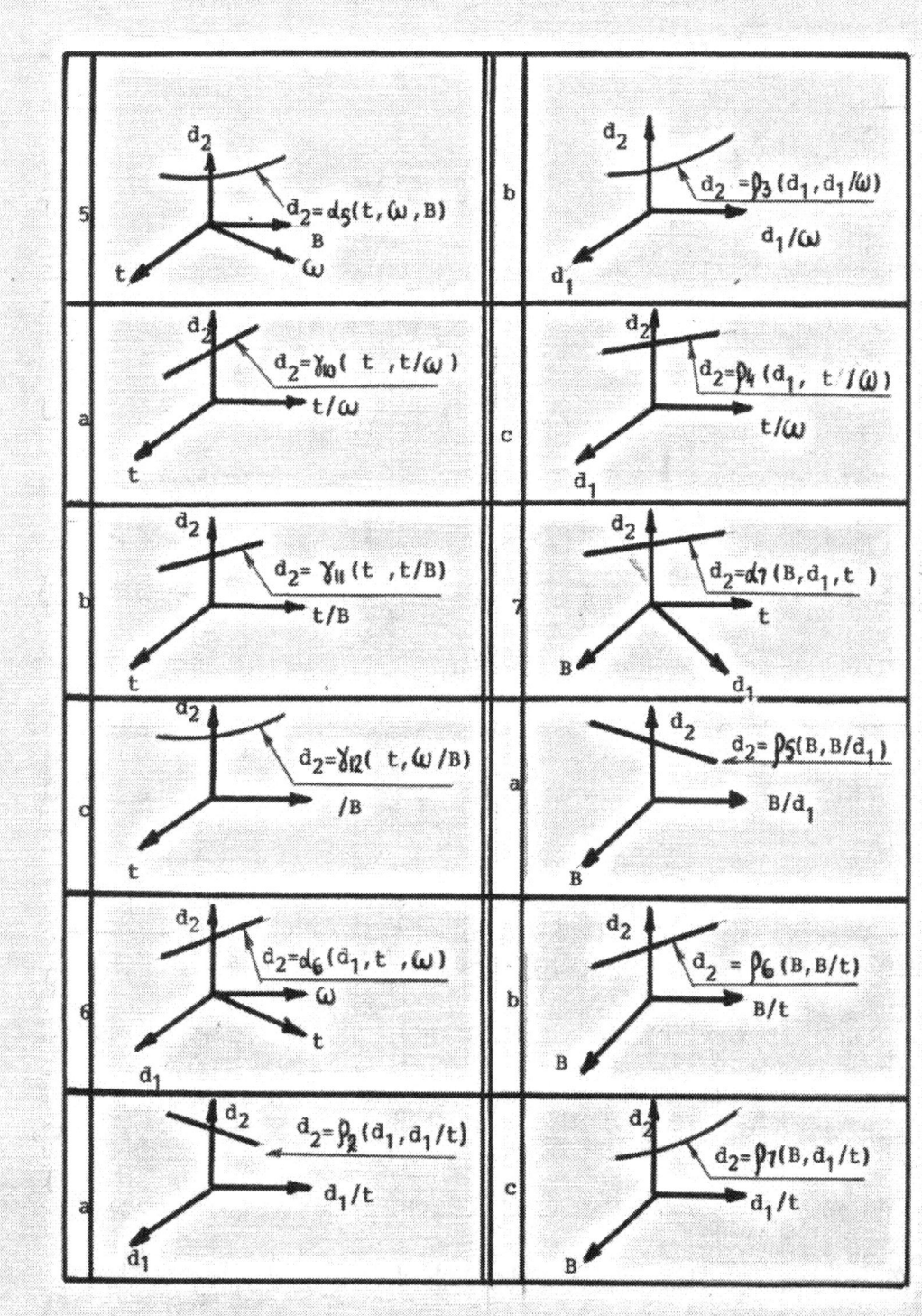

255

In Table 28 is given the Summary statistical characteristics of "Multiple Regression Analysis for external diameter".

Table 28 Summary statistical characteristics of " Multiple Regression Analysis for external diameter"

n	View function	Regression model	Empirical equation	Statistical characteristics		
				R^2	r	$S_{y/x1,x'2}/$ $S_{y/x1,x2,x3}$
1	$d_2=\alpha_1(\delta,d_1,t)$	Non-linear	$d_2=3.56-0.21\delta-0.91\delta^2+1.12d_1+3.71t$	0.950	0.974	0.203
a	$d_2=\gamma_0(\delta,d_1/t)$	Linear	$d_2=8.86+4.19\delta-0.01(d_1/t)$	0.795	0.892	0.375
b	$d_2=\gamma_1(\delta,\delta/t)$	Linear	$d_2=8.743+5.092\delta-0.297(\delta/t)$	0.799	0.894	0.369
c	$d_2=\gamma_2(\delta,\delta/d_1)$	Non-linear	$d_2=8.729+3.381\delta-0.01\delta^2+4(\delta/d_1)$	0.782	0.884	0.385
2	$d_2=\alpha_2(\delta,\omega,B)$	Linear	$d_2=7.833+4.273\delta+0.237\omega+1.47\cdot10^{-4}B$	0.802	0.896	0.404
a	$d_2=\gamma_3(\delta,\omega/B)$	Linear	$d_2=8.411+4.24\delta+0.383(\omega/B)$	0.824	0.908	0.347
b	$d_2=\gamma_4(\delta,\delta/B)$	Non-linear	$d_2=8.389+5.43\delta-3.669\delta^2+7(\delta/B)$	0.797	0.893	0.373
c	$d_2=\gamma_5(\delta,\delta/\omega)$	Linear	$d_2=8.773+2.855\delta+4(\delta/\omega)$	0.789	0.888	0.381
3	$d_2=\alpha_3(\delta,d_1,B)$	Linear	$d_2=2.32-0.76\delta+0.89d_1+0.63B$	0.917	0.958	0.261
a	$d_2=\gamma_6(\delta,\delta/d_1)$	Linear	$d_2=8.742+3.713\delta+2(\delta/d_1)$	0.790	0.890	0.30
b	$d_2=\gamma_7(\delta,d_1/B)$	Linear	$d_2=7.26+2.11\delta+2.52(d_1/B)$	0.821	0.906	0.286
4	$d_2=\alpha_4(\delta,t,B)$	Linear	$d_2=7.653+4.405\delta+2.668t+0.022B$	0.827	0.909	0.378
a	$d_2=\gamma_8(\delta,\delta/t)$	Linear	$d_2=8.743+5.077\delta-0.292(\delta/t)$	0.801	0.895	0.369
b	$d_2=\gamma_9(\delta,t/B)$	Linear	$d_2=8.68+4.03\delta+1.6(t/B)$	0.803	0.896	0.368
5	$d_2=\alpha_5(t,\omega,B)$	Non-linear	$d_2=4.09-3.069t-0.46t^2-0.02\omega+1.24B$	0.260	0.510	0.781
a	$d_2=\gamma_{10}(t,t/\omega)$	Linear	$d_2=10.97-5.84t+12.5(t/\omega)$	0.170	0.410	0.701
b	$d_2=\gamma_{11}(t,t/B)$	Linear	$d_2=11.05-2.3t-1.25(t/B)$	0.050	0.220	0.80
c	$d_2=\gamma_{12}(t,\omega/B)$	Non-linear	$d_2=12.688-0.355t-1.53t^2-3.531(\omega/B)$	0.236	0.486	0.724

Table 28 (continue)

n	View function	Regression model	Empirical equation	Statistical characteristics		
				R^2	r	$S_{y/x1,x'2}$ $S_{y/x1,x2,x3}$
6	$d_2=\alpha_6(d_1,t,\omega)$	Linear	$d_2=4.621+0.93d_1+3.769t-0.087\omega$	0.956	0.979	0.184
a	$d_2=\rho_2(d_1,d_1/t)$	Linear	$d_2=9.026-0.010d_1+0.075(d_1/t)$	0.314	0.561	0.686
b	$d_2=\rho_3(d_1,d_1/\omega)$	Non-linear	$d_2=6.29+0.75d_1+0.003d_1^2+0.05(d_1/\omega)$	0.889	0.943	0.276
c	$d_2=\rho_4(d_1,t/\omega)$	Linear	$d_2=4.47+0.85d_1+16(t/\omega)$	0.917	0.958	0.238
7	$d_2=\alpha_7(B,d_1,t)$	Linear	$d_2=1.962+0.142B+1.144d_1+5t$	0.905	0.951	0.280
a	$d_2=\rho_5(B,B/d_1)$	Linear	$d_2=9.331+0.812B-3.261(B/d_1)$	0.908	0.953	0.252
b	$d_2=\rho_6(B,B/t)$	Linear	$d_2=2.035+1.247B+0.49(B/t)$	0.264	0.514	0.710
c	$d_2=\rho_7(B,d_1/t)$	Non-linear	$d_2=7.15-0.64B+0.16B^2+0.081(d_1/t)$	0.655	0.809	0.488

SUMMARY

1. Multiple regression analysis for external diameter (d_2) of stainless chip in dependence from some its general parameters in function of $Y_i=\gamma(X_{i,1},X_{i,2},X_{i,3})$ have showed the following results:

- *Only such functions as $d_2=\alpha_1(\delta,d_1,t)$, $d_2=\gamma_2(\delta,\delta/d_1)$, $d_2=\gamma_4(\delta,\delta/B)$, $d_2=\alpha_5(t,\omega,B)$, $d_2=\gamma_{12}(t,\omega/B)$, $d_2=\rho_3(d_1,d_1/\omega)$ and $d_2=\rho_7(B,d_1/t)$ have the non-linear regression models and other functions have the linear regression models;*

- *Analysis of above-named functional dependencies indicates on that fact that statistical characteristics advantageously the coefficients of determination (R^2) and correlation (r),have the good values, besides of functions $d_2=\alpha_5(t,\omega,B), d_2=\gamma_{10}(t,t/\omega), d_2=\gamma_{11}(t,t/B), d_2=\gamma_{12}(t,\omega/B), d_2=\rho_2(d_1,d_1/t)$ and $d_2=\rho_6(B,B/t)$, where average values of coefficients determination and correlation are equal $R^2=0.28$ and $r=0.53$.*

CHAPTER SEVEN MULTIPLE REGRESSION ANAYSIS FOR INTERNAL DIAMETER AND OTHER PARAMETERS OF STAINLESS CHIP IN DEPENDENCE FROM SOM GENERAL ITS PARAMETERS IN FUNCTION OF $Y_I=(X_{i,1},X_{i,2},X_{i,3})$.

7.1 Dependence of internal diameter (d_1) from width (B) of chip, number of wraps (ω) and its thickness (δ) ,i.e $d_1=\phi_1(B,\delta,\omega)$ and also from its modifications: $d_1=\rho_8(B,B/\delta)$, $d_1=\rho_9(B,B/\omega)$, $d_1=\rho_{10}(B,\delta/\omega)$.

A. Function $d_1=\phi_1(B,\delta,\omega)$

Analysis of data and the statistical characteristics (coefficient of determination $R^2=0.869$,coefficient of correlation $r=0.932$,standard deviation $S_{y/x1,x2,x3}=0.383$), minimization of the mean square error (min MSE=0.081) and minimization of the mean absolute deviation (min MAD=0.002) show that this functional dependence $d_1=\phi_1(B,\delta,\omega)$ better submits to the linear than non-linear[1] regression model with equation view of

$$Y_c=7.968-0.846X_1+5.201X_2+0.043X_3 \text{ or } d_1=7.968-0.846B+5.201\delta+0.043\omega \ (1).$$

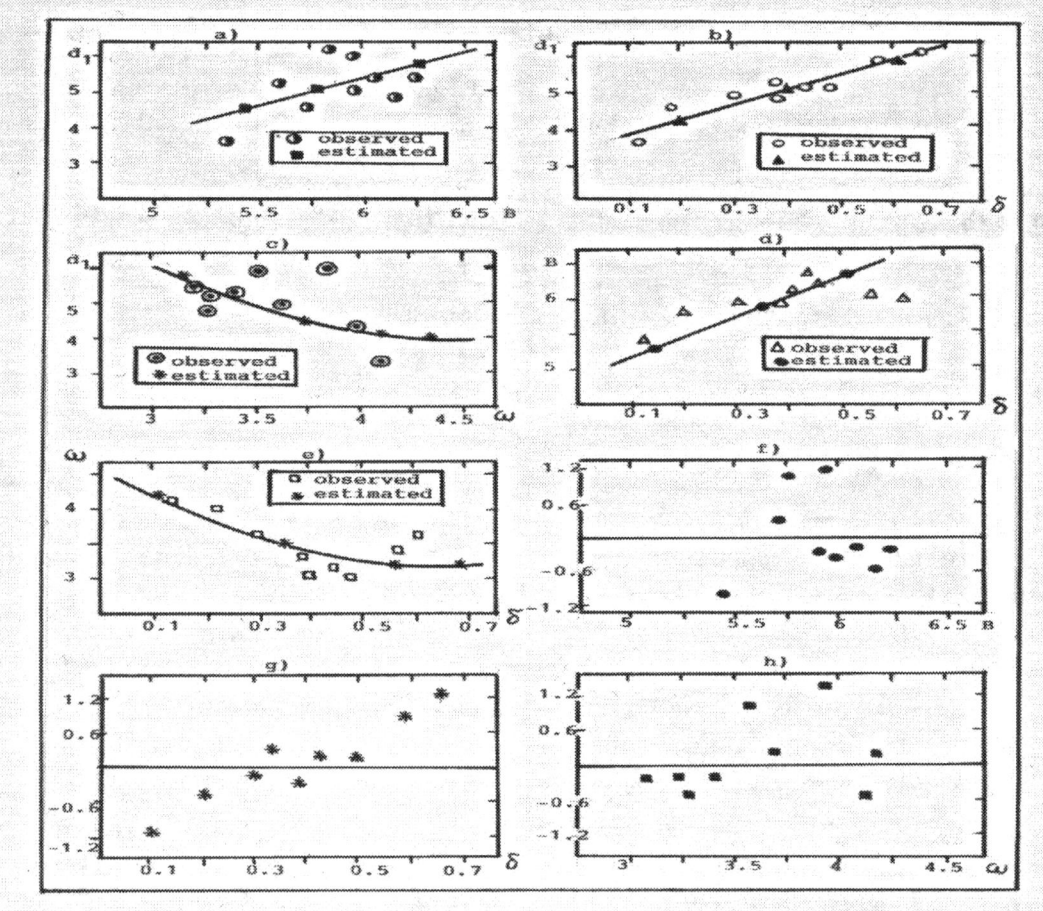

Figure 1

1- Non-linear regression model has the equation view of $d_1=551.6-186.7X_1+15.47X_1^2+13.93X_2+2.98X_3$

In Figure 1 (a) is shown the functional dependence of internal diameter (d_1) stainless chip from its width (B) ,i.e we have the functional dependence $d_1=\varphi_1(B)$.

This functional dependency has the linear regression model with equation view of $Y_c=0.06+0.863X$,where $d_1=0.06+0.0863B$.

So , from Figure 1 (a) we see that with increasing of width(B) stainless chip ,the value of internal diameter (d_1) for this chip considerably increases in accordance with the regression equation $d_1=0.06+0.863B$.

In Figure 1(b) is shown the functional dependence of internal diameter (d_1) stainless chip from its thickness (δ),i.e we have the functional dependence $d_1=\varphi(\delta)$.

This functional dependency has the linear regression model with equation view of $Y_c=3.338+4.762X$,where $d_1=3.338+4.762\delta$.

So , we see from Figure 1(b) that with increasing of thickness (δ)stainless chip ,the value of internal diameter (d_1) of this chip considerably increases in accordance with the linear regression equation $d_1=3.338+4.762\delta$.

In Figure 1 (c) is shown the functional dependence of internal diameter(d_1) stainless chip from the number of chip wraps (ω) ,i.e we have the functional dependence $d_1=\phi_3(\omega)$. This functional dependence has the non-linear regression model with equation view of $Y_c=20.32X^{-1.102}$,where $d_1=20.32\omega^{-1.102}$.

So, from Figure 1 (c) we see that with increasing of number chip wraps (ω) ,the value of internal diameter (d_1) stainless chip decreases considerably in accordance with the regression equation $d_1=20.32\omega^{-1.102}$.

In Figure 1(d) is shown the functional dependence of width (B) stainless chip from its thickness (δ) ,i.e we have the functional dependence $B=f(\delta)$.

This functional dependence has the linear regression model with equation view of $Y_c= -1.236+0.275X$ or $\delta= -1.236+0.275B$, where **$B=3.64\delta+4.5$ (2)**.

So, from Figure 1(d) we see that with increasing of thickness (δ) stainless chip, the value of width (B) this chip increases considerably in accordance with the regression equation $B=3.64\delta+4.5$.

In Figure 1 (e) is shown the functional dependence of the number chip wraps (ω) from its thickness (δ) ,i.e we have the functional dependence $\omega=\alpha(\delta)$.

This functional dependence has the non-linear regression model with equation view of $Y_c=3.09X^{-0.123}$, where $\omega=3.09\delta^{-0.123}$.

So, from Figure 1(e) we see that with increasing of thickness (δ) stainless chip, the value of number chip wraps (ω) decreases considerably in accordance with the non-linear regression equation $\omega=3.09\delta^{-0.123}$.

In Figures 1(f) , 1(g) and 1 (h) are illustrated the residual plots (residual versus B,δ,ω) of the functional dependencies $d_1=\varphi_1(B)$, $d_1=\varphi(\delta)$ and $d_1=\phi_3(\omega)$ accordingly.

All results of computations for the linear regression equation view of $Y_c=7.768-0.846X_1+5.201X_2+0.043X_3$ or $d_1=7.968-0.846B+5.201\delta+0.043\omega$ are given in Table 1.

Table 1 Evaluation of regression equation $Y_c=7.968-0.846X_1+5.201X_2+0.043X_3$

A. Mean, variance and standard deviation

Variable	Mean	Variance	Standard deviation
X_1	5.836	0.515	0.718
X_2	0.368	0.209	0.457
X_3	3.559	1.113	1.055
Y	5.098	5.612	2.369

B. results of multiple regression of Y on X_1, X_2 and X_3

Parameter	Variable	Coefficients	Standard error	T-value
β_1	X_1	−0.846	0.239	− 3.54
β_2	X_2	5.201	0.152	34.22
β_3	X_3	0.043	0.352	0.122

C. Analysis of variance results

Regression
   Degrees of freedom 3
   Sum of squares   4.876
   Mean squares   1.625
Error
   Degrees of freedom 5
   Sum of squares   0.732
   Mean squares   0.146
F-value*   11.13

* Since F=11.13> [$F_{0.05,3,5}$=5.41] we are able to reject the hypothesis that parameters β_1,β_2 and β_3 are zero.

D. Determination of residuals

Number	Observed	Estimated	Residual
1	5.36	5.15	0.21
2	5.22	5.08	0.14
3	3.52	4.14	-0.62
4	4.51	4.18	0.33
5	4.98	4.66	0.32
6	4.74	4.85	-0.11
7	5.15	5.34	-0.19
8	5.96	6.11	-0.15
9	6.44	6.39	0.05

B. Modification function $d_1=\rho_8(B,B/\delta)$

Analysis of data and the statistical characteristics (coefficient of determination R^2=0.929, coefficient of correlation r=0.964, standard deviation $S_{y/x1,x'2}$=0.257) and also the minimization of the mean square error (min MSE=0.044) and minimization of the mean absolute deviation (min MAD=0) show that this functional dependence $d_1=\rho_8(B,B/\delta)$ better submits to the linear than non-linear [2] regression model with equation view of
$$Y_c=16.963-1.773X_1-0.073X'_2 \quad \text{or} \quad Y_c=16.963-1.773X_1-0.073(X_1/X_2)$$,where
$$d_1=16.963-1.773B-0.073(B/\delta) \quad (3).$$

2- Non-linear regression model has the equation view of $d_1=14.84 – 3.463B+0.316B^2 –0.015(B/\delta)$

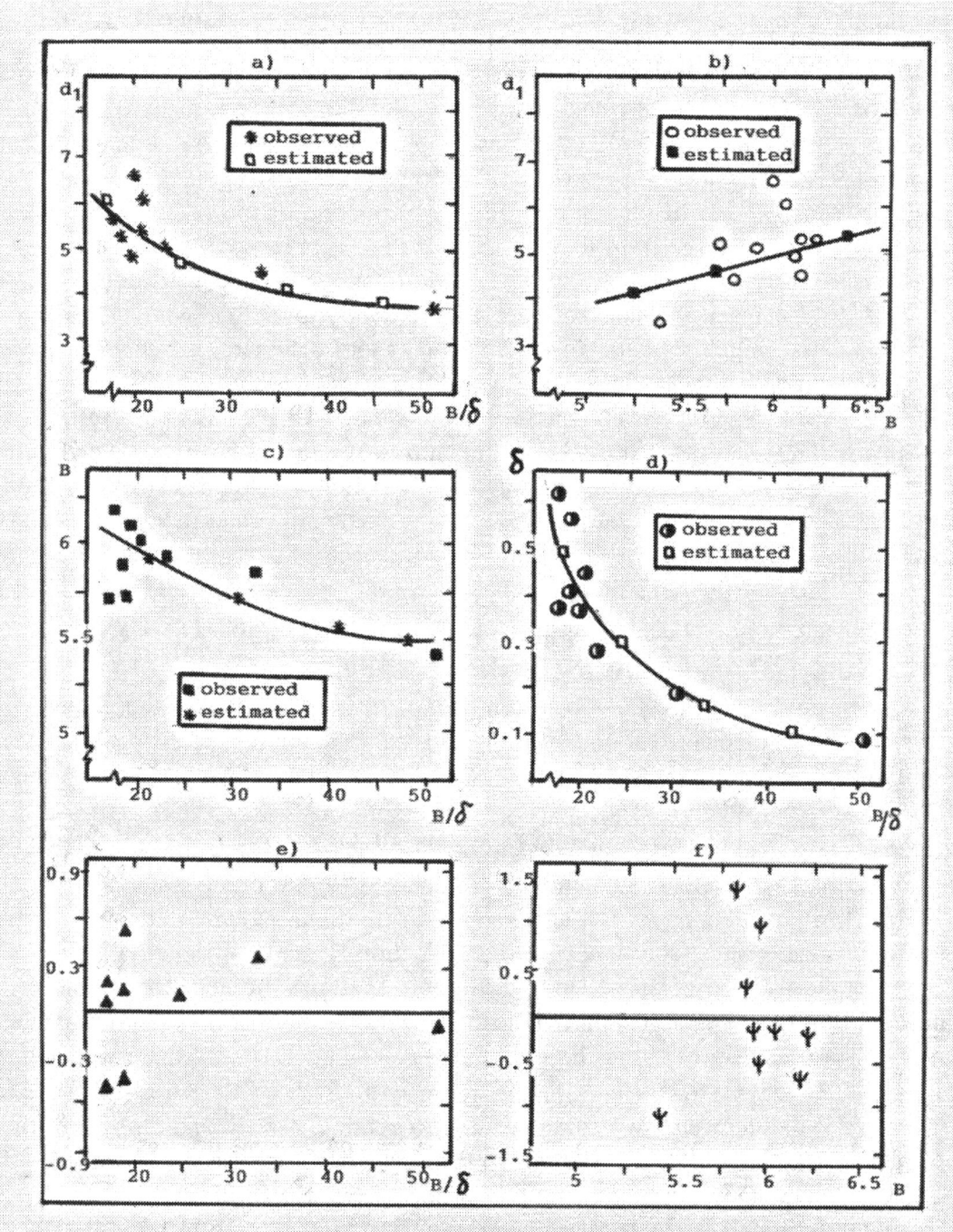

Figure 2

In Figure 2 (a) is shown the functional dependence of internal diameter (d_1) stainless chip from the ratio (B/δ) , such as width (B) of this chip to its thickness (δ) ,i.e we have the function $d_1 = \gamma_0(B/\delta)$. Analyzing of data and the statistical characteristics (coefficient of determination $R^2 = 0.894$, coefficient of correlation $r = 0.945$, standard deviation $S_{y/x'2} = 0.292$), minimization of the mean square error (min MSE= 0.066) show that this functional dependence $d_1 = \gamma_0(B/\delta)$ better submits to the non-linear than linear[3] regression

261

model with equation view of $Y_c=6.902-0.116X'_2+0.001(X'_2)^2$ or $Y_c=6.902-0.116X'_2+0.001(X_1/X_2)^2$,where $d_1=6.902-0.116(B/\delta)+$ $+0.001(B/\delta)^2$ (4) . So ,we see from Figure 2(a) that with increasing of ratio (B/δ) ,the value of internal diameter (d_1) for stainless chip decreases considerably in accordance with the non-linear regression equation view of $d_1=6.902-0.116(B/\delta)+0.001(B/\delta)^2$.

In Figure 2(b) is shown the functional dependence of internal diameter (d_1) stainless chip from its width (B). So ,we see from Figure 2(b) that this functional dependence $d_1=\varphi_1(B)$ has the linear regression model with equation view of $Y_c=0.06+0.863X$ or $d_1=0.06+0.863B$. Analysis of data shown in Figure 2(b) shows that with increasing of width (B) stainless chip, the value of internal diameter (d_1) increases considerably in accordance with the linear regression equation view of $d_1=0.06+0.863B$.

In Figure 2 (c) is shown the functional dependence of width (B) stainless chip from the ratio of values (B/δ) ,such as width (B) of this chip to its thickness (δ) ,i.e we have the function view of $B=\gamma_2(B/\delta)$. Analyzing of data and also the scatter plots shown in Figure 2 (c),and also the statistical characteristics (coefficient of determination $R^2=0.516$,coefficient of correlation r=0.718),minimization of the mean absolute deviation (min MAD=0) ,show that this functional dependence $B=\gamma_2(B/\delta)$ better submits to the non-linear than linear[4] regression model with equation view of $Y_c=6.412(X'_2)^{-0.033}$ or $Y_c=6.412(X_1/X_2)^{-0.033}$,where $B=6.412(B/\delta)^{-0.033}$ (5).

And as we see from Figure 2(c) that between two parameters ,such as width(B) and ratio (B/δ) there is the average correlation (r=0.718) and this function can be expressed in view of the non-linear regression equation $B=6.412(B/\delta)$ which shows that with increasing of ratio (B/δ) ,the value of width (B) decreases considerably.

In Figure 2 (d) is shown the functional dependence of thickness (δ) stainless chip from its ratio (B/δ) , such as width (B) of this chip to its thickness (δ) ,i.e has a place the function view of $\delta=\gamma_1(B/\delta)$. Analyzing of data and also the scatter plots shown in Figure 2 (d) ,and the statistical characteristics (coefficient of determination $R^2=1.0$,coefficient of correlation r=1.0),minimization of the mean square error (min MSE=0),minimization of the mean absolute deviation (min MAD=0) show that this functional dependence $\delta=\gamma_1(B/\delta)$ better submits to the non-linear than linear[5] regression model with equation view of $Y_c=6.457(X'_2)^{-1.035}$ or $Y_c=6.457(X_1/X_2)^{-1.035}$,where $\delta=6.457(B/\delta)^{-1.035}$ (6).And as we see from Figure 2 (d) that between two parameters, such as thickness(δ) and ratio (B/δ) there is the excellent correlation (r=1.0). And this function can be expressed in view of the non-linear regression equation $\delta=6.457(B/\delta)^{-1.035}$ which shows that with increasing of ratio (B/δ) ,the value of thickness (δ) decreases considerably.

3-Linear regression model has the equation view of $d_1=6.2-0.053(B/\delta)$
4-Linear regression model has the equation view of $B=6.07-0.01(B/\delta)$
5-Linear regression model has the equation view of $\delta=0.576-0.01(B/\delta)$

In Figures2 (e) and 2(f) are illustrated the residual plots (residual versus B/δ and B) of the functional dependencies $d_1=\gamma_0(B/\delta)$ and $d_1=\varphi_1(B)$ accordingly.

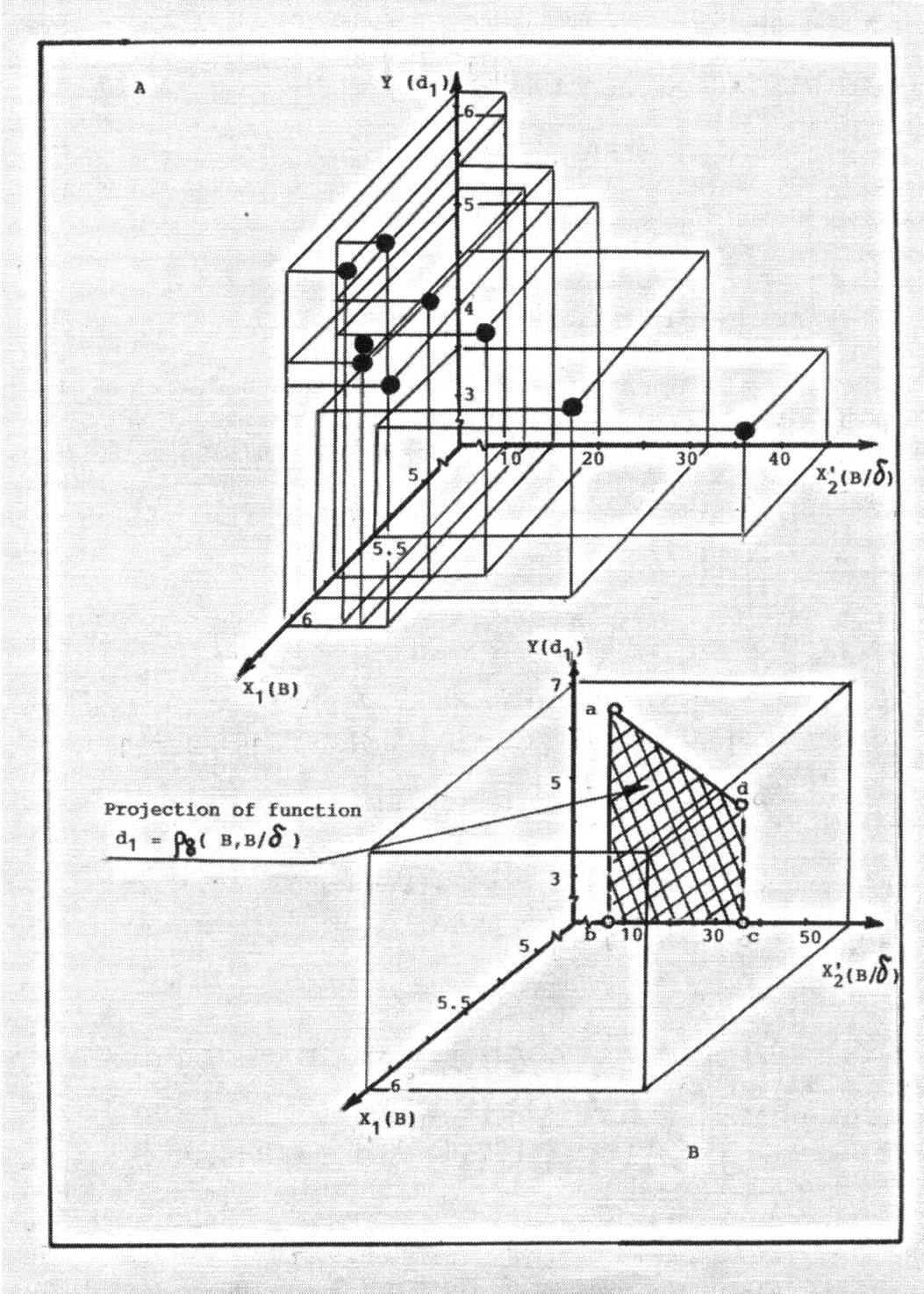

Figure 3

Analyzing the Figure 3(A) we see that functional surface $d_1=\rho_8(B,B/\delta)$ is shown in three-dimensional drawing for the linear regression model with the equation view of $Y_c=16.963-1.773X_1-0.073X'_2$,where $Y_c=16.963-1.773X_1-0.073(X_1/X_2)$,where $d_1=16.963-1.773B-0.073(B/\delta)$.

And Figure 3 (B) shows that this functional surface $d_1=\rho_8(B,B/\delta)$ has view of trapezium on the place YOX'_2 with such coordinates of their points(sizes in millimeters): a $(X_{1,1}=0,X_{2,1}=9.58,Y_1=6.07)$; b$(X_{1,2}=0,X_{2,2}=9.58,Y_2=0)$; c$(X_{1,3}=0,X_{2,3}=32.17,Y_3=0)$;d$(X_{1,4}=0,X_{2,4}=32.17,Y_4=4.35)$.

Analysis of functional surface $d_1=\rho_8(B,B/\delta)$

From Figure 3 (A) we see that functional surface $d_1=\rho_8(B,B/\delta)$,on which are situated the points of the linear regression equation view of $Y_c=16.963-1.773X_1-0.073X'_2$ or $Y_c=16.963-1.773X_1-0.073(X_1/X_2)$,where $d_1=16.963-1.773B-0.073(B/\delta)$has the following coordinates of their peaks in three- dimensional drawing (sizes in millimeters): a'($X_{1,1}=5.75$,$X_{2,1}=9.58,Y_1=6.07)$; b'$(X_{1,2}=5.75,X_{2,2}=9.58,Y_2=0)$;c'$(X_{1,3}=5.79$, $X_{2,3}=32.17$, $Y_3=0)$,d'$(X_{1,4}=5.79,X_{2,4}=32.17,Y_4=4.35)$.

1.The module of vectors are equal $|\mathbf{a'b'}|=[(X_{1,2}-X_{1,1})^2+(X_{2,2}-X_{2,1})^2+(Y_2-Y_1)^2]^{1/2}$ (7) , $|\mathbf{b'c'}|=[(X_{1,3}-X_{1,2})^2+(X_{2,3}-X_{2,2})^2+(Y_3-Y_2)^2]^{1/2}$ (8) , $|\mathbf{c'd'}|=[(X_{1,4}-X_{1,3})^2+$ $+(X_{2,4}-X_{2,3})^2+(Y_4-Y_3)^2]^{1/2}$ (9) , $|\mathbf{d'a'}|=[(X_{1,4}-X_{1,1})+(X_{2,4}-X_{2,1})^2+(Y_4-Y_1)]^{1/2}$ (10) and $\mathbf{S_{a'b'c'd'}}=0.5[|\mathbf{d'e'}|\cdot|\mathbf{a'e'}|]+|\mathbf{c'd'}|\cdot|\mathbf{b'c'}|=0.5[|\mathbf{b'c'}|\cdot(|\mathbf{a'b'}|-$ $-|\mathbf{c'd'}|)]+|\mathbf{c'd'}|\cdot|\mathbf{b'c'}|$ (11) .

At data we have $|a'b'|=6.07$, $|b'c'|=22.59, |c'd'|=4.35, |d'a'|=22.66$ and $S_{a'b'c'd'}=117.69$ mm^2.

All results of computations for the linear regression equation view of $Y_c=16.963--1.773X_1-0.073X'_2$ or $d_1=16.963-1.773B-0.073(B/\delta)$ are given in Table 2.

So ,the parameters of this functional surface $d_1=\rho_8(B,B/\delta)$ are the following:

1.Function $d_1=\rho_8(B,B/\delta)$ better submits to the linear regression model with equation view of $Y_c=16.963-1.773X_1-0.073X'_2$ or $Y_c=16.963-1.773X_1-0.073(X_1/X_2)$,where $d_1=16.963--1.773B-0.073(B/\delta)$ with such statistical characteristics:
  - Coefficients of determination $R^2=0.929$
  - Coefficient of correlation r=0.964
  - Standard deviation $S_{y/x1,x'2}=0.257$
  - Minimization of the mean square error (min MSE=0.044)
  - Minimization of the mean absolute deviation (min MAD=0).

2.The total area of functional surface $d_1=\rho_8(B,B/\delta)$ is equal $\sum S=117.69$ mm^2 with such coordinates in three-dimensional drawing for this surface:
a'($X_{1,1}=5.75$,$X_{2,1}=9.58$,$Y_1=6.07)$; b'$(X_{1,2}=5.75$,$X_{2,2}=9.58,Y_2=0)$; c' ($X_{1,3}=5.79$, $X_{2,3}=32.17,Y_3=0)$; d'$(X_{1,4}=5.79,X_{2,4}=32.17, Y_4=4.35)$.

Table 2 Evaluation of regression equation $Y_c=16.963-1.773X_1-0.073X'_2$

A. Mean , variance and standard deviation

Variable	Mean	Variance	Standard deviation
X_1	5.836	0.515	0.718
X'_2	20.792	1579.41	39.74
Y	5.098	5.612	2.369

B .Results of multiple regression of Y on X_1 and X'_2

Parameter	Variable	Coefficients	Standard error	T-value
β_1	X_1	-1.773	0.239	-7.42
β_2	X'_2	-0.073	13.24	-0.006

C. Analysis of variance results

Regression
- Degrees of freedom 2
- Sum of squares 5.211
- Mean squares 2.606

Error
- Degrees of freedom 6
- Sum of squares 0.397
- Mean squares 0.066

F-value* 39.484

* Since F=39.484>[$F_{0.05,2,6}$=5.14] we can reject the hypothesis that both β_1 and β_2 are zero.

D. Determination of residuals

Number	Observed	Estimated	Residual
1	5.36	5.68	-0.32
2	5.22	4.97	0.25
3	3.52	3. 52	0
4	4.51	4.35	0.16
5	4.98	5.13	-0.15
6	4.74	4.91	-0.17
7	5.15	5.18	-0.03
8	5.96	6.09	-0.13
9	6.44	6.07	0.37

C. Modification function $d_1=\rho_9(B,B/\omega)$

Analysis of data and the statistical characteristics (coefficient of determination R^2=0.116 ,cocfficient of correlation r=0.341,standard   deviation $S_{y/x1,x'2}$=0.908) and   also minimization of the mean square error (min MSE=0.551) and minimization of the mean absolute deviation (min MAD=0.003) show that this functional dependence $d_1=\rho_9(B,B/\omega)$ better submits to the linear than non-linear[6] regression model with equation view of
$$Y_c=8.65-1.33X_1+2.54X'_2 \quad \text{or} \quad Y_c=8.65-1.33X_1+2.54(X_1/X_2) \quad \text{,where}$$
$$d_1=8.65-1.33B+2.54(B/\omega) \quad (12)\,.$$

6-Non-linear regression model has equation view of $d_1=117.33-37.69B+3.13B^2+0.58(B/\omega)$

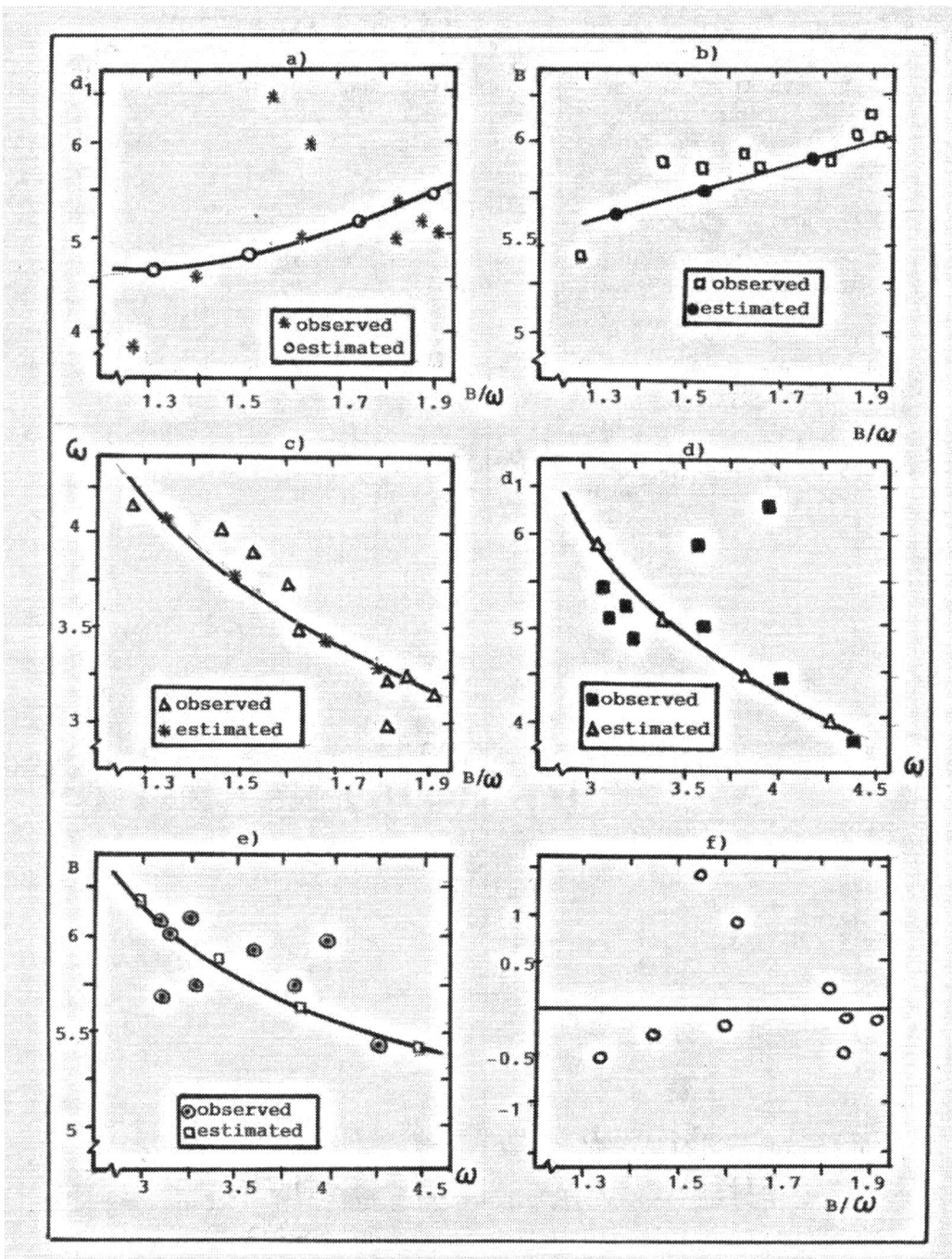

Figure 4

In Figure 4(a) is shown the functional dependence of internal diameter (d_1) stainless chip from the ratio (B/ω) , such as width(B) of this chip to its number of wraps (ω) ,i.e we have the function $d_1=\varphi_1(B/\omega)$. Analyzing of data and the statistical characteristics (coefficient of determination $R^2=0.274$,coefficient of correlation $r=0.524$,standard deviation $S_{y/x'2}=0.763$), minimization of the mean square error (min MSE $=0.452$) show that this functional dependence $d_1=\varphi_1(B/\omega)$ better submits to the non-linear than linear[7] regression model with

266

equation view of $Y_c= -3.061+9.097X'_2-2.478(X'_2)^2$ or $Y_c= -3.061+9.07(X_1/X_2)$ $-2.478(X_1/X_2)^2$, where $d_1= -3.061+9.07(B/\omega)- -2.478(B/\omega)^2$ (13).So ,we see from Figure 4(a) that with increasing of ratio (B/ω) ,the value of internal diameter (d_1) for stainless chip increases considerably in accordance with the non-linear regression equation view of $d_1= -3.061+9.097(B/\omega) - -2.478(B/\omega)^2$.

In Figure 4(b) is shown the functional dependence of width (B) stainless chip from the ratio (B/ω) ,such as width (B) of this chip to its number of wraps (ω) ,i.e we have the function $B=\varphi_2(B/\omega)$.Analyzing of data and the statistical characteristics (coefficient of determination R^2=0.746,coefficient of correlation r=0.863,standard deviation $S_{y/x'2}$=0.136) ,minimization of the mean square error(min MSE=0.015) ,minimization of the mean absolute deviation (min MAD=0) show that this functional dependence $B=\varphi_2(B/\omega)$ better submits to the linear than non-linear[8] regression model with equation view of $Y_c=4.195+0.989X'_2$ or $Y_c=4.195+0.989(X_1/X_2)$,where $B=4.195+0.989(B/\omega)$ (14) .So ,we see from Figure 4(b) that with increasing of ratio (B/ω) ,the value of width(B) for stainless chip increases considerably in accordance with the linear regression equation view of $B=4.195+ 0.989(B/\omega)$.

In Figure 4(c) is shown the functional dependence of the number wraps (ω) stainless chip from the ratio of values (B/ω) ,such as the width(B) of this chip to its number of wraps(ω) ,i.e we have the function view of $\omega=\varphi_3(B/\omega)$. Analyzing of data and also the scatter plots shown in Figure 4(c) ,and the statistical characteristics (coefficient of determination R^2=0.961 ,coefficient of correlation r=0.980,standard deviation $S_{y/x'2}$=0.078),minimization of the mean square error (min MSE=0.005), minimization of the mean absolute deviation (min MAD=0.002),show that this functional dependence $\omega=\varphi_3(B/\omega)$ better submits to the non-linear than linear[9] regression model with equation view of $Y_c= 5.152 (X'_2)^{-0.75}$ or $Y_c=5.152(X_1/X_2)^{-0.75}$, where $\omega=5.152(B/\omega)^{-0.75}$ (15) .And as we see from Figure 4 (c) that between two parameters ,such as the number of wraps (ω) stainless chip and ratio (B/ω) there is the good correlation (r=0.980). And this function $\omega=\varphi_3(B/\omega)$ can be expressed in view of the non-linear regression equation $\omega=5.152(B/\omega)^{-0.75}$ which shows that with increasing of ratio (B/ω) ,the value of number wraps (ω) decreases considerably. In Figure 4(d) is shown the functional dependence of internal diameter (d_1) stainless chip from its number of wraps (ω).So ,we see from Figure 4(d) that this functional dependence $d_1=\phi_3(\omega)$ has the non-linear regression model with equation view of $Y_c=20.32X^{-1.102}$ or $d_1=20.32\omega^{-1.102}$. Analysis of data shown in Figure 4 (d) show that with increasing of the number wraps (ω) stainless chip ,the value of internal diameter (d_1) decreases considerably in accordance with the non-linear regression equation view of $d_1=20.32\omega^{-1.102}$.

7-Linear regression model has the equation view of d_1=2.889+1.332(B/ω)
8-Non-linear regression model has the equation view of B=5.117(B/ω)$^{0.261}$
9- Linear regression model has the equation view of ω=6.271−1.635(B/ω)

In Figure 4(e) is shown the functional dependence of width (B) stainless chip from the number of chip wraps (ω). So, we see from Figure 4(e) that this functional dependence $B=\gamma_2(\omega)$ has the non-linear regression model with equation view of $Y_c=703.1X^{-3.0}$ or $\omega=703.1B^{-3.0}$,where $B=(0.001\omega)^{-0.333}$.Analysis of data shown in Figure 4(e) shows that with increasing of the number wraps (ω) stainless chip, the value of width (B) decreases considerably in accordance with the non-linear regression equation view $B=(0.001\omega)^{-0.333}$.In Figure 4(f) is illustrated the residual plot (residual versus B/ω) of functional dependency $d_1=\varphi_1(B/\omega)$.

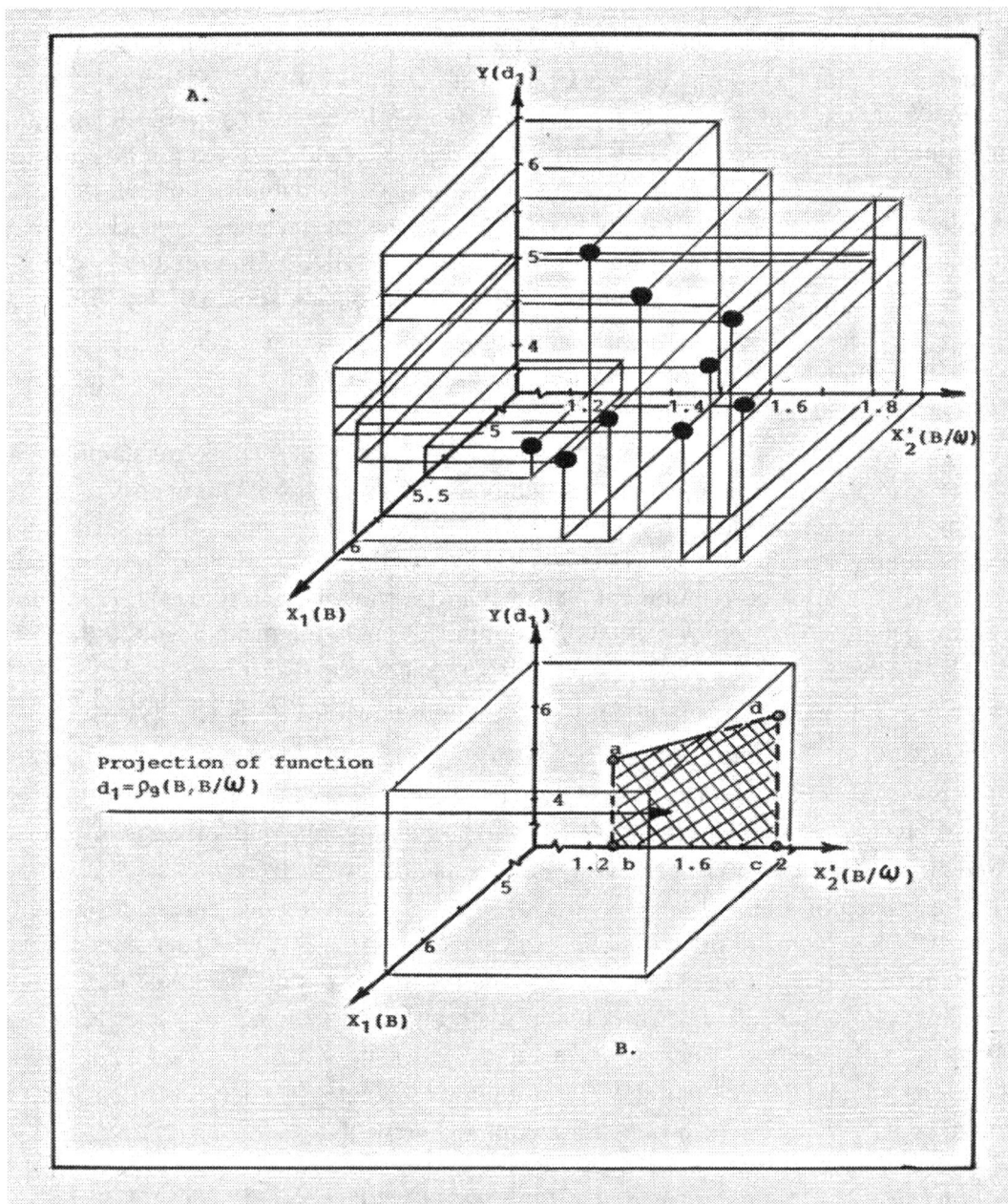

Figure 5

Analyzing the Figure 5 (A) ,we see that functional surface $d_1=\rho_9(B,B/\omega)$ is shown in three-dimensional drawing for the linear regression model with equation view of $Y_c=8.65-1.33X_1+2.54X'_2$ or $Y_c=8.65-1.33X_1+2.54(X_1/X_2)$,where $d_1=8.65-1.33B+$ $+2.54(B/\omega)$.

And Figure 5(B) shows that this functional surface $d_1=\rho_9(B,B/\omega)$ has view of trapezium on the place YOX'_2 with such coordinates of their points (sizes in millimeters):
$a(X_{1,1}=0,X_{2,1}=1.27,Y_1=4.74)$; $b(X_{1,2}=0,X_{2,2}=1.27,Y_2=0)$; $c(X_{1,3}=0,\ \ ,X_{2,3}=1.90\ \ ,Y_3=0)$; $d(X_{1,4}=0,X_{2,4}=1.90,Y_4=5.37)$.

Analysis of functional surface $d_1=\rho_9(B,B/\omega)$

From Figure 5(A) we see that functional surface $d_1=\rho_9(B,B/\omega)$,on which are situated the points of the linear regression equation view of $Y_c=8.65-1.33X_1+2.54X'_2$ or $Y_c=8.65-1.33X_1+2.54(X_1/X_2)$,where $d_1=8.65-1.33B+2.54(B/\omega)$ has the coordinates of their peaks in three- dimensional drawing(sizes in millimeters): $a'(X_{1,1}=5.37,X_{2,1}=1.27,$ $,Y_1=4.74)$; $b'(X_{1,2}=5.37,X_{2,2}=1.27,Y_2=0)$; $c'(X_{1,3}=6.10,X_{2,3}=1.90,Y_3=0)$; $d'(X_{1,4}=6.10,$ $,X_{2,4}=1.90,Y_4=5.37)$.

The module of vectors $|a'b'|,|b'c'|,|c'd'|$ and $|d'a'|$ can be calculated by formulas (7),(8), ,(9) and (10). Area of functional surface is equal $\mathbf{S_{a'b'c'd'}=0.5\{[|c'd'|-|a'b'|]\cdot|a'e'|\}+}$ $\mathbf{+|a'b'|\cdot|b'c'|}$ (**16**). At above-named data we have : $|a'b'|=4.74,|b'c'|=0.96,$ $,|c'd'|=5.37,|d'a'|=1.15$ and $S_{a'b'c'd'}=4.85mm^2$.

All results of computations for the linear regression equation view of $Y_c=8.65-1.33X_1+2.54X'_2$ or $d_1=8.65-1.33B+2.54(B/\omega)$ are given in Table 3.

So , the parameters of this functional surface $d_1=\rho_9(B,B/\omega)$ are the following:

1.Function $d_1=\rho_9(B,B/\omega)$ better submits to the linear regression model with equation view of $Y_c=8.65-1.33X_1+2.54X'_2$ or $Y_c=8.65-1.33X_1+2.54(X_1/X_2)$,where $d_1=8.65-1.33B+2.54(B/\omega)$ with such statistical characteristics :
- Coefficient of determination $R^2=0.116$
- Coefficient of correlation $r=0.341$
- Standard deviation $S_{y/x1,x'2}=0.908$
- Minimization of the mean square error (min MSE=0.551)
- Minimization of the mean absolute deviation (min MAD=0.003.

2.The total area of functional surface $d_1=\rho_9(B,B/\omega)$ is equal $\sum S=4.85mm^2$ with such coordinates in three-dimensional drawing for this surface:
a' ($X_{1,1}=5.37$, $X_{2,1}=1.27$, $Y_1=4.74$);
b' ($X_{1,2}=5.37$, $X_{2,2}=1.27$, $Y_2=0$);
c' ($X_{1,3}=6.10$, $X_{2,3}=1.90$,$Y_3=0$);
d' ($X_{1,4}=6.10$, $X_{2,4}=1.90$, $Y_4=5.37$).

Table 3 Evaluation of regression equation $Y_c=8.65-1.33X_1+2.54X'_2$

A. Mean, variance and standard deviation

Variable	Mean	Variance	Standard deviation
X_1	5.836	0.515	0.718
X'_2	1.659	0.390	0.624
Y	5.098	5.612	2.369

B. results of multiple regression of Y on X_1 and X'_2

Parameter	Variable	Coefficients	Standard error	T-value
β_1	X_1	-1.33	0.239	-5.56
β_2	X'_2	2.54	0.208	12.21

C. Analysis of variance results

Regression

Degrees of freedom 2

Sum of squares 0.621

Mean squares 0.311

Error

Degrees of freedom 6

Sum of squares 4.987

Mean squares 0.831

F-value * 0.37

* Since F=0.37 < [$F_{0.05,2,6}$=5.14] ,we can reject the hypothesis that both β_1 and β_2 are zero.

D. Determination of residuals

Number	Observed	Estimated	Residual
1	5.36	5.63	-0.27
2	5.22	5.23	-0.01
3	3.52	4.74	-1.22
4	4.51	4.64	-0.13
5	4.98	4.99	-0.01
6	4.74	5.29	-0.55
7	5.15	5.37	-0.22
8	5.96	5.17	0.79
9	6.44	4.88	1.56

D. Modification function $d_1=\rho_{10}(B,\delta/\omega)$

Analysis of data and the statistical characteristics (coefficient of determination R^2=0.650,coefficient of correlation r=0.806,standard deviation $S_{y/x',x'2}$=0.572) and also minimization of the mean square error (min MSE=0.218) shows that this functional dependence $d_1=\rho_{10}(B,\delta/\omega)$ better submits to the linear than non-linear[10] regression model with equation view of $Y_c=16.21-2.34X_1+24.10X'_2$ or $Y_c=16.21-2.34X_1 + +24.1(X_2/X_3)$, where $d_1=16.21-2.34B+24.10(\delta/\omega)$ (17).

10- *Non-linear regression model has the equation view of $d_1=3.747+0.159B+0.013B^2$*

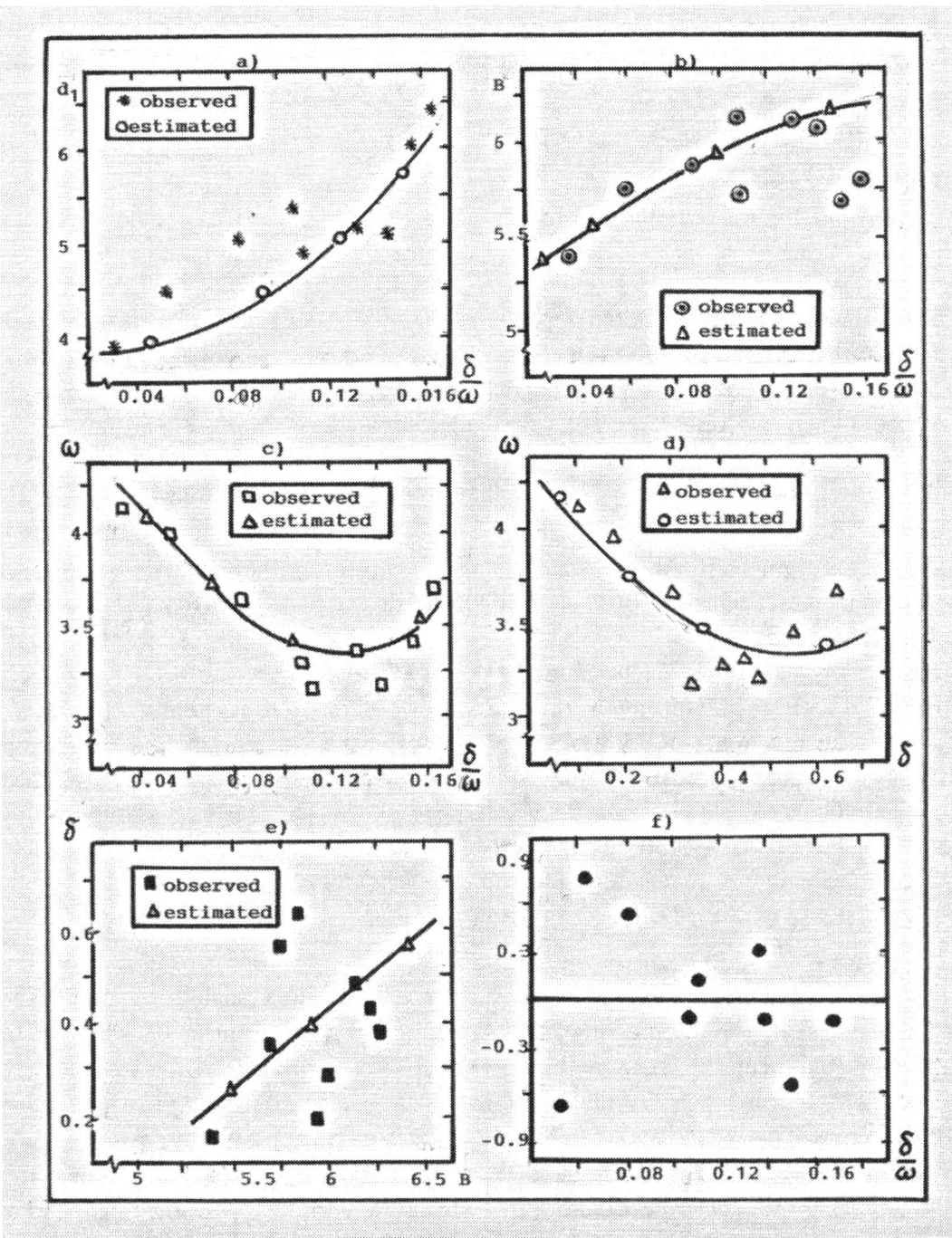

Figure 6

In Figure 6(a) is shown the functional dependence of internal diameter (d_1) stainless chip from the ratio (δ/ω),such as thickness (δ) of this chip to its number of wraps (ω) ,i.e we have the function $d_1 = \varphi_4(\delta/\omega)$. Analysis of data and the statistical characteristics (coefficient of determination $R^2 = 0.778$, coefficient of correlation $r = 0.882$, standard deviation $S_{y/x'2} = 0.422$), minimization of the mean square error (min MSE = 0.138) and

271

minimization of the mean absolute deviation (min MAD=0.004), shows that this functional dependence $d_1=\varphi_4(\delta/\omega)$ better submits to the non-linear than linear[11] regression model with equation view of $Y_c=3.232+19.932X'_2-19(X'_2)^2$ or $Y_c=3.232+19.932(X_1/X_2)-19(X_1/X_2)^2$,where $d_1=3.232+19.932(\delta/\omega)-19(\delta/\omega)^2$ (18).

So, we see from Figure 6(a) that with increasing of ratio (δ/ω) ,the value of internal diameter (d_1) for stainless chip increases considerably in accordance with the non-linear regression equation view of $d_1=3.232+19.932(\delta/\omega)-19(\delta/\omega)^2$.

In Figure 6(b) is shown the functional dependence of width(B) stainless chip from the ratio (δ/ω) ,such as thickness (δ) of this chip to its number of wraps(ω) ,i.e we have the function $B=\varphi_5(\delta/\omega)$. Analysis of data and the statistical characteristics (coefficient of determination $R^2=0.359$,coefficient of correlation r=0.599,standard deviation $S_{y/x'2}=0.217$), minimization of the mean square error (min MSE=0.037) shows that this functional dependence $B=\varphi_5(\delta/\omega)$ better submits to the non-linear than linear[12] regression model with equation view of $Y_c=6.58(X'_2)^{0.05}$ or $Y_c=6.58(X_1/X_2)^{0.05}$, where $B=6.58(\delta/\omega)^{0.05}$ (19).

So ,we see from Figure 6(b) that with increasing of ratio (δ/ω) ,the value of width(B) for stainless chip increases considerably in accordance with the non-linear regression equation view of $B=6.58(\delta/\omega)^{0.05}$.

In Figure 6(c) is shown the functional dependence of the number wraps(ω) stainless chip from the ratio of values (δ/ω) ,such as thickness (δ) of this chip to its number of wraps(ω) ,i.e we have the function view of $\omega=\varphi_6(\delta/\omega)$.Analysis of data and also the scatter plots shown in Figure 6(c) ,and the statistical characteristics (coefficient of determination $R^2=0.633$,coefficient of correlation r=0.796 ,standard deviation $S_{y/x'2}=0.241$),minimization of the mean square error (min MSE=0.045) shows that this functional dependence $\omega=\varphi_6(\delta/\omega)$ better submits to the non-linear than linear[13]regression model with equation view of $Y_c=2.624(X'_2)^{-0.126}$ or $Y_c=2.624(X_1/X_2)^{-0.126}$,where $\omega=2.624(\delta/\omega)^{-0.126}$ (20) .

And as we see from Figure 6 (c) that between two parameters ,such as the number of wraps (ω) stainless chip and ratio (δ/ω) there is the good correlation (r=0.796). And this function $\omega=\varphi_6(\delta/\omega)$ can be expressed in view of the non-linear regression equation $\omega=2.624(\delta/\omega)^{-0.126}$ which shows that with increasing of ratio (δ/ω) ,the value of the number of wraps(ω) decreases considerably.

In Figure 6(d) is shown the functional dependence of number wraps (ω) stainless chip from its thickness (δ).So ,we see from Figure 6(d) that this functional dependence has the non-linear regression model with equation view of $Y_c=3.09X^{-0.123}$ or $\omega=3.09\delta^{-0.123}$. Analysis of data shown in Figure 6(d) shows that with increasing of thickness (δ) stainless chip ,the value of number wraps (ω) decreases considerably in accordance with the non-linear regression equation view of $\omega=3.09\delta^{-0.123}$.

11-Linear regression model has the equation view of $d_1=3.23+17.63(\delta/\omega)$
12-Linear regression model has the equation view of $B=5.518+3(\delta/\omega)$
13- linear regression model has the equation view of $\omega=4.215-6.188(\delta/\omega)$

In Figure 6(e) is shown the functional dependence of thickness (δ) stainless chip from width (B) of this chip. So , we see from Figure 6(e) that this functional dependence δ=f(B) has the linear regression model with equation view of $Y_c = -1.236+0.275X$ or
$\delta = -1.236+0.275B$. Analysis of data shown in Figure 6(e) shows that with increasing of width (B) stainless chip ,the value of thickness (δ) increases considerably in accordance with the linear regression equation view of $\delta = -1.236+0.275B$.
In Figure 6(f) is illustrated the residual plot (residual versus δ/ω) of functional dependency $d_1 = \varphi_4(\delta/\omega)$.

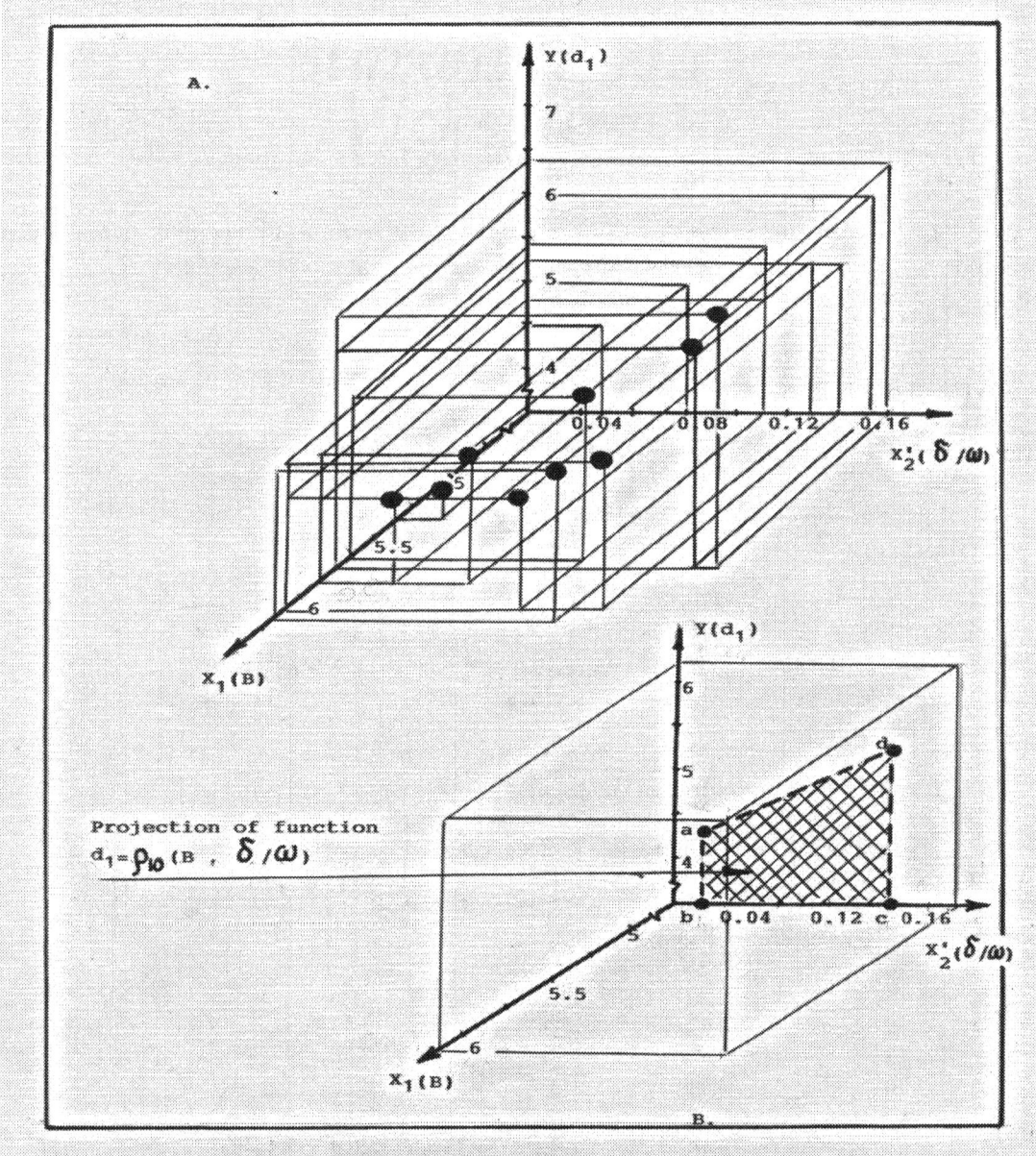

Figure 7

273

Analyzing the Figure 7(A) ,we see that functional surface $d_1=\rho_{10}(B,\delta/\omega0$ is shown in three-dimensional drawing for the linear regression model with equation view of $Y_c=16.21-2.34X_1+24.1X'_2$ or $Y_c=16.21-2.34X_1+24.1(X_2/X_3)$,where $d_1=16.21-2.34B++24.1(\delta/\omega)$. And besides the Figure 7(A) shows that this functional dependence has view of trapezium on the place YOX'$_2$ with such coordinates of their points:
$a(X_{1,1}=0,X_{2,1}=0.02,Y_1=4.22)$; $b(X_{1,2}=0$,$X_{2,2}=0.02,Y_2=0)$; $c(X_{1,3}=0,X_{2,3}=0.14,Y_3=0)$; $d(X_{1,4}=0,X_{2,4}=0.14,Y_4=5.39)$.

Analysis of functional surface $d_1=\rho_{10}(B,\delta/\omega)$

From Figure 7(A) we see that functional surface $d_1=\rho_{10}(B,\delta/\omega)$,on which are situated the points of the linear regression equation view of $Y_c=16.21-2.34X_1+24.1X'_2$ or $Y_c=16.21-2.34X_1+24.1(X_2/X_3)$,where $d_1=16.21-2.34B+24.1(\delta/\omega)$ has the coordinates of their peaks in three-dimensional drawing (sizes in millimeters): $a'(X_{1,1}=5.37,X_{2,1}=0.02,Y_1=4.22)$; $b'(X_{1,2}=5.37,X_{2,2}=0.02,Y_2=0)$; $c'(X_{1,3}=6.10,X_{2,3}=0.14,$,$Y_3=0)$; $d'(X_{1,4}=6.10,X_{2,4}=0.14,Y_4=5.39)$.
The module of vectors $|a'b'|,|b'c'|,|c'd'|,|d'a'|$ and area of functional surface $S_{a'b'c'd'}$ can be calculated by formulas (7),(8),(9),(10) and (16). So, at data we have $|a'b'|$ =4.22 ,$|b'c'|$=0.74,$|c'd'|$=5.39 ,$|d'a'|$=1.38 and $S_{a'b'c'd'}$=3.55mm^2.
All results of computations for the linear regression equation view of $Y_c=16.21-2.34X_1+24.1X'_2$ or $d_1=16.21-2.34B+24.1(\delta/\omega)$ are given in Table 4.

Table 4 Evaluation of regression equation $Y_c=16.21-2.34X_1+24.1X_2$'

A. Mean, variance and standard deviation

Variable	Mean	Variable	Standard deviation
X_1	5.836	0.515	0.718
X'_2	0.106	0.017	0.130
Y	5.098	5.612	2.369

B. Results of multiple regression of Y on X_1 and X'_2

Parameter	Variable	Coefficients	Standard error	T-value
β_1	X_1	-2.34	0.239	-9.79
β_2	X'_2	24.10	0.043	560.47

C. Analysis of variance results

Regression: Degrees of freedom 2,sum of squares 3.645, mean squares 1.823
Error: Degrees of freedom 6 ,sum of squares 1.963,mean squares 0.327
F-value *   5.574

* Since F=5.574>[F$_{0.05,2,6}$=5.14] we can reject the hypothesis that both $\beta_1$and β_2 are zero.

D. Determination of residuals

Number	Observed	Estimated	Residual
1	5.36	5.60	-0.24
2	5.22	4.89	0.33
3	3.52	4.22	-0.70
4	4.51	3.75	0.76
5	4.98	4.45	0.53
6	4.74	4.64	0.10
7	5.15	5.39	-0.24
8	5.96	6.56	-0.60
9	6.44	6.58	-0.14

So, the parameters of this functional surface $d_1=\rho_{10}(B,\delta/\omega)$ are the following:

1.Function $d_1=\rho_{10}(B,\delta/\omega)$ better submits to the linear regression model with equation view of $Y_c=16.21-2.34X_1+24.1X'_2$ or $Y_c=16.21-2.34X_1+24.1(X_2/X_3)$,where $d_1=16.21-2.34B+24.1(\delta/\omega)$ with such statistical characteristics: coefficient of determination $R^2=0.650$,coefficient of correlation $r=0.806$,standard deviation $S_{y/x1,x'2}=0.572$,minimization of the mean square error (min MSE=0.218).

2.The total area of functional surface $d_1=\rho_{10}(B,\delta/\omega)$ is equal $\sum S=3.55mm^2$ with such coordinates in three-dimensional drawing of their points for this surface (sizes in millimeters): a'($X_{1,1}=5.37,X_{2,1}=0.02,Y_1=4.22$); b'($X_{1,2}=5.37,X_{2,2}=0.02,Y_2=0$); c'($X_{1,3}=6.10,X_{2,3}=0.14,Y_3=0$); d'($X_{1,4}=6.10,X_{2,4}=0.14,Y_4=5.39$).

7.2 Dependence of internal diameter (d_1) from thickness (δ) of chip ,clearance (t) between of chip wraps(ω) and its width(B) ,i.e $d_1=\phi_2(\delta,t,B)$ and also from their modifications: $d_1=\rho_{11}(\delta,\delta/t)$, $d_1=\rho_{12}(\delta,\delta/B)$,$d_1=\rho_{13}(\delta,t/B)$.

A. Function $d_1=\phi_2(\delta,t,B)$

Analysis of data and the statistical characteristics (coefficient of determination $R^2=0.877$,coefficient of correlation $r=0.936$,standard deviation $S_{y/x1,x2,x3}=0.372$) ,minimization of the mean square error (min MSE=0.077) and minimization of the mean absolute deviation (min MAD=0) shows that this functional dependence $d_1=\phi_2(\delta,t,B)$ better submits to the linear than non-linear[14] regression model with equation view of

$$Y_c=3.802+4.629X_1-1.324X_2+0.002X_3 \text{ or } d_1=3.802+4.629\delta-1.324t+0.002B$$
(21).

In Figure 8(a) is shown the functional dependence of internal diameter (d_1) stainless chip from its thickness (δ) ,i.e we have the functional dependence $d_1=\phi(\delta)$. This dependence has the linear regression model view of $Y_c=3.338+4.762X$ or $d_1=3.338+4.762\delta$. So ,from Figure 8(a) we see that with increasing of thickness (δ) stainless chip ,the value of internal diameter(d_1) of this chip considerably increases in accordance with the linear regression equation $d_1=3.338+4.762\delta$.

In Figure 8(b) is shown the functional dependence of internal diameter (d_1) stainless chip from the clearance (t) between of chip wraps, i.e we have the functional dependence $d_1=\gamma_4(t)$. This dependence has the non-linear regression model with equation view of $Y_c=0.828X^{-0.605}$ or $t=0.828d_1^{-0.605}$,where $\mathbf{d_1=(1.207t)^{-1.653}}$ (22) . So ,we see from Figure 8(b) that with increasing of clearance (t) between of wraps for stainless chip ,the value of internal diameter (d_1) considerably decreases in accordance with the non-linear regression equation $d_1=(1.207t)^{-1.653}$.

In Figure 8(c) is shown the functional dependence of internal diameter (d_1) stainless chip from the width (B) of this chip ,i.e we have the functional dependence $d_1=\phi_1(B)$. This function has the linear regression model with equation view of $Y_c=0.06+0.863X$,where $d_1=0.06+0.863B$.

14- Non-linear regression model has the equation view of $d_1=5.158+10.98\delta-8.244\delta^2-0.412t-0.457B$

So, from Figure 8(c) we see that with increasing of width(B) stainless chip, the value of internal diameter (d_1) of this chip increases considerably in accordance with the regression equation d_1=0.06+0.863B.

In Figure 8(d) is shown the functional dependence of thickness (δ) stainless chip from its width(B) ,i.e we have the functional dependence δ=f(B). This functional dependence has the linear regression model with equation view of Y_c= −1.236+0.275X or δ=1.236+0.275B. So , from Figure 8(d) we see that with increasing of width (B) stainless chip, the value of thickness (δ) increases considerably in accordance with the regression equation view of δ= −1.236+0.275B.

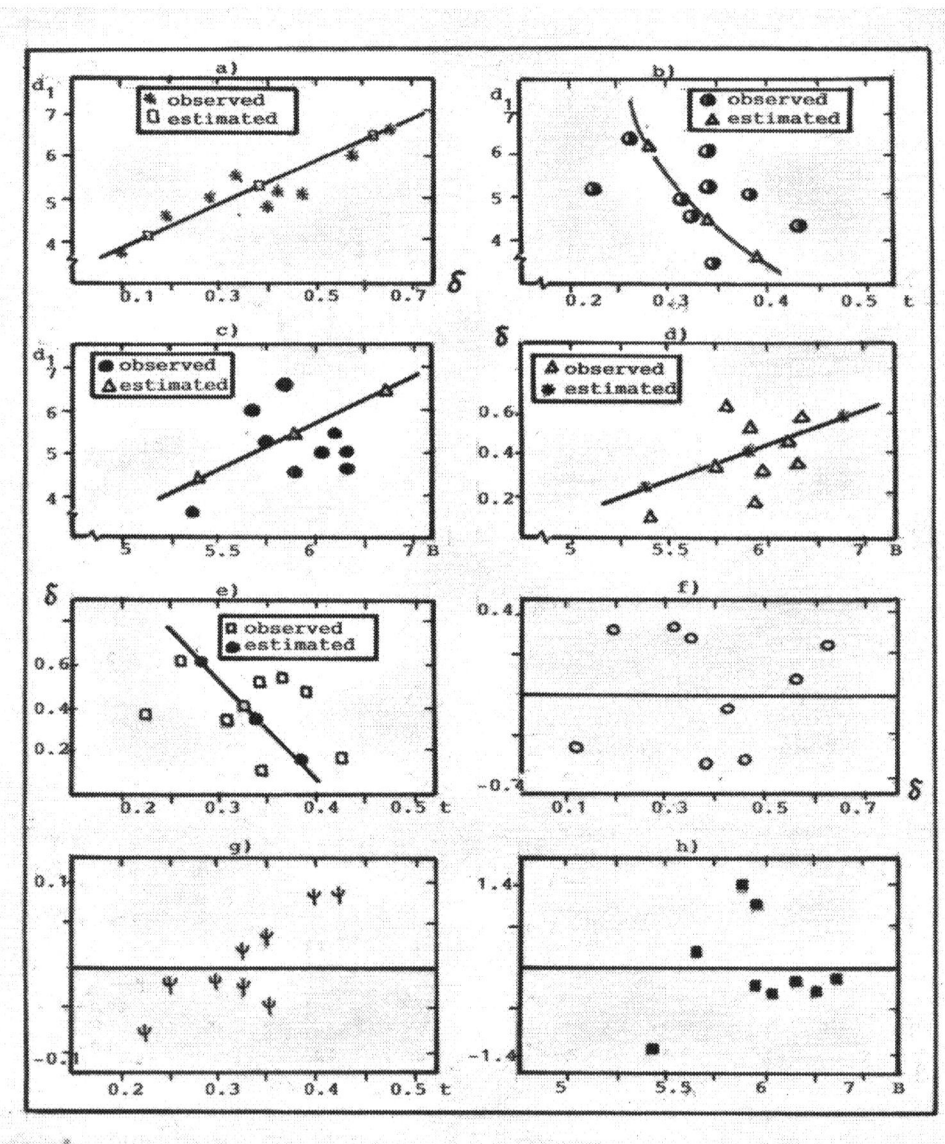

Figure 8

276

In Figure 8(e) is shown the functional dependence of thickness (δ) stainless chip from its clearance (t) between of chip wraps ,i.e we have the functional dependence $\delta=\phi(\delta)$. This function has the linear regression model with equation view of $Y_c=0.379-0.16X$ or $t=0.379-0.16\delta$,where $\delta=2.369-6.25t$. So, from Figure 8(e) we see that with increasing of clearance (t) between of chip wraps ,the value of thickness (δ) for this stainless chip decreases considerably in accordance with the linear regression equation view of $\delta=2.369-6.25t$.

In Figures 8(f),8(g) and 8(h) are illustrated the residual plots (residual versus δ,t,B) of the functional dependencies $d_1=\varphi(\delta)$, $d_1=\gamma_4(t)$ and $d_1=\varphi_1(B)$ accordingly.

All results of computations for the linear regression equation view of $Y_c=3.802+4.629X_1-1.324X_2+0.002X_3$ or $d_1=3.802+4.629\delta-1.324t+0.002B$ are given in Table 5 .

Table 5 Evaluation of regression equation $Y_c=3.802+4.629X_1-1.324X_2+0.002X_3$

A. Mean ,variance and standard deviation

Variable	Mean	Variance	Standard deviation
X_1	0.368	0.209	0.457
X_2	0.317	0.031	0.175
X_3	5.836	0.515	0.718
Y	5.098	5.612	2.369

B. Results of multiple regression of Y on X_1,X_2 and X_3

Parameter	Variable	Coefficients	Standard error	T-value
β_1	X_1	4.629	0.152	30.45
β_2	X_2	-1.324	0.058	-22.83
β_3	X_3	0.002	0.239	0.008

C. Analysis of variance results

Regression
Degrees of freedom 3
Sum of squares 4.917
Mean squares 1.639
Error
Degrees of freedom 5
Sum of squares 0.691
Mean squares 0.138
F-value* 11.88

* Since $F=11.88>[F_{0.05,3,5}=5.41]$ we are able to reject the hypothesis that parameters β_1,β_2 and β_3 are zero.

D. Determination of residuals

Number	Observed	Estimated	Residual
1	5.36	5.20	0.16
2	5.22	5.26	-0.04
3	3.52	3.83	-0.31
4	4.51	4.09	0.42
5	4.98	4.71	0.27
6	4.74	5.12	-0.38
7	5.15	5.52	-0.37
8	5.96	5.89	0.07
9	6.44	6.26	0.18

B. Modification function $d_1 = \rho_{11}(\delta, \delta/t)$

Analysis of data and the statistical characteristics (coefficient of determination R^2=0.899 ,coefficient of correlation r=0.948 ,standard deviation $S_{y/x1}$=0.308) and also minimization of the mean square error(min MSE=0.063) and minimization of the mean absolute deviation (min MAD=0) shows that this functional dependence $d_1 = \rho_{11}(\delta, \delta/t)$ better submits to the linear than non-linear[15]regression model with equation view of $Y_c=3.388+2.848X_1+0.532X'_2$ or $Y_c=3.388+2.848X_1+0.532(X_1/X_2)$, where $d_1=3.388+2.848\delta+0.532(\delta/t)$ (23).

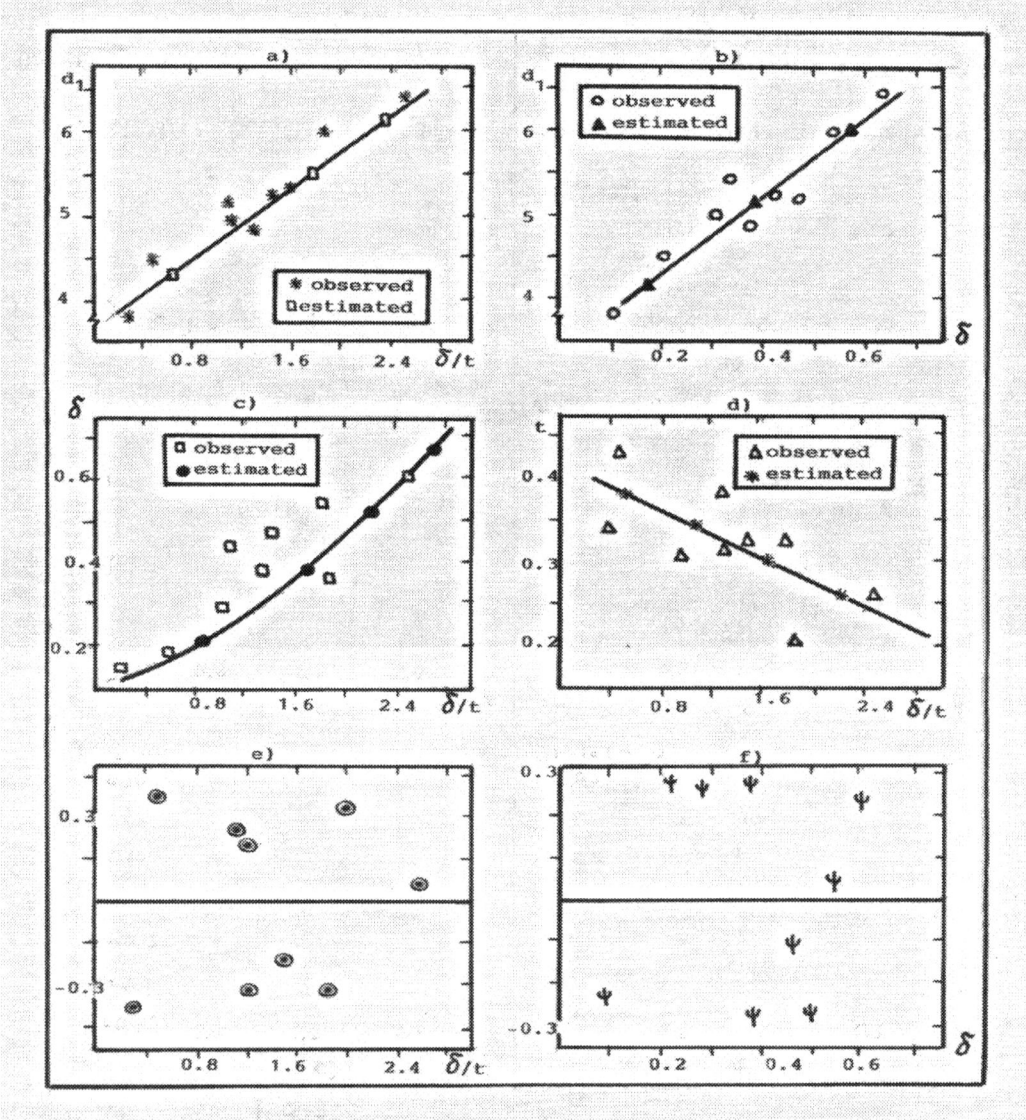

Figure 9

15-Non-linear regression model has the equation view of $d_1=3.158+4.448\delta-2.161\delta^2+0.518(\delta/t)$

In Figure 9(a) is shown the functional dependence of internal diameter (d_1) stainless chip from the ratio (δ/t) , such as thickness (δ) of this chip to its clearance (t) between of wraps ,i.e we have the function $d_1=\phi_0(\delta/t)$. Analysis of data and the statistical characteristics (coefficient of determination $R^2=0.857$,coefficient of correlation $r=0.926$,standard deviation $S_{y/x'2}=0.338$), minimization of the mean square error (min MSE=0.089) shows that this functional dependence $d_1=\phi_0(\delta/t)$ better submits to the linear than non- linear[16] regression model with equation view of **$Y_c=3.636+1.175X'_2$** or **$Y_c=3.636+1.175(X_1/X_2)$** ,where **$d_1=3.636+1.175(\delta/t)$** (24) .So ,we see from Figure 9(a) that with increasing of ratio (δ/t) ,the value of internal diameter (d_1) for stainless chip increases considerably in accordance with the linear regression equation view of $d_1=3.636+1.175(\delta/t)$.In Figure 9(b) is shown the functional dependence of internal diameter (d_1) stainless chip from its thickness (δ). This function has the linear regression model with equation view of $Y_c=3.338+4.762X$ or $d_1=3.338+4.762\delta$. Analysis of data shown in Figure (9b) shows that with increasing of thickness (δ) stainless chip ,the value of internal diameter (d_1) increases considerably in accordance with the linear regression equation view of $d_1=3.338+4.762\delta$.

In Figure 9 (c) is shown the functional dependence of thickness (δ) stainless chip from the ratio of values (δ/t),such as thickness (δ) of this chip to the clearance (t) between of chip wraps ,i.e we have the function view of $\delta=\phi_1(\delta/t)$. Analysis of data and also the scatter plots shown in Figure 9(c) ,and the statistical characteristics (coefficient of determination $R^2=0.876$,coefficient of correlation $r=0.936$,standard deviation $S_{y/x'2}=0.061$),minimization of the mean square error (min MSE=0.003) shows that this functional dependence $\delta=\phi_1(\delta/t)$ better submits to the non-linear than linear[17] regression model with equation view of **$Y_c=0.312X'_2{}^{0.807}$** or **$Y_c=0.312(X_1/X_2)^{0.807}$** , where **$\delta=0.312(\delta/t)^{0.807}$** (25).And as we see from Figure 9 (c) that between two parameters ,such as internal diameter (d_1) and ratio (δ/t) there is the good correlation ($r=0.936$). And this function $\delta=\phi_1(\delta/t)$ can be expressed in view of the non-linear regression equation $\delta=0.312(\delta/t)^{0.807}$ which shows that with increasing of ratio (δ/t) ,the value of internal diameter (d_1) increases considerably.

In Figure (9d) is shown the functional dependence of clearance (t) between of chip wraps from its ratio (δ/t) ,such as thickness (δ) of this chip to its clearance (t) between of wraps ,i.e has a place the function view of $t=\phi_2(\delta/t)$. Analysis of data and also the scatter plots shown in Figure 9(d),and the statistical characteristics (coefficient of determination $R^2=0.467$,coefficient of correlation $r=0.683$,standard deviation $S_{y/x'2}=0.048$), minimization of the mean square error (min MSE=0.002) and minimization of the mean absolute deviation (min MAD=0.002) shows that this functional dependence $t=\phi_2(\delta/t)$ better submits to the linear than non-linear[18] regression model with equation view of **$Y_c=0.403-0.069X'_2$** or **$Y_c=0.403-0.069(X_1/X_2)$** , where **$t=0.403-0.069(\delta/t)$** (26) . And as we see from Figure 9(d) that between two parameters, such as clearance (t) between of chip wraps and ratio (δ/t) there is the average correlation ($r=0.683$).

16- Non-linear regression model has the equation view of $d_1=3.516+1.419(\delta/t)-0.095(\delta/t)^2$
17-Linear regression model has the equation view of $\delta=0.087+0.226(\delta/t)$
18- Non-linear regression model has the equation view of $t=0.342(\delta/t)^{-0.193}$

And this function $t=\phi_2(\delta/t)$ can be expressed in view of the linear regression equation $t=0.403-0.069(\delta/t)$ which shows that with increasing of ratio (δ/t), the value of clearance (t) between of chip wraps decreases considerably.

In Figures 9(e) and 9(f) are illustrated the residual plots (residual versus δ/t and δ of functional dependencies $d_1=\phi_0(\delta/t)$ and $d_1=\varphi(\delta)$ accordingly.

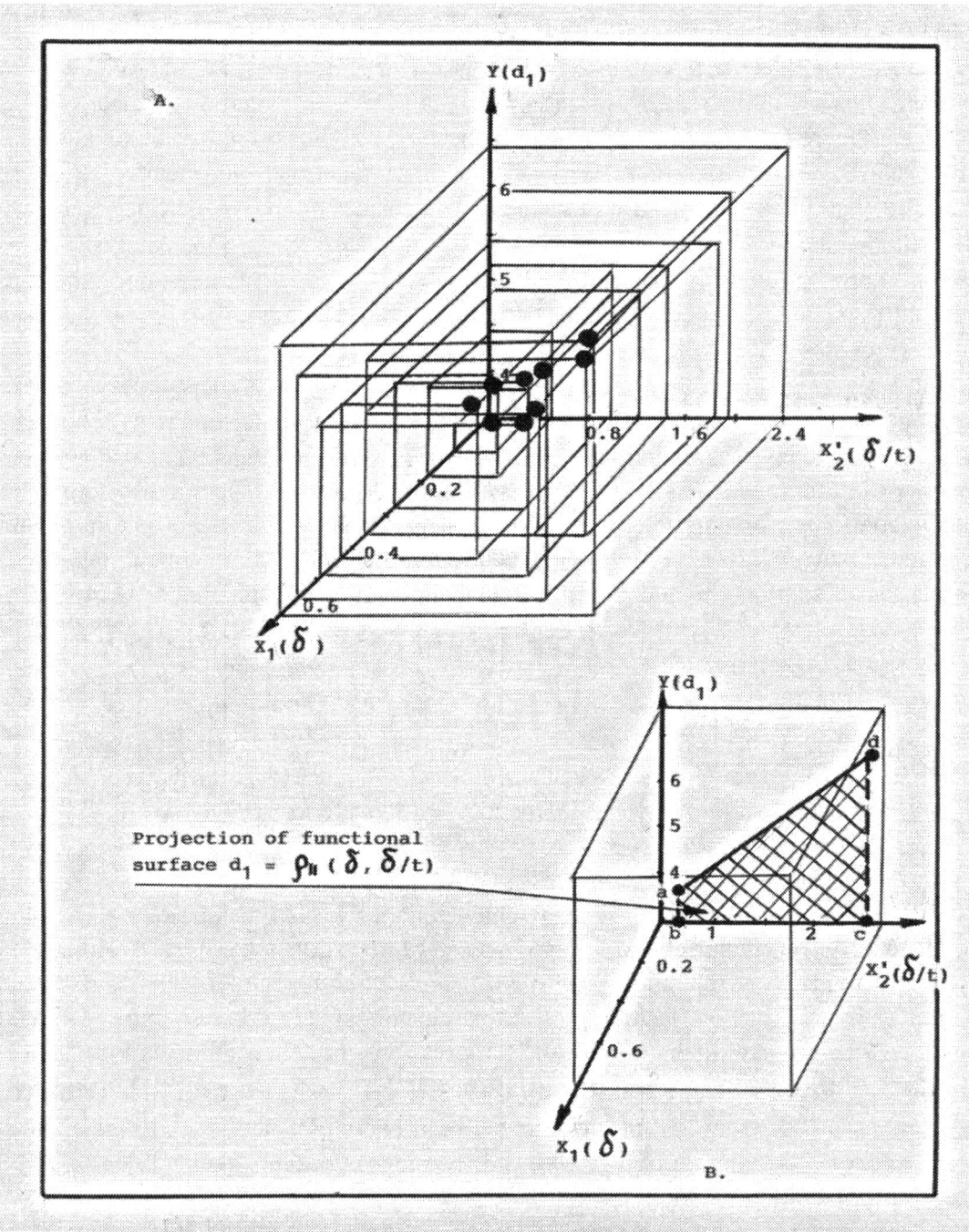

Figure 10

Analyzing the Figure 10(A), we see that functional surface $d_1=\rho_{11}(\delta,\delta/t)$ is shown in three-dimensional drawing for the linear regression model with equation view of $Y_c=3.388+2.848X_1+0.532X'_2$ or $Y_c=3.388+2.848X_1+0.532(X_1/X_2)$, where $d_1=3.388+2.848\delta+0.532(\delta/t)$.

And Figure 10(B) shows that this functional surface $d_1=\rho_{11}(\delta,\delta/t)$ has view of trapezium on the place YOX'_2 with such coordinates of their points(sizes in millimeters):a($X_{1,1}=0,X_{2,1}=0.29$,$Y_1=3.83$);b($X_{1,2}=0,X_{2,2}=0.29,Y_2=0$);c($X_{1,3}=0,X_{2,3}=2.40$, ,$Y_3=0$); d($X_{1,4}=0,X_{2,4}=2.40$,$Y_4=6.37$).

Analysis of functional surface $d_1=\rho_{11}(\delta,\delta/t)$

From Figure 10(A) we see that functional surface $d_1=\rho_{11}(\delta,\delta/t)$,on which are situated the points of the linear regression equation view of $Y_c=3.388+2.848X_1+0.532X'_2$ or $Y_c=3.388+2.848X_1+0.532(X_1/X_2)$,where $d_1=3.388+2.848\delta+0.532(\delta/t)$ has the following coordinates of their peaks in three-dimensional drawing (sizes in millimeters):
a'($X_{1,1}=0.10$,$X_{2,1}=0.29$,$Y_1=3.83$); b'($X_{1,1}=0.10,X_{2,2}=0.29,Y_2=0$),c'($X_{1,3}=0.60,X_{2,3}=2.40$, ,$Y_3=0$); d'($X_{1,4}=0.60,X_{2,4}=2.40,Y_4=6.37$).
The module of vectors $|a'b'|,|b'c'|,|c'd'|,|d'a'|$ and area of functional surface $S_{a'b'c'd'}$ can be calculated by formulas (7),(8),(9),(10) and (16).At above-shown data we have : $|a'b'|=3.83,|b'c'|=2.16,|c'd'|=6.37,|d'a'|=3.34$ and $S_{a'b'c'd'}=11.01mm^2$.
All results of computations for the linear regression equation view of $Y_c=3.388+2.848X_1+0.532X'_2$ or $Y_c=3.388+2.848X_1+0.532(X_1/X_2)$,where $d_1=3.388+2.848\delta+0.532(\delta/t)$ are given in Table 6.

So ,the parameters of this functional surface $d_1=\rho_{11}(\delta,\delta/t)$ are the following:

1.Function $d_1=\rho_{11}(\delta,\delta/t)$ better submits to the linear regression model with equation view of $Y_c=3.388+2.848X_1+0.532X'_2$ or $Y_c=3.388+2.848X_1+0.0532(X_1/X_2)$, where $d_1=3.388+2.848\delta+0.532(\delta/t)$ with such statistical characteristics:
- Coefficient of determination $R^2=0.899$
- Coefficient of correlation $r=0.948$
- Standard deviation $S_{y/x1,x'2}=0.308$
- Minimization of the mean square error (min MSE=0.063)
- Minimization of the mean absolute deviation (min MAD=0).

2. The total area of functional surface $d_1=\rho_{11}(\delta,\delta/t)$ is equal $\sum S=11.01$ mm^2 with such coordinates in three-dimensional drawing of their points for this surface (sizes in millimeters):

a' ($X_{1,1}=0.10$,$X_{2,1}=0.29$,$Y_1=3.83$);

b' ($X_{1,2}=0.10$,$X_{2,2}=0.29$,$Y_2=0$);

c' ($X_{1,3}=0.60$,$X_{2,3}=2.40$,$Y_3=0$);

d'($X_{1,4}=0.60$,$X_{2,4}=2.40$,$Y_4=6.37$).

Table 6 Evaluation of regression equation $Y_c=3.388+2.848X_1+0.532X'_2$

A. Mean , variance and standard deviation

Variable	Mean	Variance	Standard deviation
X_1	0.368	0.156	0.395
X'_2	1.244	3.478	1.865
Y	5.098	5.612	2.369

B. Results of multiple regression of Y on X_1 and X'_2

Parameter	Variable	Coefficients	Standard error	T-value
β_1	X_1	2.848	0.132	21.58
β_2	X'_2	0.532	0.622	0.86

C.   Analysis of variance results

Regression
    Degrees of freedom 2
    Sum of squares   5.039
    Mean squares   2.520
Error
    Degrees of freedom 6
    Sum of squares   0.569
    Mean squares   0.95
F-value*   26.526

*Since $F=26.526> [F_{0.05,2,6}=5.14]$ we can reject the hypothesis that both β_1 and β_2 are zero.

D.   Determination of residuals

Number	Observed	Estimated	Residual
1	5.36	5.33	0.03
2	5.22	5.17	0.05
3	3.52	3.83	-0.31
4	4.51	4.13	0.38
5	4.98	4.68	0.30
6	4.74	5.08	-0.34
7	5.15	5.46	-0.31
8	5.96	5.82	0.14
9	6.44	6.37	0.07

C. Modification function $d_1=\rho_{12}(\delta,\delta/B)$

Analysis of data and the statistical characteristics (coefficient of determination $R^2=0.871$, coefficient of correlation r=0.933, standard deviation $S_{y/x1}=0.347$) and also minimization of the mean square error (min MSE=0.080) and minimization of the mean absolute deviation (min MAD=0) shows that this functional dependence $d_1=\rho_{12}(\delta,\delta/B)$ better submits to the linear than non-linear[19] regression model with equation view of
$Y_c=3.311+4.860X_1-0.025X'_2$ or $Y_c=3.311+4.860X_1-0.025(X_1/X_2)$, where
$d_1=3.311+4.860\delta-0.025(\delta/B)$ (27) .

19- Non-linear regression model has the equation view of $d_1=3.515+3.827\delta+1.181\delta^2-0.002(\delta/B)$

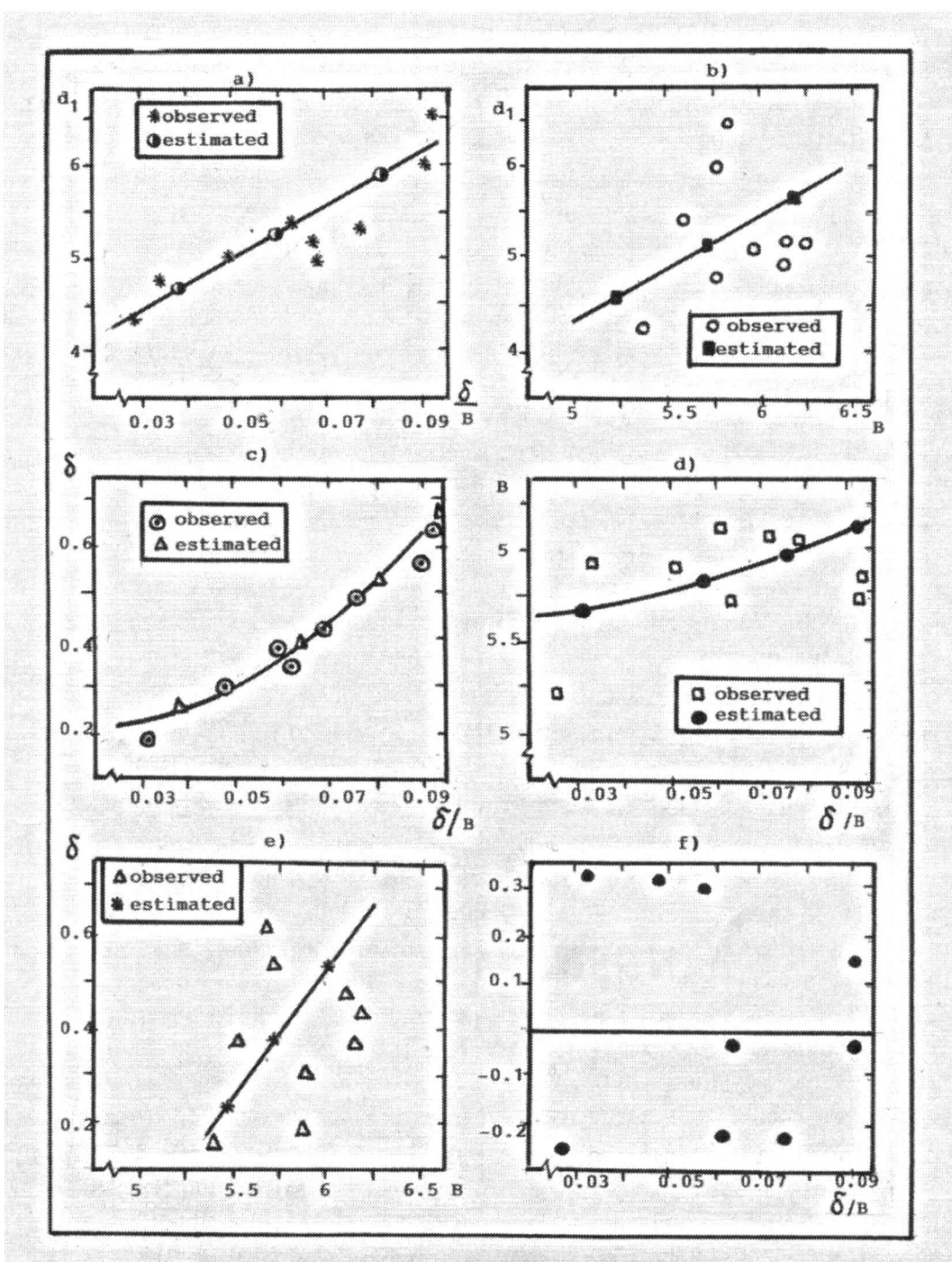

Figure 11

In Figure 11(a) is shown the functional dependence of internal diameter (d_1) stainless chip from the ratio (δ/B) ,such as thickness (δ) of this chip to its width (B) ,i.e we have the function $d_1 = \phi_3(\delta/B)$. Analysis of data and the statistical characteristics (coefficient of determination $R^2 = 0.901$,coefficient of correlation $r = 0.949$,standard deviation $S_{y/x'2} = 0.282$),minimization of the mean square error (min MSE = 0.062), minimization of the mean absolute deviation (min MAD=0.012) shows that this functional dependence $d_1=\phi_3(\delta/B)$ better submits to the linear than non-linear[20] regression model with equation view of **$Y_c=3.281+28.833X'_2$** or **$Y_c=3.281+28.833(X_1/X_2)$**, where **$d_1=3.281+28.833(\delta/B)$** (28). So, we see from Figure 11(a) that with increasing of ratio (δ/B), the value of internal diameter (d_1) for stainless chip increases considerably in accordance with the linear regression equation view $d_1=3.281+28.833(\delta/B)$.

In Figure 11(b) is shown the functional dependence of internal diameter (d_1) stainless chip from its width (B). So , we see from Figure 11(b) that this functional dependence $d_1=\phi_1(B)$ has the linear regression model with the equation view of $Y_c=0.06+0.863X$ or $d_1=0.06+0.863B$. Analysis of data given in Figure 11(b) shows that with increasing of width (B) stainless chip, the value of internal diameter (d_1) increases considerably in accordance with the linear regression equation view of $d_1=0.06+0.863B$.

In Figure 11(c) is shown the functional dependence of thickness(δ) stainless chip from the ratio of values (δ/B),such as thickness (δ) of this chip to its width (B),i.e we have the function view of $\delta=\phi_4(\delta/B)$.Analysis of data and also the scatter plots shown in Figure 11(c) ,and the statistical characteristics(coefficient of determination $R^2=1.0$,coefficient of correlation r=1.0,standard deviation $S_{y/x'2}=0$),minimization of the mean square error deviation (min MSE=0) and minimization of the mean absolute deviation (min MAD=0) show that this functional dependence $\delta=\phi_4(\delta/B)$ better submits to the non-linear than linear[21] regression model with equation view of **$Y_c=6.855(X'_2)^{1.056}$** or **$Y_c=6.855(X_1/X_2)^{1.056}$**, where **$\delta=6.855(\delta/B)^{1.056}$** (29).

And as we see from Figure 11(c) that between two parameters ,such as thickness (δ) and ratio (δ/B) there is the excellent correlation (r=1.0). And this function can be expressed in view of the non-linear regression equation $\delta=6.855(\delta/B)^{1.056}$ which shows that with increasing of ratio (δ/B) ,the value of thickness (δ) increases considerably.

In Figure 11(d) is shown the functional dependence of width(B) stainless chip from the ratio of values $(\delta?B)$,i.e has a place the function view of $B=\phi_5(\delta/B)$.Analysis of data and also the scatter plots shown in Figure 11(d) ,and the statistical characteristics (coefficient of determination $R^2=0.227$,coefficient of correlation r=0.477,standard deviation $S_{y/x'2}=0.238$),minimization of the mean square error (min MSE=0.044) and minimization of the mean absolute deviation (min MAD=0) shows that this functional dependence $B=\phi_5(\delta/B0$ better submits to the non-linear than linear[22] regression model with equation view of **$Y_c=6.73(X'_2)^{0.049}$** or **$Y_c=6.73(X_1/X_2)^{0.049}$**,where **$B=6.73(\delta/B)^{0.049}$** (30) . As we see from Figure 11(d) that between two parameters, such as width(B) stainless chip and ratio (δ/B) there is the average correlation (r=0.477).And this function $B=\phi_5(\delta/B)$ can be expressed in view of the non-linear regression equation view of $B=6.73(\delta/B)^{0.049}$ which shows that with increasing of ratio (δ/B) ,the value of width (B) for stainless chip increases considerably.

20-Non-linear regression model has the equation view of $d_1=2.782+28.833(\delta/B)+100(\delta/B)^2$

21-Linear regression model has the equation view of $\delta= -0.01+6(\delta/B)$

22- Linear regression model has the equation view of $B=5.647+3(\delta/B)$

In Figure 11(e) is shown the functional dependence of thickness (δ) stainless chip from its width (B). So , we see from Figure 11 (e) that this functional dependence $\delta=f(B)$ has the linear regression model with equation   view of $Y_c= -1.236+0.275X$ or $\delta=1.236+0.275B$. Analysis of data shown in Figure 11(e) shows that with increasing of width (B) stainless chip ,the value of thickness (δ) increases considerably in accordance with the linear regression equation view of $\delta= -1.236+0.275B$.

In Figure 11(f) is illustrated the residual plot (residual versus δ/B) of functional dependency $d_1=\phi_3(\delta/B)$.

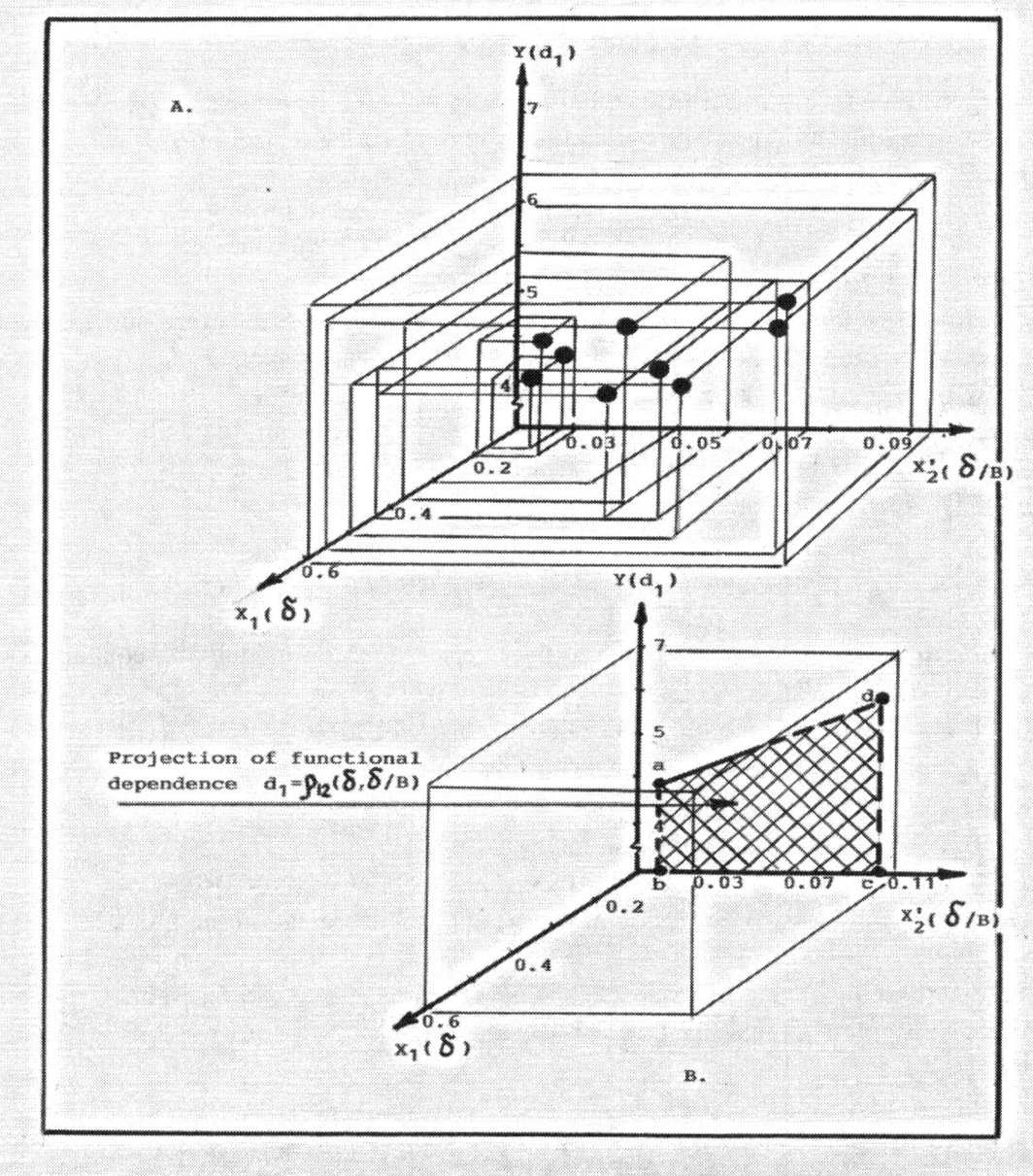

Figure 12

285

Analyzing the Figure 12(A) ,we see that functional surface $d_1=\rho_{12}(\delta,\delta/B)$ is shown in three-dimensional drawing for the linear regression model with equation view of $Y_c=3.311+4.86X_1-0.025X'_2$ or $Y_c=3.311+4.86X_1-0.025(X_1/X_2)$,where $d_1=3.311+4.86\delta--0.025(\delta/B)$.

And Figure 12(B) shows that this functional surface $d_1=\rho_{12}(\delta,\delta/B)$ has view of trapezium on the place YOX'_2 with such coordinates of their points (sizes in millimeters): $a(X_{1,1}=0$,$X_{2,1}=0.02$,$Y_1=3.79)$; $b(X_{1,2}=0,X_{2,2}=0.02,Y_2=0)$; $c(X_{1,3}=0,$,$X_{2,3}=0.10,Y_3=0)$; $d(X_{1,4}=0,X_{2,4}=0.10,Y_4=5.93)$.

Analysis of functional surface $d_1=\rho_{12}(\delta,\delta/B)$

From Figure 12(A) we see that functional surface $d_1=\rho_{12}(\delta,\delta/B)$,on which are situated the points of the linear regression equation view of $Y_c=3.311+4.86X_1-0.025X'_2$ or $Y_c=3.311+4.86X_1-0.025(X_1/X_2)$,where $d_1=3.311+4.86\delta-0.025(\delta/B)$ has the following coordinates of their peaks in three-dimensional drawing(sizes in millimeters):
$a'(X_{1,1}=0.10$,$X_{2,1}=0.02,Y_1=3.79)$; $b'(X_{1,2}=0.10,X_{2,2}=0.02,Y_2=0)$;$c'(X_{1,3}=0.54,X_{2,3}=0.10$,$Y_3=0)$;$d'(X_{1,4}=0.54,X_{2,4}=0.10,Y_4=5.93)$.
The module of vectors $\left|a'b'\right|$, $\left|b'c'\right|$, $\left|c'd'\right|$, $\left|d'a'\right|$ and area of functional surface can be calculated by formulas (7),(8),(9),(10) and (16). At above-named data we have : $|a'b'|=3.79,|b'c'|=0.44,|c'd'|=5.93,|d'a'|=2.19$ and $S_{a'b'c'd'}=2.14\text{mm}^2$.
All results of computations for the linear regression equation view of $Y_c=3.311+4.86X_1-0.025X'_2$ or $Y_c=3.311+4.86X_1-0.025(X_1/X_2)$,where $d_1=3.311+4.86\delta--0.025(\delta/B)$ are given in Table 7.

So, the parameters of this functional surface $d_1=\rho_{12}(\delta,\delta/B)$ are the following:

1.Function $d_1=\rho_{12}(\delta,\delta/B)$ better submits to the linear regression model with equation view of $Y_c=3.311+4.86X_1-0.025X'_2$ or $Y_c=3.311+4.86X_1-0.025(X_1/X_2)$,where $d_1=3.311+4.86\delta-0.025(\delta/B)$ with such statistical characteristics:
   - Coefficient of determination $R^2=0.871$
   - Coefficient of correlation $r=0.933$
   - Standard deviation $S_{y/x1,x'2}=0.347$
   - Minimization of the mean square error (min MSE=0.080)
   - Minimization of the mean absolute deviation (min MAD=0).

2.The total area of functional surface $d_1=\rho_{12}(\delta,\delta/B)$ is equal $\sum S=2.14\text{mm}^2$ with such coordinates of their points in three-dimensional drawing for this surface (sizes in millimeters):
$a'(X_{1,1}=0.10$,$X_{2,1}=0.02$,$Y_1=3.79$);
b' $(X_{1,2}=0.10$,$X_{2,2}=0.02$,$Y_2=0$);
c' $(X_{1,3}=0.54$,$X_{2,3}=0.10$, $Y_3=0)$;
d' $(X_{1,4}=0.54$,$X_{2,4}=0.10,Y_4=5.93)$.

Table 7 Evaluation of regression equation $Y_c = 3.311 + 4.86X_1 - 0.025X'_2$

A. Mean, variance and standard deviation

Variable	Mean	Variance	Standard deviation
X_1	0.368	0.156	0.395
X'_2	0.063	0.006	0.077
Y	5.098	5.612	2.369

B. Results of multiple regression of Y on X_1 and X'_2

Parameter	Variable	Coefficients	Standard error	T-value
β_1	X_1	4.86	0.132	36.82
β_2	X'_2	-0.025	0.026	-0.96

1. ### C. Analysis of variance results

      Regression

Degrees of freedom	2
Sum of squares	4.885
Mean squares	2.443

      Error

Degrees of freedom	6
Sum of squares	0.723
Mean squares	0.121
F-value*	20.19

*Since $F = 20.19 > [F_{0.05,2,6} = 5.14]$ we can reject the hypothesis that both β_1 and β_2 are zero.

D. Determination of residuals

Number	Observed	Estimated	Residual
1	5.36	5.06	0.30
2	5.22	5.35	-0.13
3	3.52	3.79	-0.27
4	4.51	4.19	0.32
5	4.98	4.67	0.31
6	4.74	5.11	-0.37
7	5.15	5.55	-0.40
8	5.96	5.93	0.03
9	6.44	6.22	0.22

D. Modification function $d_1 = \rho_{13}(\delta, t/B)$

Analysis of data and the statistical characteristics (coefficient of determination $R^2 = 0.87$, coefficient of correlation $r = 0.933$, standard deviation $S_{y/x1} = 0.348$) and minimization of the mean square error (min MSE = 0.08) shows that this functional dependence $d_1 = \rho_{13}(\delta, t/B)$ better submits to the linear than non-linear[23] regression model with equation view of **$Y_c = 3.30 + 4.86X_1 + 0.20X'_2$** or **$Y_c = 3.30 + 4.86X_1 + 0.20(X_2/X_3)$**, where **$d_1 = 3.30 + 4.86\delta + 0.2(t/B)$** (31).

23- Non-linear regression model has the equation view of $d_1 = 3.228 + 5.367\delta - 0.714\delta^2 + 0.143(t/B)$

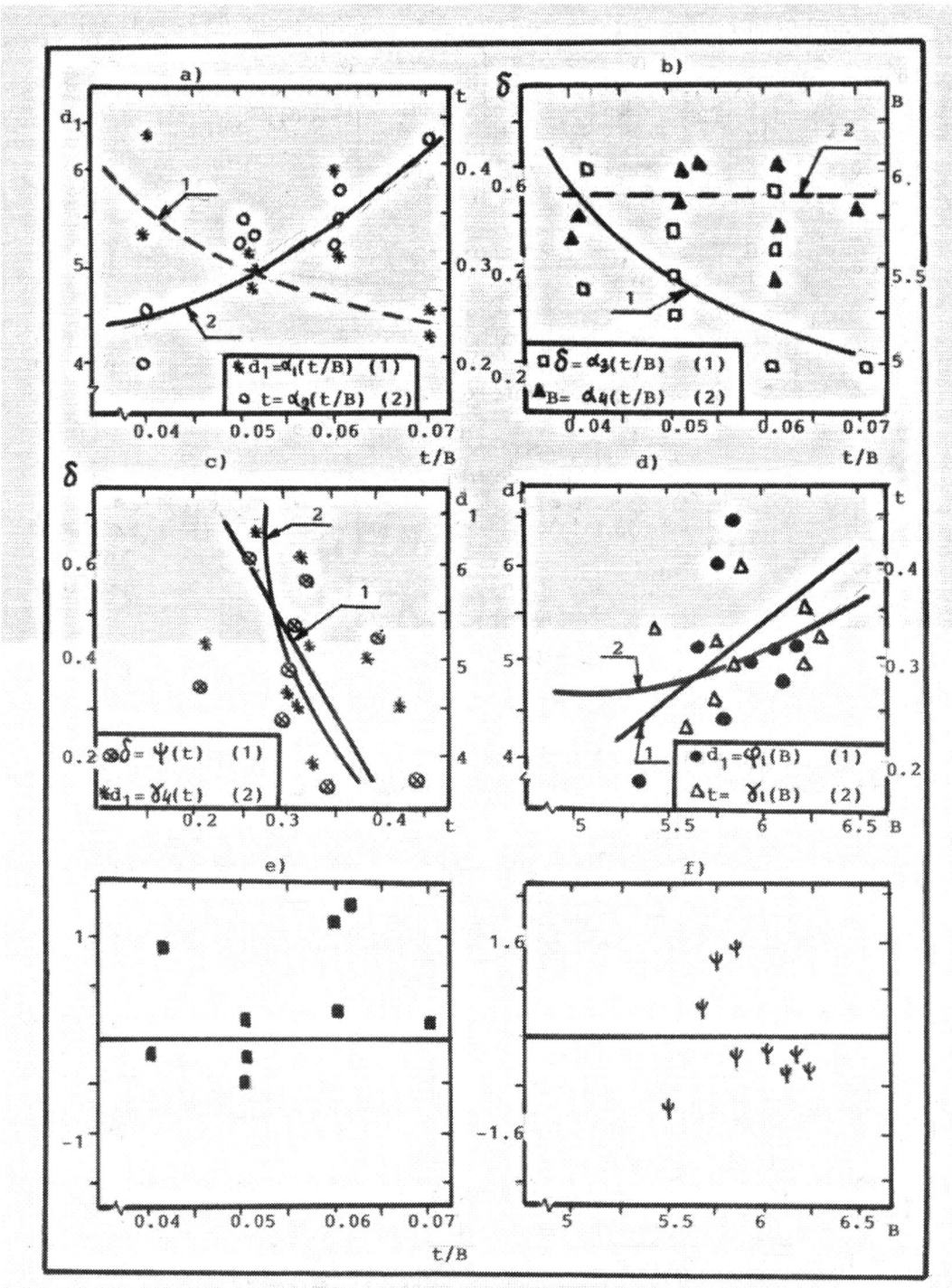

Figure 13

In Figure 13(a) is shown the functional dependence of internal diameter (d_1) stainless chip from the ratio (t/B) , such as clearance (t) between of chip wraps to its width(B) ,i.e we have the function view of $d_1=\alpha_1(t/B)$.

288

Analysis of data and the statistical characteristics (coefficient of determination R^2=0.269 ,coefficient of correlation r=0.519,standard deviation $S_{y/x'2}$=0.765,minimization of the mean square error(min MSE=0.455) shows that this functional dependence $d_1=\alpha_1(t/B)$ better submits to the non-linear than linear[24] regression model with equation view of $Y_c=1.517X'_2{}^{-0.407}$ or $Y_c=1.517(X_1/X_2)^{-0.407}$,where $d_1=1.517(t/B)^{-0.407}$ (32).

So , we see from Figure 13(a) that with increasing of ratio (t/B) ,the value of internal diameter(d_1) for stainless chip decreases considerably in accordance with the non-linear regression equation view of $d_1=1.517(t/B)^{-0.407}$.

And besides in Figure 13(a) is shown the functional dependence of clearance (t) between of chip wraps from its ratio (t/B) , such as clearance (t) between of chip wraps to its width(B) ,i.e we have the function $t=\alpha_2(t/B)$. Analysis of data and the statistical characteristics (coefficient of determination R^2 =0.871,coefficient of correlation r=0.933,standard deviation $S_{y/x'2}$=0.023),minimization of the mean square error (min MSE=0),minimization of the mean absolute deviation (min MAD=0) shows that this functional dependence $t=\alpha_2(t/B)$ better submits to the non-linear than linear[25] regression model with equation view of $Y_c=4.753X'_2{}^{0.925}$ or $Y_c=4.753(X_1/X_2)^{0.925}$,where $t=4.753(t/B)^{0.925}$ (33) . So , we see from Figure 13(a) that with increasing of ratio (t/B) ,the value of clearance (t) between of chip wraps increases considerably in accordance with the non-linear regression equation view of $t=4.753(t/B)^{0.925}$.

In Figure 13(b) is shown the functional dependence of thickness (δ) stainless chip from the ratio of values (t/B), such as clearance (t) between of chip wraps to its width(B) ,i.e we have the function view of $\delta=\alpha_3(t/B)$. Analysis of data and also the scatter plots shown in Figure 13(b) ,and the statistical characteristics (coefficient of determination R^2=0.19 ,coefficient of correlation r=0.432,standard deviation $S_{y/x'2}$=0.1560,minimization of the mean square error(min MSE=0.019) shows that this functional dependence $\delta=\alpha_3(t/B)$ better submits to the non-linear than linear[26] regression model with equation view of $Y_c=0.005(X'_2)^{-1.463}$ or $Y_c=0.005(X_1/X_2)^{-1.463}$, where $\delta=0.005(t/B)^{-1.463}$ (34) .

And as we see from Figure 13(b) that between two parameters, such as thickness (δ) and ratio (t/B) there is the average correlation (r=0.432).And this function $\delta=\alpha_3(t/B)$ can be expressed in view of the non-linear regression equation $\delta=0.005(t/B)^{-1.463}$ which shows that with increasing of ratio (t/B) ,the value of thickness decreases considerably.

In Figure 13(b) also is shown the functional dependence of width (B) stainless chip from the ratio of values (t/B) ,such as clearance (t) between of chip wraps to its width (B),i.e we have the function view of $B=\alpha_4(t/B)$. Analysis of data and also the scatter plots shown in Figure 13(b) ,and the statistical characteristics (coefficient of determination R^2=0 ,coefficient of correlation r=0 ,standard deviation $S_{y/x'2}$=0.270),minimization of the mean square error (min MSE=0.056) shows that this functional dependence $B=\alpha_4(t/B)$ better submits to the linear than non-linear[27] regression model with equation view of $Y_c=5.876-0.75X'_2$ or $Y_c=5.876-0.75(X_1/X_2)$,where $B=5.876-0.75(t/B)$(35).

24- Linear regression model has the equation view of $d_1=5.535-8.25(t/B)$

25-Linear regression model has the equation view of $t=0.264-(t/B)$

26-Linear regression model has the equation view of $\delta=0.461-1.75(t/B)$

27- Non-linear regression model has the equation view of $B=7.674(t/B)^{0.093}$

And as we see from Figure 13(b) that between two parameters, such as width (B) and ratio(t/B) there is not practically any correlation (r=o).And this function $B=\alpha_4(t/B)$ can be approximately expressed in view of the linear regression equation $B=5.876-0.75(t/B)$ which shows that with increasing of ratio (t/B) ,the value of width (B) practically does not change, and we can admit that this value of width(B) has the constant character.

In Figure 13(c) is shown the functional dependence of thickness (δ) stainless chip from its clearance (t) between of chip wraps .So , we see from Figure 13 (c) that this functional dependence $\delta=\phi(t)$ has the linear regression model with equation view of $Y_c=0.379-0.160X$ or $t=0.379-0.160\delta$,where **$\delta=2.369-6.25t$** (36).

Analysis of data shown in Figure 13(c) shows that with increasing of clearance (t) between of chip wraps ,the value of thickness (δ) stainless chip decreases considerably in accordance with the linear regression equation view of $\delta=2.369-6.25t$.

In Figure 13 (c) also is shown the functional dependence of internal diameter(d_1) stainless chip from its clearance (t) between of chip wraps. So, we see from Figure 13(c) that this functional dependence $d_1=\gamma_4(t)$ has the non-linear regression model with equation view of $Y_c=0.828X^{-0.605}$ or $t=0.828d_1^{-0.605}$,where $d_1=(1.208t)^{-1.653}$.

Analysis of data given in Figure 13(c) shows that with increasing of clearance (t) between of chip wraps ,the value of internal diameter(d_1) for stainless chip decreases considerably in accordance with the non-linear regression equation view of $d_1=(1.208t)^{-1.653}$.

In Figure 13(d) is shown the functional dependence of internal diameter (d_1) stainless chip from its width (B). So , we see from Figure 13(d) that this functional dependence $d_1=\varphi_1(B)$ has the linear regression model with the equation view of $Y_c=0.06+0.863X$ or $d_1=0.06+0.863B$.

Analysis of data shown in Figure 13(d) shows that with increasing of width(B) stainless chip ,the value of internal diameter (d_1) increases considerably in accordance with the linear regression equation view of $d_1=0.06+0.863B$.

And also in Figure 13(d) is shown the functional dependence of clearance between of chip wraps from its width (B).So ,we see from Figure 13(d) that this functional dependence $t=\gamma_1(B)$ has the non-linear regression model with equation view of $Y_c=0.032X^{1.296}$ or $t=0.032B^{1.296}$.

Analysis of data given in Figure 13(d) shows that with increasing of width (B) stainless chip , the value of clearance (t) between of chip wraps increases considerably in accordance with the non-linear regression equation view of $t=0.032B^{1.296}$.

In Figure 13(e) and 13(f) are illustrated the residual plots (residual versus t/B and B) of functional dependencies $d_1=\alpha_1(t/B)$ and $d_1=\varphi_1(B)$ accordingly.

Analyzing the Figure 14(A) , we see that functional surface $d_1=\rho_{13}(\delta,t/B)$is shown in three-dimensional drawing for the linear regression model with equation view of $Y_c=3.3+4.86X_1+0.2X_2'$ or $Y_c=3.3+4.86X_1+0.2(X_1/X_3)$,where $d_1=3.3+4.86\delta+0.2(t/B)$.

And Figure 14(B) shows that this functional surface $d_1=\rho_{13}(\delta,t/B)$ has view of trapezium on the place YOX'_2 with such coordinates of their points (sizes in millimeters):
a ($X_{1,1}=0$, $X_{2,1}=0.04$,$Y_1=6.23$) ; b ($X_{1,2}=0$,$X_{2,2}=0.04$,$Y_2=0$) ; c($X_{1,3}=0$,$X_{2,3}=0.07$,$Y_3=0$); d ($X_{1,4}=0$,$X_{2,4}=0.07$, $Y_4=4.19$).

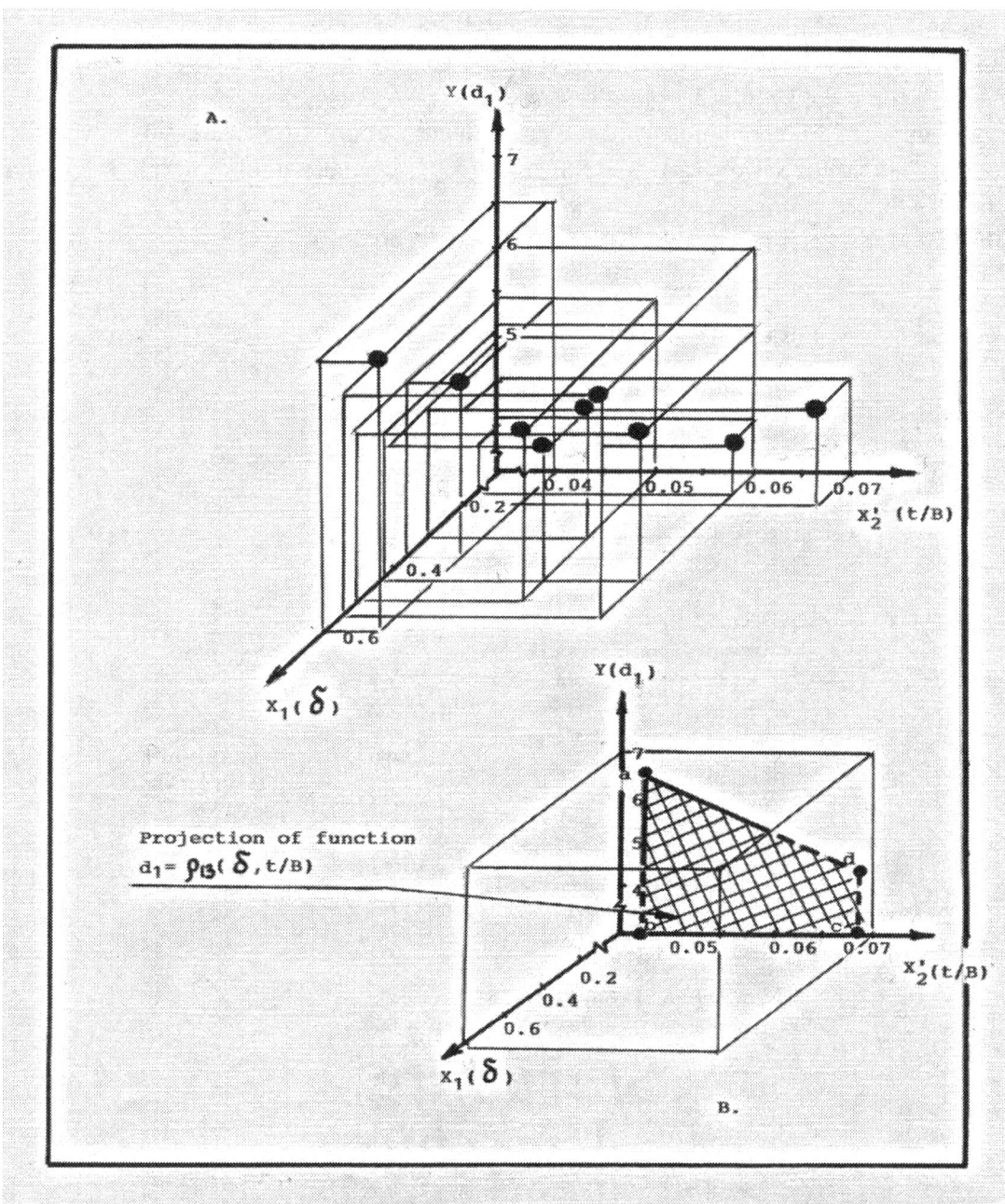

Figure 14

Analysis of functional surface $d_1=\rho_{13}(\delta,t/B)$

From Figure 14(A) we see that functional surface $d_1=\rho_{13}(\delta,t/B)$,on which are situated the points of the linear regression equation view of $Y_c=3.3+4.86X_1+0.2X'_2$ or $Y_c=3.3+4.86X_1+0.2(X_2/X_3)$,where $d_1=3.3+4.86\delta+0.2(t/B)$ has the following coordinates of their peaks in three-dimensional drawing (sizes in millimeters): a'$(X_{1,1}=0.60,X_{2,4}=0.04,Y_1=6.23)$;b'$(X_{1,2}=0.60 ,X_{2,2}=0.04,Y_2=0)$;c'$(X_{1,3}=0.18,X_{2,3}=0.07, Y_3=0)$; d'$(X_{1,4}=0.18 ,X_{2,4}=0.07,Y_4=4.19)$.

291

The module of vectors $|a'b'|,|b'c'|,|c'd'|,|d'a'|$ and area of functional surface $S_{a'b'c'd'}$ can be calculated by formulas (7),(8),(9),(10) and (11). At above-shown data we have : $|a'b'|=6.23,|b'c'|=0.42,|c'd'|=4.19,|d'a'|=2.08$ and $S_{a'b'c'd'}=2.19mm^2$.

All results of computations for the linear regression equation view of $Y_c=3.3+4.4.86X_1+0.2X'_2$ or $Y_c=3.3+4.86X_1+0.2(X_2/X_3)$,where $d_1=3.3+4.86\delta+0.2(t/B)$ are given in Table 8.

Table 8 Evaluation of regression equation $Y_c=3.3+4.86X_1+0.2X'_2$

A. Mean ,variance and standard deviation

Variable	Mean	Variance	Standard deviation
X_1	0.368	0.156	0.395
X'_2	0.053	0.001	0.032
Y	5.098	5.612	2.369

B. Results of multiple regression of Y on X_1 and X'_2

Parameter	Variable	Coefficients	Standard error	T-value
β_1	X_1	4.86	0.132	36.82
β_2	X'_2	0.20	0.011	18.19

C. Analysis of variance results

Regression
- Degrees of freedom 2
- Sum of squares 4.881
- Mean squares 2.441

Error
- Degrees of freedom 6
- Sum of squares 0.727
- Mean squares 0.121

F-value * 20.17

* Since $F=20.17>[F_{0.05,2,6}=5.14]$ we can reject the hypothesis that both β_1 and β_2 are zero.

D. Determination of residuals

Number	Observed	Estimated	Residual
1	5.36	5.15	0.21
2	5.22	5.04	0.18
3	3.52	5.04	-1.52
4	4.52	4.96	-0.45
5	4.98	5.12	-0.14
6	4.74	5.12	-0.38
7	5.15	5.12	0.03
8	5.96	5.04	0.92
9	6.44	5.15	1.29

So ,the parameters of this functional surface $d_1=\rho_{13}(\delta,t/B)$ are the following:

1.Functional surface $d_1=\rho_{13}(\delta,t/B)$ better submits to the linear regression model with equation view of $Y_c=3.3+4.86X_1+0.2X'_2$ or $Y_c=3.3+4.86X_1+0.2(X_2/X_3)$,where $d_1=3.3+4.86\delta+0.2(t/B)$ with such statistical characteristics: coefficient of determination $R^2=0.87$,coefficient of correlation r=0.933,standard deviation $S_{y/x1,x'2}=0.348$, minimization of the mean square error (min MSE=0.080).

2. The total area of functional surface $d_1 = \rho_{13}(\delta, t/B)$ is equal $\sum S = 2.19 \text{mm}^2$ with such coordinates in three-dimensional drawing of their points for this surface (sizes in millimeters): a'$(X_{1,1}=0.60, X_{2,1}=0.04, Y_1=6.23)$; b'$(X_{1,2}=0.60, X_{2,2}=0.04, Y_2=0)$; c'$(X_{1,3}=0.18, X_{2,3}=0.07, Y_3=0)$; d'$(X_{1,4}=0.18, X_{2,4}=0.07, Y_4=4.19)$.

7.3 Dependence of internal diameter (d_1) from width (B) of chip, clearance (t) between of chip wraps and its number (ω), i.e $d_1 = \phi_3(B, t, \omega)$ and also from its modifications: $d_1 = \mu_1(B, B/t), d_1 = \mu_2(B, B/\omega), d_1 = \mu_3(B, t/\omega)$.

A. Function $d_1 = \phi_3(B, t, \omega)$

Analysis of data and the statistical characteristics (coefficient of determination $R^2 = 0.370$, coefficient of correlation $r = 0.608$, standard deviation $S_{y/x_1, x_2, x_3} = 0.841$), minimization of the mean square error (min MSE=0.392) and minimization of the mean absolute deviation (min MAD=0.002) shows that this functional dependence $d_1 = \phi_3(B, t, \omega)$ better submits to the non-linear than linear[28] regression model with equation view of $\mathbf{Y_c = -5.646 + 3.436X_1 - 0.203X_1^2 - -7.351X_2 - 0.015X_3}$ or $\mathbf{d_1 = -5.646 + 3.436B - 0.203B^2 - 7.351t - 0.015\omega}$ (37).

In Figure 15(a) is shown the functional dependence of internal diameter (d_1) stainless chip from its width (B) .i.e we have the functional dependence $d_1 = \varphi_1(B)$. This dependence has the linear regression model with equation view of $Y_c = 0.06 + 0.863X$ or $d_1 = 0.06 + 0.863B$. So, from Figure 15(a) we see that with increasing of width (B) stainless chip, the value of internal diameter (d_1) of this chip considerably increases in accordance with the linear regression equation $d_1 = 0.06 + 0.863B$.

In Figure 15(a) also is shown the functional dependence of clearance (t) between of chip wraps from its width (B), i.e we have the functional dependence $t = \gamma_1(B)$. This dependence has the non-linear regression model with equation view of $Y_c = 0.032X^{1.296}$ or $t = 0.032B^{1.296}$. So, from Figure 15(a) we see that with increasing of width (B) stainless chip, the value of clearance (t) between of chip wraps increases considerably in accordance with the non-linear regression equation $t = 0.032B^{1.296}$.

In Figure 15(b) is shown the functional dependence of internal diameter (d_1) stainless chip from the clearance (t) between of chip wraps, i.e we have the functional dependence $d_1 = \gamma_4(t)$. This dependence has the non-linear regression model with equation view of $Y_c = 0.828X^{-0.605}$ or $t = 0.828d_1^{-0.605}$, where $d_1 = (1.208t)^{-1.653}$. So, from Figure 15(b) we see that with increasing of clearance (t) between of chip wraps, the value of internal diameter (d_1) decreases considerably in accordance with the non-linear regression equation $d_1 = (1.208t)^{-1.653}$.

In Figure 15(b) also is shown the functional dependence of number (ω) chip wraps from the clearance (t) between of chip wraps, i.e we have the functional dependence $\omega = \alpha_1(t)$. This functional dependency has the non-linear regression model with the equation view of $Y_c = 4.227X^{0.152}$ or $\omega = 4.227t^{0.152}$. So, from Figure 15(b) we see that with increasing of clearance (t) between of chip wraps, the value of number (ω) chip wraps increases considerably in accordance with the non-linear regression equation $\omega = 4.227t^{0.152}$.

28-Linear regression model has the equation view of $d_1 = 6.253 + 0.409B - 5.826t - 0.476\omega$

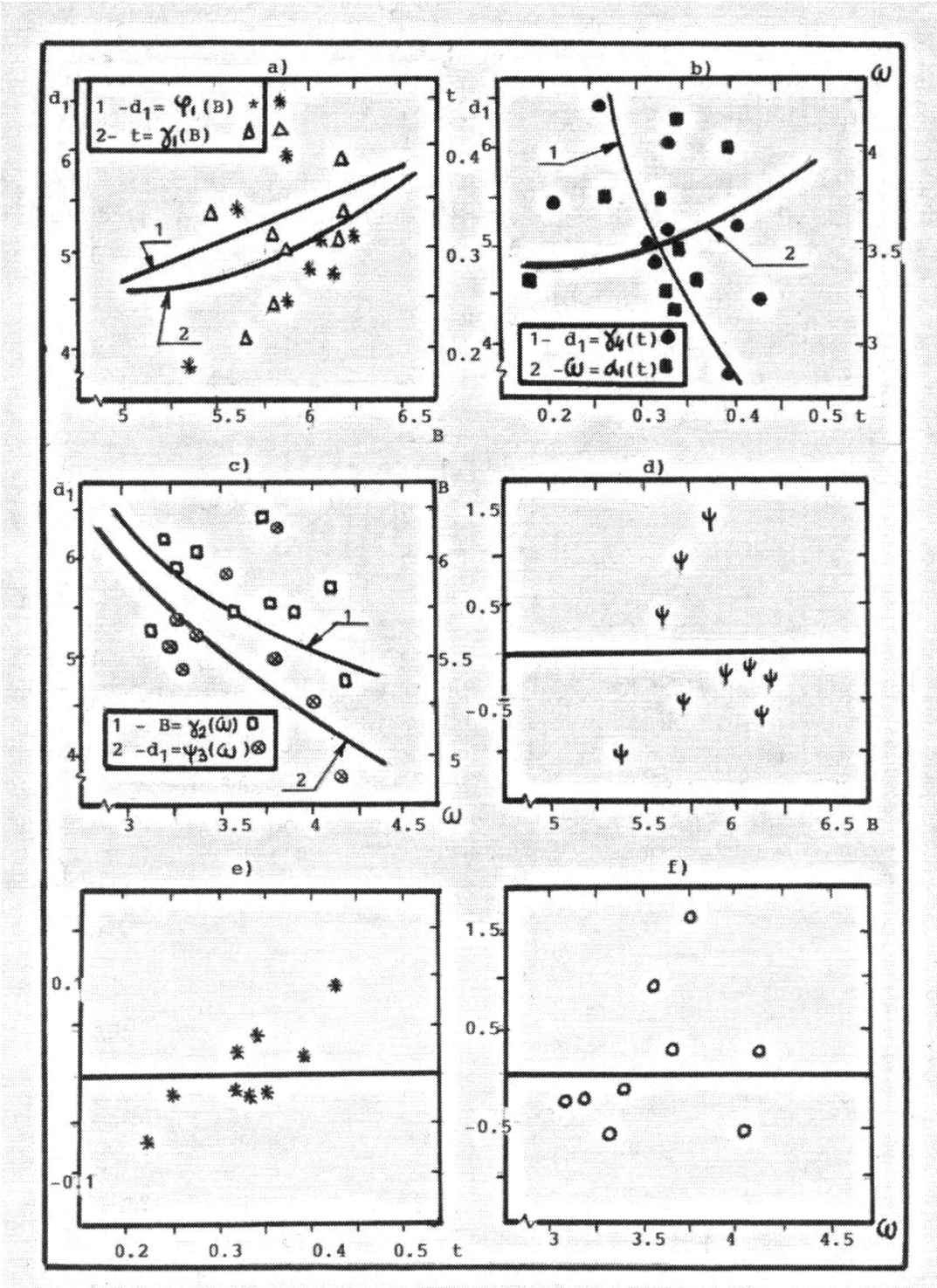

Figure 15

In Figure 15(c) is shown the functional dependence of internal diameter (d_1) stainless chip from the number (ω) chip wraps ,i.e we have the functional dependence $d_1 = \phi_3(\omega)$.

This functional dependency has the non-linear regression model with equation view of $Y_c=20.32X^{-1.102}$ or $d_1=20.32\omega^{-1.102}$. So, from Figure 15(c) we see that with increasing of number (ω) of chip wraps ,the value of internal diameter (d_1) stainless chip decreases considerably in accordance with the non-linear regression equation $d_1=20.32\omega^{-1.102}$.

In Figure 15(c) also is shown the functional dependence of width(B) stainless chip from the number (ω) of chip wraps ,i.e we have the functional dependence $B=\gamma_2(\omega)$. This dependence has the non-linear regression model with equation view of $Y_c=703.1X^{-3.0}$ or $\omega=703.1B^{-3.0}$,where $B=(0.001\omega)^{-0.333}$. So , from Figure 15 (c) we see that with increasing of number (ω) of chip wraps ,the value of width (B) stainless chip decreases considerably in accordance with non-linear regression equation view of $B=(0.001\omega)^{-0.333}$.

In Figures 15(d),15(e) and 15(f) are illustrated the residual plots (residual versus B, t,ω) of the functional dependencies $d_1=\varphi_1(B),d_1=\gamma_4(t)$ and $d_1=\phi_3(\omega)$ accordingly.

All results of computations for the non-linear regression equation view of $Y_c= -5.646+3.436X_1-0.203X_1^2-7.351X_2-0.005X_3$,where $d_1= -5.646+3.436B-0.203B^2-7.351t-0.005\omega$ are given in Table 9.

Table 9 Evaluation of regression equation $Y_c= -5.65+3.44X_1-0.203X_1^2-7.35X_2-0.01\omega$

A. Mean , variance and standard deviation			
Variable	Mean	Variance	Standard deviation
X_1	5.836	0.515	0.718
X_2	0.317	0.031	0.175
X_3	3.559	1.113	1.055
Y	5.098	5.612	2.369

B. Results of multiple regression of Y on X_1,X_2 and X_3				
Parameter	Variable	Coefficients	Standard error	T-value
β_1	X_1	3.436	0.239	14.38
β_2	X_2	-7.351	0.058	-126.74
β_3	X_3	-0.005	0.352	-0.01

C. Analysis of variance results

Regression
- Degrees of freedom 3
- Sum of squares 2.074
- Mean squares 0.691

Error
- Degrees of freedom 5
- Sum of squares 3.534
- Mean squares 0.707

F-value* 0.978

* Since F=0.978<[$F_{0.05,3,5}$=5.41] we are not able to reject the hypothesis that parameters β_1,β_2 and β_3 are zero.

D. Determination of residuals			
Number	Observed	Estimated	Residual
1	5.36	5.76	-0.40
2	5.22	4.97	0.25
3	3.52	4.39	-0.87
4	4.51	4.30	0.21
5	4.98	5.22	-0.24
6	4.74	5.45	-0.71
7	5.15	5.36	-0.21
8	5.96	4.95	1.01
9	6.44	5.50	0.94

B. Modification function $d_1=\mu_1(B,B/t)$

Analysis of data and the statistical characteristics (coefficient of determination $R^2=0.340$, coefficient of correlation $r=0.583$, standard deviation $S_{y/x1,x'2}=0.785$) and also minimization of the mean square error (min MSE=0.411) and minimization of the mean absolute deviation (min MAD=0) shows that this functional dependence $d_1=\mu_1(B,B/t)$ better submits to the linear than non-linear[29] regression model with equation view of $Y_c= -2.389+0.926X_1+0.109X'_2$ or $Y_c= -2.389+0.926X_1+0.109(X_1/X_2)$, where $d_1= -2.389+0.926B+0.109(B/t)$ (38).

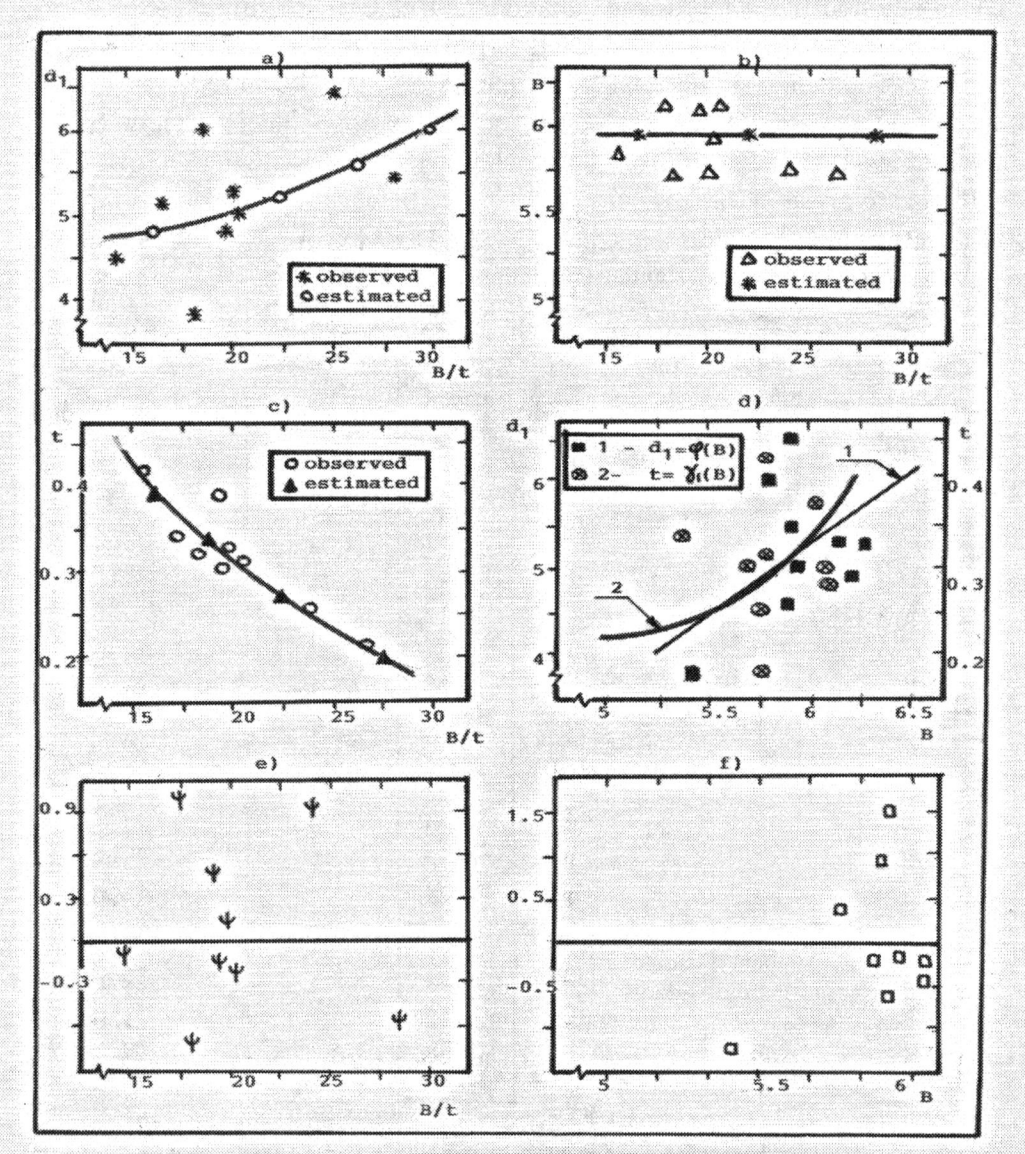

Figure 16

29- Non-linear regression model has the equation view of $d_1=675.25-221.64B+18.27B^2+0.007(B/t)$

In Figure 16(a) is shown the functional dependence of internal diameter (d_1) stainless chip from the ratio (B/t),such as width(B) of this chip to its clearance (t) between of chip wraps, i.e we have the function $d_1=\phi_1(B/t)$. Analysis of data and the statistical characteristics (coefficient of determination R^2=0.350,coefficient of correlation r=0.592,standard deviation $S_{y/x'2}$=0.721), minimization of the mean square error (min MSE=0.405) and minimization of the mean absolute deviation (min MAD=0.005) shows that this functional dependence $d_1=\phi_1(B/t)$ better submits to the non-linear than linear[30]regression model with equation view of $Y_c= -2.215+0.641X'_2-0.013(X'_2)^2$

or $Y_c= -2.215+0.641(X_1/X_2) -0.013(X_1/X_2)^2$,where $d_1= -2.215+0.641(B/t)- -0.013(B/t)^2$ (39).

And as we see from Figure 16(a) that between two parameters, such as internal diameter (d_1) and ratio (B/t) there is average correlation (r=0.592).And this function $d_1=\phi_1(B/t)$ can be expressed in view of the non-linear regression equation $d_1= -2.215+0.641(B/t)-0.013(B/t)^2$ which shows that with increasing of ratio (B/t) ,the value of internal diameter (d_1) increases considerably.

In Figure 16(b) is shown the functional dependence of width (B) stainless chip from the ratio of values (B/t) ,such as width (B) of this chip to its clearance(t) between of chip wraps ,i.e we have the function view of $B=\phi_2(B/t)$. Analysis of data and also the scatter plots shown in Figure 16(b), and the statistical characteristics (coefficient of determination R^2=0,coefficient of correlation r=0,standard deviation $S_{y/x'2}$=0.270), minimization of the mean square error (min MSE=0.056) and minimization of the mean absolute deviation (min MAD=0.002) shows that this functional dependence $B=\phi_2(B/t)$ better submits to the linear than non-linear[31] regression model with equation view of $Y_c=5.855-0.001X'_2$ or $Y_c=5.855-0.001(X_1/X_2)$,where $B=5.855-0.001(B/t)$ (40). And as we see from Figure 16(b) that between two parameters such as width (B) and ratio (B/t) practically absent any correlation (r=0). And this function $B=\phi_2(B/t)$ can be expressed in view of the linear regression equation $B=5.855-0.001(B/t)$ which shows that with increasing of ratio (B/t) ,the value of width(B) does not change ,i.e there is not correlation between two parameters of width(B) and ratio (B/t).

In Figure 16(c) is shown the functional dependence of clearance (t) between of chip wraps from the ratio of values (B/t) ,such as width(B) stainless chip to its clearance (t) between of chip wraps, i.e we have the function view of $t=\phi_3(B/t)$.

Analysis of data and also the scatter plots shown in Figure 16(c) and the statistical characteristics (coefficient of determination R^2=1.0,coefficient of correlation r=1.0,standard deviation $S_{y/x'2}$=0),minimization of the mean square error (min MSE=0),minimization of the mean absolute deviation (min MAD=0) shows that this functional dependence $t=\phi_3(B/t)$ better submits to the non-linear than linear[32] regression model with equation view of $Y_c=5.834(X'_2)^{-1.0}$ or $Y_c=5.834(X_1/X_2)^{-1.0}$, where $t=5.834(B/t)^{-1.0}$ (41).

30-Linear regression model has the equation view of $d_1=3.05+0.107(B/t)$
31-Non-linear regression model has the equation view of $B=6.982(B/t)^{-0.061}$
32-Linear regression model has the equation view of $t=0.604-0.015(B/t)$

And as we see from Figure 16(c) that between two parameters ,such as clearance (t) between of chip wraps and ratio (B/t) there is the excellent correlation (r=1.0). And this function $t=\phi_3(B/t)$ can be expressed in view of the non-linear regression equation $t=5.834(B/t)^{-1.0}$ which shows that with increasing of ratio (B/t) ,the value of clearance (t) between of chip wraps decreases considerably in accordance with the above-named non-linear regression equation.

In Figure 16(d) is shown the functional dependence of internal diameter (d_1) stainless chip from its width (B). So, we see from Figure 16(d) that this functional dependence $d_1= \varphi_1(B)$ has the linear regression model with equation view of $Y_c=0.06+0.863X$ or $d_1=0.06+0.863B$.Analysis of data shown in Figure 16(d) shows that with increasing of width(B) stainless chip ,the value of internal diameter (d_1) for this chip increases considerably in accordance with the linear regression equation view of $d_1=0.06+0.863B$.

In Figure 16(d) also is shown the functional dependence of clearance (t) between of chip wraps from the width (B) of stainless chip. So, we see from Figure 16(d) that this functional dependence $t= \gamma_1(B)$ has the non-linear regression model with equation view of $Y_c=0.032X^{1.296}$ or $t=0.032B^{1.296}$. Analysis of data given in Figure 16(d) shows that with increasing of width (B) stainless chip ,the clearance (t) between of chip wraps increases considerably in accordance with the non-linear regression equation view of $t=0.032B^{1.296}$.

In Figures 16(e) and 16(f) are illustrated the residual plots (residual versus B/t and B) of $d_1=\phi_1(B/t)$ and $d_1=\varphi(B)$ accordingly.

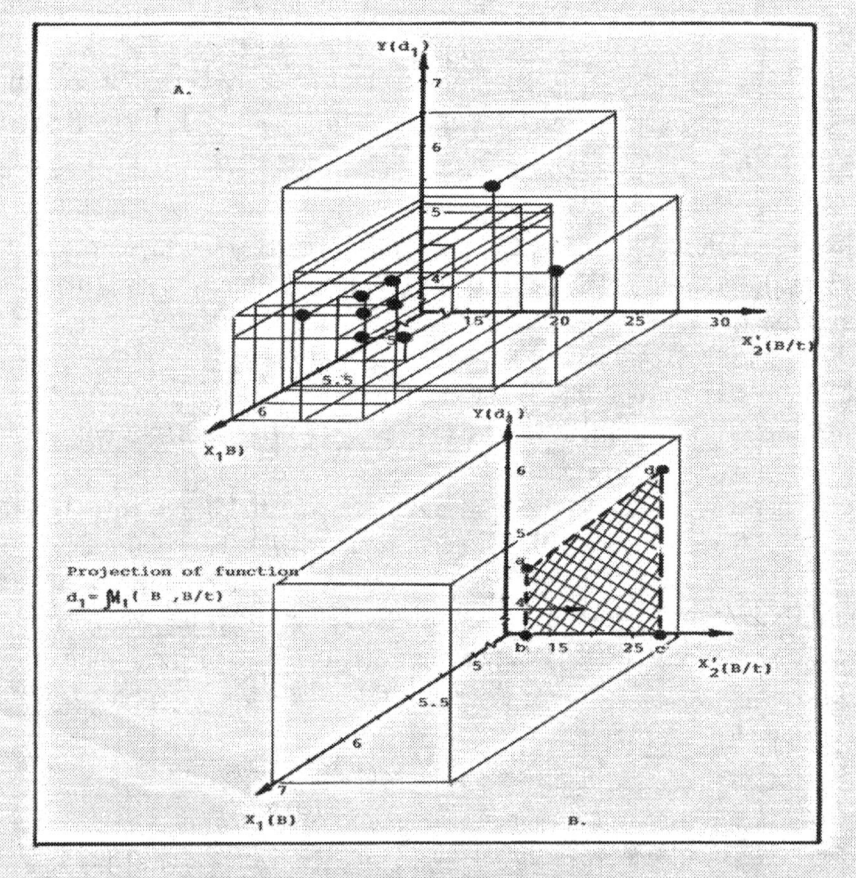

Figure 17

298

Analyzing the Figure 17(A) ,we see that functional surface $d_1 = \mu_1(B, B/t)$ is shown in three-dimensional drawing for the linear regression model with equation view of $Y_c = -2.389 + 0.926X_1 + 0.109X'_2$ or $Y_c = -2.389 + 0.926X_1 + 0.109(X_1/X_2)$, where $d_1 = -2.389 + 0.926B + 0.109(B/t)$. And Figure 17(B) shows that this functional surface $d_1 = \mu_1(B, B/t)$ has view of trapezium on the place YOX'$_2$ with such coordinates of their points (sizes in millimeters): $a(X_{1,1}=0, X_{2,1}=13.79, Y_1=4.48)$; $b(X_{1,2}=0, X_{2,2}=13.79, Y_2=0)$; $c(X_{1,3}=0, X_{2,3}=27.19, Y_3=0)$; $d(X_{1,4}=0, X_{2,4}=27.19, Y_4=5.86)$.

Analysis of functional surface $d_1 = \mu_1(B, B/t)$

From Figure 17(A) we see that functional surface $d_1 = \mu_1(B, B/t)$,on which are situated the points of the linear regression equation view of $Y_c = -2.389 + 0.926X_1 + 0.109X'_2$ or $Y_c = -2.389 + 0.926X_1 + 0.109(X_1/X_2)$,where $d_1 = -2.389 + 0.926B + 0.109(B/t)$ has the following coordinates of their peaks in three-dimensional drawing(sizes in millimeters): $a'(X_{1,1}=5.79, X_{2,1}=13.79, Y_1=4.48)$; $b'(X_{1,2}=5.79, X_{2,2}=13.79, Y_2=0)$; $c'(X_{1,3}=5.71, X_{2,3}=27.2, Y_3=0)$; $d'(X_{1,4}=5.71, X_{2,4}=27.2, Y_4=5.86)$.

The module of vectors $|a'b'|, |b'c'|, |c'd'|, |d'a'|$ and area of functional surface $S_{a'b'c'd'}$ can be calculated by formulas (7),(8),(9),(10) and (16). At above-shown data we have :$|a'b'|=4.48, |b'c'|=13.40, |c'd'|=5.86, |d'a'|=13.50$ and $S_{a'b'c'd'}=69.25mm^2$. All results of computations for the linear regression equation view of $Y_c = -2.389 + 0.926X_1 + 0.109X'_2$ or $Y_c = -2.389 + 0.926X_1 + 0.109(X_1/X_2)$,where $d_1 = -2.389 + 0.926B + 0.109(B/t)$ are given in Table 10.

Table 10 Evaluation of regression equation $Y_c = -2.389 + 0.926X_1 + 0.109X'_2$

A. Mean , variance and standard deviation			
Variable	Mean	Variance	Standard deviation
X_1	5.836	0.515	0.718
X'_2	19.11	130.326	11.416
Y	5.098	5.612	2.369

B. Results of multiple regression of Y on X_1 and X'_2				
Parameter	Variable	Coefficients	Standard error	T-value
β_1	X	0.926	0.239	3.87
β_2	X'_2	0.109	3.805	0.029

C. Analysis of variance results

Regression: Degrees of freedom 2, sum of squares 1.907, mean squares 0.954
Error: Degrees of freedom 6, sum of squares 3.701, mean squares 0.617
F-value* 1.546

* Since F=1.546< [$F_{0.05,2,6}$=5.14] we can not reject the hypothesis that both β_1 and β_2 are zero.

D. Determination of residuals			
Number	Observed	Estimated	Residual
1	5.36	5.86	-0.5
2	5.22	5.08	0.14
3	3.52	4.31	-0.79
4	4.51	4.48	0.03
5	4.98	5.12	-0.14
6	4.74	5.43	-0.69
7	5.15	5.34	-0.19
8	5.96	4.83	1.13
9	6.44	5.44	1.0

So, the parameters of this functional surface $d_1=\mu_1(B,B/t)$ are the following:

1.Functional $d_1=\mu_1(B,B/t)$ better submits to the linear regression model with equation view of $Y_c= -2.389+0.926X_1+0.109X'_2$ or $Y_c= -2.389+0.926X_1+0.109(X_1/X_2)$,where $d_1= -2.389+0.926B+0.109(B/t)$ with such statistical characteristics :

- Coefficient of determination $R^2=0.340$
- Coefficient of correlation r=0.583
- Standard deviation $S_{y/x1,x'2}=0.785$
- Minimization of the mean square error (min MSE=0.411)
- Minimization of the mean absolute deviation (min MAD=0).

2.The total area of functional surface $d_1=\mu_1(B,B/t)$ is equal $\sum S=69.25mm^2$ with such coordinates of their points in three-dimensional drawing for this surface (sizes in millimeters):
a'$(X_{1,1}=5.79,X_{2,1}=13.79,Y_1=4.48)$;b'$(X_{1,2}=5.79,X_{2,2}=13.79,Y_2=0)$;c'$(X_{1,3}=5.71,X_{2,3}=27.2,$, $Y_3=0)$;d'$(X_{1,4}=5.71,X_{2,4}=27.2,Y_4=5.86)$.

C. Modification function $d_1=\mu_2(B,B/\omega)$

Analysis of data and the statistical characteristics (coefficient of determination $R^2=0.134$,coefficient of correlation r=0.366,standard deviation $S_{y/x',x'2}=0.899$) and also minimization of the mean square error (min MSE=0.539)and minimization of the mean absolute deviation (min MAD=0) shows that this functional dependence $d_1=\mu_2(B,B/\omega)$ better submits to the linear than non-linear regression model with equation view of

$$\mathbf{Y_c=5.615-0.651X_1+1.978X'_2} \quad \text{or} \quad \mathbf{Y_c=5.615-0.651X_1+1.978(X_1/X_2),}\text{where}$$
$$\mathbf{d_1=5.615-0.651B+1.978(B/\omega)} \quad \mathbf{(42).}$$

In Figure 18(a) is shown the functional dependence of internal diameter (d_1)stainless chip from the ratio (B/ω) ,such as width (B) of this chip to its number of wraps(ω) ,i.e we have the function $d_1=\eta_1(B/\omega)$.Analysis of data and the statistical characteristics (coefficient of determination $R^2=0.13$,coefficient of correlation r=0.354,standard deviation $S_{y/x'2}=0.837$),minimization of the mean square error (min MSE=-.545) shows that this functional dependence $d_1=\eta_1(B/\omega)$ better submits to the non-linear than linear[33] regression model with equation view of $\mathbf{Y_c=3.715(X'_2)^{0.6}}$ or $\mathbf{Y_c=3.715(X_1/X_2)^{0.6}}$, where $\mathbf{d_1=3.715(B/\omega)^{0.6}}$ (43).And as we see from Figure 18(a) that between two parameters, such as internal diameter (d_1) and ratio (B/ω) there is the average correlation (r=0.354). And this function $d_1=\eta_1(B/\omega)$ can be expressed in view of the non-linear regression equation $d_1=3.715(B/\omega)^{0.6}$ which shows that with increasing of ratio (B/ω) ,the value of internal diameter (d_1) increases considerably.

In Figure 18(b) is shown the functional dependence of width (B) stainless chip from the ratio of values (B/ω) ,such as width(B) of this chip to its number of wraps (ω),i.e we have the function view of B=η_2(B/ω).Analysis of data and also the scatter plots shown in Figure 18(b) ,and the statistical characteristics (coefficient of determination $R^2=0.734$,coefficient of correlation r=0.857,standard deviation $S_{y/x'2}=0.139$), minimization of the

---

33-Linear regression model has the equation view of $d_1=2.876+1.339(B/\omega)$

mean square error (min MSE=0.015) and minimization of the mean absolute deviation (min MAD=0.007) shows that this functional dependence $B=\eta_2(B/\omega)$ better submits to the linear than non-linear[34] regression model with equation view of $\mathbf{Y_c=4.185+0.995X'_2}$ or $\mathbf{Y_c=4.185+0.995(X_1/X_2)}$, where $\mathbf{B=4.185+0.995(B/\omega)}$ (44).

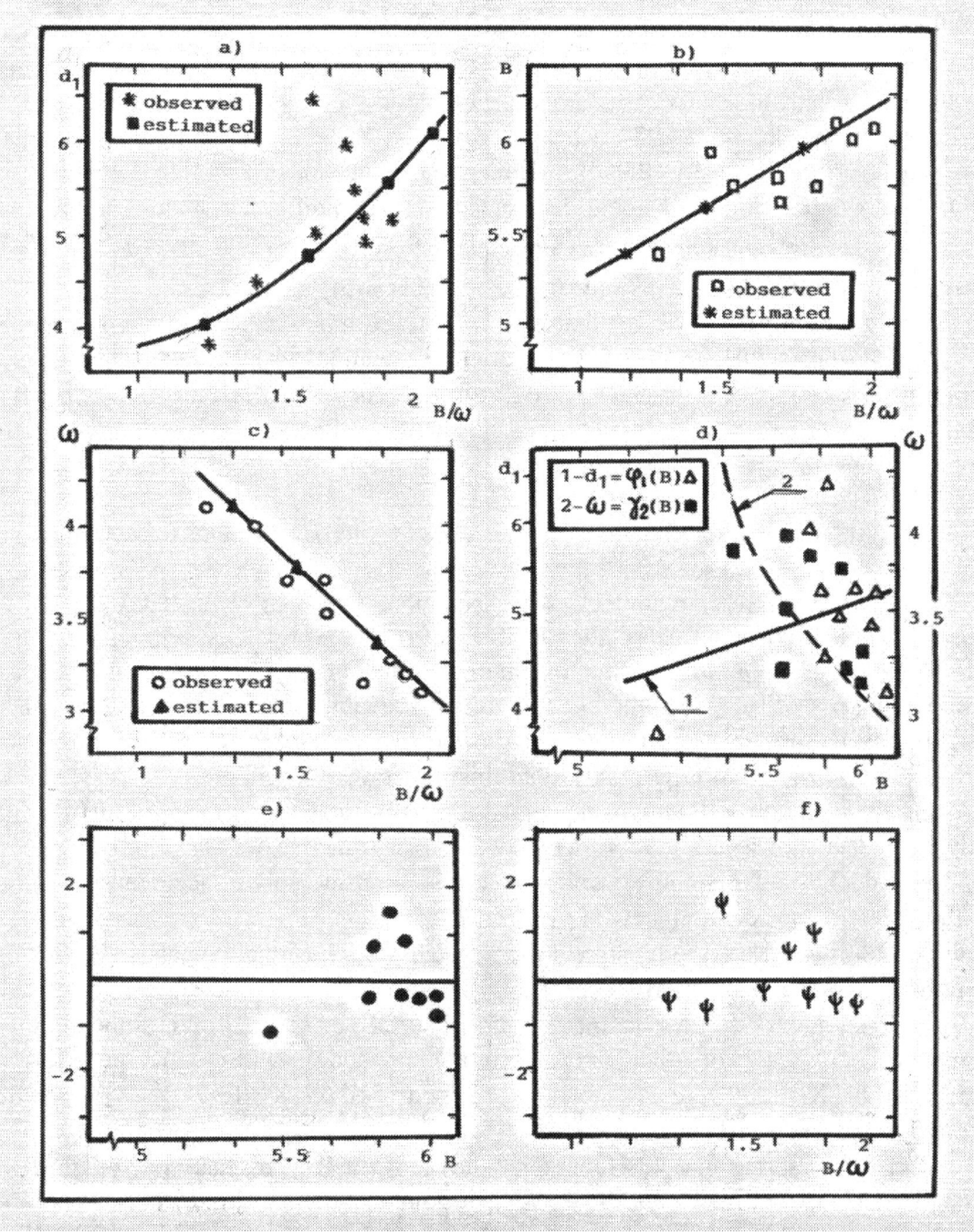

Figure 18

34- *Non-linear regression model has the equation view of* $B=3.162(B/\omega)^{0.5}$

And as we see from Figure 18(b) that between two parameters, such as width (B) and ratio (B/t) there is the good correlation (r=0.857). And this function $B=\eta_2(B/\omega)$ can be expressed in view of the linear regression equation B=4.185+0.995(B/ω) which shows that with increasing of ratio (B/ω) ,the value of width (B) increases considerably.

In Figure 18(c) is shown the functional dependence of number chip wraps (ω) from the ratio of values (B/ω) ,such as width (B) stainless chip to its number wraps (ω) ,i.e we have the function view of $\omega=\eta_3(B/\omega)$.Analysis of data and also the scatter plots shown in Figure 18(c) ,and the statistical characteristics (coefficient of determination R^2=0.949 ,coefficient of correlation r=0.974 ,standard deviation $S_{y/x'2}$=0.090) ,minimization of the mean square error (min MSE=0.006),minimization of the mean absolute deviation (min MAD=0) shows that this functional dependence $\omega=\eta_3(B/\omega)$ better submits to the linear than non-linear[35] regression model with equation view of $\mathbf{Y_c=6.268-1.633X'_2}$ or $\mathbf{Y_c=6.268-1.633(X_1/X_2)}$,where $\mathbf{\omega=6.268-1.633(B/\omega)}$ **(45)**.

And as we see from Figure 18(c) that between two parameters ,such as number of chip wraps (ω) and ratio (B/ω) there is the excellent correlation (r=0.974). And this function $\omega=\eta_3(B/\omega)$ can be expressed in view of the linear regression equation ω=6.268-1.633(B/ω) which shows that with increasing of ratio (B/ω) ,the value of number chip wraps (ω) decreases considerably in accordance with the linear regression equation view of ω=6.268-1.633(B/ω).

In Figure 18(d) is shown the functional dependence of internal diameter (d_1) stainless chip from its width (B).So, we see from Figure 18(d) that this functional dependence $d_1=\varphi_1(B)$ has the linear regression model with equation view of Y_c=0.06+0.863X or d_1=0.06+0.863B.Analysis of data indicated in Figure 18(d) shows that with increasing of width(B) stainless chip ,the value of internal diameter (d_1) for this chip increases considerably in accordance with the linear regression equation view of d_1=0.06+0.863B.

In Figure 18(d) also is shown the functional dependence of number chip wraps (ω) from its width(B).So , we see from Figure 18(d) that this functional dependence $\omega=\gamma_2(B)$ has the non-linear regression model with the equation view of $Y_c=703.1X^{-3}$ or $\omega=703.1B^{-3}$.

Analysis of data shown in Figure 18(d) shows that with increasing of width (B) stainless chip , the value of number chip wraps (ω) decreases considerably in accordance with the non-linear regression equation view of $\omega=703.1B^{-3}$.

In Figures 18(e) and 18(f) are illustrated the residual plots (residual versus B and B/ω) of functions $d_1=\varphi_1(B)$ and $d_1=\eta_1(B/\omega)$ accordingly.

Analyzing the Figure 19(A),we see that functional surface $d_1=\mu_2(B,B/\omega)$ is shown in three-dimensional drawing for the linear regression model with equation view of $Y_c=5.615-0.651X_1+1.978X'_2$ or $Y_c=5.615-0.651X_1+1.978(X_1/X_2)$,where d_1=5.615-0.615B+1.978(B/ω).

And Figure 19(B) shows that this functional surface $d_1=\mu_2(B,B/\omega)$ has view of trapezium on the place YOX_1 with such coordinates of their points (sizes in millimeters): a($X_{1,1}$=6.10 , $X_{2,1}$=0 ,Y_1=5.40); b ($X_{1,2}$=6.10,$X_{2,2}$=0,Y_2=0); c ($X_{1,3}$=5.37,$X_{2,3}$=0,Y_3=0); d ($X_{1,4}$=5.37 ,$X_{2,4}$=0,Y_4=4.63).

35-Non-linear regression model has the equation view of $\omega=2.344(B/\omega)^{1.345}$

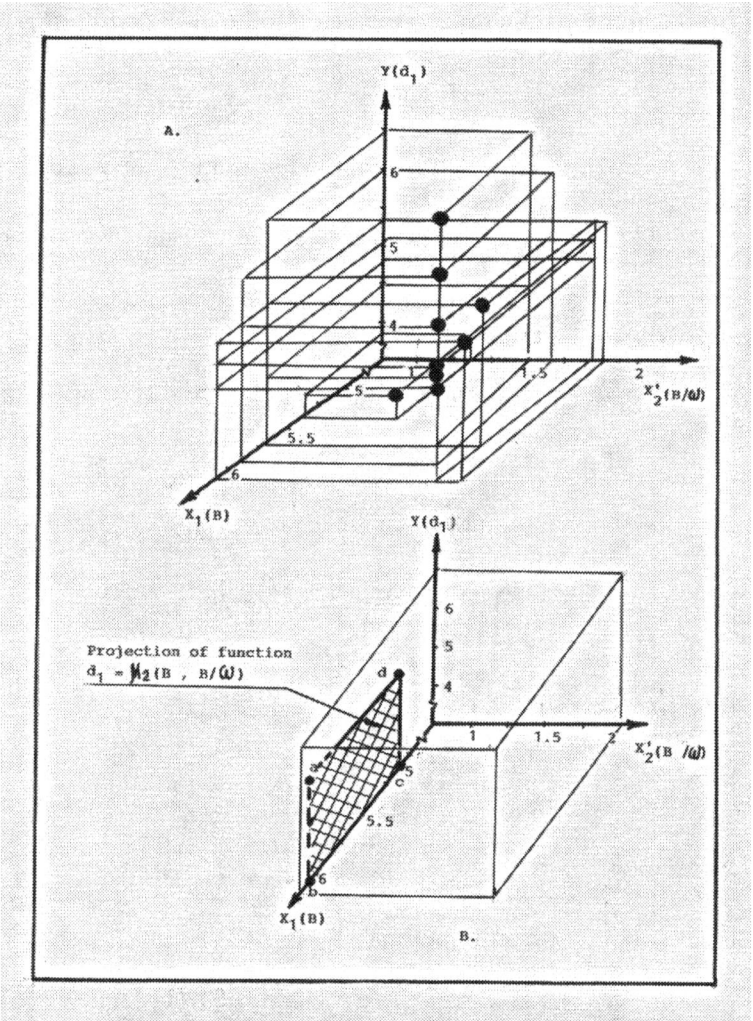

Figure 19

Analysis of functional surface $d_1=\mu_2(B,B/\omega)$

From Figure 19(A) we see that functional surface $d_1=\mu_2(B,B/\omega)$, on which are situated the points of the linear regression equation view of $Y_c=5.615-0.651X_1+1.978X'_2$ or $Y_c=5.615-0.651X_1+1.978(X_1/X_2)$,where $d_1=5.615-0.651B+1.978(B/\omega)$ has the following coordinates of their peaks in three-dimensional drawing (sizes in millimeters):
a'($X_{1,1}=6.10$, $X_{2,1}=1.90,Y_1=5.40$);b'($X_{1,2}=6.10,X_{2,2}=1.90,Y_2=0$);c'($X_{1,3}=5.37,X_{2,3}=1.27$, ,$Y_3=0$); d' ($X_{1,4}=5.37,X_{2,4}=1.27,Y_4=4.63$).
The module of vectors $|a'b'|,|b'c'|,|c'd'|,|d'a'|$ and area of functional surface $S_{a'b'c'd'}$ can be calculated by formulas (7),(8),(9),(10) and (11). At above-shown data we have : $|a'b'|=5.40,|b'c'|=0.96,|c'd'|=4.63,|d'a'|=1.23$ and $S_{a'b'c'd'}=4.80mm^2$.
All results of computations for the linear regression equation view of $Y_c=5.615-0.651X_1+1.978X'_2$ or $Y_c=5.615-0.651X_1+1.978(X_1/X_2)$,where $d_1=5.615-0.651B+1.978(B/\omega)$ are given in Table 11.

Table 11 Evaluation of regression equation $Y_c=5.615-0.651X_1+1.978X'_2$

A. Mean, variance and standard deviation

Variable	Mean	Variance	Standard deviation
X_1	5.836	0.515	0.718
X'_2	1.659	0.391	0.625
Y	5.098	5.612	2.369

B. Results of multiple regression of Y on X_1 and X'_2

Parameter	Variable	Coefficients	Standard error	T-value
β_1	X_1	-0.651	0.239	-2.720
β_2	X'_2	1.978	0.208	9.510

C. Analysis of variance results

Regression
- Degrees of freedom 2
- Sum of squares 0.752
- Mean squares 0.376

Error
- Degrees of freedom 6
- Sum of squares 4.856
- Mean squares 0.809

F-value* 0.465

* Since $F=0.465<[F_{0.05,2,6}=5.14]$ we can not reject the hypothesis that both β_1 and β_2 are zero.

D. Determination of residuals

Number	Observed	Estimated	Residual
1	5.36	5.46	-0.10
2	5.22	5.32	-0.10
3	3.52	4.63	-1.11
4	4.51	4.71	-0.2
5	4.98	5.01	-0.03
6	4.74	5.35	-0.61
7	5.15	5.40	-0.25
8	5.96	5.09	0.87
9	6.44	4.90	1.54

So, the parameters of this functional surface $d_1=\mu_2(B,B/\omega)$ are the following:

1.Functional surface $d_1=\mu_2(B,B/\omega)$ better submits to the linear regression model with equation view of $Y_c=5.615-0.651X_1+1.978X'_2$ or $Y_c=5.615-0.651X_1+1.978(X_1/X_2)$,where $d_1=5.615-0.651B+1.978(B/\omega)$ with such statistical characteristics:
- Coefficient of determination $R^2=0.134$
- Coefficient of correlation r=0.366
- Standard deviation $S_{y/x1,x'2}=0.899$

- Minimization of the mean square error (min MSE=0.539)
- Minimization of the mean absolute deviation (min MAD=0).

2.The total area of functional surface $d_1=\mu_2(B,B/\omega)$ is equal $\sum S=4.8mm^2$ with such coordinates of their points in three-dimensional drawing for this surface (sizes in millimeters): a'($X_{1,1}=6.10$,$X_{2,1}=1.90$,$Y_1=5.40$); b'($X_{1,2}=6.10$,$X_{2,2}=1.90$,$Y_2=0$); c'($X_{1,3}=5.37$,$X_{2,3}=1.27$,$Y_3=0$); d'($X_{1,4}=5.37$,$X_{2,4}=1.27$,$Y_4=4.63$).

D. Modification function $d_1=\mu_3(B, t/\omega)$

Analysis of data and minimization of the mean absolute deviation (min MAD=0.01) shows that this functional dependence $d_1=\mu_3(B, t/\omega)$ better submits to the linear than non-linear[36] regression model with equation view of $\mathbf{Y_c=11.55-1.838X_1+47X'_2}$ or $\mathbf{Y_c=11.55-1.838X_1+47(X_2/X_3)}$, where $\mathbf{d_1=11.55-1.838B+47(t/\omega)}$ (46).
In Figure 20(a) is shown the functional dependence of internal diameter (d_1) stainless chip from the ratio (t/ω), such as clearance (t) between of chip wraps to its number (ω) wraps ,i.e we have the function view of $d_1=\phi_0(t/\omega)$.Analysis of data and the statistical characteristics(coefficient of determination $R^2=0.17$,coefficient of correlation r=0.412), minimization of the mean absolute deviation (min MAD=0.0312) shows that this functional dependence $d_1=\phi_0(t/\omega)$ better submits to the linear than non-linear[37] regression model with equation view of $\mathbf{Y_c=7.53-27X'_2}$ or $\mathbf{Y_c=7.53-27(X_1/X_2)}$, where $\mathbf{d_1=7.53-27(t/\omega)}$ (47).
And as we see from Figure 20(a) that between two parameters , such as internal diameter (d_1) and ratio (t/ω) there is average correlation (r=0.412). And this function $d_1=\phi_0(t/\omega)$ can be expressed in view of the linear regression equation $d_1=7.53-27(t/\omega)$ which shows that with increasing of ratio (t/ω) , the value of internal diameter (d_1) decreases considerably.
And in Figure 20(a) also is shown the functional dependence of width(B) stainless chip from the ratio (t/ω) ,such as clearance (t) between of chip wraps to its number wraps(ω) ,i.e we have the function view of $B=\phi_1(t/\omega)$.Analysis of data and also the scatter plots shown in Figure 20(a) ,and the statistical characteristics(coefficient of determination $R^2=1.0$,coefficient of correlation r=1.0,standard deviation $S_{y/x'2}=0$),minimization of the mean square error (min MSE=0) and minimization of the mean absolute deviation (min MAD=0) show that this functional dependence $B=\phi_1(t/\omega)$ better submits to the linear

---

36-Non-linear regression model has the equation view of $d_1=3.253+2.93B-0.458B^2+0.167(t/\omega)$
37-Non-linear regression model has the equation view of $d_1=5.321(t/\omega)^{0.023}$

than non-linear[38] regression model with equation view of $\mathbf{Y_c=3.77+23(X'_2)}$ or $\mathbf{Y_c=3.77+23(X_1/X_2)}$, where $\mathbf{B=3.77+23(t/\omega)}$ (48).

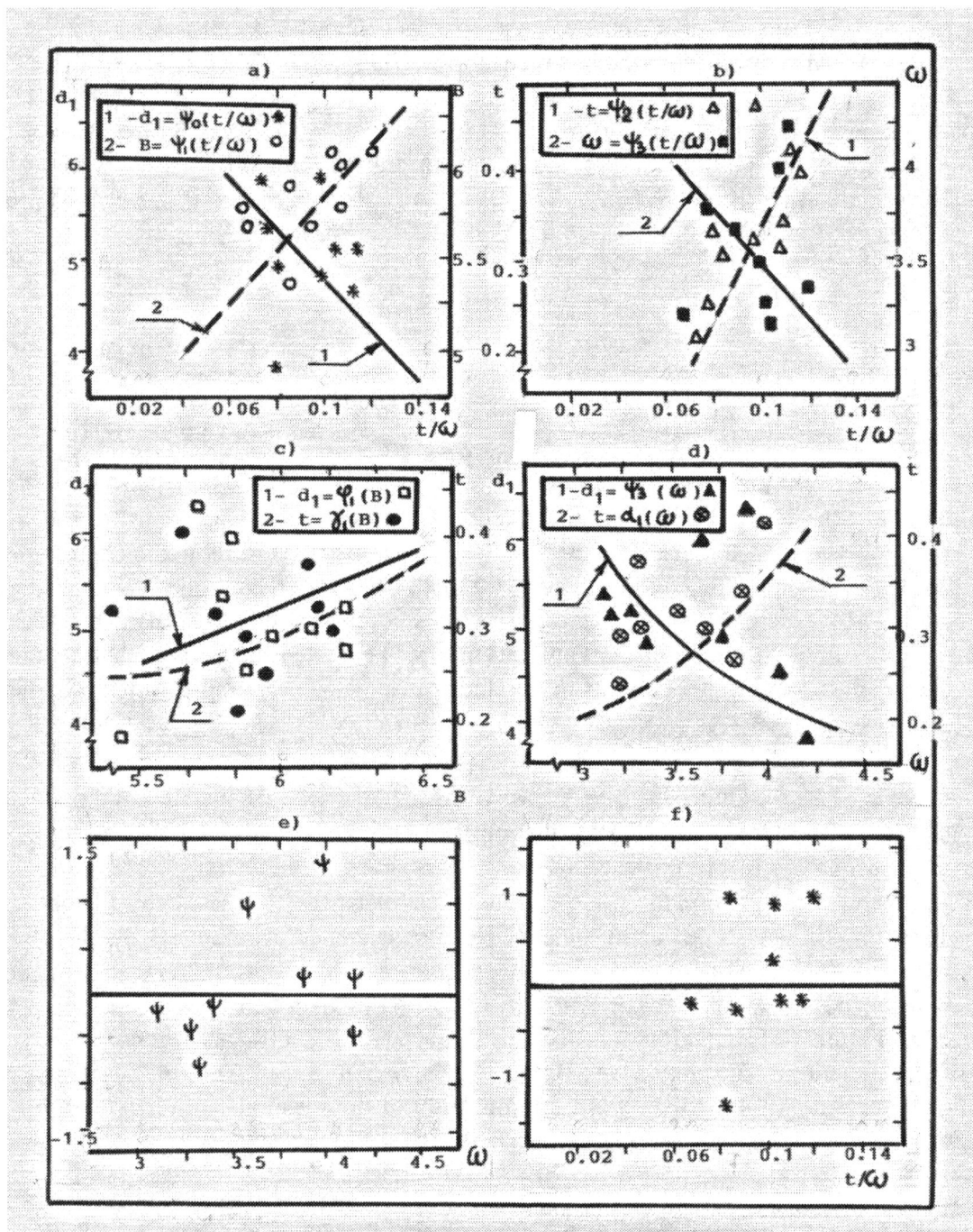

Figure 20

38-Non-linear regression model has the equation view of $d_1=5.297(t/\omega)^{-0.04}$

And as we see from Figure 20(a) that between two parameters, such as width (B) and ratio (t/ω) there is the excellent correlation (r=1.0). And this function $B=\phi_1(t/\omega)$ can be expressed in view of the linear regression equation B=3.77+23(t/ω) which shows that with increasing of ratio (t/ω) , the value of width(B)increases considerably.

In Figure 20(b) is shown the functional dependence of clearance (t) between of chip wraps from the ratio of values (t/ω), such as clearance (t) between of chip wraps to its number of wraps (ω) ,i.e we have the function view of $t=\phi_2(t/ω)$.Analysis of data and also the scatter plots shown in Figure 20(b) and also the statistical characteristics (coefficient of determination $R^2=1.0$,coefficient of correlation r=1.0)shows that this functional dependence $t=\phi_2(t/ω)$ better submits to the linear than non-linear[39]regression model with equation view of $\mathbf{Y_c= -0.403+8X'_2}$ or $\mathbf{Y_c= -0.403+8(X_1/X_2)}$,where $\mathbf{t= -0.403+8(t/ω)}$ **(49)**. And as we see from Figure 20(b) that between two parameters, such as clearance (t) between of chip wraps and ratio (t/ω) there is the excellent correlation (r=1.0).And this function $t=\phi_2(t/ω)$ can be expressed in view of the linear regression equation t= -0.403+ 8(t/ ω) which shows that with increasing of ratio (t/ω) ,the value of clearance (t) between of chip wraps increases considerably in accordance with the linear regression equation view of t= -0.403 +8(t/ω).

And Figure 20(b) also is shown the functional dependence of number chip wraps (ω) from the ratio of values (t/ω) ,such as clearance (t) between of chip wraps to its number wraps (ω) ,i.e we have the function view of $ω=\phi_3(t/ω)$.Analysis of data and also the scatter plots shown in Figure 20(b) ,and the statistical characteristics(coefficient of determination $R^2=0.10$,coefficient of correlation r=0.316),minimization of the mean absolute deviation (min MAD=0.007) shows that this functional dependence $ω=\phi_3(t/ω)$ better submits to the linear than non-linear[40] regression model with equation view of $\mathbf{Y_c=4.369-9X'_2}$ or $\mathbf{Y_c=4.369-9(X_1/X_2)}$, where $\mathbf{ω=4.369-9(t/ω)}$ **(50)**.

And as we see from Figure 20(b) that between two parameters , such as number of chip wraps (ω) and ratio (t/ω) there is small correlation (r=0.316). And this function $ω=\phi_3(t/ω)$ can be expressed in view of the linear regression equation ω=4.369-9(t/ω) which shows that with increasing of ratio (t/ω) , the value of number chip wraps(ω) decreases considerably in accordance with the linear regression equation view of ω=4.369-9(t/ω).

In Figure 20(c) is shown the functional dependence of internal diameter(d_1) stainless chip from its width(B).So ,we see from Figure 20(c) that this functional dependence $d_1=\varphi_1(B)$ has the linear regression model with equation view of $Y_c=0.06+0.863X$ or $d_1=0.06+0.863B$.Analysis of data shown in Figure 20(c) shows that with increasing of width(B) stainless chip ,the value of internal diameter (d_1) for this chip increases considerably in accordance with the linear regression equation view of $d_1=0.06+0.863B$.

In Figure 20 (c) also is shown the functional dependence of clearance (t) between of chip wraps from its width(B).So, we see from Figure 20 (c) that this functional dependence $t=\gamma_1(B)$ has the non-linear regression model with equation view of $Y_c=0.032X^{1.296}$ or $t=0.032B^{1.296}$.

39-Non-linear regression model has the equation view of $d_1=0.212(t/ω)^{-0.159}$
40-Non-linear regression model has the equation view of $ω=3.715(t/ω)^{0.02}$

Analysis of data shown in Figure 20(c) shows that with increasing of width(B) stainless chip , the value of clearance (t) between of chip wraps increases considerably in accordance with the non-linear regression equation view of $t=0.032B^{1.296}$.

In Figure 20 (d) is shown the functional dependence of internal diameter (d_1) stainless chip from its number of wraps(ω). So ,we see from Figure 20(d) that this functional dependence $d_1=\phi_3(\omega)$ has the non-linear regression model with equation view of $Y_c=20.32X^{-1.102}$ or $d_1=20.32\omega^{-1.102}$. Analysis of data shown in Figure 20(d) shows that with increasing of number wraps (ω) for stainless chip, the value of internal diameter (d_1) decreases considerably in accordance with the non-linear regression equation view of $d_1=20.32\omega^{-1.102}$.In Figure 20(d) also is shown the functional dependence of clearance (t) between of chip wraps from its number wraps (ω).So ,we see from Figure 20(d) that this functional dependence $t=\alpha_1(\omega)$ has the non-linear regression model with equation view of $Y_c=4.227X^{0.152}$ or $\omega=4.227t^{0.152}$,where $t=(0.237\omega)^{6.579}$. Analysis of data shown in Figure 20(d) shows that with increasing of number wraps (ω) stainless chip ,the value of clearance (t) between of chip wraps increases considerably in accordance with the non-linear regression equation view of $t=(0.237\omega)^{6.579}$.

In Figures 20(e) and 20(f) are illustrated the residual plots (residual versus ω and t/ω) of function $d_1=\phi_3(\omega)$ and $d_1=\phi_0(t/\omega)$ accordingly.

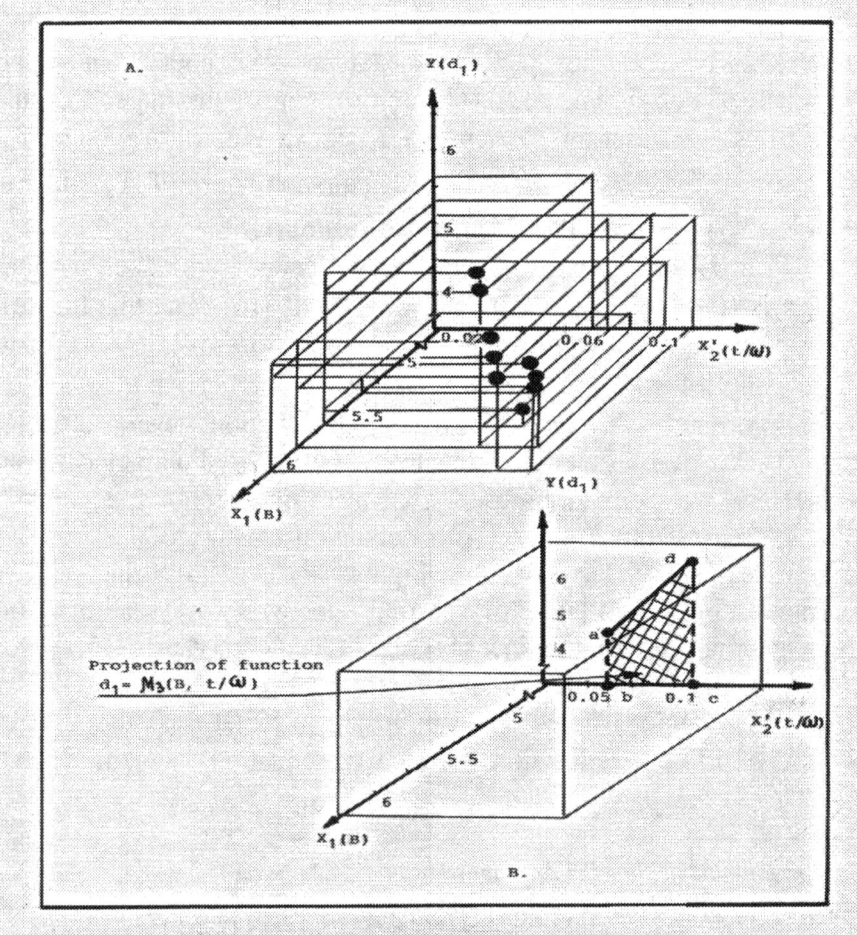

Figure 21

Analyzing the Figure 21(A) we see that functional surface $d_1=\mu_3(B, t/\omega)$ is shown in three-dimensional drawing for the linear regression model with equation view of

$Y_c=11.55-1.838X_1+47X'_2$ or $Y_c=11.55-1.838X_1+47(X_2/X_3)$, where $d_1=11.55-1.838B+$
$+47(t/\omega)$.

And Figure 21(B) shows that this functional surface $d_1=\mu_3(B, t/\omega)$ has view of trapezium on the place YOX'$_2$ with such coordinates of their points (sizes in millimeters):
a($X_{1,1}=0$,$X_{2,1}=0.07$,$Y_1=4.27$); b($X_{1,2}=0,X_{2,2}=0.07,Y_2=0$); c($X_{1,3}=0,X_{2,3}=0.11,Y_3=0$);
d($X_{1,4}=0,X_{2,4}=0.11,Y_4=6.08$).

Analysis of functional surface $d_1=\mu_3(B, t/\omega)$

From Figure 21(A) we see that functional surface $d_1=\mu_3(B, t/\omega)$, on which are situated the points of the linear regression equation view of $Y_c=11.55-1.838X_1+47X'_2$ or $Y_c=11.55-1.838X_1+47(X_2/X_3)$,where $d_1=11.55-1.838B+47(t/\omega)$ has the following coordinates of their peaks in three-dimensional drawing (sizes in millimeters):
a'($X_{1,1}=5.75,X_{2,1}=0.07$,$Y_1=4.27$); b'($X_{1,2}=5.75,X_{2,2}=0.07,Y_2=0$);c'($X_{1,3}=5.79,X_{2,3}=0.11$, ,$Y_3=0$);d'($X_{1,4}=5.79,X_{2,4}=0.11,Y_4=6.08$).
The module of vectors $|a'b'|,|b'c'|,|c'd'|,|d'a'|$ and area of functional surface can be calculated by formula (7),(8),(9),(10) and (16). At above-shown data we have : $|a'b'|=4.27,|b'c'|=0.10,|c'd'|=6.08,|d'a'|=1.80$ and $S_{a'b'c'd'}=0.52mm^2$.

All results of computations for the linear regression equation view of $Y_c=11.55-1.838X_1+47X'_2$ or $Y_c=11.55-1.838X_1+47(X_2/X_3)$,where $d_1=11.55-1.838B+$ $+47(t/\omega)$ are given in Table 12.

So , the parameters of this functional surface $d_1=\mu_3(B,t/\omega)$ are the following:

1.Functional surface $d_1=\mu_3(B, t/\omega)$ better submits to the linear regression model with equation view of $Y_c=11.55-1.838X_1+47X'_2$ or $Y_c=11.55-1.838X_1+47(X_2/X_3)$, where $d_1=11.55-1.838B+47(t/\omega)$.

2.The total area of functional surface $d_1=\mu_3(B, t/\omega)$ is equal $\sum S=0.52mm^2$ with such coordinates of their points in three-dimensional drawing for this surface (sizes in millimeters):

a'($X_{1,1}=5.75$,$X_{2,1}=0.07$,$Y_1=4.27$)
b'($X_{1,2}=5.75$,$X_{2,2}=0.07$,$Y_2=0$)
c'($X_{1,3}=5.79$,$X_{2,3}=0.11$,$Y_3=0$)
d' ($X_{1,4}=5.79$, $X_{2,4}=0.11$,$Y_4=6.08$).

Table 12 Evaluation of regression equation $Y_c = 11.55 - 1.838X_1 + 47X'_2$

A. Mean, variance and standard deviation

Variable	Mean	Variance	Standard deviation
X_1	5.836	0.515	0.718
X'_2	0.090	0.002	0.049
Y	5.098	5.612	2.369

B. Results of multiple regression of Y on X_1 and X'_2

Parameter	Variable	Coefficients	Standard error	T-value
β_1	X_1	-1.838	0.239	-7.690
β_2	X'_2	47	0.016	2937.5

C. Analysis of variance results

Regression
  Degrees of freedom 2
  Sum of squares   7.412
  Mean squares   3.706
Error
  Degrees of freedom 6
  Sum of squares   13.02
  Mean squares   2.17
F-value*   1.708

* Since $F = 1.708 < [F_{0.05,2,6} = 5.14]$ we can not reject the hypothesis that both β_1 and β_2 are zero.

D. Determination of residuals

Number	Observed	Estimated	Residual
1	5.36	4.35	1.01
2	5.22	5.85	-0.63
3	3.52	5.46	-1.94
4	4.51	6.08	-1.57
5	4.98	4.60	0.38
6	4.74	5.01	-0.27
7	5.15	5.04	0.11
8	5.96	5.30	0.66
9	6.44	4.27	2.17

7.4 Dependence of internal diameter (d_1) from clearance (t) between of chip wraps, number wraps(ω) and thickness (δ) of this stainless chip, i.e $d_1 = \phi_4(t, \omega, \delta)$ and their modifications: $d_1 = \eta_1(t, t/\omega)$, $d_1 = \eta_2(t, t/\delta)$, $d_1 = \eta_3(t, \omega/\delta)$.

A. Function $d_1 = \phi_4(t, \omega, \delta)$

Analysis of data and the statistical characteristics (coefficient of determination $R^2 = 0.820$, coefficient of correlation $r = 0.906$, standard deviation $S_{y/x1,x2,x3} = 0.448$), minimization of the mean square error (min MSE = 0.112) and minimization of the mean absolute deviation (min MAD = 0) shows that this functional dependence $d_1 = \phi_4(t, \omega, \delta)$ better submits to the linear than non-linear regression model with equation view of

$$Y_c = 5.208 - 0.471X_1 - 0.43X_2 + 4.22X_3 \text{ or } d_1 = 5.208 - 0.471t - 0.43\omega + 4.22\delta \quad (51).$$

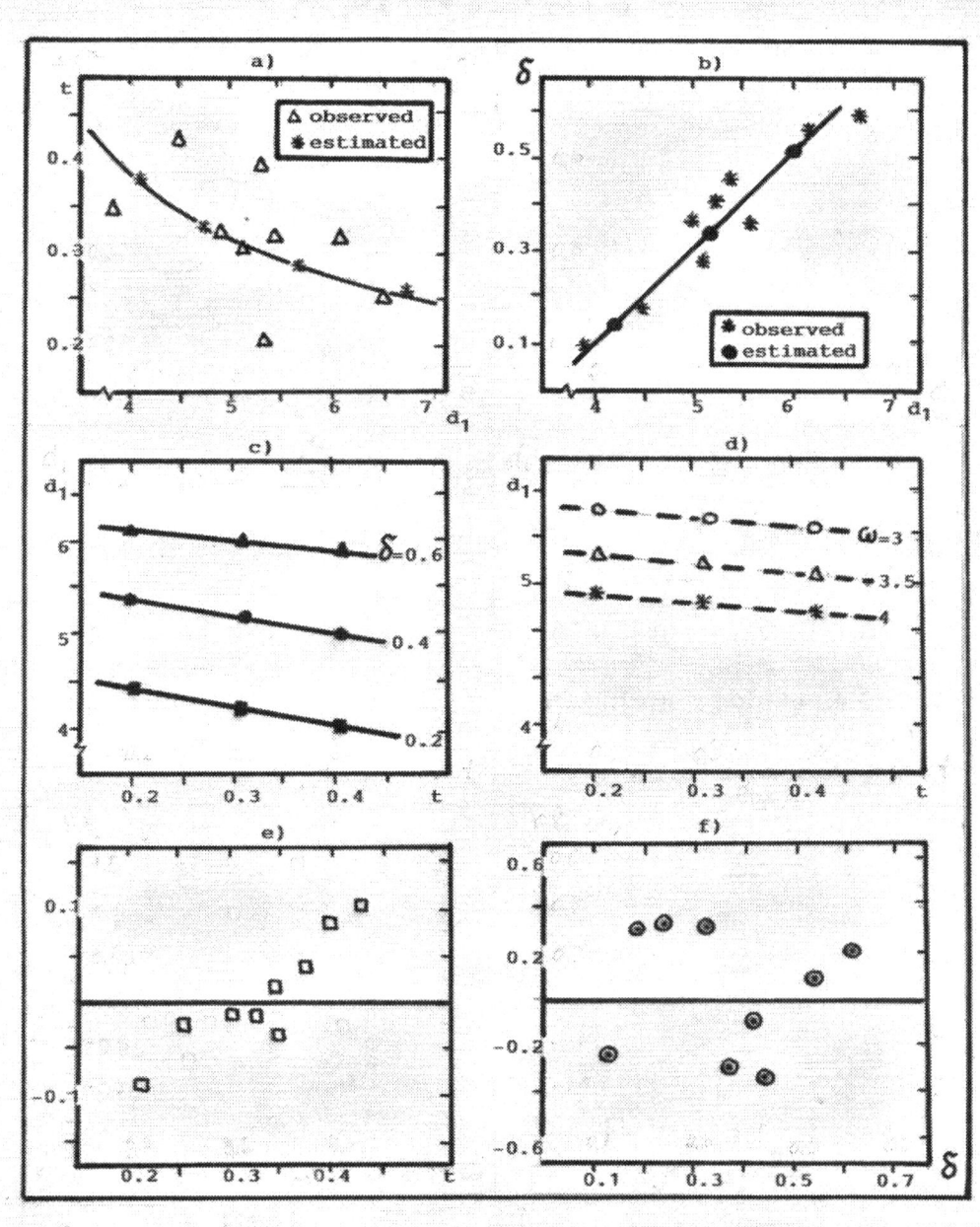

Figure 22

In Figure 22(a) is shown the functional dependence of clearance (t) between of wraps for stainless chip from its internal diameter(d_1),i.e we have the functional dependence $t=\gamma_4(d_1)$.This dependence has the non-linear regression model with equation view of $Y_c=0.828X^{-0.605}$ or $t=0.828d_1^{-0.605}$.

So , from Figure 22(a) we see that with increasing of internal diameter(d_1) stainless chip ,the value of clearance (t) between of chip wraps considerably decreases in accordance with the non-linear regression equation view of $t=0.828d_1^{-0.605}$.

In Figure 22(b) is shown the functional dependence of thickness (δ) stainless chip from its internal diameter (d_1),i.e we have the functional dependence $\delta=\varphi(d_1)$.This dependence has the linear regression model with equation view of $Y_c=3.338+4.762X$ or $d_1=3.338+4.762\delta$,where $\boldsymbol{\delta=0.21d_1-0.7}$ (52).

So, from Figure 22(b) we see that with increasing of internal diameter (d_1) stainless chip ,the value of thickness (δ) of this chip considerably increases in accordance with the linear regression equation view of $\delta=0.21d_1-0.7$.

In Figure 22(c) is shown the functional dependence of internal diameter (d_1) stainless chip from its clearance (t) between of chip wraps at the different values of thickness (δ) at constant average value of number wraps ($\omega_c=3.5$=const) ,i.e we have the functional dependence $d_1=\phi_4(t,\omega,\delta)$ at $\omega_c=3.5$=const.

This functional dependency has the linear regression model with equation view of $\boldsymbol{Y_c=5.208-0.471X_1-0.43X_2+4.22X_3}$ at $\omega_c=3.5$=const .

In Table 13 is shown the computation worksheet for Figures 22(c) and 22(d).

Table 13   Computation worksheet for determining of internal diameter (d_1) stainless chip for Figures 22(c) and 22(d).

Observed ＼ Estimated	Clearance (t)	Number wraps(ω)-Figure 22(d) At δ_c =0.37=const			Thickness (δ) – Figure 22 (c) At ω_c=3.56=const		
		3	3.5	4	0.2	0.4	0.6
Internal diameter $(d_1)^*$	0.20	5.40^*	5.20	5.00	4.50	5.30	6.10
	0.30	5.35	5.15	4.90	4.40	5.20	6.05
	0.40	5.30	5.05	4.80	4.30	5.15	6.00

So, from Figure 22 (c) we see that with increasing of clearance (t) between of chip wraps ,the value of internal diameter(d_1) considerably decreases ,but the value for internal diameter (d_1)is less for the stainless chip having the smaller value of thickness (δ) at constant value of number chip wraps (ω_c=const).

In Figure 22(d) is shown the functional dependence of internal diameter (d_1) stainless chip from its clearance (t) between of chip wraps in the different values of number wraps (ω) ,at the constant average value of thickness (δ_c=0.368=const),i.e we have also the functional dependence $d_1=\phi_4(t,\omega,\delta)$. This functional dependency $d_1=\phi_4(t,\omega,\delta)$ has the linear regression model with equation view of $Y_c=5.208-0.471X_1-0.43X_2+4.22X_3$ at δ_c=0.368=const.

So , from Figure 22(d) we see that with increasing of clearance (t) between of chip wraps ,the value of internal diameter(d_1) considerably decreases ,but the value for internal diameter(d_1) is less for stainless chip having the larger value of number chip wraps (ω) at the constant value of thickness (δ) stainless chip.

In Figures 22(e) and 22(f) are illustrated the residual plots (residual versus t and δ) of the functional dependencies $d_1=\gamma_4(t)$ and $d_1=\phi(\delta)$ accordingly.

All results of computations for the linear regression equation view of $Y_c=5.208-0.471X_1-0.43X_2+4.22X_3$ or $d_1=5.208-0.471t-0.43\omega+4.22\delta$ are given in Table 14.

Table 14 Evaluation of regression equation $Y_c=5.208-0.471X_1-0.43X_2+4.22X_3$

A. Mean ,variance and standard deviation

Variable	Mean	Variance	Standard deviation
X_1	0.317	0.031	0.175
X_2	3.559	1.113	1.055
X_3	0.368	0.209	0.457
Y	5.098	5.612	2.369

B. Results of multiple regression of Y on X_1,X_2 and X_3

Parameter	Variable	Coefficients	Standard error	T-value
β_1	X_1	-0.471	0.058	-8.12
β_2	X_2	-0.43	0.352	-1.21
β_3	X_3	4.22	0.152	27.74

C. Analysis of variance results

Regression
  Degrees of freedom 3
  Sum of squares   4.60
  Mean squares   1.533
Error
  Degrees of freedom 5
  Sum of squares   1.008
  Mean squares   0.202
F-value*   7.59

* Since $F=7.59>[F_{0.05,3,5}=5.41]$ we are able to reject the hypothesis that parameters β_1,β_2 and β_3 are zero.

D. Determination of residuals

Number	Observed	Estimated	Residual
1	5.36	5.28	0.08
2	5.22	5.41	-0.19
3	3.52	3.68	-0.16
4	4.51	4.07	0.44
5	4.98	4.71	0.27
6	4.74	5.24	-0.50
7	5.15	5.63	-0.48
8	5.96	5.83	0.13
9	6.44	6.02	0.42

B. Modification function $d_1=\eta_1(t,t/\omega)$

Analysis of data and the statistical characteristics (coefficient of determination $R^2=0.18$,coefficient of correlation r=0.420,standard deviation $S_{y/x1}=1.141$) and also minimization of the mean square error (min MSE=0.512) ,and minimization of the mean absolute deviation (min MAD=0.018) shows that this functional dependence $d_1=\eta_1(t,t/\omega)$ better submits to the non-linear than linear[41]regression model with equation view of

$$Y_c=5.03+9.01X_1-26.8X_1^2+0.08X'_2 \qquad \text{or} \qquad Y_c=5.03+9.01X_1-26.8X_1^2+$$
$$+0.08(X_1/X_2) , \text{ where } d_1=5.03+9.01t-26.8t^2+0.08(t/\omega) \quad (53) .$$

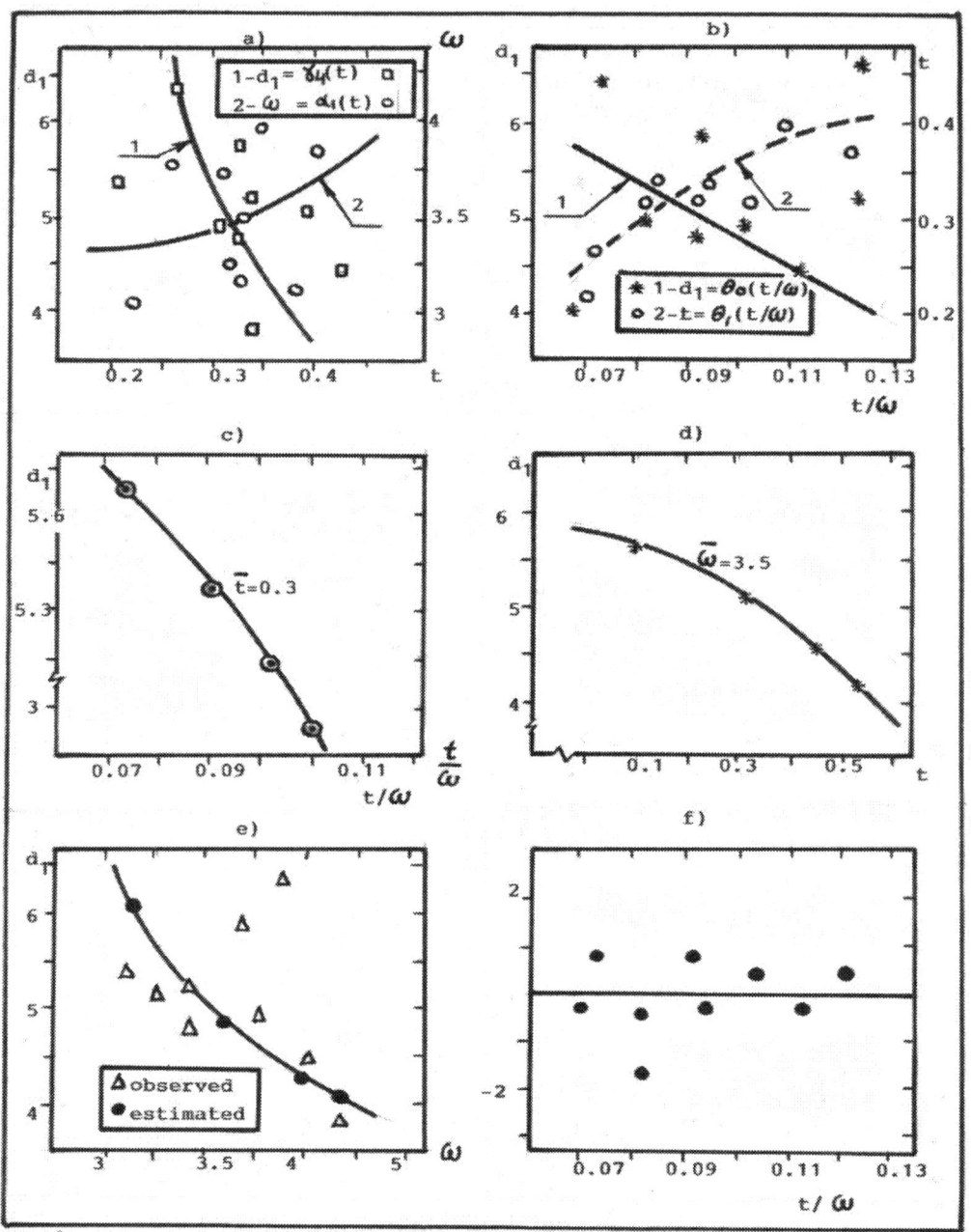

Figure 23

41- Linear regression model has the equation view of $d_1=7.378+1.61t-31(t/\omega)$

In Figure 23(a) is shown the functional dependence of internal diameter (d_1) stainless chip from its clearance (t) between of chip wraps. So , we see from Figure 23(a) that this functional dependence $d_1=\gamma_4(t)$ has the non- linear regression model with equation view of $Y_c=0.828X^{-0.605}$ or $t=0.828d_1^{-0.605}$, where $d_1=(1.208t)^{-1.653}$. Analysis of data shown in Figure 23(a) shows that with increasing of clearance (t) between of chip wraps for stainless chip, the value of internal diameter (d_1) decreases considerably in accordance with the non-linear regression equation view of $d_1=(1.208t)^{-1.653}$.

In Figure 23(a) also is shown the functional dependence of number chip wraps (ω) from its clearance (t) between of chip wraps. So, we see from Figure 23(a) that this functional dependence $\omega=\alpha_1(t)$ has the non-linear regression model with equation view of $Y_c=4.227X^{0.152}$ or $\omega=4.227t^{0.152}$. Analysis of data given in Figure 23(a) shows that with increasing of clearance (t) between of chip wraps, the value of number chip wraps (ω) increases considerably in accordance with the non-linear regression equation view of $\omega=4.227t^{0.152}$.

In Figure 23(b) is shown the functional dependence of internal diameter (d_1) stainless chip from the ratio of values (t/ω), such as clearance (t) between of chip wraps to its number wraps(ω), i.e we have the function view of $d_1=\theta_0(t/\omega)$. Analysis of data and also the scatter plots shown in Figure 23(b), and the statistical characteristics(coefficient of determination $R^2=0.10$, coefficient of correlation r=0.29, standard deviation $S_{y/x'2}=0.914$), minimization of the mean square error (min MSE=0.650)shows that this functional dependence $d_1=\theta_0(t/\omega)$ better submits to the linear than non-linear[42]regression model with equation view of **$Y_c=7.17-23X'_2$ or $Y_c=7.17-23(X_1/X_2)$** ,where **$d_1=7.17-23(t/\omega)$ (54)**.And as we see from Figure 23(b) that between two parameters, such as internal diameter (d_1) and ratio (t/ω) there is small correlation (r=0.29). And this function $d_1=\theta_0(t/\omega)$ can be expressed in view of the linear regression equation $d_1=7.17-23(t/\omega)$ which shows that with increasing of ratio (t/ω) ,the value of internal diameter (d_1) decreases considerably.

In Figure 23(b) also is shown the functional dependence of clearance (t) between of chip wraps from the ratio of values (t/ω) , such as clearance (t) between of chip wraps to the number wraps (ω) ,i.e we have the function view of $t=\theta_1(t/\omega)$. Analysis of data and also the scatter plots shown in Figure 23(b),and minimization of the mean square error (min MSE=0.003) and minimization of the mean absolute deviation (min MAD=0) shows that this functional dependence $t=\theta_1(t/\omega)$ better submits to the non-linear than linear[43] regression model with equation view of **$Y_c=0.241(X'_2)^{-0.114}$** or **$Y_c=0.241(X_1/X_2)$** ,where **$t=0.241(t/\omega)^{-0.114}$ (55)**.And as we see from Figure 23(b) that between two parameters, such as clearance (t) between of chip wraps and ratio (t/ω) there is small correlation . And this function $t=\theta_1(t/\omega)$ can be expressed in view of the non-linear regression equation view of $t=0.241(t/\omega)^{-0.114}$ which shows that with increasing of ratio (t/ω) ,the value of clearance (t) between of chip wraps increases considerably in accordance with the non-linear regression equation view of $t=0.241(t/\omega)^{-0.114}$.

42- Non-linear regression model has the equation view of $d_1=5.89(t/\omega)^{0.062}$

43-Linear regression model has the equation view of $t= -0.4+8(t/\omega)$

In Figure 23(c) is shown the functional dependence of internal diameter (d_1) stainless chip from the ratio of values (t/ω) ,such as clearance (t) between of chip wraps to the number of wraps(ω)at constant average value of clearance (t) between of chip wraps(t_c=0.3=const),i.e we have the functional dependence $d_1=\eta_1(t,t/\omega)$.

Analysis of data given in Figure 23(c) shows that with increasing of ratio (t/ω) for stainless chip ,the value of internal diameter (d_1) decreases considerably in accordance with the non-linear regression equation view of $d_1=5.03+9.01t-26.8t^2+0.08(t/\omega)$ at t_c=0.3=const.

In Figure 23(d) is shown the functional dependence of internal diameter (d_1) stainless chip from the clearance (t) between of chip wraps at constant average value of number of chip wraps (ω_c=3.5=const),i.e we have the functional dependence $d_1=\eta_1(t,t/\omega)$.

Analysis of data given in Figure 23(d) shows that with increasing of clearance (t) between of chip wraps for stainless chip, the value of internal diameter (d_1) decreases considerably in accordance with the non-linear regression equation view of d_1=5.03+9.01t−26.8t^2+0.08(t/ω) at ω_c=3.5=const. In Table 15 is shown the computation worksheet for Figures 23(c) and 23(d).

Table 15 Computation worksheet for determining of internal diameter (d_1)

Observed / Estimated	Clearance (t)	Figure 23(c) –ratio (t/ω)			Figure 23(d) –number of chip wraps(ω).		
		0.07	0.09	0.11	3.0	3.5	4.0
Internal diameter (d_1)*	0.10	5.66*	5.66	5.66	5.66	5.66	5.66
	0.30	5.32	5.32	5.32	5.32	5.32	5.32
	0.50	2.84	2.84	2.84	2.84	2.84	2.84

In Figure 23(e) is shown the functional dependence of the internal diameter (d_1) stainless chip from the number chip wraps (ω).So , we see from Figure 23(e) that this functional dependence $d_1=\phi_3(\omega)$ has the non-linear regression model with equation view of $Y_c=20.32X^{-1.102}$ or $d_1=20.32\omega^{-1.102}$. Analysis of data given in Figure 23(e) shows that with increasing of the number chip wraps (ω) the value of internal diameter (d_1) decreases considerably in accordance with the non-linear regression equation view of $d_1=20.32\omega^{-1.102}$.In Figure 23(f) is illustrated the residual plot (residual versus t/ω) of function $d_1=\phi_0(t/\omega)$.

Analyzing the Figure 24(A) ,we see that functional surface $d_1=\eta_1(t, t/\omega)$ is shown in view of three-dimensional drawing for the non-linear regression model with equation view of $Y_c=5.03+9.01X_1-26.8X_1{}^2+0.08X'_2$ or $d_1=5.03+9.01t-26.8t^2+0.08(t/\omega)$.

And Figure 24(B) shows that functional surface $d_1=\eta_1(t, t/\omega)$ has view of trapezium on the face YOX'$_2$ with such coordinates of their points (sizes in millimeters):

a($X_{1,1}$=0,$X_{2,1}$=0.07 ,Y_1=5.75);b($X_{1,2}$=0,$X_{2,2}$=0.07,Y_2=0); c($X_{1,3}$=0 ,$X_{2,3}$=0.11,Y_3=0), d($X_{1,4}$=0 ,$X_{2,4}$=0.11,Y_4=4.10).

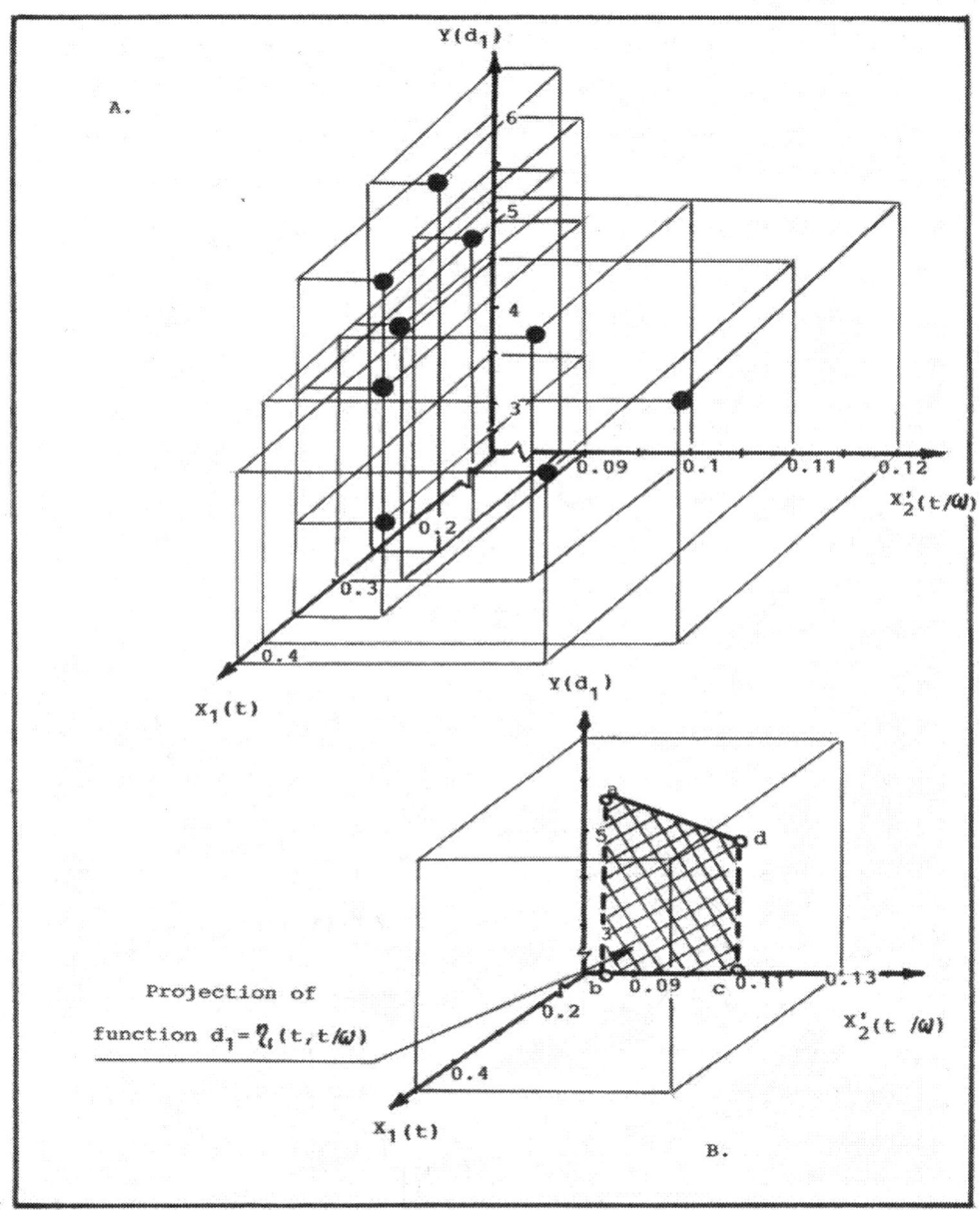

Figure 24

Analysis of functional surface $d_1 = \eta_1(t, t/\omega)$

From Figure 24(A) we see that functional surface $d_1 = \eta_1(t, t/\omega)$, on which are situated the points of the non-linear regression equation view of $Y_c = 5.03 + 9.01X_1 - 26.8X_1^2 + 0.08X'_2$ or $d_1 = 5.03 + 9.01t - 26.8t^2 + 0.08(t/\omega)$ has the following coordinates of their peaks in three-dimensional drawing (sizes in millimeters): a'($X_{1,1} = 0.21, X_{2,1} = 0.07, Y_1 = 5.75$); b'($X_{1,2} = 0.21$, $X_{2,2} = 0.07, Y_2 = 0$); c'($X_{1,3} = 0.42, X_{2,3} = 0.11, Y_3 = 0$); d'($X_{1,4} = 0.42, X_{2,4} = 0.11, , Y_4 = 4.10$).

317

In Figure 25 schematically is shown the functional surface $d_1=\eta_1(t\,,t/\omega)$ and graph for calculation of module for vectors $|b'c'|$ and $|a'd'|$.

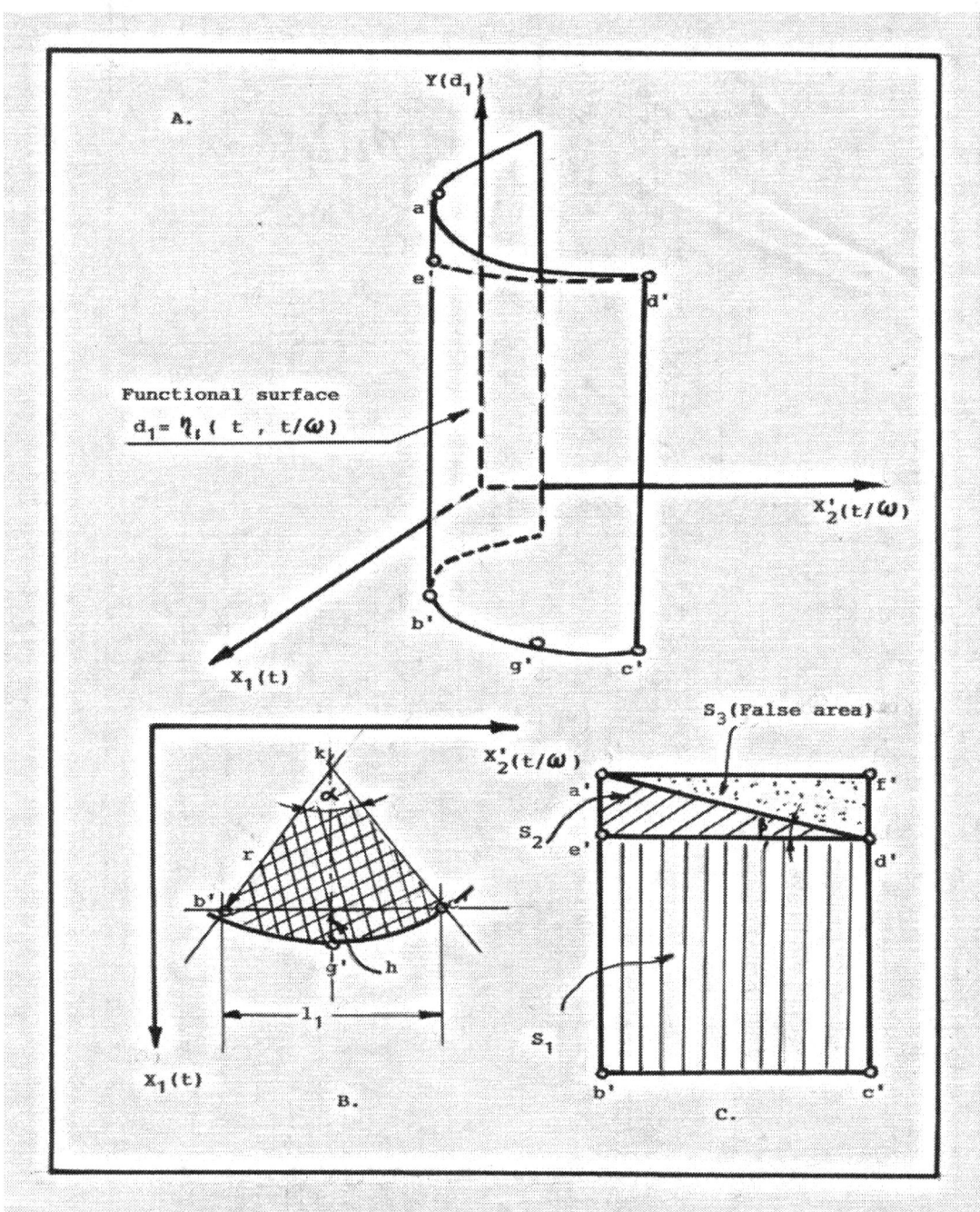

Figure 25 Graph for calculation of module vectors $|b'c'|$,$|a'd'|$ and area of functional surface $d_1=\eta_1(\,t\,,t/\omega)$: A=total view of functional surface; B=graph for calculation of module vector $|b'c'|$; C=graph for calculation of module vector $|a'd'|$ and area of this surface.

1.The module of vector |a'b'| and |c'd'| can be calculated from the projection of this functional surface $d_1=\eta_1(t,t/\omega)$ on the plane YOX'_2(Figure 24 B and 25A) by formulas (7) and (9).So, the module of vectors |a'b'|=5.75 and |c'd|=4.10.

2.The module of vector |b'c'| or it length can be calculated by formulas $L=|b'c'|=0.01745r\alpha$, where L=length of arc|b'c'|, r=radius of sector, α=angle of sector. So ,the module of vector |k'g'|is equal: $r=|k'g'|=[(X_{1,6}-X_{1,5})^2+(X_{2,6}-X_{2,5})^2]^{1/2}=0.35$, where the coordinates of points g' and k: $g'(X_{1,6}=0.34,X_{2,6}=0.08)$,$k(X_{1,5}=0,X_{2,5}=0)$. At data $\alpha=82°$,r=0.35 we have the module of vector |b'c'|=0.50 and other parameters of circular segment which are equal: height $h=r[1-\cos(\alpha/2)]=0.08$,the length of chord $l_1=2[h(2r-h)]^{1/2}=0.44$.
The graph for calculation of module of vectors |b'c'| ,|a'd'| and area of functional surface $d_1=\eta_1(t,t/\omega)$ is shown in Figure 25.

3.The module of vector |a'd'| and area of this functional surface , as shown in Figure 25 (C) ,can be calculated the following way: the module of length for curve |d'e'|is equal |d'e'|=|b'c'|,i.e we have |d'e'|=0.50 and module of length for |a'e'|=|a'b'|-|c'd'|=1.65.

4.For calculation of area S_1,S_2 and S_3 we find the angle β which is equal: $\tan\beta=|a'e'|/|d'e'|=3.30$,where $\beta=73°3'$ and then the module of length for curve |a'd'|is equal |a'd'|=|a'e'|/sin73°3'=1.72.The coordinates of point f' is equal: $f'(X_{1,3}=0.42,$ $,X_{2,3}=0.11,Y_1=5.75)$ and module of length |f'a'|=|b'c'|=0.50. Area of $S_2=S_3=0.5S=$ =0.5[|a'e'|·|d'e'|=0.41mm^2. For these conditions, we have the area S_1 which is equal: S_1=|c'd'|·|b'c'|=2.05mm^2 and total area $\sum S$=2.46mm^2.
All results of computations for the non-linear regression equation view of $Y_c=5.03+9.01X_1-26.8X_1{}^2+0.08X'_2$ or $Y_c=5.03+9.01X_1-26.8X_1{}^2+0.08(X_1/X_2)$,where $d_1=5.03+9.01t-26.8t^2+0.08(t/\omega)$ are given in Table 16.

So ,the parameters of the functional surface $d_1=\eta_1(t,t/\omega)$ are the following:

1.Function $d_1=\eta_1(t,t/\omega)$ better submits to non-linear regression model in view of curvilinear surface ,as shown in Figure 24(A) and 25(A),and describes by equation view of $Y_c=5.03+9.01X_1-26.8X_1{}^2+0.08X'_2$ or $d_1=5.03+9.01t-26.8t^2+0.08(t/\omega)$ with such statistical characteristics:
- Coefficient of correlation R^2=0.18
- Coefficient of correlation r=0.42
- Standard deviation $S_{y/x',x'2}$=1.141
- Minimization of the mean square error (min MSE=0.512)
- Minimization of the mean absolute deviation (min MAD=0.018).

2.The total area of functional surface $d_1=\eta_1(t,t/\omega)$ is equal $\sum S$=2.46 mm^2 with such coordinates of their points in three- dimensional drawing for this surface(sizes in millimeters):
$a'(X_{1,1}=0.21,X_{2,1}=0.07,Y_1=5.75)$;$b'(X_{1,2}=0.21,X_{2,2}=0.07,Y_2=0)$; $c'(X_{1,3}=0.42,X_{2,3}=0.11,$ $,Y_3=0)$;$d'(X_{1,4}=0.42,X_{2,4}=0.11,Y_4=4.10)$.

Table 16 Evaluation of regression equation $Y_c=5.03+9.01X_1-26.8X_1^2+0.08X'_2$

A. Mean, variance and standard deviation

Variable	Mean	Variance	Standard deviation
X_1	0.317	0.031	0.176
X'_2	0.090	0	0
Y	5.098	5.612	2.369

B. Results of multiple regression of Y on X_1 and X'_2

Parameter	Variable	Coefficients	Standard error	T-value
β_1	X_1	9.01	0.06	150.16
β_2	X'_2	0.08	0	0

C. Analysis of variance results

Regression
- Degrees of freedom 2
- Sum of squares   0.997
- Mean squares   0.50

Error
- Degrees of freedom 6
- Sum of squares   4.611
- Mean squares   0.769

F-value*   0.65

* Since $F=0.65<[F_{0.05,2,6}=5.14]$ we can not reject the hypothesis that both β_1 and β_2 are zero.

D. Determination of residuals

Number	Observed	Estimated	Residual
1	5.36	5.75	-0.39
2	5.22	4.60	0.62
3	3.52	5.00	-1.48
4	4.51	4.10	0.41
5	4.98	5.33	-0.35
6	4.74	5.29	-0.55
7	5.15	5.19	-0.04
8	5.96	5.18	0.78
9	6.44	5.60	0.84

C. Modification function $d_1=\eta_2(\,t\,,t/\delta\,)$

Analysis of data and the statistical characteristics (coefficient of determination $R^2=0.748$, coefficient of correlation $r=0.865$, standard deviation $S_{y/x1,x'2}=0.486$) and also minimization of the mean square error (min MSE=0.157) and minimization of the mean absolute deviation (min MAD=0) shows that this functional dependence $d_1=\eta_2(t\,,t/\delta)$ better submits to the linear than non-linear[44] regression model with equation view of
$$Y_c=6.033-0.229X_1-0.718X'_2 \quad \text{or} \quad Y_c=6.033-0.229X_1-0.073(X_1/X_2)\ ,\ \text{where}$$
$$d_1=6.033-0.229t-0.073(t/\delta) \quad (\,56\,)\,.$$

44-Non-linear regression model has the equation view of $d_1=25.1-140.49t+251t^2-1.31(t/\delta)$

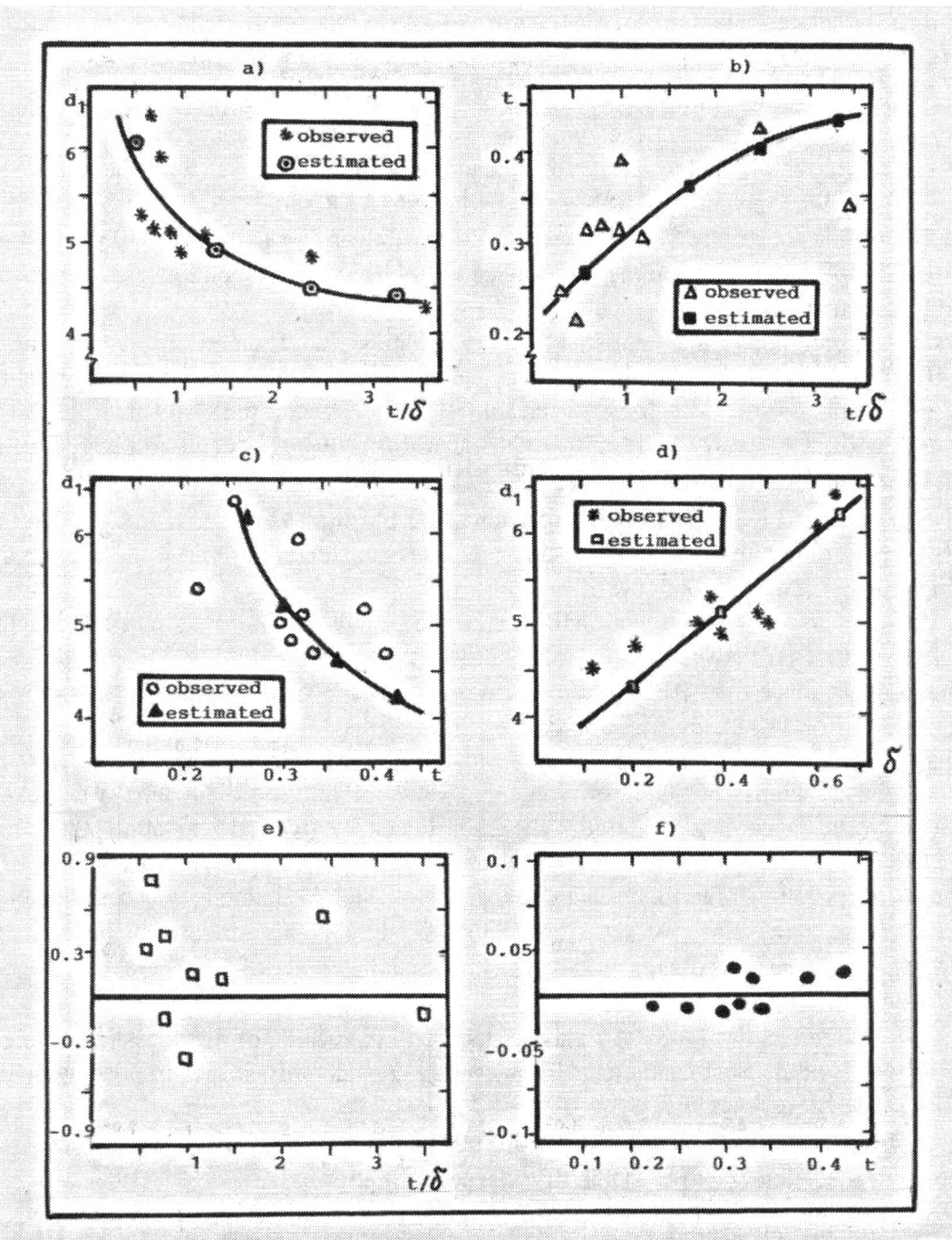

Figure 26

In Figure 26(a) is shown the functional dependence internal diameter (d_1) stainless chip from the ratio (t/δ) such as clearance (t) between of chip wraps to its thickness (δ) ,i.e we have the function $d_1=\rho_0(t/\delta)$.Analysis of data and the statistical characteristics (coefficient of determination $R^2=0.554$,coefficient of correlation r=0.744 ,standard deviation $S_{y/x'2}=0.598$),minimization of the mean square error (min MSE=0.278) and minimization of the mean absolute deviation (min MAD=0.114) shows that this functional dependence $d_1=\rho_0(t/\delta)$ better submits to non-linear than linear[45] regression model with equation view of

$$Y_c=4.98(X'_2)^{-0.237} \text{ or } Y_c=4.98 (X_1/X_2)^{-0.237}, \text{ where } d_1=4.98(t/\delta)^{-0.237} \quad (57).$$

So, we see from Figure 26(a) that with increasing of ratio (t/δ), the value of internal diameter (d_1) for stainless chip decreases considerably in accordance with the non-linear regression equation view of $Y_c=4.98X'_2{}^{-0.237}$, where $d_1=4.98(t/\delta)^{-0.237}$.

In Figure 26(b) is shown the functional dependence of clearance (t) between of chip wraps from its ratio (t/δ), such as clearance (t) between of chip wraps to its thickness (δ), i.e we have the function $t=\rho_1(t/\delta)$.

Analysis of data and the statistical characteristics (coefficient of determination $R^2=0.467$, coefficient of correlation r=0.683, standard deviation $S_{y/x'2}=0.048$), minimization of the mean square error (min MSE=0) and minimization of the mean absolute deviation (min MAD=0) better submits to the non-linear than linear[46] regression model with equation view of $Y_c=0.32X'_2{}^{0.19}$ or $Y_c=0.32(X_1/X_2)^{0.19}$, where $t=0.32(t/\delta)^{0.19}$ (58). So, we see from Figure 26(b) that with increasing of ratio (t/δ), the value of clearance (t) between of chip wraps for stainless chip increases considerably in accordance with the non-linear regression equation view of $t=0.32(t/\delta)^{0.19}$.

In Figure 26(c) is shown the functional dependence of internal diameter(d_1) stainless chip from its clearance (t) between of chip wraps. So, we see from Figure 26(c) that this functional dependence $d_1=\gamma_4(t)$ has the non-linear regression model with equation view of $Y_c=0.828X^{-0.605}$ or $t=0.828d_1{}^{-0.605}$, where $d_1=(1.208t)^{-1.653}$.

Analysis of data shown in Figure 26(c) shows that with increasing of clearance (t) between of chip wraps for stainless chip, the value of internal diameter(d_1) decreases considerably in accordance with the non-linear regression equation view of $d_1=(1.208t)^{-1.653}$.

In Figure 26(d) is shown the functional dependence of internal diameter (d_1) stainless chip from its thickness (δ). So, we see from Figure 26(d) that this functional dependence $d_1=\varphi(\delta)$ has the linear regression model with the equation view of $Y_c=3.338+4.762X$ or $d_1=3.338+4.762\delta$.

Analysis of data given in Figure 26(d) shows that with increasing of thickness (δ) stainless chip, the value of internal diameter (d_1) increases considerably in accordance with linear regression equation view of $d_1=3.338+4.762\delta$.

In Figures 26(e) and 26(f) are illustrated the residual plots (residual versus t/δ and t) of functional dependencies $d_1=\rho_0(t/\delta)$ and $d_1=\gamma_4(t)$ accordingly.

Analyzing the Figure 27(A), we see that functional dependence $d_1=\eta_2(t, t/\delta)$ is shown in three-dimensional drawing for the linear regression model with equation view of $Y_c=6.033-0.229X_1-0.718X'_2$ or $Y_c=6.033-0.229X_1-0.718(X_1/X_2)$, where $d_1=6.033-0.229t-0.718(t/\delta)$.

45-Linear regression model has the equation view of $d_1=5.971-0.726(t/\delta)$
46-Linear regression model has the equation view of $t=0.276-0.034(t/\delta)$

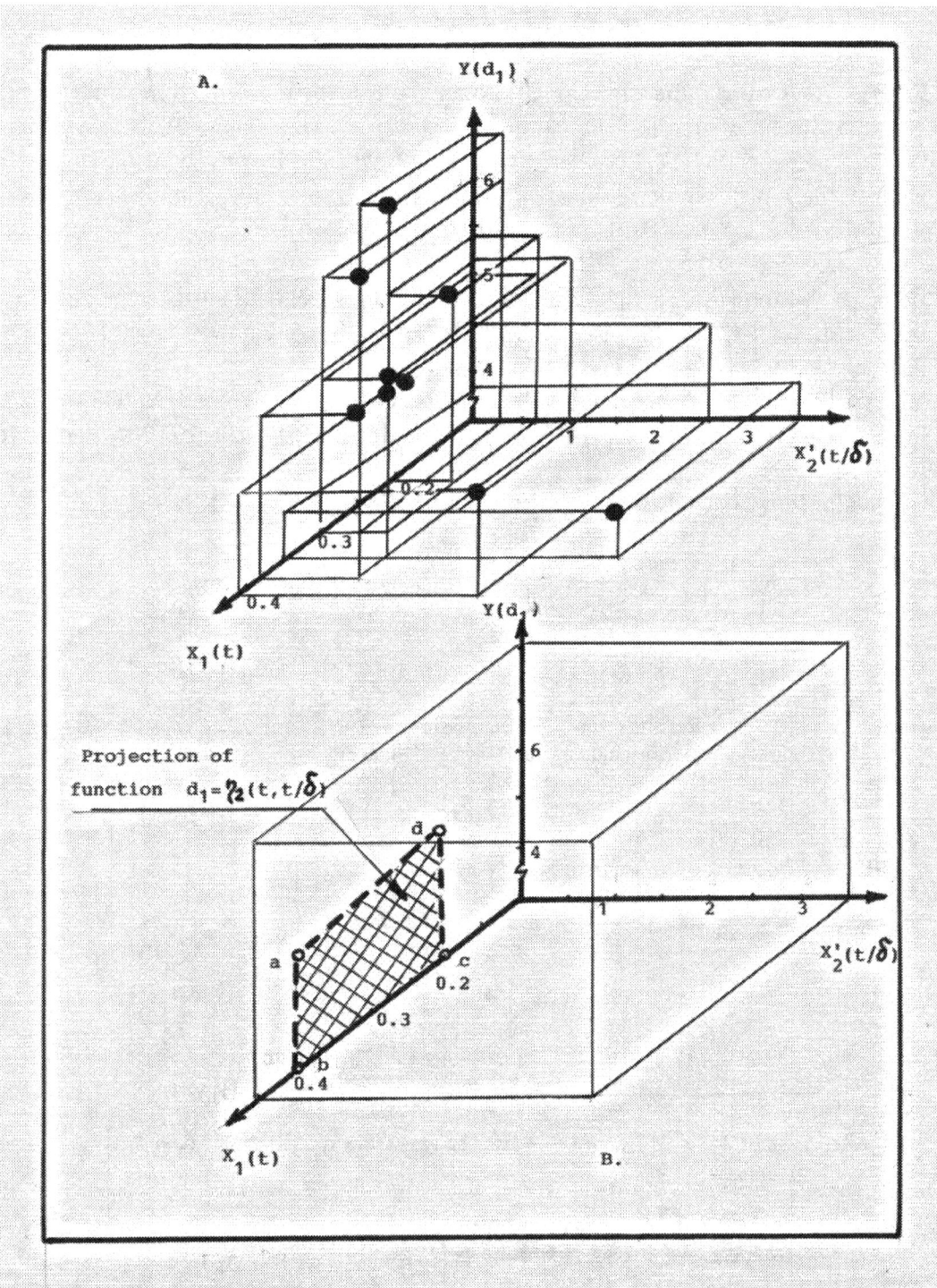

Figure 27

And Figure 27(B) shows that this functional surface $d_1=\eta_2(t,t/\delta)$ has view of trapezium on the place YOX_1 with such coordinates of their points in three-dimensional drawing (sizes in millimeters): $a(X_{1,1}=0.38, X_{2,1}=0, Y_1=5.30)$; $b(X_{1,2}=0.38, X_{2,2}=0, Y_2=0)$; $c(X_{1,3}=0.21, X_{2,3}=0, Y_3=0)$; $d(X_{1,4}=0.21, X_{2,4}=0, Y_4=5.57)$.

Analysis of functional surface $d_1=\eta_2(t,t/\delta)$

From Figure 27(A) we see that functional surface $d_1=\eta_2(t,t/\delta)$ on which are situated the points of the linear regression equation view of $Y_c=6.033-0.229X_1-0.718X'_2$ or $Y_c=6.033-0.229X_1-0.718(X_1/X_2)$,where $d_1=6.033-0.229t-0.718(t/\delta)$ has the following coordinates of their peaks in three-dimensional drawing(sizes in millimeters):

a'$(X_{1,1}=0.38,X_{2,1}=0 ,Y_1=5.30)$; b'$(X_{1,2}=0.38,X_{2,2}=0,Y_2=0)$;c'$(X_{1,3}=0.21,X_{2,3}=0,Y_3=0)$; d'$(X_{1,4}=0.21,X_{2,4}=0,Y_4=5.57)$.

The module of vectors $|a'b'|,|b'c'|,|c'd'|,|d'a'|$ and area of functional surface $S_{a'b'c'd'}$ can be calculated by formulas (7),(8),(9),(10) and (16).At above-shown data we have : $|a'b'|=5.30$, $|b'c'|=0.17,|c'd'|=5.57,|d'a'|=0.32$ and $S_{a'b'c'd'}=0.93mm^2$.

All results of computation for the linear regression equation view of $Y_c=6.033-0.229X_1-0.718X_2'$ or $Y_c=6.033-0.229X_1-0.718(X_1/X_2)$,where $d_1=6.033-0.229t-0.718(t/\delta)$ are given in Table 17.

Table 17 Evaluation of regression equation $Y_c=6.033-0.229X_1-0.718X'_2$

A. Mean , variance and standard deviation

Variable	Mean	Variance	Standard deviation
X_1	0.317	0.029	0.170
X_2	1.202	7.978	2.824
Y	5.098	5.612	2.369

B. Results of multiple regression of Y on X_1 and X'_2

Parameter	Variable	Coefficients	Standard error	T-value
β_1	X_1	-0.229	0.057	- 4.02
β_2	X'_2	-0.718	0.941	- 0.76

C. Analysis of variance results

Regression
Degrees of freedom 2
Sum of squares 4.192
Mean squares 2.096
Error
Degrees of freedom 6
Sum of squares 1.416
Mean squares 0.236
F-value * 8.88

*Since $F=8.88>[F_{0.05,2,6}=5.14]$ we can reject the hypothesis that both β_1 and β_2 are zero.

D. Determination of residuals

Number	Observed	Estimated	Residual
1	5.36	5.57	- 0.21
2	5.22	5.30	- 0.08
3	3.52	3.51	0.01
4	4.51	4.26	0.25
5	4.98	5.20	-0.22
6	4.74	5.36	-0.62
7	5.15	5.47	- 0.32
8	5.96	5.54	0.42
9	6.44	5.67	0.77

So ,the parameters of this functional surface $d_1=\eta_2(t,t/\delta)$ are the following:

1.Function $d_1=\eta_2(t,t/\delta)$ better submits to the linear regression model with equation view of $Y_c=6.033-0.229X_1-0.718X'_2$ or $Y_c=6.033-0.229X_1-0.718(X_1/X_2)$,where $Y_c=6.033-0.229t-0.718(t/\delta)$ with such statistical characteristics: coefficients of determination $R^2=0.748$,coefficient of correlation $r=0.865$,standard deviation $S_{y/x1,x'2}=0.486$, minimization of the mean square error (min MSE=0.157), minimization of the mean absolute deviation (min MAD=0).

2. The total area of functional surface $d_1=\eta_2(t, t/\delta)$ is equal $\sum S=0.93mm^2$ with such coordinates of their points in three-dimensional drawing for this surface (sizes in millimeters): a'($X_{1,1}=0.38,X_{2,1}=0,Y_1=5.30$);b'($X_{1,2}=0.38,X_{2,2}=0$,$Y_2=0$);c'($X_{1,3}=0.21$, ,$X_{2,3}=0,Y_3=0$),d'($X_{1,4}=0.21,X_{2,4}=0,Y_4=5.57$).

D. Modification function $d_1=\eta_3(t,\omega/\delta)$

Analysis of data and the statistical characteristics (standard deviation $S_{y/x1,x'2}=1.75$) and minimization of the mean square error (min MSE=2.03) shows that this functional dependence $d_1=\eta_3(t,\omega/\delta)$ better submits to the linear than non-linear[47] regression model with equation view of **$Y_c=5.61+6.65X_1-0.19X'_2$** or **$Y_c=5.61+6.65X_1-0.19(X_2/X_3)$** , where **$d_1=5.61+6.65t-0.19(\omega/\delta)$** (59).

In Figure 28(a) is shown the functional dependence of internal diameter (d_1) stainless chip from the ratio (ω/δ) , such as number of chip wraps (ω) to its thickness (δ) ,i.e we have the function $d_1=\theta_0(\omega/\delta)$. Analysis of data and the statistical characteristics (coefficient of determination $R^2=0.565$,coefficient of correlation $r=0.752$,standard deviation $S_{y/x'2}=0.59$), minimization of the mean square error (min MSE=0.271) and minimization of the mean absolute deviation (min MAD=0.038) shows that this functional dependence $d_1=\theta_0(\omega/\delta)$ better submits to the non-linear than linear[48] regression model with equation view of **$Y_c=6.61(X'_2)^{-0.113}$** or **$d_1=6.61(\omega/\delta)^{-0.113}$** (60). So ,we see from Figure 28(a) that with increasing of ratio (ω/δ) ,the value of internal diameter (d_1) for stainless chip decreases considerably in accordance with the non-linear regression equation view of $d_1=6.61(\omega/\delta)^{-0.113}$.

And besides in Figure 28(a) is shown the functional dependence of clearance (t) between of chip wraps from the ratio (ω/δ) ,such as number of chip wraps (ω) to its thickness (δ) ,i.e we have the function view of $t=\theta_1(\omega/\delta)$.Analysis of data and the statistical characteristics (coefficient of determination $R^2=0.29$,coefficient of correlation $r=0.54$,standard deviation $S_{y/x'2}=0.067$),minimization of the mean square error(min MSE=0.003) and minimization of the mean absolute deviation (min MAD=0.005) shows that this functional dependence $t=\theta_1(\omega/\delta)$ better submits to the non-linear than linear[49] regression model with equation view of **$Y_c=0.229(X'_2)^{0.128}$** or **$t=0.229(\omega/\delta)^{0.129}$** (61).

47- Non-linear regression model has the equation view of $d_1=5.93+3.67t+6.46\ t^2-0.196(\omega/\delta)$
48- Linear regression model has the equation view of $d_1=4.83+0.02(\omega/\delta)$
49-Linear regression model has the equation view of $t=0.29+0.002(\omega/\delta)$

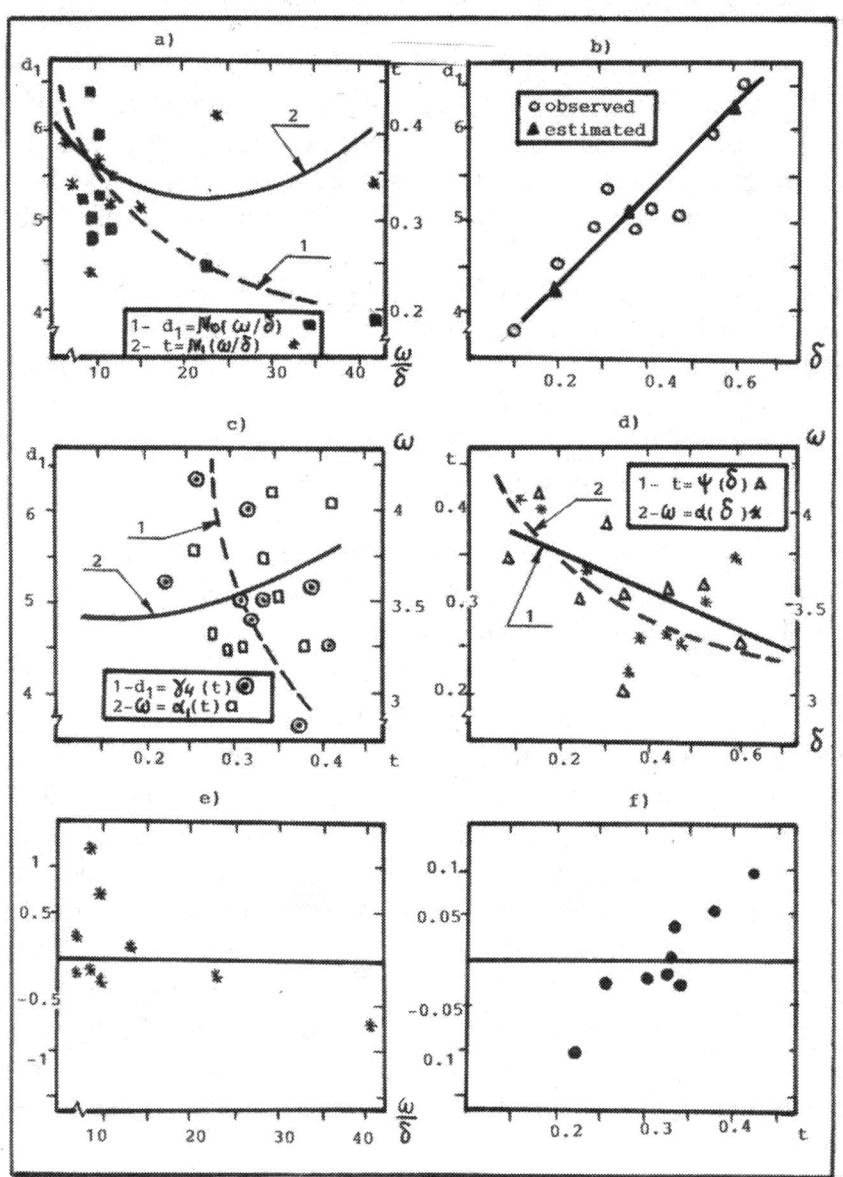

Figure 28

So , we from Figure 28(a) that with increasing of ratio (ω/δ) ,the value of clearance (t) between of chip wraps increases considerably in accordance with the non-linear regression equation view of $t=0.229(\omega/\delta)^{0.128}$.

In Figure 28(b) is shown the functional dependence of internal diameter (d_1) from thickness (δ) ,i.e we have the function view of $d_1=\varphi(\delta)$.So ,we see from Figure 28(b) that this functional dependence has the linear regression model with equation view of $Y_c=3.338+4.762X$ or $d_1=3.338+4.762\delta$.

Analysis of data shown in Figure 28(b) shows that with increasing of thickness (δ) stainless chip, the value of internal (d_1) increases considerably in accordance with the linear regression equation view of $d_1=3.338+4.762\delta$.

In Figure 28 (c) is shown the functional dependence of internal diameter(d_1) from clearance (t) between of chip wraps. This function $d_1=\gamma_4(t)$ has the non-linear regression model with equation view of $Y_c=0.828X^{-0.605}$ or $t=0.828d_1^{-0.605}$, where $d_1=(1.208t)^{-1.653}$.

Analyzing of data shown in Figure 28(c), we see that with increasing of clearance (t) between of chip wraps for stainless chip, the value of internal diameter(d_1) decreases considerably in accordance with the non-linear regression equation view of $d_1=(1.208t)^{-1.653}$.

In Figure 28 (c) also is shown the functional dependence of number (ω) chip wraps from the clearance (t) between of chip wraps ,i.e we have the function view of $\omega=\alpha_1(t)$. So ,we see from Figure 28 (c) that this functional dependence has the non-linear regression model with equation view of $Y_c=4.227X^{0.152}$ or $\omega=4.227t^{0.152}$. Analyzing of data shown in Figure 28(c), we see that with increasing of clearance (t) between of chip wraps ,the value of number (ω) chip wraps increases considerably in accordance with the non-linear regression equation view of $\omega=4.227t^{0.152}$.

In Figure 28(d) is shown the functional dependence of clearance (t) between chip wraps from thickness (δ) of this stainless chip ,i.e we have the function view of $t=\phi(\delta)$. So ,we see from Figure 28(d) that this functional dependence has the linear regression model with equation view of $Y_c=0.379-0.16X$ or $t=0.379-0.16\delta$. Analyzing of data shown in Figure 28(d) ,we see that with increasing of thickness (δ) stainless chip , the value of clearance (t) between chip wraps decreases considerably in accordance with the linear regression equation view of $t=0.379-0.16\delta$.

In Figure 28(d) also is shown the functional dependence of number (ω) chip wraps ,i.e we have the function view of $\omega=\alpha(\delta)$. So ,we see from Figure 28(d) that this functional dependence has the non-linear regression model with equation view of $Y_c=3.09X^{-0.123}$, where $\omega=3.09\delta^{-0.123}$.

Analyzing of data shown in Figure 28(d) ,we see that with increasing of thickness (δ) stainless chip, the value of number (ω) chip wraps decreases considerably in accordance with the non-linear regression model with equation view of $\omega=3.09\delta^{-0.123}$.

In Figures 28(e) and 28(f) are illustrated the residual plots (residual versus ω/δ and t) of functional dependencies $d_1=\theta_0(\omega/\delta)$ and $d_1=\gamma_4(t)$ accordingly.

Analyzing the Figure 29(A) ,we see that functional dependence $d_1=\eta_3(t,\omega/\delta)$ is shown in three-dimensional drawing for the linear regression model with equation view of $Y_c=5.61+6.65X_1-0.19X'_2$ or $Y_c=5.61+6.65X_1-0.19(X_2/X_3)$,where $d_1=5.61+6.65t-0.19(\omega/\delta)$.

And Figure 29(B) shows that this functional surface $d_1=\eta_3(t,\omega/\delta)$ has view of trapezium on the place YOX_1 with such coordinates of their points in three-dimensional drawing (sizes in millimeters):

$a(X_{1,1}=0.21,X_{2,1}=0,Y_1=5.36)$;$b(X_{1,2}=0.21,X_{2,2}=0,Y_2=0)$; $\quad c(X_{1,3}=0.38,X_{2,3}=0,Y_3=0)$; $d(X_{1,4}=0.38,X_{2,4}=0,Y_4=5.22)$.

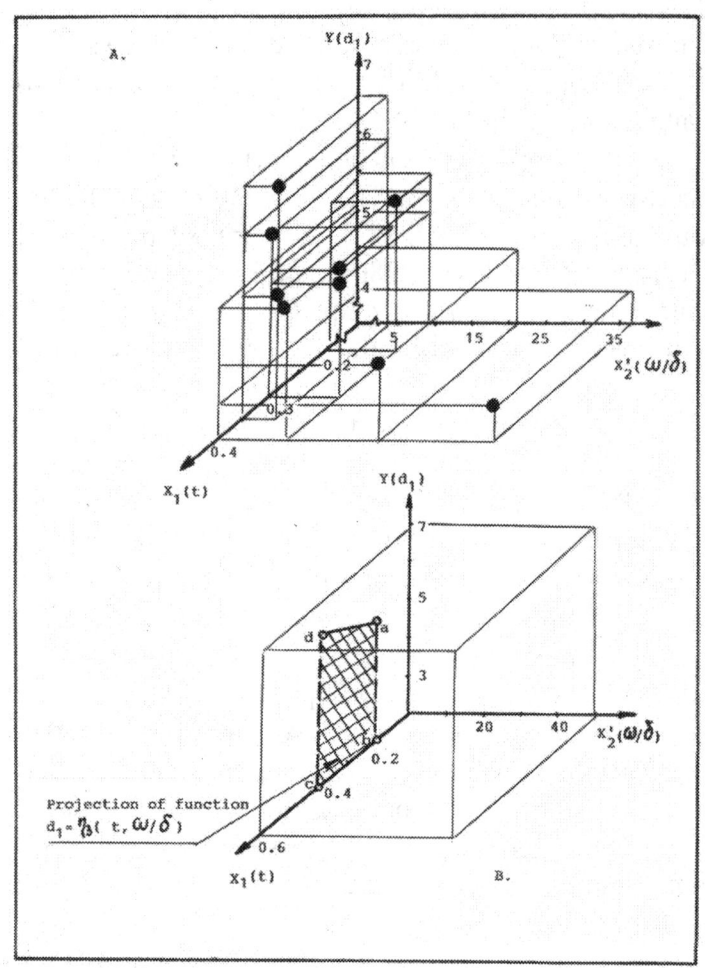

Figure 29

Analysis of functional surface $d_1 = \eta_3(t, \omega/\delta)$

From Figure 29(A) we see that functional surface $d_1 = \eta_3(t, \omega/\delta)$, on which are situated the points of the linear regression equation view of $Y_c = 5.61 + 6.65X_1 - 0.19X'_2$ or $Y_c = 5.61 + 6.65X_1 - 0.19(X_2/X_3)$, where $d_1 = 5.61 + 6.65t - 0.19(\omega/\delta)$ has the following coordinates of their peaks in three-dimensional drawing(sizes in millimeters):
a'$(X_{1,1}=0.21, X_{2,1}=8.81, Y_1=5.36)$;b'$(X_{1,2}=0.21, X_{2,2}=8.81, Y_2=0)$;c'$(X_{1,3}=0.38, X_{2,3}=7.81, Y_3=0)$;d'$(X_{1,4}=0.38, X_{2,4}=7.81, Y_4=5.22)$.
The module of vectors $|a'b'|, |b'c'|, |c'd'|, |d'a'|$ and area of functional surface $S_{a'b'c'd'}$ can be calculated by formulas (7),(8),(9),(10) and (11). At above-shown data we have : $|a'b'|=5.36, |b'c'|=1.0, |c'd'|=5.22, |d'a'|=1.03$ and $S_{a'b'c'd'}=5.29mm^2$.

All results of computation for the linear regression equation view of $Y_c = 5.61 + 6.65X_1 - 0.19X_2'$ or $Y_c = 5.61 + 6.65X_1 - 0.19(X_2/X_3)$, where $d_1 = 5.61 + 6.65t - 0.19(\omega/\delta)$ are given in Table 18.

Table 18 Evaluation of regression equation $Y_c=5.61+6.65X_1-0.19X'_2$

A. Mean , variance and standard deviation

Variable	Mean	Variance	Standard deviation
X_1	0.317	0.029	0.170
X'_2	13.609	1119.01	33.452
Y	5.098	5.612	2.369

B. Results of multiple regression of Y on X_1 and X'_2

Parameter	Variable	Coefficients	Standard error	T-value
β_1	X	6.65	0.057	116.67
β_2	X'_2	-0.19	11.15	-0.02

C. Analysis of variance results

Regression
 Degrees of freedom 2
 Sum of squares   12.682
 Mean squares   6.341
Error
 Degrees of freedom 6
 Sum of squares   18.29
 Mean squares   3.048
 F- value *   2.08

* Since F=2.08< [$F_{0.05,2,6}$=5.14] we can not reject the hypothesis that both β_1 and β_2 are zero.

D. Determination of residuals

Number	Observed	Estimated	Residual
1	5.36	5.33	0.03
2	5.22	6.66	-1.44
3	3.52	0	3.52
4	4.51	4.19	0.32
5	4.98	5.15	-0.17
6	4.74	6.04	-1.30
7	5.15	6.41	-1.26
8	5.96	6.50	-0.54
9	6.44	6.08	0.36

So , the parameters of this functional surface $d_1=\eta_3(t,\omega/\delta)$ are the following:

1. Function $d_1=\eta_3(t,\omega/\delta)$ better submits to the linear regression model with equation view of $Y_c=5.61+6.65X_1-0.19X'_2$ or $Y_c=5.61+6.65X_1-0.19(X_2/X_3)$,where $d_1=5.61+6.65t-0.19(\omega/\delta)$ with such statistical characteristics: standard deviation $(S_{y/x1,x'2}=1.75)$,minimization of the mean square error (min MSE=2.03).

2.The total area of functional surface $d_1=\eta_3(t,\omega/\delta)$ is equal $\sum S=5.29mm^2$ with such coordinates of their points in three-dimensional drawing for this surface (sizes in millimeters):

a'$(X_{1,1}=0.21,X_{2,1}=8.81,Y_1=5.36)$;b'$(X_{1,2}=0.21,X_{2,2}=8.81,Y_2=0)$;c'$(X_{1,3}=0.38,X_{2,3}=7.81,$,$Y_3=0)$;d'$(X_{1,4}=0.38,X_{2,4}=7.81,Y_4=5.22)$.

7.5 Dependence of clearance (t) between of chip wraps from thickness(δ) of this chip and also width(B),number of chip wraps(ω),i.e $t=\theta(\delta,B,\omega)$ and their modifications: $t=\gamma_1(\delta, \delta/B)$, $t=\gamma_2(\delta,\delta/\omega)$,$t=\gamma_3(\delta,B/\omega)$.

A. Function $t=\theta(\delta,B,\omega)$

Analysis of data and the statistical characteristics (coefficient of determination $R^2=0.60$,coefficient of correlation r=0.775,standard deviation $S_{y/x1,x'2}=0.049$) ,minimization of the mean square error (min MSE=0) and minimization of the mean absolute deviation (min MAD=0.004) shows that this functional dependence $t=\theta(\delta,B,\omega)$ better submits to the linear than non-linear [50] regression model with equation view of

$$\mathbf{Y_c= -0.99-0.16X_1+0.17X_2+0.10X_3} \text{ or } \mathbf{t= -0.99-0.16\delta+0.17B+0.10\omega} \ (62).$$

In Figure 30 (a) is shown the functional dependence of clearance (t) between of chip wraps from its thickness (δ) ,i.e we have the functional dependence $t=\phi(\delta)$. This functional dependency has the linear regression model with equation view of $Y_c=0.379--0.16X$,where $t=0.379-0.16\delta$.So, from Figure 30(a) we see that with increasing of thickness (δ) stainless chip, the value of clearance (t) between of chip wraps increases considerably in accordance with linear regression equation $t=0.379-0.16\delta$.
And besides in Figure 30(a) is shown the functional dependence of width (B) stainless chip from its thickness (δ),i.e we have the functional dependence $B=f(\delta)$.
This function has the linear regression model with equation view of $Y_c= -1.236+0.275X$ or $\delta= -1.236+ 0.275B$, where $\mathbf{B=(3.64\delta+4.49)}$ (63). So, from Figure 30(a) we see that with increasing of thickness (δ) stainless chip, the value of width(B) increases considerably in accordance with the linear regression equation $B=(3.64\delta+4.49)$.
In Figure 30(b) is shown the functional dependence of clearance (t) between of chip wraps from its number wraps (ω) ,i.e we have the function view of $t=\alpha_1(\omega)$. This functional dependency has the non-linear regression model with equation view of $Y_c=4.227X^{0.152}$ or $\omega=4.227t^{0.152}$,where $\mathbf{t=(0.237\omega)^{6.579}}$ (64).

50- Non-linear regression model has the equation view $t= -0.1-0.576\delta+0.527\delta^2+0.084B+0.017\omega$

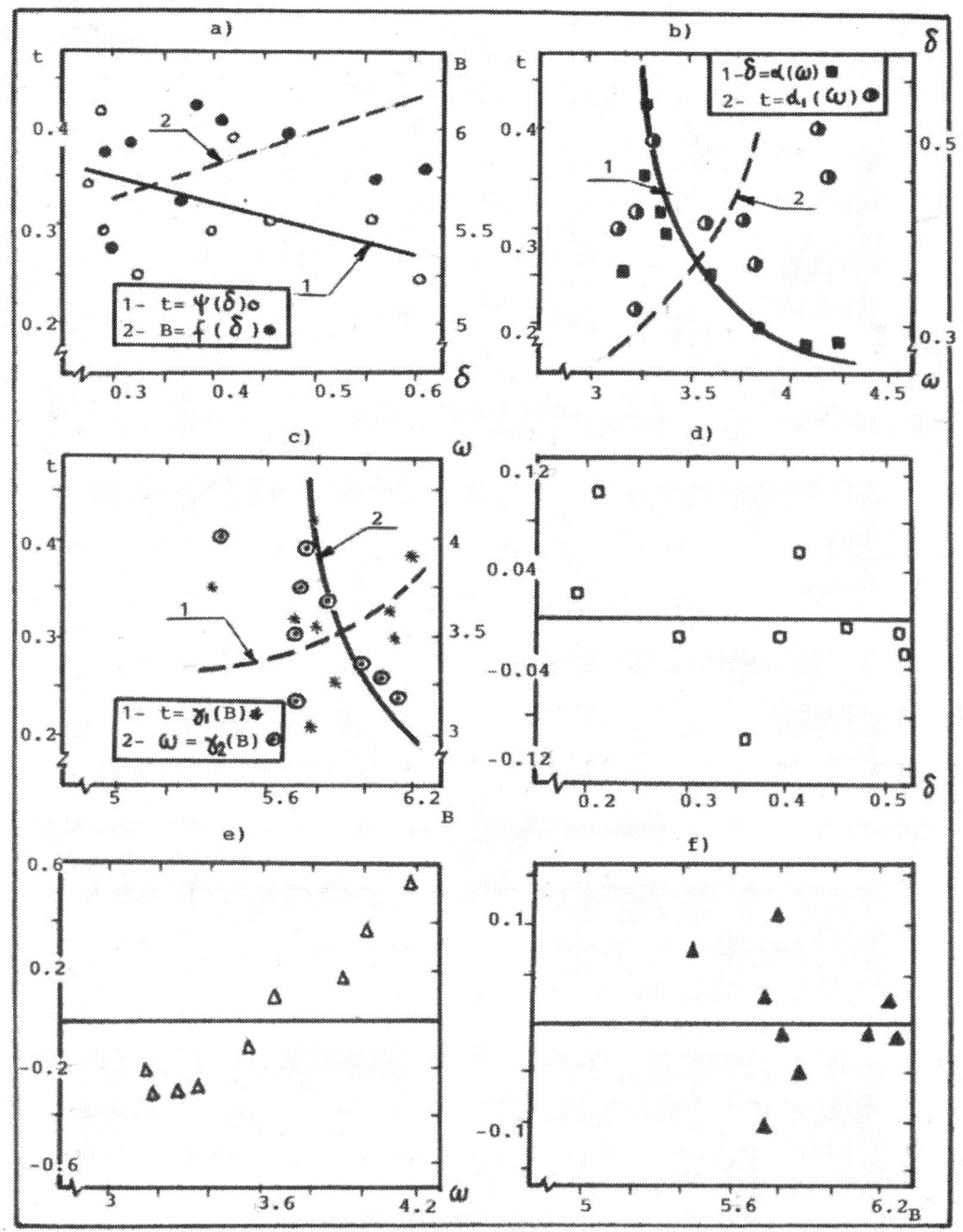

Figure 30

So , from Figure 30(b) we see that with increasing of number (ω) chip wraps ,the value of clearance (t) between of chip wraps increases considerably in accordance with the non-linear regression equation view of $t=(0.237\omega)^{6.579}$.

In Figure 30(b) also is shown the functional dependence of thickness (δ) stainless chip from the number (ω) of chip wraps ,i.e we have the function view of $\delta=\alpha(\omega)$.

This functional dependency has the non-linear regression model with equation view of $Y_c=3.09X^{-0.123}$ or $\omega=3.09\delta^{-0.123}$, where $\delta=(0.324\omega)^{-8.13}$. So, from Figure 30(b) we see that with increasing of number (ω) chip wraps , the value of thickness (δ) decreases considerably in accordance with the non-linear regression equation view of $\delta=(0.324\omega)^{-8.13}$.

In Figure 30(c) is shown the functional dependence of clearance(t) between of chip wraps from width(B) of this stainless chip ,i.e we have the function view of $t=\gamma_1(B)$. This functional dependency has the non-linear regression model with equation view of $Y_c=0.032X^{1.296}$,where $t= 0.032B^{1.296}$.

So ,we see from Figure 30 (c) that with increasing of width(B) stainless chip, the value of clearance (t) between of chip wraps increases considerably in accordance with the non-linear regression equation view of $t=0.032B^{1.296}$.

In Figure 30 (c) also is shown the functional dependence of number(ω)chip wraps from the width(B) of stainless chip, i.e we have the function view of $\omega=\gamma_2(B)$.This functional dependency has the non-linear regression model with equation $Y_c=703.1X^{-3.0}$,where $\omega=703.1B^{-3.0}$.

So, from Figure 30(c) we see that with increasing of width(B) stainless chip , the value of number (ω)chip wraps decreases considerably in accordance with the non-linear regression equation view of $\omega=703.1B^{-3.0}$.

In Figure 30(d) ,30(e) and 30(f) are illustrated the residual plots(residual versus δ,ω and B) of functional dependencies $t=\phi(\delta)$,$t=\alpha_1(\omega)$ and $t=\gamma_1(B)$ accordingly.

All results of computations for the linear regression equation view of $Y_c=$ -0.991-$-0.161X_1+0.171X_2+0.104X_3$ are given in Table 19.

So, the functional dependence $t=\theta(\delta,B,\omega)$ has the following characteristics:

1. *Such dependent parameter of stainless chip as clearance (t) between of chip wraps mainly depends from such independent variables as thickness (δ), width (B) and number(ω) chip wraps;*

2. *The most influence on increasing of clearance (t) between of chip wraps plays such parameters as number (ω) of chip wraps and width (B) of this stainless chip;*

3. *The most influence on decreasing of clearance (t) between of chip wraps plays such parameters as the thickness (δ) of this stainless chip.*

Table 19 Evaluation of regression equation $Y_c = -0.991 - 0.161X_1 + 0.171X_2 + 0.104X_3$

A. Mean, variance and standard deviation

Variable	Mean	Variance	Standard deviation
X_1	0.368	0.209	0.457
X_2	5.836	0.515	0.718
X_3	3.559	1.113	1.055
Y	0.317	0.031	0.175

B. Results of multiple regression of Y on X_1, X_2 and X_3

Parameter	Variable	Coefficients	Standard error	T-value
β_1	X_1	-0.161	0.152	-1.06
β_2	X_2	0.171	0.239	0.715
β_3	X_3	0.104	0.352	0.295

C. Analysis of variance results

Regression
- Degrees of freedom 3
- Sum of squares 0.018
- Mean squares 0.006

Error
- Degrees of freedom 5
- Sum of squares 0.012
- Mean squares 0.002

F-value *   3.0

* Since F=3.0< [$F_{0.05,3,5}$=5.41] we are not able to reject the hypothesis that parameters β_1, β_2 and β_3 are zero.

D. Determination of residuals

Number	Observed	Estimated	Residual
1	0.21	0.26	-0.05
2	0.38	0.34	0.04
3	0.34	0.35	-0.01
4	0.42	0.39	0.03
5	0.30	0.34	-0.04
6	0.31	0.34	-0.03
7	0.32	0.31	0.01
8	0.32	0.27	0.05
9	0.25	0.29	-0.04

B. Modification function $t = \gamma_1(\delta, \delta/B)$

Analysis of data and the statistical characteristics such as minimization of the mean absolute deviation (min MAD=0.002) shows that this functional dependence $t = \gamma_1(\delta, \delta/B)$ better submits to the linear than non-linear[51] regression model with equation view of

$$Y_c = 0.379 - 0.336X_1 + X'_2 \text{, where } t = 0.379 - 0.336\delta + \delta/B \quad (65).$$

51- *Non-linear regression model has the equation view of $t = 0.378 - 0.494\delta - 0.004\delta^2 + 1.964(\delta/B)$*

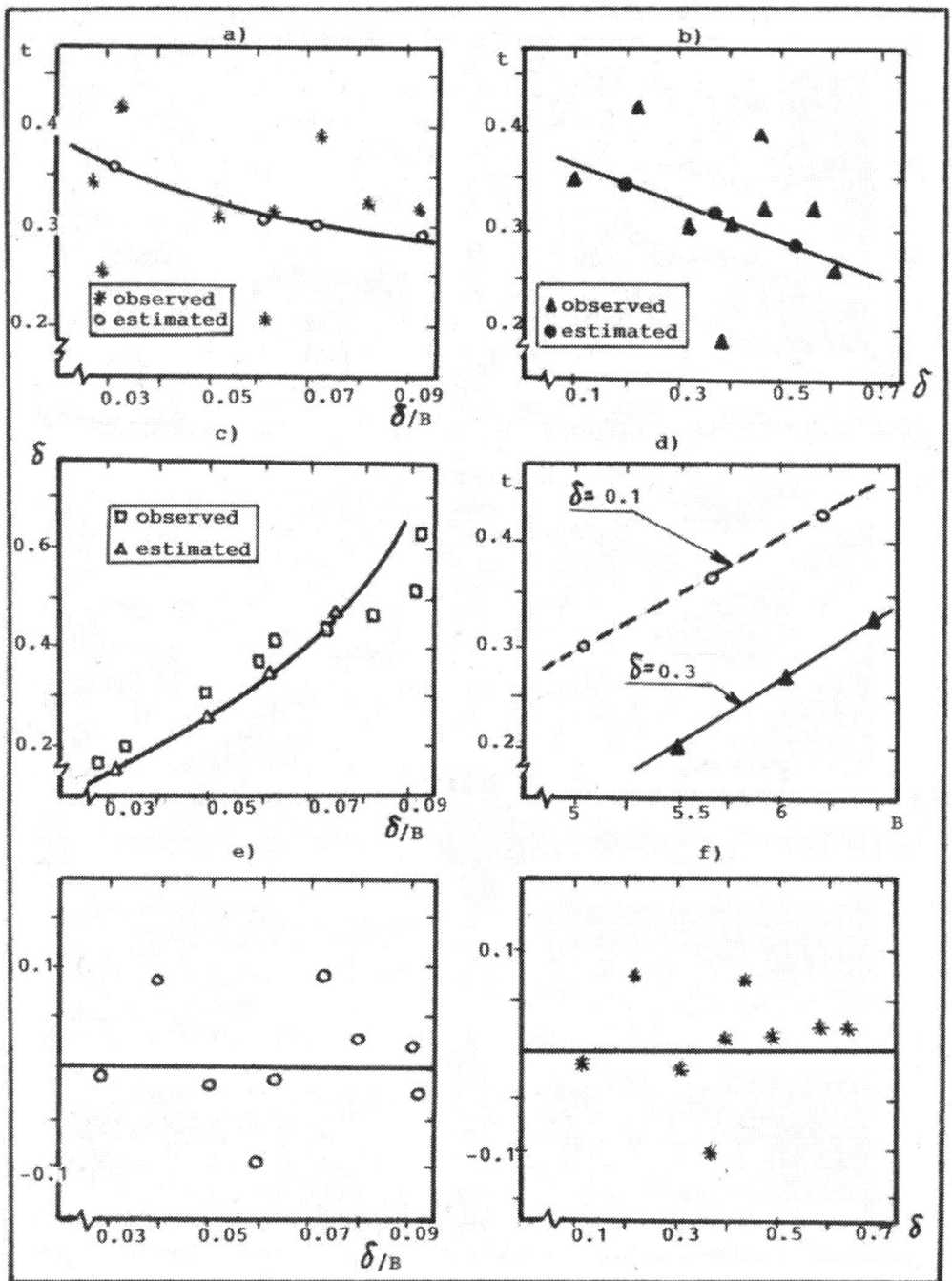

Figure 31

In Figure 31(a) is shown the functional dependence of clearance (t) between chip wraps for stainless chip from the ratio (δ/B), such as thickness (δ) to its width (B) ,i.e we have the function $t=\rho_0(\delta/B)$.Analysis of data and also the statistical characteristics(coefficient of determination $R^2=0.267$,coefficient of correlation $r=0.516$,standard deviation $S_{y/x'2}=0.056$),minimization of the mean square error (min MSE=0.002) and minimization

334

of the absolute deviation (min MAD=0.006) shows that this functional dependence $t=\rho_0(\delta/B)$ better submits to the non-linear than linear[52] regression model with equation view of $\mathbf{Y_c=0.2(X'_2)^{-0.153}}$,where $\mathbf{t=0.2(\delta/B)^{-0.153}}$ (66) .

So , we see from Figure 31(a) that with increasing of ratio $(\delta/B0$,the value of clearance (t) between chip wraps for stainless chip decreases considerably in accordance with the non-linear regression equation view of $t=0.2(\delta/B)^{-0.153}$.

In Figure 31(b) is shown the functional dependence of clearance (t) between chip wraps from thickness (δ) of stainless chip, i.e we have the function view of $t=\phi(\delta)$. So ,we see from Figure 319b) that this functional dependence has the linear regression model with equation view of $Y_c=0.379-0.16X$ or $t=0.379-0.16\delta$. Analysis of data shown in Figure 31(b) shows that with increasing of thickness (δ) stainless chip ,the value of clearance (t) between chip wraps decreases considerably in accordance with the linear regression equation view of $t=0.379-0.16\delta$.

In Figure 31(c) is shown the functional dependence of thickness (δ) stainless chip from the ratio of values (δ/B),such as thickness (δ) to its width(B) ,i.e we have the function $\delta=\rho_1(\delta/B)$.Analysis of data and also the statistical characteristics (coefficient of determination $R^2=0.884$,coefficient of correlation r=0.94,standard deviation $S_{y/x'2}=0.062$),minimization of the mean square error (min MSE=0.003),minimization of the absolute deviation (min MAD=0.024) shows that this functional dependence $\delta=\rho_1(\delta/B)$better submits to the non-linear than linear[53] regression model with equation view of $\mathbf{Y_c=16.63(X'_2)^{1.363}}$ or $\mathbf{\delta=16.63(\delta/B)^{1.363}}$ (67).

So ,we see from Figure 31(c) that with increasing of ratio (δ/B) ,the value of thickness (δ) stainless chip increases considerably in accordance with the non-linear regression equation view of $\delta=16.63(\delta/B)^{1.363}$.In Table 20 is shown the statistical data for Figure 31(d).

In Figure 31(d) is shown the functional dependence of clearance (t) between of chip wraps for stainless chip from the width (B) of this chip at the different values of its thickness (δ) ,i.e we have the function view of $t=\varphi_0(B,\delta)$ at δ=0.1mm.

 Analysis of data and the statistical characteristics (coefficients of determination $R^2=0.11$, coefficient of correlation r=0.33 ,standard deviation $S_{y/x}=0.115$) ,minimization of the mean square error (min MSE=0.011) and minimization of the mean absolute deviation (min MAD=0.029) shows that this functional dependence $t=\varphi_0(B,\delta)$ at δ=0.1mm better submits to the linear regression than non-linear[54] regression model with equation view of

$\mathbf{Y_c=-0.084+0.071X}$ or $\mathbf{t=-0.084+0.071B}$ (68).

In Figure 31(d) also is shown the functional dependence of clearance (t) between of chip wraps for stainless chip from the width (B) of this chip at the different values of its thickness (δ) ,i.e we have the function view of $t=\varphi_1(B,\delta)$ at δ=0.30mm.Analysis of data and the statistical characteristics(coefficient of determination $R^2=0.213$,coefficient of correlation r=0.461 ,standard deviation $S_{y/x}=0.148$),minimization of the mean square error

---

52-Linear regression model has the equation view of $t=0.353-0.6(\delta/B)$
53- Linear regression model has the equation view of $\delta=0.137+3(\delta/B)$
54-Non-linear regression model has the equation view of $t=0.05B^{0.967}$

(min MSE =0.017) and minimization of the mean absolute deviation (min MAD=0) shows that this functional dependence $t=\varphi_1(B,\delta)$ better submits to the linear than non-linear[55] regression model with equation view of $Y_c= -0.48+0.12X$ or $t= -0.48+0.12B$ (**69**) at $\delta=0.30$ mm .

So, we see from Figure 31(d) that with increasing of width (B) and thickness (δ) ,the value of clearance (t) between of chip wraps decreases considerably in accordance with the linear regression equation view of t= -0.48+0.12B for $\delta=0.30$mm.

Table 20 Computation worksheet for Figure 31(d)

Thickness ,δ	Observed parameters(sizes in millimeters)		
	Width, B	Ratio ,δ/B	Clearance (t)
0.10	4.7	0.02	0.3
	4.9	0.02	0.3
	5.0	0.02	0.3
	5.1	0.02	0.3
	5.2	0.02	0.1,0.3,0.4
	5.4	0.02	0.1
	5.5	0.02	0.2
	5.8	0.02	0.4
	6.0	0.02	0.4
	6.2	0.02	0.4,0.5
	6.3	0.02	0.2
0.30	5.4	0.06	0.1
	5.5	0.05	0.2
	5.6	0.05	0.3
	6.0	0.05	0.4
	6.3	0.05	0.2
	6.5	0.05	0.1,0.2
	6.6	0.05	0.2
	6.9	0.04	0.6
	7.1	0.04	0.4
0.50	5.6	0.09	0.4
	5.7	0.09	0.1
	5.8	0.09	0.1,0.2,0.3
	6.0	0.08	0.2,0.3
	6.3	0.08	0.3
	6.5	0.08	0.2,0.4,0.5
	7.0	0.07	0.4

55-Non-linear regression model has the equation view of $t=0.007B^{1.90}$

In Figure 31(e) and 31(f) are illustrated the residual plots (residual versus δ/B and δ)of functional dependencies $t=\rho_0(\delta/B)$ and $t=\phi(\delta)$ accordingly.

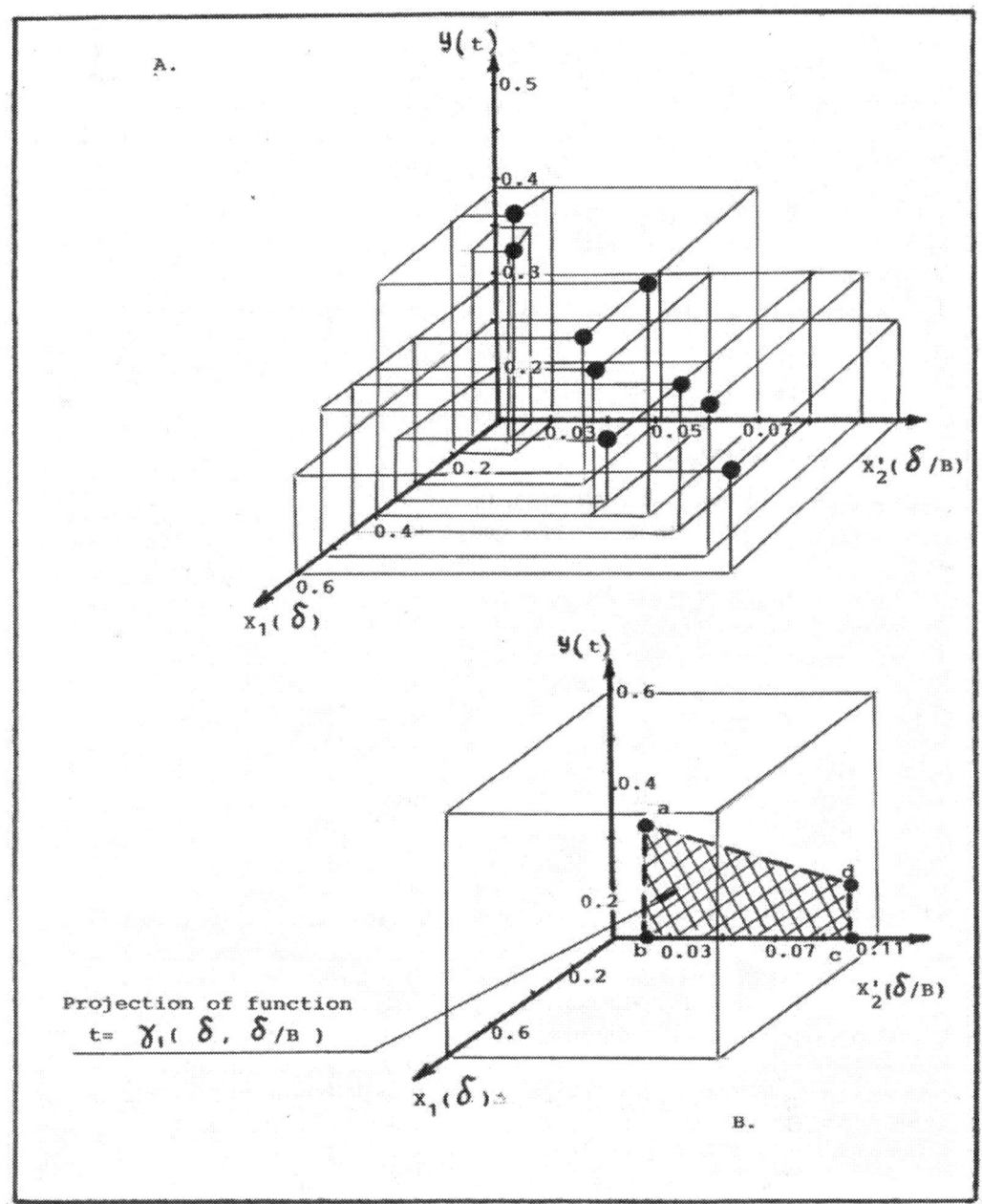

Figure 32

Analyzing the Figure 32(A) ,we see that functional dependence $t=\gamma_1(\delta,\delta/B)$ is shown in three-dimensional drawing for the linear regression model with equation view of $Y_c=0.379-0.336X_1+X'_2$ or $Y_c=0.379-0.336X_1+(X_1/X_2)$,where $t=0.379-0.336\delta+\delta/B$.

And Figure 32(B) shows that this functional surface $t=\gamma_1(\delta,\delta/B)$ has view of trapezium on the place YOX'_2 with such coordinates of their points (sizes in millimeters):
$a(X_{1,1}=0,X_{2,1}=0.02,Y_1=0.37)$; $\quad b(X_{1,2}=0,X_{2,2}=0.02,Y_2=0)$; $\quad c(X_{1,3}=0,X_{2,3}=0.10,Y_3=0)$; $d(X_{1,4}=0, X_{2,4}=0.10,Y_4=0.28)$.

Analysis of functional surface $t=\gamma_1(\delta, \delta/B)$

From Figure 32(A) we see that functional surface $t=\gamma_1(\delta,\delta/B)$,on which are situated the points of the linear regression equation view of $Y_c=0.379-0.336X_1+X'_2$ or $Y_c=0.379-0.336X_1+(X_1/X_2)$,where $t=0.379-0.336\delta+(\delta/B)$ has the following coordinates of their peaks in three-dimensional drawing(sizes in millimeters): $a'(X_{1,1}=0.10,X_{2,1}=0.02,Y_1=0.37)$; $\quad b'(X_{1,2}=0.10,X_{2,2}=0.02,Y_2=0)$,$c'(X_{1,3}=0.60,X_{2,3}=0.10,Y_3=0)$,$d'(X_{1,4}=0.60,X_{2,4}=0.10,Y_4=0.28)$.

The module of vectors $|a'b'|,|b'c'|,|c'd'|,|d'a'|$ and area of functional surface $S_{a'b'c'd'}$ can be calculated by formulas (7),(8),(9),(10) and (11). At above-shown data we have :$|a'b'|=0.37,|b'c'|=0.50,|c'd'|=0.28,|d'a'|=0.52$ and $S_{a'b'c'd'}=0.16mm^2$. All results of computations for the linear regression equation view of $Y_c=0.379-0.336X_1+X'_2$ or $t=0.379-0.336\delta+\delta/B$ are given in Table 21.

Table 21 Evaluation of linear regression equation $Y_c=0.379-0.336X_1+X'_2$

A. Mean , variance and standard deviation			
Variable	Mean	Variance	Standard deviation
X_1	0.368	0.209	0.457
X'_2	0.06	0.006	0.077
Y	0.317	0.029	0.170

B. Results of multiple regression of Y on X_1 and X'_2				
Parameter	Variable	Coefficients	Standard error	T-value
β_1	X_1	-0.336	0.152	-2.21
β_2	X'_2	1.0	0.026	38.46

C. Analysis of variance results

Regression
    Degrees of freedom 2
    Sum of squares 0.006
    Mean squares 0.003
Error
    Degrees of freedom 6
    Sum of squares 0.024
    Mean squares 0.004
F-value * 0.75

*Since F=0.75< [$F_{0.05,2,6}$=5.14] we can not reject the hypothesis that both β_1 and β_1 are zero.

D. Determination of residuals			
Number	Observed	Estimated	Residual
1	0.21	0.32	-0.11
2	0.38	0.31	0.07
3	0.34	0.37	-0.03
4	0.42	0.35	0.07
5	0.30	0.34	-0.04
6	0.31	0.32	0.01
7	0.32	0.30	0.02
8	0.32	0.29	0.03
9	0.25	0.28	-0.03

1.Function $t=\gamma_1(\delta,\delta/B)$ better submits to the linear regression model with equation view of $Y_c=0.379-0.336X_1+X'_2$ or $Y_c=0.379-0.336X_1+(X_1/X_2)$, where $t=0.379-0.336\delta+(\delta/B)$ with such statistical characteristics: coefficients of determination $R^2=0.20$,coefficient of correlation r=0.45,standard deviation $S_{y/x1,x'2}=0.63$,minimization of the mean square error (min MSE=0.003) and minimization of the mean absolute deviation (min MAD=0.002).

2.The total area of functional surface $t=\gamma_1(\delta,\delta/B)$ is equal $\sum S=0.16mm^2$ with such coordinate of their points in three- dimensional drawing for this surface (sizes in millimeters): a'$(X_{1,1}=0.10,X_{2,1}=0.02,Y_1=0.37)$; b'$(X_{1,2}=0.10,X_{2,2}=0.02$,$Y_2=0)$; c'$(X_{1,3}=0.60,X_{2,3}=0.10,Y_3=0)$; d'$(X_{1,4}=0.60,X_{2,4}=0.10,Y_4=0.28)$.

C. Modification function $t=\gamma_2(\delta,\delta/\omega)$

Analysis of data and the statistical characteristic, such as minimization of the mean absolute deviation (min MAD=0.003) shows that this functional dependence $t=\gamma_2(\delta,\delta/\omega)$ better submits to the linear than non-linear[56] regression model with equation view of **$Y_c=0.422+0.022X_1-X'_2$** or **$Y_c=0.422+0.022X_1-(X_1/X_2)$** , where **$t=0.422+0.022\delta-(\delta/\omega)$ (70).**

In Figure 33(a) is shown the functional dependence of clearance (t) between chip wraps for stainless chip from the ratio (δ/ω) , such as thickness (δ) to its number of wraps(ω),i.e we have the function $t=\theta_1(\delta/\omega)$. Analysis of data and also the statistical characteristics (coefficient of determination $R^2=0.23$, coefficient of correlation r=0.48,standard deviation $S_{y/x'2}=0.058$),minimization of the mean square error (min MSE=0.003)and minimization of the absolute deviation (min MAD=0) shows that this functional dependence $t=\theta_1(\delta/\omega)$ better submits to the linear than non-linear[57] regression model with equation view of **$Y_c=0.39-0.69X'_2$** or **$Y_c=0.39-0.69(X_1/X_2)$** , where **$t=0.39--0.69(\delta/\omega)$ (71).**So , we see from Figure 33(a) that with increasing of (δ/ω) ,the value of clearance (t) between chip wraps for stainless chip decreases considerably in accordance with the linear regression equation view of $t=0.39-0.69(\delta/\omega)$.

In Figure 33(b) is shown the functional dependence of number chip wraps (ω) from the ratio (δ/ω) ,such as thickness (δ) to its number of chip wraps(ω) ,i.e we have the function view of $\omega=\theta_2(\delta/\omega)$.Analysis of data and also the statistical characteristics (coefficient of determination $R^2=0.626$,coefficient of correlation r=0.791,standard deviation $S_{y/x'2}=0.244$),minimization of the mean square error (min MSE=0.046) and minimization of the absolute deviation (min MAD=0.007) shows that this functional dependence $\omega=\theta_2(\delta/\omega)$ better submits to the non-linear than linear[58] regression model with equation view of **$Y_c=2.57(X'_2)^{-0.134}$** ,where **$\omega=2.57(\delta/\omega)^{-0.134}$ (72).**

56-Non-linear regression model has the equation view $t=0.477-0.577\delta+\delta^2-(\delta/\omega)$
57- Non-linear regression model has the equation view of $t=0.179(\delta/\omega)^{-0.23}$
58-Linear regression model has the equation view of $\omega=4.23-6.31(\delta/\omega)$

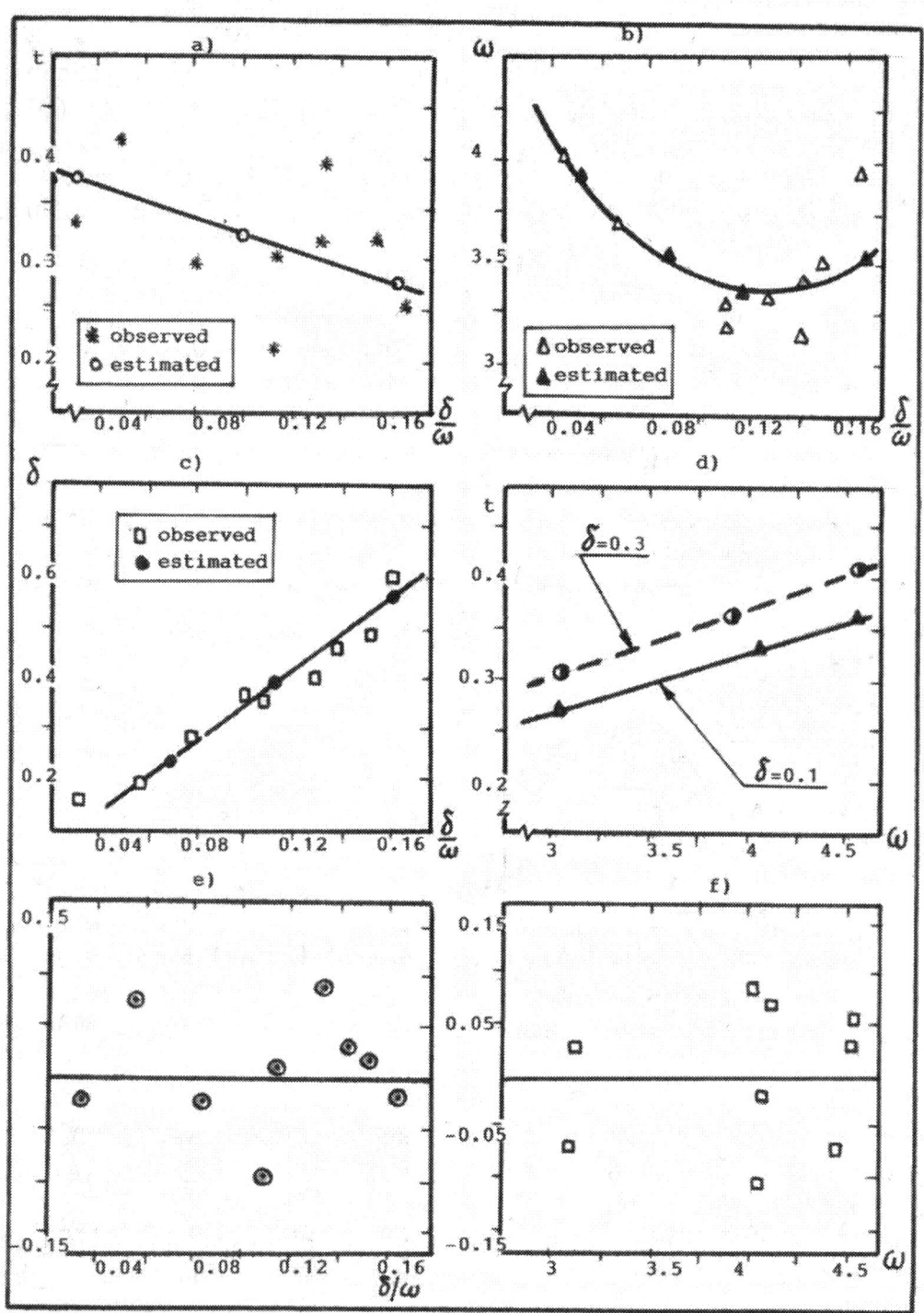

Figure 33

So, we see from Figure 33(b) that with increasing of (δ/ω) ,the value of number chip wraps(ω) for stainless chip decreases considerably in accordance with the non-linear regression equation view of $\omega=2.57(\delta/\omega)^{-0.134}$.

In Figure 33(c) is shown the functional dependence of thickness (δ) stainless chip from the ratio (δ/ω),such as thickness(δ) to its number wraps(ω) ,i.e we have the function view of $\delta=\theta_3(\delta/\omega)$. Analysis of data and also the statistical characteristics (coefficient of determination $R^2=0.962$,coefficient of correlation r=0.981,standard deviation $S_{y/x'2}=0.034$), minimization of the mean square error (min MSE=0) and minimization of the absolute deviation (min MAD=0.004) show that this functional dependence $\delta=\theta_3(\delta/\omega)$ better submits to the linear than non-linear[59] regression model with equation view of **$Y_c= -0.04+3.81X'_2$** or **$Y_c= -0.04+3.81(X_1/X_2)$** , where $\delta= -0.04+3.81(\delta/\omega)$ (73).

So, we see from Figure 33(c) that with increasing of ratio (δ/ω),the value of thickness (δ) for stainless chip increases considerably in accordance with the linear regression equation view of $\delta= - 0.04+3.81(\delta/\omega)$.In Table 22 is given the statistical data for Figure 33(d).

Table 22 Computation worksheet for Figure 33(d)

Thickness,δ (in mm)	Observed parameters [sizes in millimeters]	
	Number of wraps (ω)	**Clearance (t)** (in mm)
0.10	3	0.2
	3	0.3
	4	0.2
	4	0.3
	4	0.4
	4	0.4
	5	0.3
	5	0.3
	5	0.4
	5	0.4
0.30	1	0.2
	2	0.4
	3	0.1
	3	0.3
	4	0.1
	4	0.2
	4	0.4
	4	0.4
	4	0.5
	5	0.4

In Figure 33(d) is shown the functional dependence of clearance (t) between of chip wraps for stainless chip from the number of chip wraps (ω) at the different values of its thickness (δ) ,i.e we have the function view of $t=\eta_0(\delta,\omega)$ at $\delta=0.10$mm and $\delta=0.30$mm.

Analysis of data and the statistical characteristics (coefficient of determination $R^2=0.11$,

---

59- Non-linear regression model has the equation view of $\delta=0.04(\delta/\omega)^{-0.89}$

coefficient of correlation r=0.33,standard deviation $S_{y/x}=0.126$),minimization of the mean square error(min MSE=0) and minimization of the mean absolute deviation (min MAD=0) shows that this functional dependence $t=\eta_0(\delta,\omega)$ at $\delta=0.10$mm better submits to the linear than non-linear[60] regression model with equation view of **$Y_c=0.125+0.046X$** or **$t=0.125+0.046\omega$** at $\delta=0.10$ mm **(74).**

In Figure 33(d) also is shown the functional dependence of clearance (t) between of chip wraps for stainless chip from the number of wraps (ω) at the different values of its thickness ,i.e we have the function $t=\eta_1(\delta,\omega)$ at $\delta=0.30$mm.Analysis of data and the statistical characteristics (coefficient of determination $R^2=0.11$,coefficient of correlation $r=0.33$,standard deviation $S_{y/x}=0.126$),minimization of the mean square error(min MSE=0) and minimization of the mean absolute deviation (min MAD=0) shows that this functional dependence $t=\eta_1(\delta,\omega)$ better submits to the linear than non-linear[61] regression model with equation view of **$Y_c=0.19+0.032X$** or **$t=0.19+0.032\omega$** at $\delta=0.30$mm **(75).** So ,we see from Figure 33(d) that with increasing of number chip wraps (ω) ,the value of clearance (t) between of chip wraps increases considerably in accordance with the linear regression equation view of $t=0.19+0.032\omega$ at $\delta=0.30$mm.

And besides from Figure 33(d) we see that with increasing of number chip wraps (ω) and decreasing of thickness (δ) stainless chip ,the value of clearance (t) between of chip wraps decreases considerably in accordance with the linear regression equation view of $t=0.125+0.046\omega$ at $\delta=0.10$mm and on the contrary- the value of clearance (t) between of chip wraps increases considerably with increasing of number chip wraps (ω) and thickness (δ) in accordance with the linear regression equation view of $t=0.19+0.032\omega$ at $\delta=0.30$mm.

In Figure 33(e) and 33(f) are illustrated the residual plots (residual versus δ/ω and ω) of functional dependencies $t=\theta_1(\delta/\omega)$ and $t=\eta_0(\delta,\omega)$ at $\delta=0.10$ mm accordingly.

Analyzing the Figure 34(A) ,we see that functional dependence $t=\gamma_2(\delta,\delta/\omega)$ is shown in three-dimensional drawing for the linear regression model with equation view of $Y_c=0.422+0.022X_1-X'_2$ or $Y_c=0.422+0.022X_1-(X_1/X_2)$,where $t=0.422+0.022\delta-(\delta/\omega)$.

And Figure 34(B) shows that this functional surface $t=\gamma_2(\delta,\delta/\omega)$ has view of trapezium on the place YOX'$_2$ with such coordinates of their points (sizes in millimeters):
$a(X_{1,1}=0,X_{2,1}=0.02,Y_1=0.40)$; $b(X_{1,2}=0,X_{2,2}=0.02,Y_2=0)$;$c(X_{1,3}=0,X_{2,3}=0.16,Y_3=0)$;
$d(X_{1,4}=0,X_{2,4}=0.16,Y_4=0.28)$.

Analysis of functional surface $t=\gamma_2(\delta,\delta/\omega)$

From Figure 34(A) we see that functional surface $t=\gamma_2(\delta,\delta/\omega)$,on which are situated the points of the linear regression equation view of $Y_c=0.422+0.022X_1-X'_2$ or $Y_c=0.422+0.022X_1-(X_1/X_2)$,where $t=0.422+0.022\delta-(\delta/\omega)$ has the following coordinates of their peaks in three-dimensional drawing(sizes in millimeters):
$a'(X_{1,1}=0.10,X_{2,1}=0.34,Y_1=0.40)$;$b'(X_{1,2}=0.10,X_{2,2}=0.34,Y_2=0)$; $c'(X_{1,3}=0.60,X_{2,3}=0.16,Y_3=0)$; $d'(X_{1,4}=0.60,X_{2,4}=0.16,Y_4=0.28)$.

60-Non-linear regression model has the equation view of $t=0.113\omega^{0.714}$
61-Non-linear regression model has the equation view of $t=0.2\omega^{0.196}$

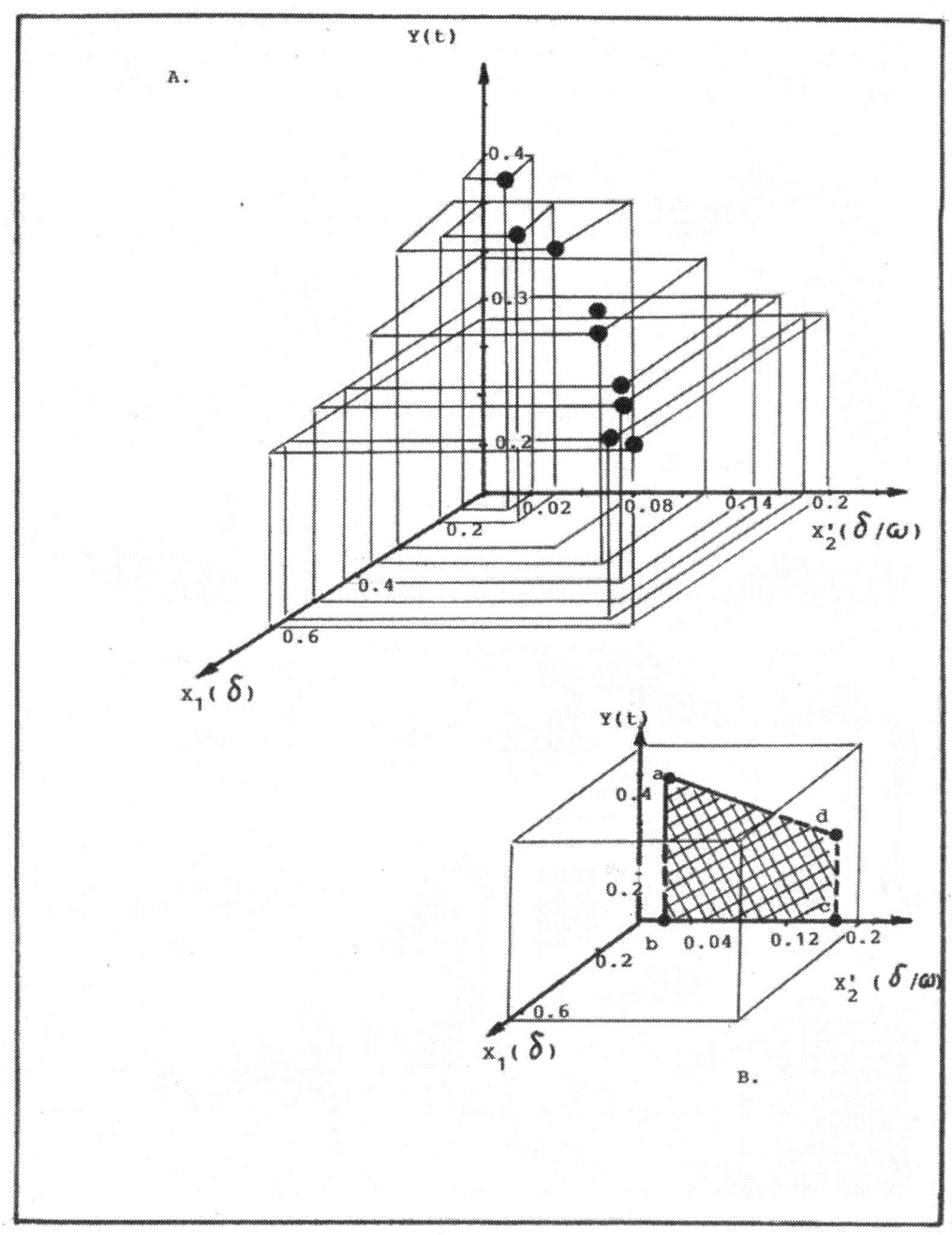

Figure 34

The module of vectors |a'b'|,|b'c'|,|c'd'|,|d'a'| and area of functional surface $S_{a'b'c'd'}$ be calculated by formulas (7),(8),(9),(10) and (11). At above-shown data we have : |a'b'|=0.40, |b'c'|=0.52 ,|c'd'|=0.28 ,|d'a'|=0.55 and $S_{a'b'c'd'}$=0.18mm^2.

All results of computations for the linear regression equation view of Y_c=0.422+0.022X_1- $-X'_2$ or t=0.422+0.022δ–(δ/ω) are given in Table 23.

Table 23 Evaluation of linear regression equation $Y_c=0.422+0.022X_1-X'_2$

A. Mean ,variance and standard deviation

Variable	Mean	Variance	Standard deviation
X_1	0.368	0.209	0.457
X'_2	0.106	0.018	0.134
Y	0.317	0.029	0.170

B. Results of multiple regression of Y on X_1 and X'_2

Parameter	Variable	Coefficients	Standard error	T-value
β_1	X_1	0.022	0.153	0.14
β_2	X'_2	-1.0	0.045	-22.22

C. Analysis of variance results

Regression
    Degrees of freedom 2
    Sum of squares   0.002
    Mean squares   0.001
Error
    Degrees of freedom 6
    Sum of squares   0.029
    Mean squares   0.005
F-value *   0.20

* Since F=0.20< [$F_{0.05,2,6}$=5.14] we can not reject the hypothesis that both β_1 and β_2 are zero.

D. Determination of residuals

Number	Observed	Estimated	Residual
1	0.21	0.32	-0.11
2	0.38	0.30	0.08
3	0.34	0.40	-0.06
4	0.42	0.38	0.04
5	0.30	0.35	-0.05
6	0.31	0.32	-0.01
7	0.32	0.29	0.03
8	0.32	0.28	0.04
9	0.25	0.28	-0.03

So, the parameters of this functional surface $t=\gamma_2(\delta,\delta/\omega)$ are the following:

1.Function $t=\gamma_2(\delta,\delta/\omega)$ better submits to the linear regression model with equation view of $Y_c=0.422+0.022X_1-X'_2$ or $Y_c=0.422+0.022X_1-(X_1/X_2)$,where $t=0.422+0.022\delta-(\delta/\omega)$ with such statistical characteristics : coefficient of determination R^2=0.06,coefficient of correlation r=0.25,standard   deviation $S_{y/x1,x'2}$=0.069,minimization of the mean square error(min MSE=0.003,minimization of the mean absolute deviation (min MAD=0.008).

2.The total area of functional surface $t=\gamma_2(\delta,\delta/\omega)$ is equal $\sum S$=0.18mm^2 with such coordinate of their points in three-dimensional drawing (sizes in millimeters):
a'($X_{1,1}$=0.10 ,$X_{2,1}$=0.34 ,Y_1=0.40);b'($X_{1,2}$=0.10,$X_{2,2}$=0.34,Y_2=0);c'($X_{1,3}$=0.60,$X_{2,3}$=0.16, Y_3=0);d'($X_{1,4}$=0.60,$X_{2,4}$=0.16,Y_4=0.28).

D. Modification function t=γ₃(δ,B/ω)

Analysis of data and the statistical characteristics (coefficient of determination R^2=0.322 ,coefficient of correlation r=0.568,standard deviation $S_{y/x1,x'2}$=0.0590,minimization of the mean square error (min MSE=0.002) and minimization of the mean absolute deviation (min MAD=0.002) shows that this function better submits to the linear than non-linear[62] regression equation view of **Y_c=0.427-0.164X_1-0.03X'_2** or **Y_c=0.427−0.164X_1− −0.03 (X_2/X_3)** , where **t=0.427−0.164δ−0.03(B/ω)** (76).

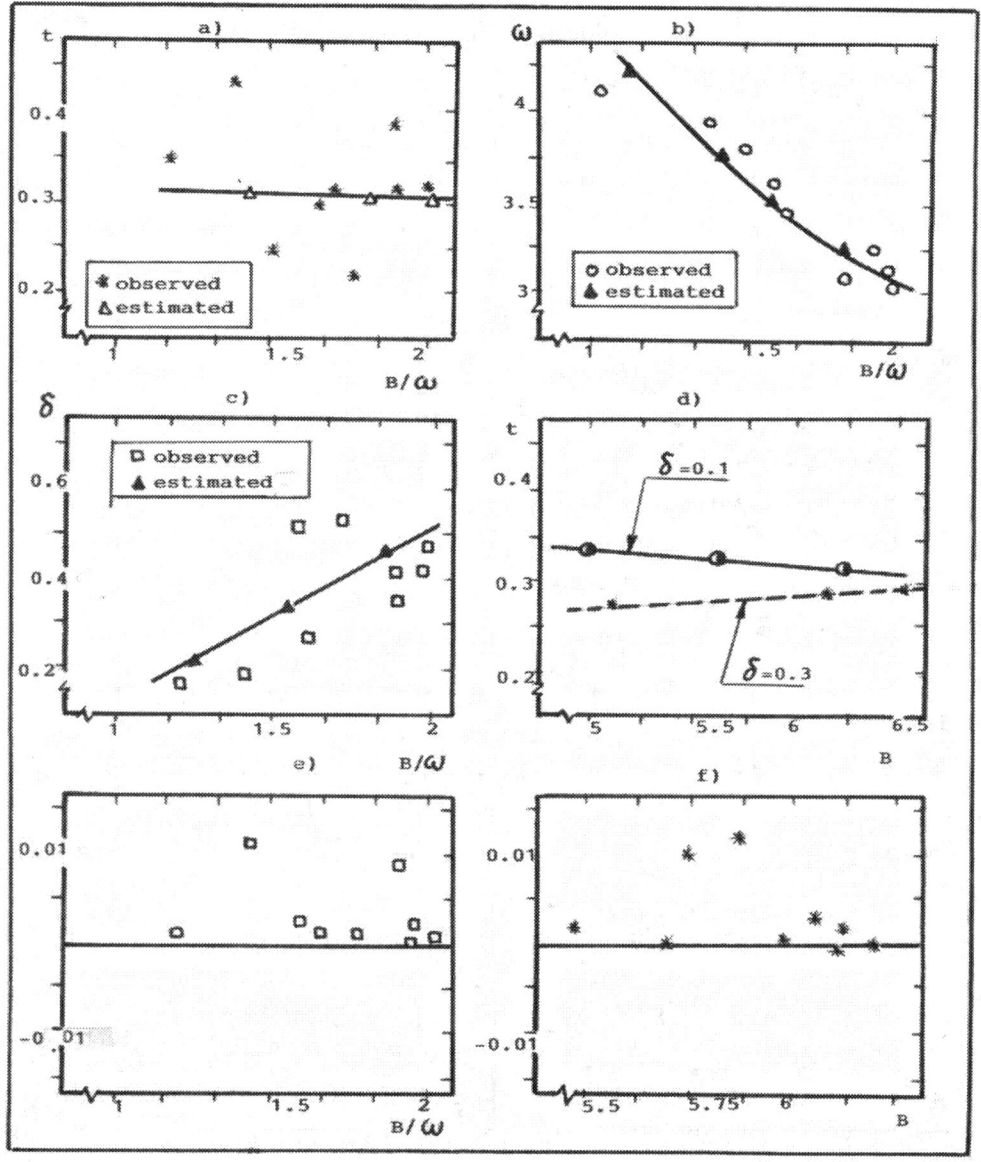

Figure 35

62- *Non-linear regression model has the equation view of* $t=0.191+1.179δ-1.97δ^2+0.003(B/ω)$

In Figure 35(a) is shown the functional dependence of clearance (t) between chip wraps for stainless chip from the ratio (B/ω),such as width(B) to its number chip wraps (ω) ,i.e we have the function t=ϕ_0(B/ω).

Analysis of data and also the statistical characteristics (coefficients of determination and correlation practically are equal zero), minimization of the mean square error (min MSE=0.004) and minimization of the absolute deviation (min MAD=0) shows that this functional dependence t=ϕ_0(B/ω) better submits to the linear than non-linear[63] regression model with equation view of $\mathbf{Y_c=0.407-0.054X'_2}$ or $\mathbf{Y_c=0.407-0.054(X_1/X_2)}$, where $\mathbf{t=0.407-0.054(B/\omega)}$ (77).So ,we see from Figure 35(a) that between of clearance (t) and ratio (B/ω) the correlation is absents, but this function can be evaluated approximately by formula t=0.407--0.054(B/ω).

In Figure 35(b) is shown the functional dependence of number chip wraps (ω) from the ratio(B/ω) ,such as width(B) to its number of chip wraps (ω) ,i.e we have the function ω=ϕ_1(B/ω).Analysis of data and also the statistical characteristics (coefficient of determination R^2=0.959,coefficient of correlation r=0.980,standard deviation $S_{y/x'2}$=0.0710,minimization of the mean square error (min MSE=0.005),minimization of the absolute deviation (min MAD=0.006) shows that this functional dependence ω=ϕ_1(B/ω) better submits to the non-linear than linear[64] regression model with equation view of $\mathbf{Y_c=5.093X'_2{}^{-0.733}}$, where $\mathbf{\omega=5.093(B/\omega)^{-0.733}}$ (78).

So ,we see from Figure 35(b) that with increasing of (B/ω) ,the value of number chip wraps(ω) for stainless chip decreases considerably in accordance with the non-linear regression equation view of ω=5.093(B/ω)$^{-0.733}$.

In Figure 35(c) is shown the functional dependence of thickness (δ) stainless chip from the ratio (B/ω) ,such as width (B) to its number of wraps (ω) ,i.e we have the function δ=ϕ_2(B/ω).

Analysis of data and also the statistical characteristics(coefficient of determination R^2=0.134,coefficient of correlation r=0.366,standard deviation $S_{y/x'2}$ = =0.161),minimization of the mean square error (min MSE=0.020) and minimization of the absolute deviation (min MAD=0) shows that this functional dependence δ=ϕ_2(B/ω) better submits to the linear than non-linear[65] regression model with equation view of $\mathbf{Y_c=-0.269+0.384X'_2}$ or $\mathbf{Y_c= - 0.269+0.384(X_1/X_2)}$, where $\mathbf{\delta= -0.269+}$ $\mathbf{+0.384(B/\omega)}$ (79).

So , we see from Figure 35(c) that with increasing of ratio (B/ω) ,the value of thickness (δ) for stainless chip increases considerably in accordance with the linear regression equation view of Y_c= -0.269+0.384X'$_2$,where δ= -0.269+0.384(B/ω).

In Table 24 is given the statistical data for Figure 35(d). In Figure 35(d) is shown the functional dependence of clearance (t) between of chip wraps for stainless chip from width (B) at the different values of its thickness (δ) ,i.e we have the function t=θ_1(B) at δ=0.10 mm.

63-*Non-linear regression model has the equation view of t= -0.015+0.48(B/ω) –0.166(B/ω)^2*
64- *Linear regression model has the equation view of ω= 6.289-1.648(B/ω)*
65-*Non-linear regression model has the equation view δ=0.012(B/ω)$^{6.615}$*

Analysis of data and the statistical characteristics such as minimization of the mean square error(min MSE=0.005) and minimization of the mean absolute deviation (min MAD=0.003) shows that this functional dependence t= θ_1(B) better submits to the linear than non-linear [66] regression model with equation view of **Y_c=0.42-0.02X** or **t=0.42-−0.02B** at δ=0.10mm (80). So ,we see from Figure 35(d) that with increasing of width(B) chip ,the value of clearance (t) between of chip wraps decreases considerably in accordance with the linear regression equation view of t=0.42-0.02B at δ=0.10mm.

Table 24 Computation worksheet for data of Figure 35(d)

Thickness,δ	Observed parameters (sizes in millimeters)	
	Width ,B	Clearance ,t
0.1	4.7	0.3
	4.9	0.4
	5.0	0.3
	5.1	0.3
	5.2	0.3
	5.2	0.4
	5.5	0.2
	5.8	0.4
	6.2	0.4
	6.3	0.2
0.3	5.2	0.1
	5.2	0.5
	5.6	0.3
	5.7	0.1
	6.0	0.5
	6.0	0.4
	6.1	0.5
	6.1	0.2
	6.5	0.4
	6.5	0.1

In Figure 35(d) also is shown the functional dependence of clearance (t) between of chip wraps for stainless chip from the width (B) at the different values of its thickness (δ) ,i.e we have the function t=θ_2(B) at δ=0.3mm. Analysis of data and the statistical characteristics (standard deviation $S_{y/x}$=0.038,minimization of the mean square error(min MSE=0.027) and minimization of the mean absolute deviation (min MAD=0.030) shows that this functional dependence t=θ_2(B) better submits to the linear than non-linear[67] regression model with equation view of **Y_c=0.186+0.016X** or **t=0.186+0.016B** at δ =0.3mm (81). So, we see from Figure 35(d) that with increasing of width (B) chip, the value of clearance (t) between of chip wraps increases considerably in accordance with the linear regression equation view of t=0.186+0.016B at δ=0.30 mm.

66- Non-linear regression model has the equation view of $t=0.811B^{-0.435}$
67-Non-linear regression model has the equation view of $t=0.099B^{0.583}$

And besides from Figure 35(d) we see that with increasing of width(B) chip and thickness (δ) of stainless chip ,the value of clearance (t) between of chip wraps increases considerably in accordance with the linear regression equation view of t=0.186+0.016B at δ=0.30mm. And on the otherwise, the value of clearance (t) between of chip wraps decreases considerably in accordance with the linear regression equation view of t=0.42- −0.02B at δ=0.10mm.In Figures 35(e) and 35(f) are illustrated the residual plots (residual versus B/ω and B) of functional t= ϕ_0(B/ω) and t=γ_1(B) accordingly.

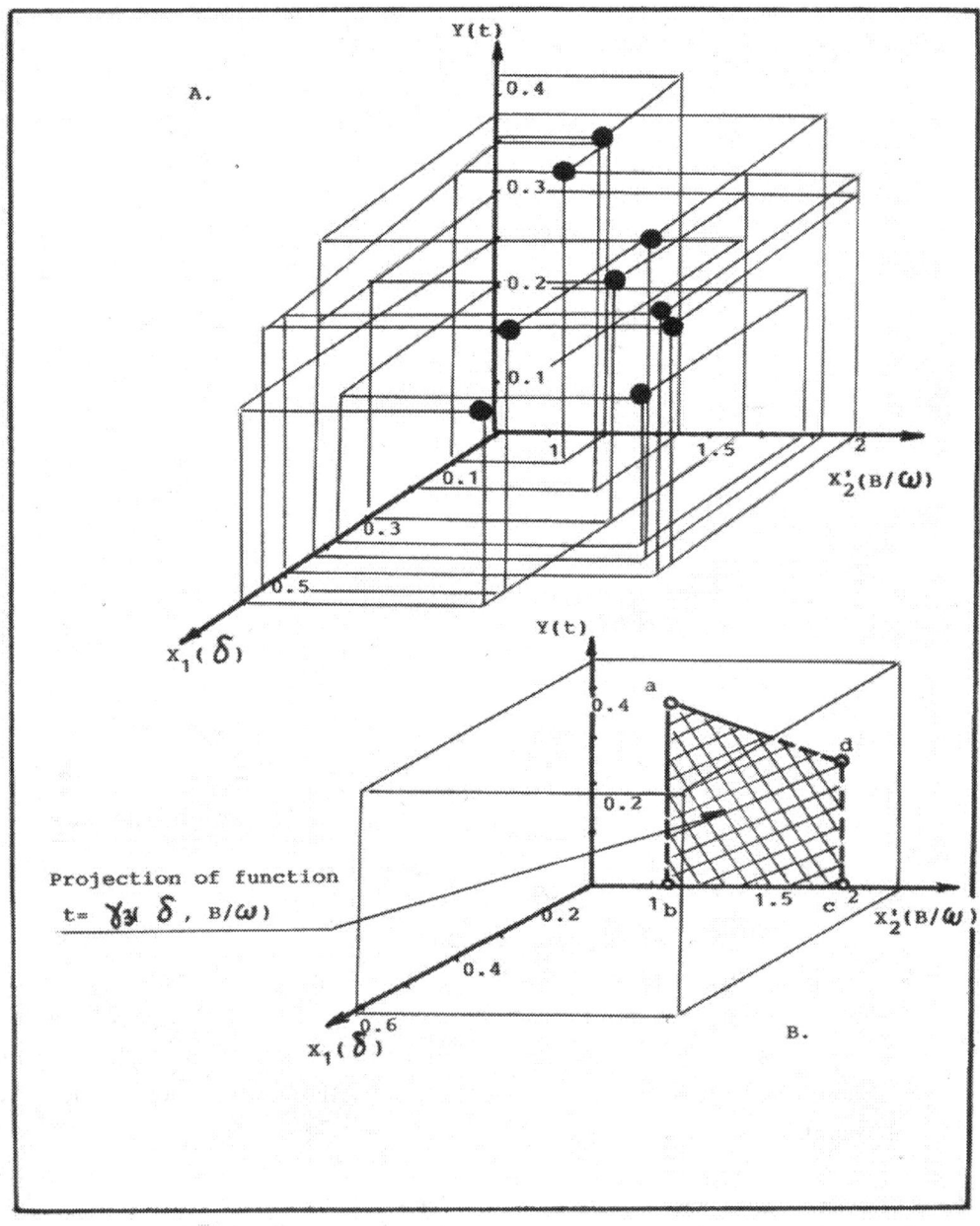

Figure 36

348

Analyzing the Figure 36(A) we see that functional dependence $t=\gamma_3(\delta,B/\omega)$ is shown in three-dimensional drawing for the linear regression model with equation view of $Y_c=0.427-0.164X_1-0.03X'_2$ or $Y_c=0.427-0.164X_1-0.03(X_2/X_3)$,where $t=0.427-0.164\delta--0.03(B/\omega)$. And Figure 36(B) shows that this functional surface $t=\gamma_3(\delta,B/\omega)$ has view of trapezium on the place YOX'_2 with such coordinates of their points (sizes in millimeters): $a(X_{1,1}=0,X_{2,1}=1.27$,$Y_1=0.37)$; $b(X_{1,2}=0,X_{2,2}=1.27,Y_2=0)$;$c(X_{1,3}=0,X_{2,3}=1.61$, ,$Y_3=0)$; $d(X_{1,4}=0,X_{2,4}=1.61,Y_4=0.29)$.

Analysis of functional surface $t=\gamma_3(\delta,B/\omega)$

From Figure 36(A) we see that functional surface $t=\gamma_3(\delta,B/\omega)$,on which are situated the points of the linear regression equation view of $Y_c=0.427-0.164X_1-0.03X'_2$ or $Y_c=0.427-0.164X_1-0.03(X_2/X_3)$,where $t=0.427-0.164\delta-0.03(B/\omega)$ has the following coordinate of their peaks in three-dimensional drawing: $a'(X_{1,1}=0.10,X_{2,1}=1.27,Y_1=0.37)$;$b'(X_{1,2}=0.10$, $X_{2,2}=1.27$,$Y_2=0)$; $c'(X_{1,3}=0.54,X_{2,3}=1.61,Y_3=0)$;$d'(X_{1,4}=0.54,X_{2,4}=1.61,Y_4=0.29)$.
The module of vectors $|a'b'|,|b'c'|,|c'd'|,|d'a'|$ and area of functional surface $S_{a'b'c'd'}$ can be calculated by formulas (7),(8),(9),(10) and (11). At above-shown data we have : $|a'b'|=0.37,|b'c'|=0.56,|c'd'|=0.29,|d'a'|=0.58$ and $S_{a'b'c'd'}=0.18mm^2$.

All results of computations for the linear regression equation view of $Y_c=0.427-0.164X_1--0.03X'_2$ or $Y_c=0.427-0.164X_1-0.03(X_2/X_3)$,where $t=0.427-0.164\delta-0.03(B/\omega)$ are given in Table 25.

So, the parameters of this functional surface $t=\gamma_3(\delta,B/\omega)$ are the following:

1. Function $t=\gamma_3(\delta,B/\omega)$ better submits to the linear regression model with equation view of $Y_c=0.427-0.164X_1-0.03X'_2$ or $Y_c=0.427-0.164X_1-0.03(X_2/X_3)$,where $t=0.427--0.164\delta-0.03(B/\omega)$ with such statistical characteristics:
   - Coefficient of determination $R^2=0.322$
   - Coefficient of correlation $r=0.568$
   - Standard deviation $S_{y/x1,x'2}=0.059$
   - Minimization of the mean square error (min MSE=0.002)
   - Minimization of the mean absolute deviation (min MAD=0.002).

2. The total area of functional surface $t=\gamma_3(\delta,B/\omega)$ is equal $\sum S=0.18mm^2$ with such coordinate of their points in three-dimensional drawing for this surface (sizes in millimeters):
$$a'(X_{1,1}=0.10,X_{2,1}=1.27,Y_1=0.37)$$
$$b'(X_{1,2}=0.10,X_{2,2}=1.27,Y_2=0)$$
$$c'(X_{1,3}=0.54,X_{2,3}=1.61,Y_3=0)$$
$$d'(X_{1,4}=0.54,X_{2,4}=1.61,Y_4=0.29).$$

Table 25 Evaluation of linear regression equation $Y_c=0.427-0.164X_1-0.03X'_2$

A. Mean ,variance and standard deviation

Variable	Mean	Variance	Standard deviation
X_1	0.368	0.209	0.457
X'_2	1.659	0.391	0.625
Y	0.317	0.029	0.170

B. Results of multiple regression of Y on X_1 and X'_2

Parameter	Variable	Coefficients	Standard error	T-value
β_1	X_1	-0.164	0.152	-1.08
β_2	X'_2	-0.03	0.208	-0.14

C. Analysis of variance results

Regression
 Degrees of freedom 2
 Sum of squares   0.10
 Mean squares   0.05
Error
 Degrees of freedom 6
 Sum of squares   0.021
 Mean squares   0.004
F-value*   12.50

* Since F=12.50>[$F_{0.05,2,6}$=5.14] we can reject the hypothesis that both β_1 and β_2 are zero.

D. Determination of residuals

Number	Observed	Estimated	Residual
1	0.21	0.31	-0.10
2	0.38	0.30	0.08
3	0.34	0.37	-0.03
4	0.42	0.35	0.07
5	0.30	0.33	-0.03
6	0.31	0.31	0
7	0.32	0.29	0.03
8	0.32	0.29	0.03
9	0.25	0.28	-0.03

In Table 26 is given the summary graphical functions *"Multiple Regression Analysis for internal diameter of stainless chip"* in dependence from some its parameters in function of $Y_c=\gamma(X_{i,1},X_{i,2},X_{i,3})$.

Table26Graphical function of " Multiple Regression Analysis for internal diameter"

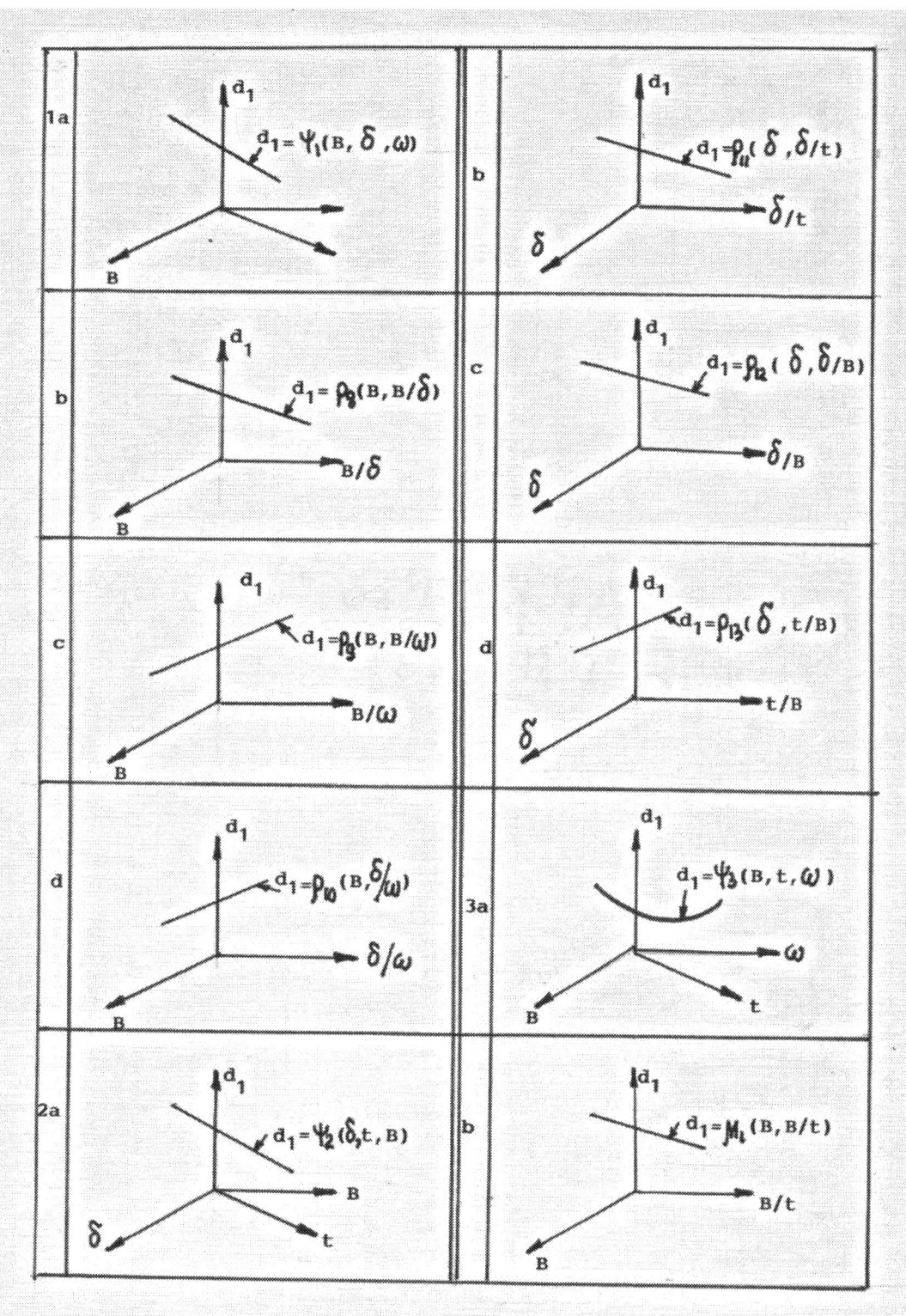

Table 26 (continue)

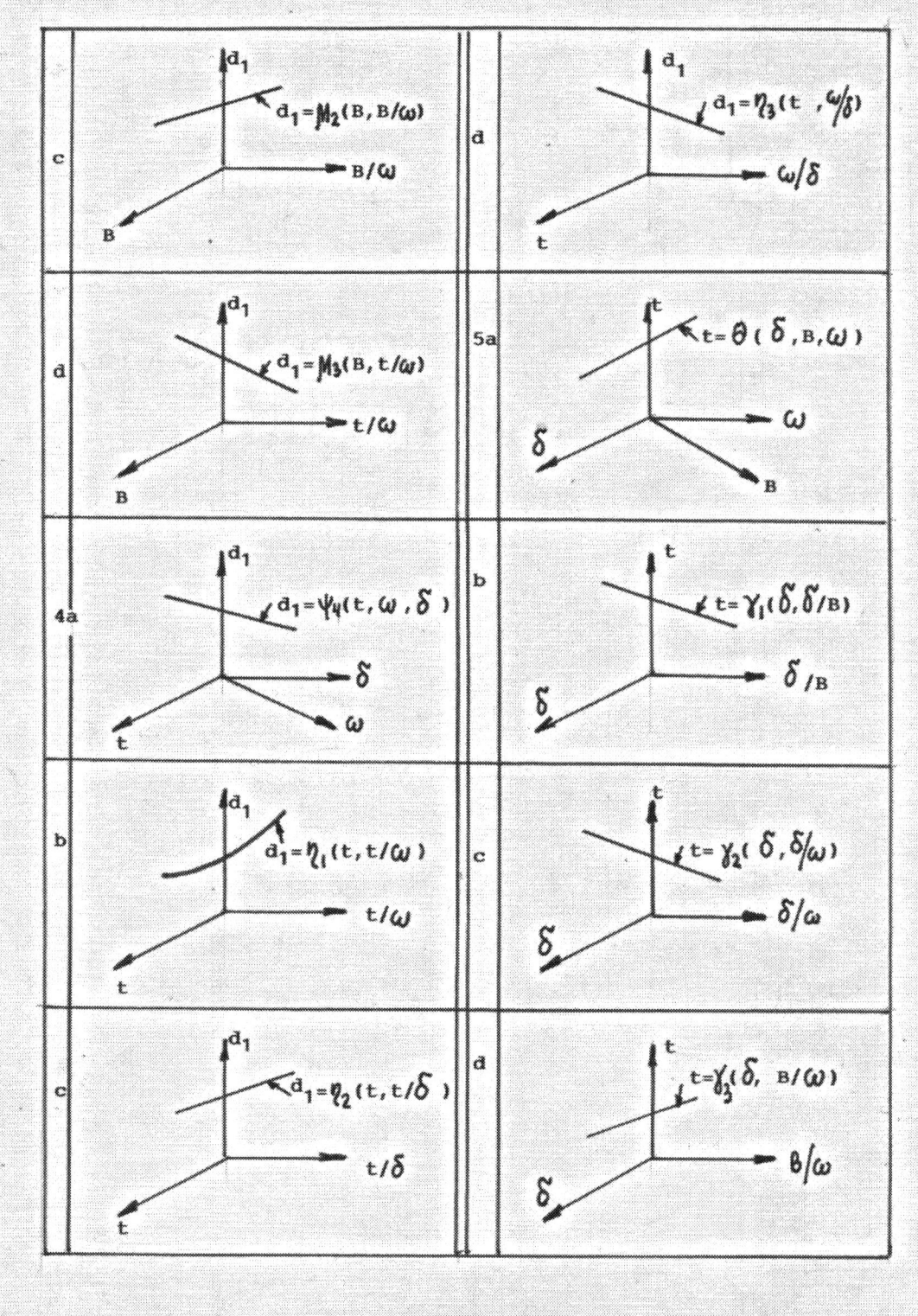

In Table 27 is given the summary statistical characteristics of *"Multiple Regression Analysis of internal diameter"*.

Table 27 Summary statistical characteristics of "Multiple Regression Analysis of internal diameter" for stainless chip

N	View function	Regression model	Empirical equation	Statistical characteristics		
				R^2	r	$S_{y/x1,x'2}$
1a	$d_1=\phi_1(B,\delta,\omega)$	Linear	$d_1=7.968-0.846B+5.201\delta+0.043\omega$	0.869	0.932	0.383
b	$d_1=\rho_8(B,B/\delta)$	Linear	$d_1=16.963-1.773B-0.073(B/\delta)$	0.929	0.964	0.257
c	$d_1=\rho_9(B,B/\omega)$	Linear	$d_1=8.65-1.33B+2.54(B/\omega)$	0.116	0.341	0.908
d	$d_1=\rho_{10}(B,\delta/\omega)$	Linear	$d_1=16.21-2.34B+24.1(\delta/\omega)$	0.650	0.806	0.572
2a	$d_1=\phi_2(\delta,t,B)$	Linear	$d_1=3.802+4.629\delta-1.324t+0.002B$	0.877	0.936	0.372
b	$d_1=\rho_{11}(\delta,\delta/t)$	Linear	$d_1=3.388+2.848\delta+0.532(\delta/t)$	0.899	0.948	0.308
c	$d_1=\rho_{12}(\delta,\delta/B)$	Linear	$d_1=3.311+4.860\delta-0.025(\delta/B)$	0.871	0.933	0.347
d	$d_1=\rho_{13}(\delta,t/B)$	Linear	$d_1=3.30+4.86\delta+0.2(t/B)$	0.870	0.933	0.348
3a	$d_1=\phi_3(B,t,\omega)$	Non-linear	$d_1=-5.65+3.44B-0.20B^2-7.35t-0.02\omega$	0.370	0.608	0.841
b	$d_1=\mu_1(B,B/t)$	Linear	$d_1=-2.389+0.926B+0.109(B/t)$	0.340	0.583	0.785
c	$d_1=\mu_2(B,B/\omega)$	Linear	$d_1=5.615-0.651B+1.970(B/\omega)$	0.134	0.366	0.899
d	$d_1=\mu_3(B,t/\omega)$	Linear	$d_1=11.55-1.838B+47(t/\omega)$	0.170	0.412	0.712
4a	$d_1=\phi_4(t,\omega,\delta)$	Linear	$d_1=5.208-0.471t-0.43\omega+4.22\delta$	0.820	0.906	0.448
b	$d_1=\eta_1(t,t/\omega)$	Non-linear	$d_1=5.03+9.01t-26.8t^2+0.08(t/\omega)$	0.180	0.420	1.141
c	$d_1=\eta_2(t,t/\delta)$	Linear	$d_1=6.033-0.229t-0.073(t/\delta)$	0.748	0.865	0.486
d	$d_1=\eta_3(t,\omega/\delta)$	Linear	$d_1=5.61+6.65t-0.19(\omega/\delta)$	0.565	0.750	1.750
5a	$t=\theta(\delta,B,\omega)$	Linear	$t=-0.99-0.16\delta+0.17B+0.10\omega$	0.60	0.775	0.049
b	$t=\gamma_1(\delta,\delta/B)$	Linear	$t=0.379-0.336\delta+\delta/B$	0.20	0.45	0.63
c	$t=\gamma_2(\delta,\delta/\omega)$	Linear	$t=0.422+0.022\delta-\delta/\omega$	0.06	0.25	0.069
d	$t=\gamma_3(\delta,B/\omega)$	Linear	$t=0.427-0.164\delta-0.03(B/\omega)$	0.322	0.568	0.059

SUMMARY

1.Multiple regression analysis for internal diameter (d_1) of stainless chip in dependence from some its general parameters in function of $Y_i = \gamma(X_{i,1}, X_{i,2}, X_{i,3})$ have showed the following results:

- Only such functions as $d_1 = \phi_3(B, t, \omega)$ and $d_1 = \eta_1(t, t/\omega)$ have the non-linear regression models and other functions have the linear regression models;

- Analysis of above-named functional dependencies in Table 27 indicates on fact that statistical characteristics such as coefficients of determination (R^2) and correlation (r) have the good values besides of such functions as $d_1 = \rho_9(B, B/\omega)$, $d_1 = \phi_3(B \quad t, \quad \omega)$, $d_1 = \mu_1(B, B/t), d_1 = \mu_2(B, B/\omega), d_1 = \mu_3(B, \quad t/\omega)$, $d_1 = \eta_1(t, t/\omega), t = \gamma_1(\delta, \delta/B)$, $t = \gamma_2(\delta, \delta/\omega)$ and $t = \gamma_3(\delta, B/\omega)$, where average values of coefficients determination and correlation are equal $R^2 = 0.21$ and $r = 0.46$.

CHAPTER EIGHT   MULTIPLE REGRESSION ANALYSIS OF THE MAIN PARAMETERS OF STAINLESS CHIP

8.1 Dependence of external diameter (d_2) of stainless chip from all its parameters for the multiple regression model view of $Y_i=\phi(X_{i,1},X_{i,2},X_{i,3},X_{i,4},X_{i,5})$

The illustration of this paragraph is shown on the relationships between of external diameter (Y) and internal diameter(X_1),number of chip wraps (X_2),clearance (X_3) between of chip wraps ,width (X_4) of this chip and also the thickness (X_5). External diameter (d_2) of stainless chip is the dependent variable and other five parameters are the independent variables.

We can designate our variables in such manner:

Dependent variable:

External diameter....................Y

Independent   variables:

Internal diameter..................... X_1

Number of chip wraps............... X_2

Clearance between of chip wraps...X_3

Width of chip...........................X_4

Thickness of chip.....................X_5

Computation Procedure

Since this multiple regression model view of $Y=\phi(X_1,X_2,X_3,X_4,X_5)$ requires a considerable number of measures of relationship between the five variables , it will be

convenient to complete at one time all values that are needed in the different formulas. To put the matter concretely in our problem, we may write directly the multiple regression model view of $Y=b_0+b_1X_1+b_2X_2+b_3X_3+b_4X_4+b_5X_5$ (1) ,where b_0,b_1,b_2,b_3,b_4 and b_5 unknown coefficients which can be calculated from the six normal equations:

$$\Sigma Y=nb_0 +b_1\Sigma X_1+b_2\Sigma X_2+b_3\Sigma X_3 +b_4\Sigma X_4 +b_5\Sigma X_5 \qquad (2)$$

$$\Sigma X_1Y=b_0\Sigma X_1+b_1\Sigma X_1^{2}+b_2\Sigma X_1X_2 +b_3\Sigma X_1X_3 +b_4\Sigma X_1X_4+b_5\Sigma X_1X_5 \qquad (3)$$

$$\Sigma X_2Y=b_0\Sigma X_2 +b_1\Sigma X_1X_2 +b_2\Sigma X_2^{2} +b_3\Sigma X_3X_2 +b_4\Sigma X_4X_2 +b_5\Sigma X_5X_2 \qquad (4)$$

$$\Sigma X_3Y=b_0\Sigma X_3 +b_1\Sigma X_1X_3 +b_2\Sigma X_2X_3 +b_3\Sigma X_3^{2}+b_4\Sigma X_4X_3 +b_5\Sigma X_5X_3 \qquad (5)$$

$$\Sigma X_4Y=b_0\Sigma X_4 +b_1\Sigma X_1X_4 +b_2\Sigma X_2X_4 +b_3\Sigma X_3X_4 +b_4\Sigma X_4^{2} +b_5\Sigma X_5X_4 \qquad (6)$$

$$\Sigma X_5Y=b_0\Sigma X_5 +b_1\Sigma X_1X_5 +b_2\Sigma X_2X_5 +b_3\Sigma X_3X_5 +b_4\Sigma X_4X_5 +b_5\Sigma X_5^{2} \qquad (7)$$

Considering the statistical data given in **Appendix 1** , we can determine some parameters of the above-shown equations (2),(3),(4),(5),(6) and (7) which are indicated in Tables1,2.3 and 4.

Table 1 Computation worksheet for determining sum parameters of Multiple linear regression equation

Number of observation n	External diameter Y	Internal diameter X_1	Number of chip wraps X_2	Clearance between of chip wraps X_3	Width of chip X_4	Thickness of chip X_5	Check column
1	10.03	5.36	3.17	0.21	5.71	0.36	24.84
2	10.36	5.22	3.28	0.38	6.16	0.42	25.82
3	8.69	3.52	4.22	0.34	5.37	0.10	22.24
4	10.13	4.51	3.99	0.42	5.79	0.18	25.02
5	10.22	4.98	3.62	0.30	5.82	0.28	25.22
6	10.07	4.74	3.25	0.31	6.12	0.37	24.86
7	10.49	5.15	3.21	0.32	6.10	0.46	25.73
8	11.06	5.96	3.53	0.32	5.70	0.54	27.11
9	11.18	6.44	3.76	0.25	5.75	0.60	27.98
Total:	92.23	45.88	32.03	2.85	52.52	3.31	228.82

Table 2 Computation worksheet for determining sum parameters of Multiple linear regression equation

n	X_1Y	X_2Y	X_3Y	X_4Y	X_5Y	X_1^2	X_1X_2	X_1X_3	Check column
1	53.76	31.79	2.11	57.27	3.61	28.73	16.99	1.13	195.39
2	54.08	33.98	3.94	63.82	4.35	27.25	17.12	1.98	206.52
3	30.59	36.67	2.95	46.67	0.87	12.39	14.85	1.19	146.18
4	45.69	40.42	4.25	58.65	1.82	20.34	17.99	1.89	191.05
5	50.89	36.99	3.07	59.48	2.86	24.80	18.03	1.49	197.61
6	47.73	32.73	3.12	61.63	3.73	22.47	15.41	1.47	188.29
7	54.02	33.67	3.36	63.99	4.83	26.52	16.53	1.65	204.57
8	65.92	39.04	3.54	63.04	5.97	35.52	21.04	1.91	235.98
9	71.99	42.04	2.79	64.29	6.71	41.47	24.21	1.61	255.11
Total:	474.67	327.33	29.13	538.84	34.75	239.49	162.17	14.32	1820.7

Table 3 Computation worksheet for determining sum parameters of Multiple linear regression equation

n	X_1X_4	X_1X_5	X_2^2	X_3X_2	X_4X_2	X_5X_2	Check column
1	30.61	1.93	10.05	0.67	18.10	1.14	62.50
2	32.16	2.19	10.76	1.25	20.20	1.38	67.94
3	18.90	0.35	17.81	1.43	22.66	0.42	61.57
4	26.11	0.81	15.92	1.68	23.10	0.72	68.34
5	28.98	1.39	13.10	1.09	21.07	1.01	66.64
6	29.00	1.75	10.56	1.08	19.89	1.20	63.48
7	31.42	2.37	10.30	1.03	19.58	1.48	66.18
8	33.97	3.22	12.46	1.13	20.12	1.91	72.81
9	37.03	3.86	14.14	0.94	21.62	2.26	79.85
Total:	268.18	17.87	115.1	10.30	186.34	11.52	609.31

Table 4 Computation worksheet for determining sum parameters of Multiple linear regression equation

n	X_3^2	X_4X_3	X_5X_3	X_4^2	X_5X_4	X_5^2	Check column
1	0.04	1.19	0.08	32.60	2.06	0.13	36.10
2	0.14	2.34	0.16	37.95	2.59	0.18	43.36
3	0.12	1.83	0.03	28.84	0.54	0.01	31.37
4	0.18	2.43	0.08	33.52	1.04	0.03	37.28
5	0.09	1.75	0.08	33.87	1.63	0.08	37.50
6	0.10	1.89	0.11	37.45	2.26	0.14	41.95
7	0.10	1.95	0.15	37.21	2.81	0.21	42.43
8	0.10	1.82	0.17	32.49	3.08	0.29	37.95
9	0.06	1.44	0.15	33.06	3.45	0.36	38.52
Total:	0.93	16.64	1.01	306.99	19.46	1.43	346.46

So, from above-named Tables 1,2,3 and 4 we have the following system of normal equations for determining of coefficients b_0, b_1, b_2, b_3, b_4 and b_5 :

$92.23 = 9b_0 + 45.88b_1 + 32.03b_2 + 2.85b_3 + 52.52b_4 + 3.31b_5$
$474.67 = 45.88b_0 + 239.49b_1 + 162.17b_2 + 14.32b_3 + 268.18b_4 + 17.87b_5$
$327.33 = 32.03b_0 + 162.17b_1 + 115.10b_2 + 10.30b_3 + 186.34b_4 + 11.52b_5$
$29.13 = 2.85b_0 + 14.32b_1 + 10.30b_2 + 0.93b_3 + 16.64b_4 + 1.01b_5$

$538.84 = 52.52b_0 + 268.18b_1 + 186.34b_2 + 16.64b_3 + 306.99b_4 + 19.46b_5$

$34.75 = 3.31b_0 + 17.87b_1 + 11.52b_2 + 1.01b_3 + 19.46b_4 + 1.43b_5$

After of solving the above-shown system of normal equations , we have the following value of coefficients: **$b_0 = 0.91$, $b_1 = 0.42$, $b_2 = -0.04$, $b_3 = -0.01$, $b_4 = 1.29$, $b_5 = -0.51$** and *the Multiple Regression Linear Equation for External diameter(d_2)of stainless chip has view of*

$$Y_c = 0.91 + 0.42X_1 - 0.04X_2 - 0.01X_3 + 1.29X_4 - 0.51X_5 \quad (8)$$

or $\quad d_2 = 0.91 + 0.42d_1 - 0.04\omega - 0.01t + 1.29B - 0.51\delta \quad (8a)$

In Table 5 is shown the residuals from data shown in Table 1 and also the statistical characteristics for the *Multiple Regression Linear equation* for external diameter (d_2) of stainless chip view of $Y_c = 0.91 + 0.42X_1 - 0.04X_2 - 0.01X_3 + 1.29X_4 - 0.51X_5$.

Table 5 Computation worksheet for determining of residuals

Var. n	Y	X_1	X_2	X_3	X_4	X_5	Y_c	$(Y-Y_c)$	$(Y-Y_c)^2$	$(Y-\overline{Y})$	$(Y-\overline{Y})^2$
1	10.03	5.36	3.17	0.21	5.71	0.36	10.21	-0.18	0.03	-0.22	0.05
2	10.36	5.22	3.28	0.38	6.16	0.42	10.70	-0.34	0.12	0.11	0.01
3	8.69	3.52	4.22	0.34	5.37	0.10	9.09	-0.40	0.16	-1.56	2.43
4	10.13	4.51	3.99	0.42	5.79	0.18	10.01	0.12	0.01	-0.12	0.01
5	10.22	4.98	3.62	0.30	5.82	0.28	10.21	0.01	0	-0.03	0
6	10.07	4.74	3.25	0.31	6.12	0.37	10.47	-0.40	0.16	-0.18	0.03
7	10.49	5.15	3.21	0.32	6.10	0.46	10.58	-0.09	0.01	0.24	0.06
8	11.06	5.96	3.53	0.32	5.70	0.54	10.35	0.71	0.50	0.81	0.66
9	11.18	6.44	3.76	0.25	5.75	0.60	10.57	0.61	0.37	0.93	0.86
Total:	92.23	45.88	32.03	2.85	52.52	3.31		0.04	1.36	-0.02	4.11

	The main formulas for calculation of statistical characteristics :
Mean: $\overline{X_1}=5.098, \overline{X_2}=3.559,$ $\overline{X_3}=0.317, \overline{X_4}=5.836,$ $\overline{X_5}=0.368, \overline{Y}=10.250$	$R^2 = [\sum(Y-\overline{Y})^2 - \sum(Y-Y_c)^2] / \sum(Y-\overline{Y})^2$ (9) $r = \pm (R^2)^{1/2}$ (10)
Coefficient of determination $R^2=0.669$	$S_{y/x1,x2,x3,x4,x5} = [\sum(Y-Y_c)^2/(n-6)]^{1/2}$ (11) $\min MSE = \sum(Y-Y_c)^2/n$ (12) $\min MAD = \sum|Y-Y_c)/n$ (13)
Coefficient of correlation $r=0.818$	
Standard deviation $S_{y/x1,x2,x3,x4,x5}=0.673$	
min MSE=0.151	
min MAD=0.004	

All results of computation for the multiple regression linear equation view of $Y_c=0.091+0.42X_1-0.04X_2-0.01X_3+1.29X_4-0.51X_5$ or $d_2=0.91+0.42d_1-0.04\omega-0.01t+ +1.29B-0.51\delta$ are given in Table 6.

Table 6 Evaluation of multiple regression equation $Y_c = 0.91 + 0.42X_1 - 0.04X_2 - 0.01X_3 + +1.29X_4 - 0.51X_5$

A. Mean ,variance and standard deviation

Variable	Mean	Variance	Standard deviation
X_1	5.098	5.609	2.368
X_2	3.559	1.113	1.055
X_3	0.317	0.030	0.173
X_4	5.836	0.515	0.718
X_5	0.368	0.209	0.457
Y	10.250	4.110	2.027

B. Results of multiple regression of Y on X_1,X_2,X_3,X_4 and X_5

Parameter	Variable	Coefficients	Standard error	T-value
β_1	X_1	0.42	0.789	0.532
β_2	X_2	-0.04	0.352	-0.114
β_3	X_3	-0.01	0.058	-0.172
β_4	X_4	1.29	0.239	5.397
β_5	X_5	-0.51	0.152	-3.355

C. Analysis of variance results

Regression
- Degrees of freedom   5
- Sum of squares   2.75
- Mean square   0.55

Error
- Degrees of freedom 3
- Sum of squares   1.36
- Mean square   0.453

F-value*   1.21

* Since F=1.21< [$F_{0.05,5,3}$=9.01] we can not reject the hypothesis that parameters $\beta_1,\beta_2,\beta_3,\beta_4$ and β_5 are zero.

D. Determination of residuals

Number	Observed	Estimated	Residual
1	10.03	10.21	-0.18
2	10.36	10.70	-0.34
3	8.69	9.09	-0.40
4	10.13	10.01	0.12
5	10.22	10.21	0.01
6	10.07	10.47	-0.40
7	10.49	10.58	-0.09
8	11.06	10.35	0.71
9	11.18	10.57	0.61

8.2 Dependence of internal diameter(d_1) of stainless chip from all its parameters for the multiple regression model view of $Y_i = \phi(X_{i,1}, X_{i,2}, X_{i,3}, X_{i,4}, X_{i,5})$

The illustration of this paragraph is shown on the relationship between of internal diameter (Y) and external diameter (X_1) , number of chip wraps (X_2) ,clearance (X_3) between of chip wraps ,width(X_4) of chip, thickness (X_5) of stainless chip.

Internal diameter (d_1) of stainless chip is the dependent variable and other five parameters are the independent variables. In this case we can designate our variables in such manner:

Dependent variable:

Internal diameter Y

Independent variables:

External diameter X_1

Number of chip wraps X_2

Clearance between of chip wraps X_3

Width of chip X_4

Thickness of chip X_5

Using of above-shown formulas (1),(2),(3),(4),(5),(6) and (7) we can determine the sum of parameters for multiple linear regression equation. Computation worksheets for these data are given in Tables 7,8,9 and 10.

Table 7 Computation worksheet for determining sum parameters for multiple linear regression equation

Number of average observation n	Internal diameter Y	External diameter X_1	Number of chip wraps X_2	Clearance between of chip wraps X_3	Width of chip X_4	Thickness of chip X_5	Check column
1	5.36	10.03	3.17	0.21	5.71	0.36	24.84
2	5.22	10.36	3.28	0.38	6.16	0.42	25/82
3	3.52	8.69	4.22	0.34	5.37	0.10	22.24
4	4.51	10.13	3.99	0.42	5.79	0.18	25.02
5	4.98	10.22	3.62	0.30	5.82	0.28	25.22
6	4.74	10.07	3.25	0.31	6.12	0.37	24.86
7	5.15	10.49	3.21	0.32	6.10	0.46	25.73
8	5.96	11.06	3.53	0.32	5.70	0.54	27.11
9	6.44	11.18	3.76	0.25	5.75	0.60	27.98
Total:	45.88	92.23	32.03	2.85	52.52	3.31	228.82

Table 8 Computation worksheet for determining sum of parameters for multiple linear regression equation

n	X_1Y	X_2Y	X_3Y	X_4Y	X_5Y	X_1^2	X_1X_2	X_1X_3	Check column
1	53.76	16.99	1.13	30.61	1.93	100.60	31.79	2.11	238.92
2	54.08	17.12	1.98	32.16	2.19	107.33	33.98	3.94	252.78
3	30.59	14.85	1.19	18.90	0.35	75.52	36.67	2.95	181.02
4	45.69	17.99	1.89	26.11	0.81	102.62	40.42	4.25	239.78
5	50.89	18.03	1.49	28.98	1.39	104.45	36.99	3.07	245.29
6	47.73	15.41	1.47	29.01	1.75	101.40	32.73	3.12	232.62
7	54.02	16.53	1.65	31.42	2.37	110.04	33.67	3.36	253.06
8	65.92	21.04	1.91	33.97	3.22	122.32	39.04	3.54	290.96
9	71.99	24.21	1.61	37.03	3.86	124.99	42.04	2.79	308.52
Total:	474.67	162.17	14.32	268.19	17.87	949.27	327.33	29.13	2242.95

Table 9 Computation worksheet for determining sums of parameters for multiple linear regression equation

n	X_1X_4	X_1X_5	X_2^2	X_3X_2	X_4X_2	X_5X_2	Check column
1	57.27	3.61	10.05	0.67	18.10	1.14	90.84
2	63.82	4.35	10.76	1.25	20.20	1.38	101.76
3	46.67	0.87	17.81	1.43	22.66	0.42	89.86
4	58.65	1.82	15.92	1.68	23.10	0.72	101.89
5	59.48	2.86	13.10	1.09	21.07	1.01	98.61
6	61.63	3.73	10.56	1.01	19.89	1.20	98.02
7	63.99	4.83	10.30	1.03	19.58	1.48	101.21
8	63.04	5.97	12.46	1.13	20.12	1.91	104.63
9	64.29	6.71	14.14	0.94	21.62	2.26	109.96
Total:	538.84	34.75	115.10	10.23	186.34	11.52	896.78

Table 10 Computation worksheet for determining sums of parameters for multiple linear regression equation

n	$X_3{}^2$	X_4X_3	X_5X_3	$X_4{}^2$	X_5X_4	$X_5{}^2$	Check column
1	0.044	1.199	0.076	32.604	2.056	0.129	36.108
2	0.144	2.341	0.159	37.946	2.587	0.176	43.353
3	0.116	1.826	0.034	28.837	0.537	0.010	31.360
4	0.176	2.432	0.076	33.524	1.042	0.032	37.282
5	0.090	1.746	0.084	33.872	1.629	0.078	37.499
6	0.096	1.897	0.115	37.454	2.264	0.137	41.963
7	0.102	1.952	0.147	37.210	2.806	0.212	42.429
8	0.102	1.824	0.173	32.490	3.078	0.292	37.959
9	0.063	1.438	0.150	33.063	3.450	0.360	38.524
Total:	0.933	16.655	1.014	307	19.449	1.426	346.477

At data from above-named Tables 7,8,9 and 10 we have the following system of normal equations for determining of coefficients b_0, b_1, b_2, b_4 and b_5 :

$45.88 = 9b_0 + 92.23b_1 + 32.03b_2 + 2.85b_3 + 52.52b_4 + 3.31b_5$

$474.67 = 92.23b_0 + 949.27b_1 + 327.33b_2 + 29.13b_3 + 538.84b_4 + 34.75b_5$

$162.17 = 32.03b_0 + 327.33b_1 + 115.10b_2 + 10.23b_3 + 186.34b_4 + 11.52b_5$

$14.32 = 2.85b_0 + 29.13b_1 + 10.23b_2 + 0.933b_3 + 16.655b_4 + 1.014b_5$

$268.19 = 52.52b_0 + 538.84b_1 + 186.34b_2 + 16.655b_3 + 307b_4 + 19.449b_5$

$17.87 = 3.31b_0 + 34.75b_1 + 11.52b_2 + 1.014b_3 + 19.449b_4 + 1.426b_5$

After of solving the above-shown system of normal equations, we have the following value of coefficients: $b_0 = -5.435$, $b_1 = 1.07$, $b_2 = -0.144$, $b_3 = 0.30$, $b_4 = 0.004$, $b_5 = -0.017$ and Multiple regression equation for determining of internal diameter(d_1) for stainless chip has view of $Y_c = -5.435 + 1.07X_1 - 0.144X_2 + 0.3X_3 + 0.004X_4 - 0.017X_5$ (14),

where $d_1 = -5.435 + 1.07d_2 - 0.144\omega + 0.3t + 0.004B - 0.017\delta$ (14 a).

In Table 11 is shown the residuals from data of Table 7 and statistical characteristics for Multiple linear regression equation $Y_c = -5.44 + 1.07X_1 - 0.144X_2 + 0.3X_3 + 0.004X_4 - 0.017X_5$.

Table 11 Computation worksheet for determining of residuals

n	Y	X_1	X_2	X_3	X_4	X_5	Y_c	$(Y-Y_c)$	$(Y-Y_c)^2$	$(Y-\bar{Y})$	$(Y-\bar{Y})^2$
1	5.36	10.03	3.17	0.21	5.71	0.36	4.92	0.44	0.194	0.262	0.069
2	5.22	10.36	3.28	0.38	6.16	0.42	5.31	-0.09	0.008	0.122	0.015
3	3.52	8.69	4.22	0.34	5.37	0.10	3.38	0.14	0.019	-1.578	2.490
4	4.51	10.13	3.99	0.42	5.79	0.18	4.98	-0.47	0.221	-0.588	0.346
5	4.98	10.22	3.62	0.30	5.82	0.28	5.09	-0.11	0.012	-0.118	0.014
6	4.74	10.07	3.25	0.31	6.12	0.37	4.98	-0.24	0.058	-0.358	0.128
7	5.15	10.49	3.21	0.32	6.10	0.46	5.44	-0.29	0.084	0.052	0.003
8	5.96	11.06	3.53	0.32	5.70	0.54	6.00	-0.04	0.002	0.862	0.743
9	6.44	11.18	3.76	0.25	5.75	0.60	6.08	0.36	0.129	1.342	1.801
Total:	45.88	92.23	32.03	2.85	52.52	3.31		-0.30	0.727		5.608

Mean:

$\bar{X_1}=10.248$, $\bar{X_2}=3.559$

$\bar{X_3}=0.317$, $\bar{X_4}=5.836$

$\bar{X_5}=0.368$, $\bar{Y}=5.098$

Coefficient of determination
$R^2=0.870$

Coefficient of correlation
$r=0.933$

Standard deviation
$S_{y/x1,x2,x3,x4,x5}=0.492$

Minimization of mean square min MSE=0.081

Minimization of the mean absolute deviation
min MAD=0.033

The main formulas for calculation of statistical characteristics :

$$R^2= [\ \Sigma(Y-\bar{Y})^2 - \Sigma(Y-Y_c)^2]/\ \Sigma(Y-\bar{Y})^2$$

$$r = \pm (R^2)^{1/2}$$

$$S_{y/x1,x2,x2,x3,x4,x5}= [\ \Sigma(Y-Y_c)^2 /(n-6)\ ]^{1/2}$$

$$\min MSE= \Sigma(Y-Y_c)^2/n$$

$$\min MAD = \Sigma|Y-Y_c|/n$$

All results of computation for the multiple regression equation view of Y_c= **-5.435 + +1.07X$_1$−0.144X$_2$+0.3X$_3$+0.004X$_4$−0.017X$_5$** or **d$_1$= −5.435+1.07d$_2$− −0.144ω+ 0.3t+0.004B−0.017δ** are given in Table 12.

Table 12 Evaluation of multiple regression equation Y_c= **−5.435+1.07X$_1$−0.144X$_2$+ +0.3X$_3$+0.004X$_4$−0.017X$_5$**

A. Mean ,variance and statistical deviation

Variable	Mean	Variance	Standard deviation
X_1	10.248	4.12	2.029
X_2	3.559	1.113	1.055
X_3	0.317	0.030	0.173
X_4	5.836	0.515	0.718
X_5	0.368	0.209	0.457
Y	5.098	5.608	2.368

B. Results of multiple regression of Y on X$_1$,X$_2$,X$_3$,X$_4$ and X$_5$

Parameter	Variable	Coefficients	Standard error	T-value
β_1	X_1	1.07	0.676	1.583
β_2	X_2	-0.144	0.352	-0.409
β_3	X_3	0.3	0.058	5.172
β_4	X_4	0.004	0.239	0.017
β_5	X_5	-0.017	0.152	-0.112

C. Analysis of variance results

Regression
    Degrees of freedom 5
    Sum of squares   4.881
    Mean square   0.976
Error
    Degrees of freedom 3
    Sum of squares   0.727
    Mean square   0.242
F-value*   4.033

* Since F=4.033 <[F$_{0.05,5,3}$=9.01] we can not reject the hypothesis that parameters β$_1$,β$_2$,β$_3$,β$_4$ and β$_5$ are zero.

D. Determination of residuals

Number	Observed	Estimated	Residual
1	5.36	4.92	0.44
2	5.22	5.31	-0.09
3	3.52	3.38	0.14
4	4.51	4.98	-0.47
5	4.98	5.09	-0.11
6	4.74	4.98	-0.24
7	5.15	5.44	-0.29
8	5.96	6.00	-0.04
9	6.44	6.08	0.36

REFERENCES

[1] Oberg E.,Johes F.D., HortonH.L., Ryffel H.H. Machinery's Handbook.24th Edition. – Industrial Press Inc.(New York,1992,p.373,p.58)

[2] Rozenblat A.I." *Criteria of evaluation of resistance circular saws in during cutting of billets"* . Technology and Organization of Production . Monthly Scientific magazine of the Ukrainian Scientific Research Institute of the Technical Information (Kiev,1974,pp.40-42)

[3] Metalcutting: Today's techniques for Engineers and Shop Personnel by the editors of American Machinist (American Machinist, 1977,pp.130-144)

[4] Rozenblat A.I .,The Russian Inventor and Scientist brings the new technologies to USA.- 1st Books Library (Bloomington ,USA ,2001 ,pp.3-8)

[5] Справочник по технологии резания материалов ,том 2.Под редакцией Г.Шпура ,Т.Штеферле .-Москва,Машиностроение ,1985 ,стр.98)

[6] Rozenblat A.I. *"Analysis of resistance of segmental circular saws".* The collections of papers . Machinery Technology: Scientific technical information. Central Scientific Research Institute of the Technical Information of the Light and Food Machine-Building, issue 3,1975(Moscow ,pp.16-18)

[7] Резников Н.И .,Производительная обработка нержавеющих и жаропрочных сталей. – Москва,Издательство"Машиностроение",1960

[8] Amstead B.H., Ostwald P.F., Begeman M.L., Metal cutting process . Manufacturing process. –John Wiley & Sons, New York ,1977

[9] Бронштейн И.Н.,Семендяев К.А.,Справочник по математике.- Москва "Наука" Главная редакция физико-математической литературы, 1980,стр.184.

[10] Rozenblat A.I. Advanced machining problem solving.-1stBooks Library (Bloomington ,USA ,2001)

[11] Справочник по технологии резания материалов ,том 1 .Под редакцией Г.Шпура ,Т.Штеферле. – Москва"Машиностроение",1985,стр.77)

[12] Roger C., Pfaffenberger J. ,Patterson H. Statistical methods for business and economics.- Richard D. Irwing Inc.(Homewood,USA,1977)

[13] Дроздов Ф., Лебедевич В.,Рубежин В. Справочное пособие по отрезным станкам. –Минск. Издательство" Беларусь" ,1968

[14] Rozenblat A.I. Author's Certificate # 1220856,USSR MKI 14B 23B 27/00.Combination Lathe tool of A.I.Rozenblat

[15] Rozenblat A.I., 30 Innovations of the Russian Engineer.-1st Books Library (Bloomington ,USA ,2003)

[16] Rozenblat A.I , Author's Certificate #1131634,USSR,MKI B23Q 11/02.Tool Cleaning Device. Rozenblat A.I.USSR# 3435998/25-8.Applied on 5-10-82.Published on 12-30-84,Bulletin#48"Discoveries.Inventions", 1984,#48,p.35

[17] Rozenblat A.I *"The Russian Inventor and Scientist brings the new technology to USA".-* Proceeding of the 27th Israel Conference on Mechanical Engineers.Haifa,Israel,1998.

[18] Applied General Statistics by Croxton F.E and Cowden.D.J.-New York.Prentice-Hall,Inc.1939

[19] Byrkit D.R. Elements of Statistics.3rd Edition. –D.Van Nostrand Company.1980

[20] Ефимов Н.В. Краткий курс аналитической геометрии.-Издательство"Наука" .Москва,1969 ,стр.155

[21] Х.Шенк.Теория Инженерного эксперимента.-Издательство"Мир".Москва,1972

APPENDIX 1 Short observed average data (n_s=9) are used for determining of above-shown general statistical functional parameters and all multiple regression models

n_s / Parameters	1	2	3	4	5	6	7	8	9
Thickness δ	0.36	0.42	0.10	0.18	0.28	0.37	0.46	0.54	0.60
Internal diameter (d_1)	5.36	5.22	3.52	4.51	4.98	4.74	5.15	5.96	6.44
External diameter (d_2)	10.03	10.36	8.69	10.13	10.22	10.07	10.49	11.06	11.18
Number of chip wraps (ω)	3.17	3.28	4.22	3.99	3.62	3.25	3.21	3.53	3.76
Clearance between of chip wraps (t)	0.21	0.38	0.34	0.42	0.30	0.31	0.32	0.32	0.25
Width of chip (B)	5.71	6.16	5.37	5.79	5.82	6.12	6.10	5.70	5.75

APPENDIX 2 Full observed statistical data ($n_f=154$) are collected in the industrial experiments for stainless chip formed by the cold segmental circular saws

n	1	2	3	4	5	6	7	8	9	10	11	12	13	14
δ	0.1	0.3	0.3	0.4	0.4	0.4	0.5	0.5	0.5	0.4	0.4	0.3	0.3	0.2
d_1	3.6	2.5	6.0	3.5	2.7	4.8	3.1	4.8	2.3	7.0	8.5	7.0	5.5	6.9
d_2	9.0	10.3	11.0	7.8	10.1	10.5	10.8	10.3	10.2	10.0	11.0	11.0	11.0	10.0
ω	5	4	3	3	4	4	4	4	4	3	2	1	3	3
t	0.3	0.1	0.1	0.3	0.4	0.2	0.1	0.2	0.1	0.2	0.1	0.2	0.3	0.3
B	5.2	5.4	6.5	5.6	5.4	5.7	5.8	5.8	5.7	5.5	5.7	5.5	5.6	5.8

n	15	16	17	18	19	20	21	22	23	24	25	26	27	28
δ	0.2	0.2	0.5	0.5	0.5	0.5	0.4	0.3	0.4	0.4	0.3	0.5	0.6	0.2
d_1	7.0	6.5	7.5	7.2	6.5	6.5	3.0	3.5	2.0	7.5	8.0	5.8	7.5	3.8
d_2	7.5	11.0	10.0	10.0	10.0	11.0	9.0	9.0	9.0	11.0	10.0	10.0	12.0	10.0
ω	2.0	4.0	1.0	3.0	3.0	3.0	4.0	4.0	4.0	3.0	2.0	3.0	3.0	4.0
t	0.2	0.4	0.1	0.2	0.2	0.4	0.5	0.5	0.3	0.2	0.4	0.3	0.4	0.7
B	6.0	6.0	5.7	5.8	6.0	5.6	5.8	6.0	7.0	6.8	6.5	5.8	5.0	5.5

n	29	30	31	32	33	34	35	36	37	38	39	40	41	42
δ	0.2	0.3	0.5	0.5	0.5	0.5	0.5	0.5	0.1	0.1	0.1	0.1	0.1	0.1
d_1	3.0	5.0	5.1	6.5	5.2	4.3	5.5	5.2	3.0	3.4	4.1	3.5	3.1	3.8
d_2	10.0	10.0	11.0	10.0	11.0	10.0	11.5	12.0	8.2	9.4	10.0	7.9	9.7	7.5
ω	4.0	4.0	3.0	3.0	3.0	3.0	3.0	3.0	4.0	5.0	4.0	4.0	4.0	4.0
t	0.5	0.4	0.4	0.2	0.4	0.3	0.3	0.5	0.3	0.4	0.4	0.3	0.2	0.3
B	6.0	6.0	6.5	6.5	7.0	6.3	6.0	6.5	5.0	5.8	6.0	4.7	5.5	4.9

n	43	44	45	46	47	48	49	50	51	52	53	54	55	56
δ	0.1	0.1	0.1	0.1	0.1	0.1	0.1	0.1	0.1	0.1	0.1	0.1	0.1	0.1
d_1	3.0	2.7	3.5	3.8	4.1	3.8	3.7	4.2	3.1	3.8	3.0	3.8	3.7	3.6
d_2	6.3	7.4	8.9	9.0	11.0	9.2	9.5	10.3	8.5	7.9	8.0	8.7	9.0	9.0
ω	3.0	3.0	3.0	4.0	5.0	5.0	5.0	3.0	5.0	5.0	5.0	5.0	5.0	5.0
t	0.3	0.2	0.3	0.4	0.4	0.3	0.4	0.5	0.4	0.3	0.3	0.4	0.4	0.3
B	5.1	6.3	5.0	5.2	6.2	5.2	5.4	6.2	4.9	4.7	5.3	5.3	5.0	5.2

n	57	58	59	60	61	62	63	64	65	66	67	68	69	70
δ	0.2	0.2	0.2	0.2	0.2	0.2	0.2	0.2	0.2	0.2	0.2	0.2	0.2	0.2
d_1	3.8	3.0	6.9	7.0	6.5	6.3	5.9	5.0	3.8	3.9	4.3	3.7	3.4	3.3
d_2	10.0	10.0	10.0	7.5	11.0	11.2	10.1	10.5	11.5	11.7	11.5	10.6	9.8	9.3
ω	4.0	4.0	3.0	2.0	4.0	4.2	4.0	4.3	3.0	3.6	3.8	4.2	4.3	4.4
t	0.7	0.5	0.3	0.2	0.4	0.5	0.3	0.4	0.3	0.4	0.6	0.2	0.5	0.6
B	5.5	6.0	5.8	6.0	6.0	6.2	6.0	6.2	5.8	6.0	6.3	5.4	5.1	5.8

n	71	72	73	74	75	76	77	78	79	80	81	82	83	84
δ	0.2	0.2	0.2	0.2	0.2	0.2	0.3	0.3	0.3	0.3	0.3	0.3	0.3	0.3
d_1	3.6	3.5	3.9	4.3	4.8	5.1	5.0	2.5	6.0	5.5	3.5	8.0	7.5	7.3
d_2	9.5	10.1	10.7	8.8	10.0	11.0	10.0	10.3	11.0	11.0	9.0	10.0	11.5	10.0
ω	5.1	4.0	4.0	4.0	3.6	3.8	4.0	4.0	3.0	3.0	4.0	2.0	4.0	3.0
t	0.5	0.4	0.3	0.6	0.2	0.5	0.4	0.1	0.1	0.3	0.5	0.4	0.2	0.2
B	6.0	6.0	6.1	6.3	6.3	6.5	6.0	5.4	6.5	5.6	6.0	6.5	6.1	6.2

n	85	86	87	88	89	90	91	92	93	94	95	96	97	98
δ	0.3	0.3	0.3	0.3	0.3	0.3	0.3	0.3	0.3	0.3	0.3	0.3	0.4	0.4
d_1	5.0	3.4	3.6	5.0	5.7	3.6	3.9	7.3	5.1	4.8	4.6	5.8	3.5	2.7
d_2	10.3	11.7	9.5	9.3	9.7	10.2	10.8	11.3	11.8	9.3	10.5	10.7	7.8	10.1
ω	3.0	4.5	4.8	5.1	2.8	2.5	3.0	3.0	4.0	2.0	3.0	3.5	3.0	4.0
t	0.1	0.1	0.5	0.4	0.4	0.1	0.2	0.2	0.2	0.6	0.5	0.4	0.3	0.4
B	5.7	5.2	5.2	4.7	4.3	6.2	6.3	6.5	6.6	6.9	7.0	7.1	5.6	5.4

n	99	100	101	102	103	104	105	106	107	108	109	110	111	112
δ	0.4	0.4	0.4	0.4	0.4	0.4	0.4	0.4	0.4	0.4	0.4	0.4	0.4	0.4
d_1	4.8	7.0	8.5	3.0	2.0	7.5	4.3	3.8	3.9	2.8	7.6	7.0	8.0	3.6
d_2	10.5	10.0	11.0	9.0	9.0	11.0	11.2	10.5	7.5	9.3	8.7	8.5	10.3	10.8
ω	4.0	3.0	2.0	4.0	4.0	3.0	3.0	4.0	2.0	4.0	3.8	3.5	3.1	4.0
t	0.2	0.2	0.1	0.5	0.3	0.2	0.3	0.4	0.2	0.3	0.5	0.4	0.2	0.4
B	5.7	5.5	5.7	5.8	7.0	6.8	5.1	5.3	5.8	6.1	7.2	7.0	6.8	4.3

n	113	114	115	116	117	118	119	120	121	122	123	124	125	126
δ	0.4	0.4	0.4	0.4	0.5	0.5	0.5	0.5	0.5	0.5	0.5	0.5	0.5	0.5
d_1	4.2	4.8	5.1	2.9	5.1	6.5	5.2	4.3	5.5	5.2	3.1	4.8	2.3	7.5
d_2	11.0	11.3	10.6	10.9	11.0	10.0	11.0	10.0	11.5	12.0	10.8	10.3	10.2	10.0
ω	4.3	2.0	2.2	3.8	3.0	3.0	3.0	3.0	3.0	3.0	4.0	4.0	4.0	1.0
t	0.3	0.3	0.5	0.6	0.4	0.2	0.4	0.3	0.3	0.5	0.1	0.2	0.1	0.1
B	5.0	5.5	5.0	7.2	6.5	6.5	7.0	6.3	6.0	6.5	5.8	5.8	5.7	5.7

n	127	128	129	130	131	132	133	134	135	136	137	138	139	140
δ	0.5	0.5	0.5	0.5	0.5	0.5	0.5	0.5	0.5	0.5	0.6	0.6	0.6	0.6
d_1	7.2	6.5	6.5	5.8	3.0	3.4	2.5	7.3	6.4	5.0	7.5	6.3	6.0	6.2
d_2	10.0	10.0	11.0	10.00	11.3	11.8	12.3	12.0	11.8	11.5	12.0	11.5	12.2	11.0
ω	3.0	3.0	3.0	3.0	3.2	4.1	5.2	4.3	3.2	4.5	3.0	2.8	2.3	2.0
t	0.2	0.2	0.4	0.3	0.1	0.3	0.2	0.4	0.3	0.4	0.4	0.3	0.3	0.4
B	5.8	6.0	5.6	5.8	5.6	5.8	6.2	5.0	6.3	5.8	5.0	5.2	5.5	6.1

n	141	142	143	144	145	146	147	148	149	150	151	152	153	154
δ	0.6	0.6	0.6	0.6	0.6	0.6	0.6	0.6	0.6	0.6	0.6	0.6	0.6	0.6
d_1	7.1	7.2	6.3	7.0	7.0	7.8	5.3	5.4	6.8	7.1	7.3	6.3	6.2	6.0
d_2	10.2	10.3	9.5	10.7	12.1	12.0	11.5	11.6	11.0	12.1	10.3	10.5	11.0	9.8
ω	4.3	4.5	5.2	3.0	3.0	2.0	4.2	4.7	4.0	3.3	3.8	4.2	4.0	5.0
t	0.3	0.3	0.4	0.5	0.5	0.4	0.3	0.3	0.5	0.4	0.4	0.3	0.4	0.3
B	6.3	5.0	5.1	6.5	6.4	6.0	4.8	4.3	6.2	6.3	6.7	6.0	4.9	5.3

Устройство для очистки зубьев дисковой сегментной пилы
[Invention: Cleaning tool for the cold segmental circular saw]

СОЮЗ СОВЕТСКИХ СОЦИАЛИСТИЧЕСКИХ РЕСПУБЛИК

ГОСУДАРСТВЕННЫЙ КОМИТЕТ СССР
ПО ДЕЛАМ ИЗОБРЕТЕНИЙ И ОТКРЫТИЙ

АВТОРСКОЕ СВИДЕТЕЛЬСТВО

№ 1131634

На основании полномочий, предоставленных Правительством СССР, Государственный комитет СССР по делам изобретений и открытий выдал настоящее авторское свидетельство на изобретение: "Устройство для очистки инструмента"

Автор (авторы): Розенблат Анатолий Исаакович

Заявитель: он же

Заявка № 3435996 Приоритет изобретения 10 мая 1982г.
Зарегистрировано в Государственном реестре изобретений СССР

1 сентября 1984г.
Действие авторского свидетельства распространяется на всю территорию Союза ССР.

Председатель Комитета

Начальник отдела

www.ingramcontent.com/pod-product-compliance
Lightning Source LLC
Chambersburg PA
CBHW081104170526
45165CB00008B/2320